T0290854

IET ENERGY ENGINEERING SERIES 142B

# Wind Turbine System Design

**Other volumes in this series:**

| | |
|---|---|
| Volume 1 | **Power Circuit Breaker Theory and Design** C.H. Flurscheim (Editor) |
| Volume 4 | **Industrial Microwave Heating** A.C. Metaxas and R.J. Meredith |
| Volume 7 | **Insulators for High Voltages** J.S.T. Looms |
| Volume 8 | **Variable Frequency AC Motor Drive Systems** D. Finney |
| Volume 10 | **$SF_6$ Switchgear** H.M. Ryan and G.R. Jones |
| Volume 11 | **Conduction and Induction Heating** E.J. Davies |
| Volume 13 | **Statistical Techniques for High Voltage Engineering** W. Hauschild and W. Mosch |
| Volume 14 | **Uninterruptible Power Supplies** J. Platts and J.D. St Aubyn (Editors) |
| Volume 15 | **Digital Protection for Power Systems** A.T. Johns and S.K. Salman |
| Volume 16 | **Electricity Economics and Planning** T.W. Berrie |
| Volume 18 | **Vacuum Switchgear** A. Greenwood |
| Volume 19 | **Electrical Safety: A guide to causes and prevention of hazards** J. Maxwell Adams |
| Volume 21 | **Electricity Distribution Network Design, 2nd Edition** E. Lakervi and E.J. Holmes |
| Volume 22 | **Artificial Intelligence Techniques in Power Systems** K. Warwick, A.O. Ekwue and R. Aggarwal (Editors) |
| Volume 24 | **Power System Commissioning and Maintenance Practice** K. Harker |
| Volume 25 | **Engineers' Handbook of Industrial Microwave Heating** R.J. Meredith |
| Volume 26 | **Small Electric Motors** H. Moczala *et al.* |
| Volume 27 | **AC-DC Power System Analysis** J. Arrillaga and B.C. Smith |
| Volume 29 | **High Voltage Direct Current Transmission, 2nd Edition** J. Arrillaga |
| Volume 30 | **Flexible AC Transmission Systems (FACTS)** Y-H. Song (Editor) |
| Volume 31 | **Embedded Generation** N. Jenkins *et al.* |
| Volume 32 | **High Voltage Engineering and Testing, 2nd Edition** H.M. Ryan (Editor) |
| Volume 33 | **Overvoltage Protection of Low-Voltage Systems, Revised Edition** P. Hasse |
| Volume 36 | **Voltage Quality in Electrical Power Systems** J. Schlabbach *et al.* |
| Volume 37 | **Electrical Steels for Rotating Machines** P. Beckley |
| Volume 38 | **The Electric Car: Development and future of battery, hybrid and fuel-cell cars** M. Westbrook |
| Volume 39 | **Power Systems Electromagnetic Transients Simulation** J. Arrillaga and N. Watson |
| Volume 40 | **Advances in High Voltage Engineering** M. Haddad and D. Warne |
| Volume 41 | **Electrical Operation of Electrostatic Precipitators** K. Parker |
| Volume 43 | **Thermal Power Plant Simulation and Control** D. Flynn |
| Volume 44 | **Economic Evaluation of Projects in the Electricity Supply Industry** H. Khatib |
| Volume 45 | **Propulsion Systems for Hybrid Vehicles** J. Miller |
| Volume 46 | **Distribution Switchgear** S. Stewart |
| Volume 47 | **Protection of Electricity Distribution Networks, 2nd Edition** J. Gers and E. Holmes |
| Volume 48 | **Wood Pole Overhead Lines** B. Wareing |
| Volume 49 | **Electric Fuses, 3rd Edition** A. Wright and G. Newbery |
| Volume 50 | **Wind Power Integration: Connection and system operational aspects** B. Fox *et al.* |
| Volume 51 | **Short Circuit Currents** J. Schlabbach |
| Volume 52 | **Nuclear Power** J. Wood |
| Volume 53 | **Condition Assessment of High Voltage Insulation in Power System Equipment** R.E. James and Q. Su |
| Volume 55 | **Local Energy: Distributed generation of heat and power** J. Wood |
| Volume 56 | **Condition Monitoring of Rotating Electrical Machines** P. Tavner, L. Ran, J. Penman and H. Sedding |
| Volume 57 | **The Control Techniques Drives and Controls Handbook, 2nd Edition** B. Drury |
| Volume 58 | **Lightning Protection** V. Cooray (Editor) |
| Volume 59 | **Ultracapacitor Applications** J.M. Miller |
| Volume 62 | **Lightning Electromagnetics** V. Cooray |

Volume 63 **Energy Storage for Power Systems, 2nd Edition** A. Ter-Gazarian
Volume 65 **Protection of Electricity Distribution Networks, 3rd Edition** J. Gers
Volume 66 **High-Voltage Engineering Testing, 3rd Edition** H. Ryan (Editor)
Volume 67 **Multicore Simulation of Power System Transients** F.M. Uriate
Volume 68 **Distribution System Analysis and Automation** J. Gers
Volume 69 **The Lightening Flash, 2nd Edition** V. Cooray (Editor)
Volume 70 **Economic Evaluation of Projects in the Electricity Supply Industry, 3rd Edition** H. Khatib
Volume 72 **Control Circuits in Power Electronics: Practical issues in design and implementation** M. Castilla (Editor)
Volume 73 **Wide Area Monitoring, Protection and Control Systems: The enabler for smarter grids** A. Vaccaro and A. Zobaa (Editors)
Volume 74 **Power Electronic Converters and Systems: Frontiers and applications** A.M. Trzynadlowski (Editor)
Volume 75 **Power Distribution Automation** B. Das (Editor)
Volume 76 **Power System Stability: Modelling, analysis and control** A.A. Sallam And B. Om P. Malik
Volume 78 **Numerical Analysis of Power System Transients and Dynamics** A. Ametani (Editor)
Volume 79 **Vehicle-to-Grid: Linking electric vehicles to the smart grid** J. Lu and J. Hossain (Editors)
Volume 81 **Cyber-Physical-Social Systems and Constructs in Electric Power Engineering** S. Suryanarayanan, R. Roche and T.M. Hansen (Editors)
Volume 82 **Periodic Control of Power Electronic Converters** F. Blaabjerg, K.Zhou, D. Wang and Y. Yang
Volume 86 **Advances in Power System Modelling, Control and Stability Analysis** F. Milano (Editor)
Volume 87 **Cogeneration: Technologies, optimisation and implementation** C.A. Frangopoulos (Editor)
Volume 88 **Smarter Energy: From smart metering to the smart grid** H. Sun, N. Hatziargyriou, H.V. Poor, L. Carpanini and M.A. Sánchez Fornié (Editors)
Volume 89 **Hydrogen Production, Separation and Purification for Energy** A. Basile, F. Dalena, J. Tong and T.N.Veziroğlu (Editors)
Volume 90 **Clean Energy Microgrids** S. Obara and J. Morel (Editors)
Volume 91 **Fuzzy Logic Control in Energy Systems with Design Applications in MATLAB®/Simulink®** İ.H. Altaş
Volume 92 **Power Quality in Future Electrical Power Systems** A.F. Zobaa and S.H.E.A. Aleem (Editors)
Volume 93 **Cogeneration and District Energy Systems: Modelling, analysis and optimization** M.A. Rosen and S. Koohi-Fayegh
Volume 94 **Introduction to the Smart Grid: Concepts, technologies and evolution** S.K. Salman
Volume 95 **Communication, Control and Security Challenges for the Smart Grid** S.M. Muyeen and S. Rahman (Editors)
Volume 96 **Industrial Power Systems with Distributed and Embedded Generation** R Belu
Volume 97 **Synchronized Phasor Measurements for Smart Grids** M.J.B. Reddy and D.K. Mohanta (Editors)
Volume 98 **Large Scale Grid Integration of Renewable Energy Sources** A. Moreno-Munoz (Editor)
Volume 100 **Modeling and Dynamic Behaviour of Hydropower Plants** N. Kishor and J. Fraile-Ardanuy (Editors)
Volume 101 **Methane and Hydrogen for Energy Storage** R. Carriveau and D.S-K. Ting
Volume 104 **Power Transformer Condition Monitoring and Diagnosis** A. Abu-Siada (Editor)
Volume 106 **Surface Passivation of Industrial Crystalline Silicon Solar Cells** J. John (Editor)
Volume 107 **Bifacial Photovoltaics: Technology, applications and economics** J. Libal and R. Kopecek (Editors)

| | |
|---|---|
| Volume 108 | **Fault Diagnosis of Induction Motors** J. Faiz, V. Ghorbanian and G. Joksimović |
| Volume 109 | **Cooling of Rotating Electrical Machines: Fundamentals, modelling, testing and design** D. Staton, E. Chong, S. Pickering and A. Boglietti |
| Volume 110 | **High Voltage Power Network Construction** K. Harker |
| Volume 111 | **Energy Storage at Different Voltage Levels: Technology, integration, and market aspects** A.F. Zobaa, P.F. Ribeiro, S.H.A. Aleem and S.N. Afifi (Editors) |
| Volume 112 | **Wireless Power Transfer: Theory, technology and application** N.Shinohara |
| Volume 114 | **Lightning-Induced Effects in Electrical and Telecommunication Systems** Y. Baba and V.A. Rakov |
| Volume 115 | **DC Distribution Systems and Microgrids** T. Dragičević, F.Blaabjerg and P. Wheeler |
| Volume 116 | **Modelling and Simulation of HVDC Transmission** M. Han (Editor) |
| Volume 117 | **Structural Control and Fault Detection of Wind Turbine Systems** H.R. Karimi |
| Volume 118 | **Modelling and Simulation of Complex Power Systems** A. Monti and A. Benigni |
| Volume 119 | **Thermal Power Plant Control and Instrumentation: The control of boilers and HRSGs, 2nd Edition** D. Lindsley, J. Grist and D. Parker |
| Volume 120 | **Fault Diagnosis for Robust Inverter Power Drives** A. Ginart (Editor) |
| Volume 121 | **Monitoring and Control Using Synchrophasors in Power Systems with Renewables** I. Kamwa and C. Lu (Editors) |
| Volume 123 | **Power Systems Electromagnetic Transients Simulation, 2nd Edition** N. Watson and J. Arrillaga |
| Volume 124 | **Power Market Transformation** B. Murray |
| Volume 125 | **Wind Energy Modeling and Simulation, Volume 1: Atmosphere and plant** P.Veers (Editor) |
| Volume 126 | **Diagnosis and Fault Tolerance of Electrical Machines, Power Electronics and Drives** A.J.M. Cardoso |
| Volume 127 | **Lightning Electromagnetics, Volume 1: Return stroke modelling and electromagnetic radiation, 2nd Edition** V. Cooray, F. Rachidi, M. Rubinstein (Editors) |
| Volume 127 | **Lightning Electromagnetics, Volume 2: Electrical processes and effects, 2nd Edition** V. Cooray, F. Rachidi, M. Rubinstein (Editors) |
| Volume 128 | **Characterization of Wide Bandgap Power Semiconductor Devices** F. Wang, Z. Zhang and E.A. Jones |
| Volume 129 | **Renewable Energy from the Oceans: From wave, tidal and gradient systems to offshore wind and solar** D. Coiro and T. Sant (Editors) |
| Volume 130 | **Wind and Solar Based Energy Systems for Communities** R. Carriveau and D. S-K. Ting (Editors) |
| Volume 131 | **Metaheuristic Optimization in Power Engineering** J. Radosavljević |
| Volume 132 | **Power Line Communication Systems for Smart Grids** I.R.S Casella and A. Anpalagan |
| Volume 134 | **Hydrogen Passivation and Laser Doping for Silicon Solar Cells** B. Hallam and C. Chan (Editors) |
| Volume 139 | **Variability, Scalability and Stability of Microgrids** S.M. Muyeen, S.M. Islam and F. Blaabjerg (Editors) |
| Volume 142A | **Wind Turbine System Design, Volume 1: Nacelles, drive trains and verification** J. Wenske (Editor) |
| Volume 143 | **Medium Voltage DC System Architectures** B. Grainger and R.D. Doncker (Editors) |
| Volume 145 | **Condition Monitoring of Rotating Electrical Machines** P. Tavner, L. Ran, C. Crabtree |
| Volume 146 | **Energy Storage for Power Systems, 3rd Edition** A.G. Ter-Gazarian |
| Volume 147 | **Distribution Systems Analysis and Automation, 2nd Edition** J. Gers |
| Volume 151 | **SiC Power Module Design: Performance, robustness and reliability** A. Castellazzi and A. Irace (Editors) |
| Volume 152 | **Power Electronic Devices: Applications, failure mechanisms and reliability** F Iannuzzo (Editor) |

| | |
|---|---|
| Volume 153 | **Signal Processing for Fault Detection and Diagnosis in Electric Machines and Systems** M. Benbouzid (Editor) |
| Volume 155 | **Energy Generation and Efficiency Technologies for Green Residential Buildings** D. Ting and R. Carriveau (Editors) |
| Volume 156 | **Lithium-Ion Batteries Enabled by Silicon Anodes** C. Ban and K. Xu (Editors) |
| Volume 157 | **Electrical Steels, 2 Volumes** A. Moses, K. Jenkins, Philip Anderson and H. Stanbury |
| Volume 158 | **Advanced Dielectric Materials for Electrostatic Capacitors** Q Li (Editor) |
| Volume 159 | **Transforming the Grid Towards Fully Renewable Energy** O. Probst, S. Castellanos and R. Palacios (Editors) |
| Volume 160 | **Microgrids for Rural Areas: Research and case studies** R.K. Chauhan, K. Chauhan and S.N. Singh (Editors) |
| Volume 161 | **Artificial Intelligence for Smarter Power Systems: Fuzzy logic and neural networks** M.G. Simoes |
| Volume 165 | **Digital Protection for Power Systems, 2nd Edition** Salman K Salman |
| Volume 166 | **Advanced Characterization of Thin Film Solar Cells** N. Haegel and M Al-Jassim (Editors) |
| Volume 167 | **Power Grids with Renewable Energy** Storage, integration and digitalization A. A. Sallam and B. Om P. Malik |
| Volume 169 | **Small Wind and Hydrokinetic Turbines** P. Clausen, J. Whale and D. Wood (Editors) |
| Volume 170 | **Reliability of Power Electronics Converters for Solar Photovoltaic Applications** F. Blaabjerg, A.l Haque, H. Wang, Z. Abdin Jaffery and Y. Yang (Editors) |
| Volume 171 | **Utility-Scale Wind Turbines and Wind Farms** A. Vasel-Be-Hagh and D.S.-K. Ting |
| Volume 172 | **Lighting Interaction with Power Systems, 2 volumes** A. Piantini (Editor) |
| Volume 174 | **Silicon Solar Cell Metallization and Module Technology** T. Dullweber (Editor) |
| Volume 175 | **n-Type Crystalline Silicon Photovoltaics: Technology, applications and economics** D. Munoz and R. Kopecek (Editors) |
| Volume 178 | **Integrated Motor Drives** Xu Deng and B. Mecrow (Editors) |
| Volume 180 | **Protection of Electricity Distribution Networks, 4th Edition** J. Gers and E. Holmes |
| Volume 182 | **Surge Protection for Low Voltage Systems** A. Rousseau (Editor) |
| Volume 184 | **Compressed Air Energy Storage: Types, systems and applications** D. Ting and J. Stagner |
| Volume 186 | **Synchronous Reluctance Machines: Analysis, optimization and applications** N. Bianchi, C. Babetto and G. Bacco |
| Volume 190 | **Synchrophasor Technology: Real-time operation of power networks** N. Kishor and S.R. Mohanty (Editors) |
| Volume 191 | **Electric Fuses: Fundamentals and new applications, 4th Edition** N. Nurse, A. Wright and P.G. Newbery |
| Volume 193 | **Overhead Electric Power Lines: Theory and practice** S. Chattopadhyay and A. Das |
| Volume 194 | **Offshore Wind Power Reliability, Availability and Maintenance, 2nd Edition** P. Tavner |
| Volume 196 | **Cyber Security for Microgrids** S. Sahoo, F. Blaajberg and T. Dragicevic |
| Volume 198 | **Battery Management Systems and Inductive Balancing** A. Van den Bossche and A. Farzan Moghaddam |
| Volume 199 | **Model Predictive Control for Microgrids: From power electronic converters to energy management** J. Hu, J.M. Guerrero and S. Islam |
| Volume 201 | **Design, Control and Monitoring of Tidal Stream Turbine Systems** M. Benbouzid (Editor) |
| Volume 204 | **Electromagnetic Transients in Large HV Cable Networks: Modeling and calculations** Ametani, Xue, Ohno and Khalilnezhad |
| Volume 208 | **Nanogrids and Picogrids and Their Integration with Electric Vehicles** S. Chattopadhyay |
| Volume 210 | **Superconducting Magnetic Energy Storage in Power Grids** M.H. Ali |

Volume 211 **Blockchain Technology for Smart Grids: Implementation, management and security** H.L Gururaj, K.V. Ravi, F. Flammini, H. Lin, B. Goutham, Kumar B.R. Sunil and C. Sivapragash
Volume 212 **Battery State Estimation: Methods and models** S. Wang
Volume 215 **Industrial Demand Response: Methods, best practices, case studies, and applications** H.H. Alhelou, A. Moreno-Muñoz and P. Siano (Editors)
Volume 213 **Wide Area Monitoring of Interconnected Power Systems, 2nd Edition** A.R. Messina
Volume 214 **Distributed Energy Storage in Urban Smart Grids** P.F. Ribeiro and R.S. Salles (Editors)
Volume 217 **Advances in Power System Modelling, Control and Stability Analysis, 2nd Edition** F. Milano (Editor)
Volume 225 **Fusion–Fission Hybrid Nuclear Reactors: For enhanced nuclear fuel utilization and radioactive waste reduction** W.M. Stacey
Volume 227 **Processing and Manufacturing of Electrodes for Lithium-Ion Batteries** J. Li and C. Jin (Editors)
Volume 228 **Digital Technologies for Solar Photovoltaic Systems: From general to rural and remote installations** S. Motahhir (Editor)
Volume 236 **Green HV Switching Technologies for Modern Power Networks,** K. Niayesh (Editor)
Volume 237 **High Voltage Power Transformers: State of the art and technological innovations** M. Sforna, S.Zunino and F.Ferrari
Volume 238 **AI for Status Monitoring of Utility Scale Batteries** S. Wang, K. Liu, Y. Wang, D. Stroe, C. Fernandez and J.M. Guerrero
Volume 239 **Intelligent Control of Medium and High Power Converters** M. Bendaoud, Y. Maleh and S. Padmanaban (Editors)
Volume 247 **Power System Strength Evaluation Methods, Best Practice, Case Studies, and Applications** H. Alhelou, N. Hosseinzadeh and B. Bahrani (Editors)
Volume 250 **Interactions of Wind Turbines with Aviation Radio and Radar Systems** Alan Collinson
Volume 905 **Power System Protection, 4 volumes**

# Wind Turbine System Design

## Volume 2: Electrical systems, grid integration, control and monitoring

Edited by
Jan Wenske

The Institution of Engineering and Technology

Published by The Institution of Engineering and Technology, London, United Kingdom

The Institution of Engineering and Technology is registered as a Charity in England & Wales (no. 211014) and Scotland (no. SC038698).



First published 2023



The Institution of Engineering and Technology
Futures Place
Kings Way, Stevenage
Hertfordshire, SG1 2UA, United Kingdom

www.theiet.org



**British Library Cataloguing in Publication Data**
A catalogue record for this product is available from the British Library

**ISBN 978-1-78561-858-1 (hardback) vol 2**
**ISBN 978-1-78561-859-8 (PDF) vol 2**

**ISBN 978-1-78561-856-7 (hardback) vol 1**
**ISBN 978-1-78561-857-4 (PDF) vol 1**

**ISBN 978-1-78561-864-2 2 vol set (hardback)**

Typeset in India by MPS Limited

Cover Image: ZU_09/E+ via Getty Images

# Contents

# About the editor

**Jan Wenske** is a professor at the University of Bremen and deputy director of the Fraunhofer-Institute for Wind Energy Systems (IWES), Germany, since 2011. He studied mechanical engineering and received his Ph.D. in 1999 from the Institute of Electrical Engineering at the TU Clausthal. Professional assignments include the Department of Drivetrain Development and the Pre-development for forklift trucks at STILL GmbH from 2000 to 2004 as well as the Department Power Electronics Development at Jenoptik Defense & Civil Systems from 2005 to 2010. He works intensively with major industry players in wind energy technology.

# Preface

## Synopsis

Research and development in the field of wind energy utilization and the advancement of modern onshore and offshore wind turbines have entered a unique, intricate, and multidisciplinary realm. Thanks to significant progress in modeling, simulation, experimental validation, and measurements, it is now feasible to intricately model the subsystems of a wind turbine for comprehensive multi-domain analysis. Conversely, when translating this model into an actual turbine design, a thorough assessment of the complex interactions among the subsystems is imperative. This acquired knowledge must then be seamlessly integrated into the design process. While turbine manufacturers have established and optimized their internal processes, they have sometimes aligned these processes primarily to satisfy the requirements of certifying authorities. Presently, at least for onshore turbines, which remain highly complex products, there exists a constant risk of them devolving into commodities. This means they could be perceived by customers as nearly identical across manufacturers, with price being the sole easily discernible differentiator, despite significant differences existing in intricate details. Manufacturers are already under immense pressure to cut costs, as evidenced by the swift consolidation trend among both manufacturers and equipment suppliers. This pressure is further highlighted by the introduction of new platforms, modifications, and variants in increasingly compressed timeframes. Within the wind industry, the need for continuous innovation and optimization is paramount in order to break free from this situation.

Research institutions and universities are actively striving to make valuable contributions by offering specialized expertise across a diverse range of fields. This proficiency encompasses wind physics, delving into the intricate nature of wind with its intricate and stochastic turbulences, as well as atmospheric dynamics. Furthermore, these institutions are dedicated to exploring turbine technology and controls, aerodynamics, civil and ocean engineering, grid integration, and even material science. This collective effort significantly bolsters the advancement of the industry. The offshore sector has undergone rapid development, optimizing all facets of the value chain within a relatively short time span, particularly in logistics, foundation, and construction. The continuous trend to reduce costs and achieve even lower levelized cost of energy (LCoE) remains the predominant catalyst for technical innovations. A few years ago, operators of new offshore wind farms held expectations that the rated output of next-generation offshore wind turbines would nearly double to approximately 15 MW within the subsequent five years, allowing for operation in a subsidy-free electricity generation market. This anticipation has indeed materialized, reflecting the dynamic pace of market evolution. Given these rapid market advancements, there is a compelling case for synergizing industrial, application-oriented research on the one hand, and academic research in wind power on the other, to ensure effective and efficient progress in this field.

Having access to appropriate specialized literature is essential for students, Ph.D. candidates, industry experts, and individuals interested in career changes. It serves as a fundamental catalyst, inspiring a deeper exploration of the fascinating realm of wind turbines. An extensive array of standard literature is accessible for wind energy utilization and turbine technology, often focused to offer readers an initial comprehensive introduction and overarching insight into the world of wind energy. Furthermore, specialized literature is readily available across specific domains, encompassing simulation, modeling, control systems, aerodynamics, hydro- and aeroelasticity, fiber-based materials, and general structural dynamics. These resources provide an extensive collection of in-depth knowledge tailored to those seeking specialized expertise.

So why create a small book series dedicated to turbine technology, particularly emphasizing the power mechatronics of the drivetrain, nacelle systems, and grid integration? The motivation behind this endeavor lies in addressing existing gaps within the literature, specifically targeting application-oriented readers in engineering, mechatronics, controls, and the energy system sector. This series is addressed to individuals at various stages of their professional careers, including novices, engineers, scientists, and students, all engaged in the realm of applied research within wind energy systems. With the overarching aim of bridging the gap between foundational knowledge and practical application, while accounting for pertinent boundary conditions (ranging from construction-related aspects to system considerations and economic factors), these books are designed to provide insightful answers to more practical related questions such as:

- How lightning protection is managed within a wind turbine?
- What's about personal safety against electrical hazards – a matter of system design?
- What is essential for designing generator-converter systems for wind turbines?
- How control of a wind turbine was and will be realized from the hardware and software perspective?
- Which are the boundary conditions to configure a direct-drive generator?
- What are the secrets of a proper DFIG system design?
- What is a modern, generic controller design for a wind turbine?
- What is the status of current monitoring technologies?
- Is an extensive test and validation program for the electrical system characteristics reasonable?

- X-in-loop testing, what does this mean for wind turbines and the grid integration
- Which control challenges coming next, corresponding to future requirements of large wind farms?
- What kind of grid integration topics will be the next big challenges for the wind turbines?
- ...

This book aims to cultivate the reader's practical design acumen, instilling a tangible understanding of the intricacies involved. Naturally, this book cannot delve into every design process down to the minutiae. Instead, its objective is to illustrate these processes through examples while judiciously referencing additional literature or established requirements and standards as necessary. Upon completing this book, readers will possess the ability to comprehend the design intricacies of specific components and subsystems within modern wind turbines and even delve into their more detailed specifications. They will also gain a robust grasp of boundary conditions, best practices, pitfalls to avoid, system interaction, and the requisites inherent in the design journey of nacelle systems and complete drive-trains. While some chapters may offer a more application-oriented overview of the current technological landscape and ongoing research, such as in the realms of monitoring, controls, signals, and test and validation concepts, others will delve into the practical design process.

Hence, each chapter has been authored by established experts or small, select groups. These teams consist of industry professionals, seasoned researchers, or a blend of the two, with a consistent presence of authors possessing industry insight to ensure the pragmatic applicability of the content. Every chapter is designed to provide concise foundational knowledge, often including a historical perspective, illustrative design process instances, requisites, encountered challenges, avenues for optimization, and prospects for future research.

## Readership, who might be interested

This book series is geared towards scientists and engineers who have an interest in or are actively engaged in wind turbine design, particularly within the realm of drivetrain, mechatronics, and electrical power systems. Additionally, the series holds significant value for students studying wind energy systems, particularly those specializing in mechanical design, simulation, mechatronics, or electrical engineering. These volumes (Volume 1, Volume 2) aim to foster a comprehensive and practical grasp of the intricacies of the entire wind turbine as a complex system. When utilized in conjunction with established foundational literature on wind energy systems and other titles within the IET wind energy series (such as "Modelling and Simulation"), this book serves as a valuable practical supplement, enriching the reader's understanding and expertise.

In this manner, engineers and professionals within the wind industry can gain a comprehensive insight into this subject area. This is particularly beneficial for those whose daily responsibilities encompass interconnected subsystems, such as tower or blade experts, as well as drivetrain specialists from other sectors who welcome fresh inspiration and perspectives from the wind industry. This series also caters to specialized students aiming to grasp the intricate demands of turbine and subsystem design. The purpose of this concise book series is to make a valuable contribution towards effectively addressing essential prerequisites and navigating the intricate challenges within the realms of electromagnetic transient (EMT) and multi-body simulation (MBS), finite element method (FEM) calculations, advanced fatigue analysis for mechanical components and systems, controls, generator design, and power electronics development. These volumes are intended to facilitate the successful advancement of even more sophisticated wind turbines, as well as to equip stakeholders for forthcoming challenges in system integration.

## Wind Turbine System Design and its authors

### *Volume 2: Electrical systems, grid integration, control, and monitoring*

When it comes to new product developments, ensuring the safety of human life and property is always paramount. Modern wind turbines are advanced electrical power generation units, boasting multi-megawatt output power and substantial mechanical dimensions, which inherently present a fundamentally high-risk potential. The contents of this volume (Volume 2) primarily delve into electro-technical subjects, focusing on the electrical drivetrain, including the generator, main converter, and grid integration. It also encompasses control technology, measurement technology, and fieldbus systems for data exchange within the turbine. The electrical systems of wind turbines are exposed to specific hazards based on their sites and characteristics, such as lightning strikes, and can, in turn, be sources of potential dangers due to high electrical voltages and currents. This encompasses risks like electric shock for operating and service personnel, as well as the possibility of spark-induced fire hazards. The central objective is to manage and, if necessary, minimize these potential risks. Therefore, the initial chapter of this book is dedicated to electrical safety. A team of authors from specialized companies, Bender GmbH and DEHN SE, provides an expert and practical perspective. They detail the design boundary conditions, criteria, and technical solutions essential for mitigating these electrical hazards effectively within wind turbines.

The author team of Chapter 1 consists of **Benjamin Moosburger**, he is the Global Business Development Manager at DEHN SE with a focus on the energy segment (Wind, PV). He completed an apprenticeship as an electronics technician at DEHN before going on to study electrical engineering at the University of Applied Science OTH Regensburg. After graduation, he returned to DEHN as an R&D Engineer for information technology. Later, in his role as Project Manager for R&D projects, he undertook further training to become a certified Project Management Professional (PMP) and Scrum Master (PSM). Since 2022, he has held the position of Global Business Development Manager at DEHN SE with a focus on the energy segment (wind, PV, and grid). **Christian Vögerl** is the head of Sales Region South and West Europe at DEHN SE. After graduating as a Bachelor Professional in Electrical Technology, he started his career at DEHN as a market manager for photovoltaics. He subsequently deepened his expertise in the field of renewable energies in a number of different positions: Solution Manager Photovoltaics, Business Development Manager Energy Generation (wind & PV), and Global Account Manager Energy Generation. From February 2021, he led the Business Development Infrastructure and Renewables team at DEHN SE before becoming Head of Sales Region South and West Europe in mid-2022.

Dipl.-Ing. **Tilo Püschel** studied electrical engineering and communications engineering and was first employed for many years in an international company in the Data Centre Infrastructure division. He then worked on the industrial utilization of direct current which at the same time integrates renewables (photovoltaics, wind energy). During this time, he was active in working groups of the IEC (International Electrotechnical Commission) and the German VDE (Association for Electrical, Electronic & Information Technologies). He then focused on the combination of renewable energy with electrolysis processes for the production of hydrogen. His general interest in the operation of electrical installations that are both fail-safe and in compliance with the applicable laws and standards led him to Bender GmbH & Co. KG in Grünberg (Germany), where he presently works as market segment manager. Moreover, Dipl.-Ing. **Jörg Irzinger** studied electrical engineering at the University of Applied Sciences Gießen Friedberg with a focus on automation technology. He then worked in the field of electrical drive systems and automation technology as an application and sales engineer for a manufacturer of linear synchronous motors and handling systems. At Bender GmbH & Co. KG, he then focused on the insulation monitoring of IT systems working for 10 years as a product manager for insulation monitoring devices and insulation fault locators. Currently, he works in the same company's Standards & Innovation department focusing on industrial topics in the field of electrical standardization.

The author of Chapter 2 delves deeper into the subject of the electrical drivetrain and focuses on one of its key components, the generator. As previously introduced in Volume 1 (Chapter 5), there is a wide array of technical variants for wind turbine drivetrain solutions. In this chapter, Henri Arnold provides a comprehensive overview of suitable generator concepts, encompassing both asynchronous and synchronous machines. The chapter digs deeper into the theoretical, mathematical but also practical aspects of these concepts except direct drive applications. Consequently, this chapter primarily concentrates on generator solutions tailored for geared turbine variations, which are sometimes referred to as high- and medium-speed generator applications. Dr.-Ing. **Henri Arnold** studied mechanical engineering with an emphasis on mechatronic at the Technical University Bergakademie Freiberg, where he got his Diploma in 2002. In the same year, he started there to work at the Institute of Electrical Engineering as a scientific employee. His main interest was field-oriented control, simulation, and parameter identification of electrical machines. He finalized his doctoral thesis on parameter identification of asynchronous machines in 2006. In the year 2007, he joined the VEM Sachsenwerk GmbH. There he was responsible for the electromagnetic design of wind power generators, the development of calculation and simulation software, and later the technical lead for drive systems within the VEM group. Since 2020, he has worked as a freelance engineer, mainly doing electromagnetic calculation, simulation, teaching, and programming.

Chapter 3 supplements the introduction to generator technology from the previous chapter with regard to the special variant of the Direct Drives in gearless wind turbines. Two industry experts, Tobias Muik and Stephan Jöckel, with a lot of theoretical and practical experience in Direct Drive development, give us a dedicated insight into concepts, development, and wind energy application, especially for permanent magnet synchronous generators. M.Sc. **Tobias Muik** graduated in electrical engineering in 2009 at the Darmstadt University of Technology with a focus of designing electrical machines, power electronics, and motion control. Since 2009, he has been working at wind-direct GmbH, a wind engineering consultant located in Mannheim, Germany. At wind-direct, he has been dealing with electrical machines as well as the whole drivetrain design for Direct Drive wind turbines. Besides designing he had been involved in prototyping and commissioning as well. In 2019, he shifted to the research and development department at ENERCON where he is currently working as a Senior Expert on various topics in the field of electrical machines and drivetrain design. Dr.-Ing. **Stephan Jöckel** studied Mechanical Engineering and received his Ph.D. on "Generator Systems for Wind Turbines" in 2002 at the Institute of Electrical Energy Conversion at the Darmstadt University of Technology. Professional assignments since 2000 included wind turbine OEMs like GE Wind Energy and Vensys Energiesysteme AG. Between 2008 and 2019, he worked as Managing Director of wind-direct GmbH, a wind engineering consultant located in Mannheim/Germany. In 2019, he shifted to the research and development department at ENERCON where he is currently engaged as chief engineer in the pre-development team.

The main converter, along with its hardware-related firmware, constitutes the second pivotal element within the electrical drivetrain of a wind turbine. Often, it holds the determining influence over the system's behavior and attributes at the common point of coupling (PCC). The rapid evolution of power electronics technology over recent decades has endowed these components with a significant role in facilitating grid integration and characterization of the reliability of a turbine's electrical system. This trajectory, particularly in the context of wind turbine grid integration and supplementary services, is expected to gain further momentum in the forthcoming years. Chapter 4 describes the team of authors from science, Holger Groke, and the industry, Nobert Hennchen, the development and trends for the use of power electronics in wind turbines. Dipl.-Phys. **Norbert Hennchen** is the managing director and owner of Freqcon GmbH. He pursued his studies in Physics at the University of Bremen and has amassed extensive experience as a designer and leader of development teams for wind turbine control and converter systems at various wind turbine original equipment manufacturers (OEMs) since 1988. In 2005, he established his own company, where he continues to drive advancements in new technologies and concepts for wind energy converter designs and grid-connected battery systems. His efforts are dedicated to assisting industrial clients in addressing the evolving demands stemming from the transformation of power supply systems. FreqCon's offerings span the international market. Dr.-Ing. **Holger Groke**'s academic journey led him to specialize in electrical engineering at the University of Bremen. His expertise lies in converter control for a wide array of electrical machines, encompassing both conventional types such as doubly fed induction machines and more unconventional varieties like transversal flux machines. Since 2013, he has assumed the role of Chief Scientist at the Institute for Electrical Drives, Power Electronics, and Semiconductors (IALB) at the University of Bremen. Within this capacity, he oversees the control of the electrical drives group. Holger's exceptional focus centers on wind energy applications, a commitment evidenced by his involvement in numerous public research projects in collaboration with industry partners. His diverse background, which includes prior training as a professional radio and television technician, equips him not only with a profound understanding of control theory and software expertise but also with practical, hands-on experience, and a keen interest in the electronic and communications engineering facets of control implementation.

As the proportion of renewable energy sources integrated into the electrical power supply system continues to rise, the demands for future grid operation are correspondingly escalating. The structures, requirements, and limitations built upon the existing principles of conventional large-scale power plants are undergoing rapid transformation. A reliable and stable electrical energy supply is also the top priority for future power system operation and a basic requirement for competitive industrial locations and broad social acceptance. This underscores the pivotal role of Chapter 5 – "Grid compliance and electrical system characterization" – within this book, even in the context of individual turbine design. It is evident that the imperative expansion of wind energy, both locally and globally, is intricately linked to the prospective grid operation. The author Torben Jersch, who possesses over a decade of expertise in this field, collaborates with his team to streamline solutions and rationalizes processes for evaluating the electrical characteristics of wind turbine and wind farm grid integration.

M.Sc. **Torben Jersch** completed his studies in electrical engineering at the University of Applied Sciences Hamburg. Early in his career, he served as a development engineer for power electronics at the company Jenoptik from 2008 to 2011. During this time, his focus was on grid-forming inverters within high-power onboard electrical systems of specialized vehicles. In 2011, Torben joined Fraunhofer IWES as a research assistant. In this role, he undertook responsibilities for designing, constructing, and overseeing the commissioning of the electrical systems for the 10 MW nacelle test rig located at IWES's DyNaLab. Progressing in his career, he took on the role of group leader for electrical systems and grid integration, and in 2018, he assumed the position of Head of the Systems Technology Department, a role he continues to hold. Since 2015, he has been a driving force in advancing the field of grid integration testing for wind turbines through power electronic grid emulators on a international scale. His contributions extend to active participation in working groups responsible for developing

Technical Guideline 3 – focused on determining the electrical attributes of generation units, plants, storage facilities, and their components within medium, high, and extra-high voltage grids. Additionally, Torben is an engaged member of the Technical Committee for IEC 61400-21, which pertains to the measurement and assessment of electrical characteristics of wind turbines.

As previously mentioned, the electrical drivetrain, and more notably, the main converter, plays a pivotal role in determining the electrical characteristics of the wind turbine. Given the intricate control structures and dynamic, nonlinear system behavior, validation is crucial, as elucidated in the V-model introduced earlier (refer to Volume 1, Chapter 9), ideally employing a fitting testing methodology. In recent decades, x-in-the-loop methods have gained prominence. These methods involve integrating the device under examination into a virtual test environment, which subsequently emulates physically absent components, such as the wind rotor, mechanical system, electrical power network, or control sensors, through test bench interfaces. This approach expedites the validation of component and system properties, offering notable time and cost savings. While this methodology might appear straightforward, its implementation constitutes a scientific challenge of its own. Thus, Chapter 6 assembles a sizeable team of authors to comprehensively describe the extensive range of variations and applications of this testing method, particularly in the context of wind energy.

Dr.-Ing. **Florian Hans** is an electrical engineer holding a doctoral degree in control engineering from the Technische Universität Braunschweig, Germany. Since joining the Department of Power Electronics and Grid Integration at Fraunhofer IWES in 2021, he has been responsible for the power electronic grid simulator of the nacelle test (DyNaLab) rig in Bremerhaven. In this context, he is also involved in the simulation of large power networks on real-time platforms and their integration at the test bench level. In addition, he is working on projects related to direct current technologies and grid connection systems for offshore wind farms. M.Sc. **Mohsen Neshati** works as a lead engineer and is active in the wind industry, in the field of grid connection of wind energy converters, too. His background is in electrical and controls, and he started as a development engineer at Fraunhofer IWES and worked for 8 years at the dynamic nacelle-testing laboratory in Bremerhaven, Germany. During this time, he worked on his Ph.D. on mHiL test systems for full-scale wind turbine nacelles. Dr.-Ing. **Oliver Feindt** is an engineer with a background in aeronautical and systems engineering studies. He has worked on the development of the Ariane 5 and 6 launchers at ArianeGroup and later joined the Institute of Automation at the University of Bremen for research focused on control and optimization of district heating networks. Since 2023, he has been working at Fraunhofer IWES, where he specializes in mHiL control of nacelle and generator/converter test benches. Dipl.-Ing. **Adam Zuga** is an automation engineer with a background in computer engineering studies. After graduation, he worked for Siemens Wind Power for 3 years in the area of mobile and IT solutions before joining Fraunhofer IWES in 2009. There he first worked in the field of wind turbine simulation and load calculation. Since 2014, he has been working in the field of systems engineering at the Dynamic Nacelle Testing Laboratory (DyNaLab). Since

2018, he leads the group Process Automation and Real-Time Systems, which is responsible for the automation of test rigs for wind turbine systems and components as well as the integration of real-time simulation models using hardware-in-the-loop (HIL) methodologies mainly in the context of grid compliance testing. Last but not least M.Sc. Nils Johannsen, who is introduced below as the main author of Chapter 7.

As previously indicated, in Chapter 7 Nils Johannsen from Beckhoff Automation provides a comprehensive overview of his extensive expertise and insights into practical concepts and designs pertaining to wind turbine controls and their software architecture. M.Sc. **Nils Johannsen** pursued his studies in electrical and information engineering at the West Coast University of Applied Sciences in Heide and the Distance-Learning University of Hagen. His bachelor thesis from 2010 on the development of a fuzzy controller for the pitch angle of wind turbines was awarded an industry prize for outstanding final theses. Subsequently, he embarked on a journey within the wind energy industry, assuming a managerial role at Beckhoff Automation. His involvement in numerous end-to-end development projects aimed at automating wind turbines has been significant. Focusing on information technology and applied computer science, his achievements encompass a comprehensive software framework for wind turbine automation, coupled with a master's thesis centered on integrating and interfacing simulation environments with real-time systems. Over time, Nils Johannsen has transitioned into the position of technical manager, overseeing development, research, and application endeavors within the wind industry.

Sensors and monitoring play an increasingly pivotal role within wind turbines. They serve as indispensable components for enhancing turbine reliability and fostering cost-efficient maintenance strategies. Ongoing advancements in measurement and evaluation methods continuously refine the capability to directly or indirectly assess plant conditions through intelligent measurements. In the context of wind turbines, a clear distinction is drawn between condition monitoring for rotating machine elements (such as bearings and gears) and structural health monitoring for various structural components (including rotor blades, support structures, and foundations). In Chapter 8, experts from Wölfel Engineering and the University of Siegen provide a comprehensive overview of the current state-of-the-art sensor technology and evaluation methods within the realm of structural health monitoring systems for wind turbines. The following contributors have lent their expertise to this chapter.

Dr.-Ing. **Carles Colomer Segura** is the Head of Development of Monitoring Systems at Wölfel Engineering GmbH Co. + KG. He graduated as a Civil Engineer in 2007 from the UPV Valencia, Spain. He obtained his Ph.D. in 2016 from the RWTH Aachen, Germany, where he specialized in structural dynamics, wind engineering, and software development. In 2016, he joined Wölfel and worked in different roles as a Data Scientist (2016–2017), Head of Data Processing and Analytics (2017–2022), and Head of Development of Monitoring Systems (since 2022). Prof. Dr.-Ing. **Peter Kraemer** has been a Professor of Mechanics with a focus on Structural Health Monitoring at the University of Siegen since 2018.

Between 2016 and 2018, he was a Professor of Machine Diagnostics and Machine Dynamics at the Bochum University of Applied Sciences. His professional experience is shaped by his collaboration with Wölfel Engineering (2011–2016), where he was responsible for the business unit "Signal and Data Analysis" and developed methods and algorithms for different SHM products, especially with applications in wind energy. Dr.-Ing. **Carsten Ebert** is the CTO of the Wölfel Group and is responsible for the Structure Monitoring business unit. He studied civil engineering at the HTWK Leipzig and completed his diploma and master's degree. He received his Ph.D. from the University of Siegen for a thesis on vibration-based damage detection in bridge structures. He has been working for Wölfel since 2008 and is also active in several working groups for the preparation of standards, including VDI4551 for structure monitoring and assessment.

Chapter 9 that follows; extends the discussion on controls but adopts a broader perspective – referred to as a meta-system approach – encompassing wind turbine control in its current state (as detailed in Chapter 7) and delving into the optimization of wind farm control within the context of diverse, interconnected, and multi-domain control objectives. While this may be considered commonplace with the state-of-the-art hardware (such as PLCs, field bus communication, 5G technology, etc.), it remains less standardized and common within the wind energy industry and is a topic of greater emphasis within academia, but with a lot of potential practical benefits in term of wind farm control. Niklas Requate and Tobias Meyer, who have dedicated several years to this endeavor at Fraunhofer-IWES, share their insights and general concepts in this chapter, offering a valuable perspective on this evolving field. M.Sc. **Niklas Requate** completed his M.Sc. studies in industrial mathematics at the University of Bremen in 2018. Afterward, he joined the Advanced Control Systems group at Fraunhofer-IWES as a research assistant. He works on intelligent methods for the operational management of wind energy systems considering technical reliability. The systems considered can be individual wind turbines, entire wind farms, or, more recently, the coupled operation of wind turbines with an electrolyzer for hydrogen production. For his work, the interaction of different areas from control to load calculation to economic evaluation is essential. The topic is also part of his Ph.D. work, which he is working on. Dr.-Ing. **Tobias Meyer** heads the group Advanced Control Systems at Fraunhofer-Institute for Wind Energy Systems (IWES). His research focuses on cross-system control loops and their use to optimally operate wind turbines and wind farms. He is particularly interested in applying control-engineering methods to problems that have not yet been considered as control loops. His main interest is resource conservation over the entire system and its entire lifetime. To this end, he and his team are developing methods for optimizing the operation of wind turbines that allow for the production of as much energy as possible from an existing structure. Previously, he was a research associate at the Paderborn University, where he conducted research on the autonomous adaptation of self-optimizing systems. Tobias studied mechanical engineering.

Finally, in Chapter 10, M.Eng. Christoph Kaufmann and the team at the Fraunhofer Application Center ILES (Integration of Local Energy Systems) discuss the future challenges of transforming today's largely centrally organized energy power supply system. According to the status, the operation of a converter-dominated grid with highly decentralized generator structures from renewable energy sources, storage systems, and intelligent consumers seems to be possible, but still requires extensive research and adapted standards and advanced grid codes for grid integration.

**Christoph Kaufmann** received his B.Sc. in Industrial Engineering and Management from the University of Hamburg in 2016 and worked in different fields related to renewable energies. In 2018, he did a research internship at the research and technology transfer center CITCEA-UPC during an exchange semester at the Polytechnic University of Catalonia. After that, he received an M.Eng. from the Hamburg University of Applied Sciences in 2019. Since then, he has been working for the Fraunhofer Society, where he is currently working for the Fraunhofer Institute for Wind Energy Systems IWES at the Application Center for Integration of Local Energy Systems in Hamburg. In 2021, he started a Ph.D. related to power system analysis using multilinear models at the Polytechnic University of Catalonia together with the Hamburg University of Applied Sciences. His research concerns converter-dominated power systems, its modeling, and analysis, with a focus on grid-forming converters and the integration of electrolyzer systems. In this book chapter, he received great support from the ILES expert team, whose members are briefly introduced below. **Aline Luxa** received a B.Sc. and an M.Sc. degree in energy and process engineering from the Technical University of Berlin in 2017 and 2020, respectively. Since 2020, she has been part of the Application Center ILES at the Fraunhofer Institute for Wind Energy Systems. Her research topics include modeling and control of grid-supporting hydrogen systems with a focus on electrolyzers. Since 2023, she has been enrolled in the Ph.D. program for electrical engineering at the Leibniz University of Hannover in cooperation with the Hamburg University of Applied Sciences. M.Sc. **Marina Nascimento Souza** has an academic background in process engineering. Marina's research focus lies on the detailed modeling and simulation of electrolyzers, where she aims on contributing to the energy revolution towards decarbonization. Also with it, M. Eng. **Carlos Cateriano Yáñez**, born in Arequipa, Peru. Carlos received a B.Eng. degree in industrial and systems engineering from the University of Piura, Lima, Peru, in 2012 and an M.Eng. in renewable energy systems from the Hamburg University of Applied Sciences, in 2017. He is currently pursuing a Ph.D. degree in automation, robotics, and industrial computer science at the Universitat Politècnica de València, Valencia, Spain, in cooperation with the Hamburg University of Applied Sciences. Since 2017, Carlos is working at the Fraunhofer-Gesellschaft zur Förderung der angewandten Forschung e.V. in Hamburg, Germany. His research interest includes model predictive control and its applications in the field of power systems integrated with renewable energy sources. And, last but not least an expert in advanced control of energy systems, Dr.-Ing. **Georg Pangalos** received the Dipl.-Ing. degree in electrical engineering and his Ph.D. in control engineering from the Hamburg University of Technology in 2011 and 2015, respectively. Since 2015, he has been working at the Fraunhofer-IWES Application Center ILES in

Hamburg, Germany. His research interest includes the areas of renewable energy integration and energy systems transition.

## *Volume 1: Nacelles, drivetrains and verification*

The content of Volume 1 will be succinctly outlined in this section. A comprehensive breakdown of chapters and the contributing authors can be found in the corresponding preface of Volume 1 of this book series. Unlike Volume 2, Volume 1 primarily delves into mechanical loads, the mechanical drivetrain, auxiliary mechanical and hydraulic systems, and approaches to validation and verification (including drivetrain, pitch and yaw systems, simulation, load calculation, and mechanical testing) within the framework of wind turbine systems and components. A thorough analysis of technical constraints and intricate interplays among mechanical subsystems within the nacelle system is expounded upon. The essential assumptions underlying design considerations, such as those for pitch and yaw drives, rotor main bearings, and gearboxes, are meticulously outlined. Furthermore, the authors place significant emphasis on their specific practical expertise in system design throughout this volume. The examples provided there for designs and dimensioning calculations are mostly related to a generic 7.5 MW wind turbine or its corresponding turbine class as defined by Fraunhofer-IWES. Within the intricate process of developing complex technical products like a wind turbine, it has proven effective to contemplate the validation and verification strategy for planned or developed components and systems during the developmental phase. This strategy is then integrated into a comprehensive test and validation plan. Volume 1 introduces this well-established approach, following the general V-model, and elaborates on it through the lens of the structural and mechanical core components and systems of the wind turbine. In this regard, Volume 2 seamlessly carries forward this methodology, expanding its application to encompass electrical and control systems. Additionally, it introduces novel methodologies such as component-based testing for intricate electrical systems.

The editor and the entire team of authors, driven by a profound commitment and shared passion for wind energy, have meticulously written this book for those who hold an interest in this subject. We sincerely hope that all readers find pleasure in perusing its contents and extend our good wishes for your prosperous endeavors in the captivating realm of wind energy systems.

With best regards
**Jan Wenske (Ed.)**

# Acknowledgments

On behalf of all the authors of this book, the editor would like to extend sincere gratitude to the following companies and institutions for their generous support in facilitating the publication of this work. The information and images provided by these companies are invaluable for enhancing the comprehension and elucidation of the intricate domains of knowledge explored within this book.

Beckhoff Automation GmbH & Co. KG
Bender GmbH & Co. KG
Bundesverband WindEnergie e.V.
DEHN SE
ECO 5 GmbH
Flender GmbH
Fraunhofer-IWES
Fraunhofer-Anwendungszentrum ILES
FREQCON GmbH
Henri Arnold – Ingenieur für elektrische Antriebstechnik
OPAL-RT TECHNOLOGIES, Inc.
PIETRO CARNAGHI S.P.A.
Typhoon HIL, Inc.
University Bremen IALB
University Siegen, Department of Mechanical Engineering
VEM Sachsenwerk GmbH (VEM Group)
Yantai Dongxing Magnetic Materials Inc (YSM)
Wind-Direct GmbH
Wölfel Engineering GmbH + Co. KG

# Abbreviations and terminologies

| | |
|---|---|
| **4QC** | Four-Quadrant-Circuit |
| **AC** | Alternating Current |
| **AEP** | Annual Energy Production |
| **AI** | Artificial Intelligence |
| **ANN** | Artificial Neural Network |
| **API** | Application Programming Interface |
| **B2B** | Back-to-Back |
| **BDF** | Backward Differentiation |
| **BMS** | Building Management System |
| **BOM** | Bill of Material |
| **C** | Capacitor |
| **CAPEX** | Capital Expenditure |
| **CB** | Converter Unit |
| **CC** | Central Controller |
| **CCM** | Converter Control Module |
| **CE** | Conformite Europeenne |
| **CHIL** | Controller-HIL |
| **CI** | Continuous Integration |
| **CIG** | Converter-Interconnected Generator |
| **CM** | Condition Monitoring |
| **CMS** | Condition Monitoring System |
| **$CO_2$eq.** | Carbon Dioxide Equivalent |
| **CPU** | Central Processing Unit |
| **CSI** | Current Source Inverters |
| **CSV** | Complex Space Vectors |
| **CSV** | Comma-Separated Values |
| **DC** | Direct Current |
| **DC-I/O** | DC-Input and DC-Output |
| **DD** | Direct-Drive |
| **DER** | Distributed Energy Resource |
| **DERs** | Distributed Energy Resources |

| | |
|---|---|
| **DES** | Decentralized Energy System |
| **DFIG** | Doubly-Fed Induction Generator |
| **DG** | Distributed Generation |
| **DIM** | Damping Impedance Method |
| **DLC** | Design Load Case |
| **DNV** | Det Norske Veritas |
| **DOL** | Direct On-Line |
| **DRESs** | Decentralized Renewable Energy Systems |
| **DSO** | Distribution System Operator |
| **DUT** | Device Under Test |
| **dVOC** | Dispatchable Virtual Oscillator Control |
| **DyNaLab** | Dynamic Nacelle Testing Laboratory |
| **EAP** | EtherCAT Automation Protocol |
| **ECD** | Equivalent Circuit Diagram |
| **ECU** | Electronic Control Units |
| **EDT** | Electrical Drive Train |
| **EESG** | Electrical Excited Synchronous Generator |
| **EFC** | Emergency Feather Command |
| **EMC** | Electromagnetic Compatibility |
| **EMF** | Electromotive Force |
| **EMI** | Electromagnetic Interference |
| **EMT** | Electromagnetic Transient |
| **EOC** | Environmental Operating Condition |
| **EOG** | Extreme Operating Gust |
| **ESS** | Energy Storage Systems |
| **EtherCAT** | Ethernet for Control Automation Technology |
| **EZE** | (de.) Erzeugereinheit |
| **FBG** | Fiber Bragg Grating |
| **FCF** | Feedback Current Filtering |
| **FE** | Feature Extraction |
| **FE** | Finite Element |
| **FFR** | Fast Frequency Response |
| **FFT** | Fast Fourier Transform |
| **FGW** | (de.) Fördergesellschaft Windenergie |
| **FMEA** | Failure Mode and Effects Analysis |
| **FPGA** | Field-Programmable Gate Array |
| **FRT** | Fault Ride Through |
| **FTIM** | Filtered Ideal Transformer Method |

| | |
|---|---|
| **GC** | Grid Codes |
| **GDT** | Gas Discharge Tube |
| **GFL** | Grid-Following |
| **GFM** | Grid-Forming |
| **GPS** | Global Positioning System |
| **GPU** | Graphics Processing Unit |
| **GRP** | Glass Reinforced Plastic |
| **HDD** | Hard Disk Drive |
| **HIL** | Hardware-in-the-Loop |
| **HiL-GridCop** | Hardware in the Loop Grid Compliance |
| **HMI** | Human Machine Interface |
| **HSE** | Health Safety Environmental |
| **HTS** | High Temperature Superconductors |
| **HUT** | Hardware Under Test |
| **HVAC** | Heating, Ventilation, Air-Conditioning and Cooling |
| **HVDC** | High-voltage Direct Current |
| **HVRT** | High Voltage Ride Through |
| **I/O** | Input/Output |
| **IA** | Interfacing Algorithm |
| **IACS** | Industrial Automation and Control Systems |
| **ICCP** | Impresses Current Cathodic Protection |
| **IDE** | Integrated Development Environments |
| **IDS** | Intrusion Detection System |
| **IEC** | International Electrotechnical Commission |
| **IEEE** | Institute of Electrical and Electronics Engineers |
| **IG** | Induction Generator (with squirrel-cage rotor) |
| **IGBT** | Insulated-Gate Bipolar Transistor |
| **IGCT** | Integrated Gate-Commuted Thyristor |
| **IMD** | Insulation Monitoring Devices |
| **IoT** | Internet of Things |
| **IPC** | Individual Pitch Control |
| **IPM** | Intelligent Power Module ®SEMIKRON |
| **IRENA** | International Renewable Energy Agency |
| **IT** | (fr.) Isolé Terre |
| **IT** | Information Technology |
| **ITM** | Ideal Transformer Model or Method |
| **ITU-T** | International Telecommunication Union Telecommunication Standardization Sector |

| | |
|---|---|
| **IWES** | Institute for Wind Energy Systems |
| **KPI** | Key Performance Indicator |
| **L** | Inductor |
| **LAS** | Load Application System |
| **LC** | Local Controller |
| **LCOE** | Levelized Cost of Energy |
| **LEMP** | Lightning Electromagnetic Pulse |
| **LES** | Local Energy System |
| **LiDAR** | Light Detection And Ranging |
| **LMS** | Linear Multistep |
| **LORC** | Lindø Offshore Renewables Center |
| **LPL** | Lightning Protection Level |
| **LPS** | Lightning Protection System |
| **LPSP** | Loss of Power Supply Probability |
| **LPZ** | Lightning Protection Zone |
| **LSC** | Line Side Converter |
| **MBD** | Model-Based Design |
| **MBS** | Multi-Body Simulation |
| **MC** | Main Converter |
| **MEM** | Micro-Electro-Mechanical |
| **MGOe** | Mega-Gauss-Oersted |
| **MHIL** | Mechanical-HIL |
| **MIL** | Model-in-the-Loop |
| **ML** | Machine Learning |
| **M-UVRT** | Multiple Under-Voltage-Ride-Through |
| **MOSFET** | Metal-Oxide-Semiconductor Field-Effect Transistor |
| **MoWiT** | Modelica Library for Wind Turbines |
| **MPC** | Model-Predictive Control |
| **MPPT** | Maximum Power Point Tracking |
| **MSC** | Machine Side Converter |
| **MTPA** | Max Torque per Ampere |
| **MV** | Medium Voltage |
| **MVGE** | Medium Voltage Grid Emulator |
| **N** | Neutral |
| **NCS** | Naturally Coupled System |
| **NdFeB** | Neodymium Magnets |
| **NPC** | Neutral Point Clamped |
| **NREL** | National Renewable Energy Laboratory |

| | |
|---|---|
| **NTV** | Negative Temperature Coefficient |
| **O&M** | Operation & Maintenance |
| **ODE** | Ordinary Differential Equation |
| **OEM** | Original Equipment Manufacturer |
| **OPC** | Open Platform Communications |
| **OPC UA** | OPC Unified Architecture |
| **OPEX** | Operational Expenditure |
| **ORE** | Offshore Renewable Energy |
| **OS** | Operating System |
| **OT** | Operational Technology |
| **PC** | Personal Computer |
| **PCC** | Point of Common Coupling |
| **PCD** | Partial Circuit Duplication |
| **PE** | Protective Earth |
| **PEC** | Power Electronic Controller |
| **PEN** | Protective Earth Neutral |
| **PFC** | Power Factor Correction |
| **PHIL** | Power-HIL |
| **PI** | Power Interface |
| **PI** | Proportional-Integral |
| **PID** | Proportional Integral Derivative |
| **PIL** | Processor-in-the-Loop |
| **PL** | Performance Level |
| **PLC** | Programmable Logic Controller |
| **PLL** | Phase-Locked Loop |
| **PMG** | Permanent Magnet Generator |
| **PMSG** | Permanent Magnet Synchronous Generator |
| **PMSM** | Permanent Magnet Synchronous Motor |
| **PMSM/G** | Permanent Magnet Synchronous Machine/Generator |
| **PPM** | Power Park Modules |
| **PQ4Wind** | Power Quality for Wind Energy |
| **PSD** | Power Spectral Density |
| **PWM** | Pulse-Width Modulation |
| **Q(U)** | Apparent Power Dependent on Inst. Voltage |
| **R** | Resistor |
| **RAM** | Random Access Memory |
| **RBDO** | Reliability-Based Design Optimization |
| **RCA** | Root Cause Analysis |

| | |
|---|---|
| **RCM** | Residual-Current Monitor |
| **RCP** | Rapid Control Prototyping |
| **RE** | Renewable Energy |
| **RES** | Renewable Energy Sources |
| **RK** | Runge-Kutta |
| **RMS** | Root Mean Square |
| **ROCOF** | Rate of Change of Frequency |
| **RT** | Real-Time |
| **RTOS** | Real-Time Operating System |
| **RTS** | Real-Time Simulator |
| **RWTH** | Rheinisch-Westfälische Technische Hochschule |
| **SCADA** | Supervisory Control and Data Acquisition |
| **SEMP** | Switching Electromagnetic Pulse |
| **SG** | Synchronous Generator |
| **SGRE** | Siemens Gamesa Renewable Energy |
| **SHM** | Structural Health Monitoring |
| **SIC** | Silicon Carbide |
| **SIL** | Software-in-the-Loop |
| **SLD** | Single Line Diagram |
| **SLES** | Smart Local Energy System |
| **SMC** | Simulink Motor Control |
| **SPSG** | Salient Pole Synchronous Generator |
| **SQL** | Structured Query Language |
| **SW** | Software |
| **TAB** | (de.) Technische Anschlussbedingungen |
| **TBDT** | Test Bench Drive Train |
| **TBM** | Tolerance-Band-Modulation |
| **TCP** | Transmission Control Protocol |
| **TFA** | Time-Variant First-order Approximation |
| **THD** | Total Harmonic Distortion |
| **TLM** | Transmission Line Model |
| **TLS** | Transport Layer Security |
| **TN** | (fr.) Terre Neutre |
| **TN-C** | (fr.) Terre Neutre Combiné |
| **TN-S** | (fr.) Terre Neutre Séparé |
| **TPU** | Tensor Processing Unit |
| **TR** | (de.) Technische Richtlinie |
| **TRL** | Technology Readiness Level |

| | |
|---|---|
| **TSN** | Time-Sensitive Networking |
| **TSO** | Transmission System Operator |
| **TT** | (fr.) Terre Terre |
| **TwinCAT** | The Windows Control and Automation Technology |
| **UPS** | Uninterruptible Power Supply |
| **UVRT** | Under-Voltage-Ride-Through |
| **VDE** | Verband der Elektrotechnik Elektronik Informationstechnik e. V. |
| **VDI** | Verein Deutscher Ingenieure e. V. |
| **VFD** | Variable Frequency Drive |
| **VOC** | Virtual Oscillator Control |
| **VPI** | Vacuum Pressure Impregnation |
| **VPP** | Virtual Power Plant |
| **VSC** | Voltage Source Converter |
| **VSI** | Voltage Source Inverters |
| **VSM** | Virtual Synchronous Machine |
| **WT** | Wind Turbine |

# Chapter 1
# Electrical safety

*Benjamin Moosburger[1], Christian Vögerl[1], Tilo Püschel[2] and Jörg Irzinger[2]*

## 1.1 Electrical safety

The value of feed-in compensation for renewable energy is falling around the world, putting the wind energy sector, both onshore and offshore, under increasing pressure. Despite low rates of reimbursement, the high costs of investment for a wind turbine generator system should be recouped over just a few years. Therefore, in order to increase the efficiency of these installations, operating and maintenance costs must be reduced and downtimes as a result of unexpected electrical faults such as insulation faults or damage caused by lighting and surges must be prevented. When it comes to energy supply, the security of supply in the event of a fault and protection against electric shock are central tasks. An undetected insulation fault is synonymous with unexpected downtimes or increases fire hazards. Unscheduled service work resulting from insulation faults or even lightning strikes and surges also cause high time and cost expenditure.

Insurers of wind turbines also demand effective condition monitoring systems so that all the relevant data of the facility can be kept centrally in view. Modern detection devices for recording lightning events and evaluating them in detail enable the systematic planning of maintenance and servicing work after lightning events. This can prevent consequential damage and reduce downtime. In wind turbines, extensive, permanent monitoring by a variety of sensors is required in order to identify the developing deterioration of sensors in good time, thereby ensuring safety and a high level of availability.

Reliable solutions, products, and coordinated services for wind turbines with the aim of optimum safety and efficiency are therefore paramount.

### *1.1.1 Lightning and surge protection for wind turbines*

Lightning protection standard

IEC 61400-24 [1] must be followed for the lightning protection of wind turbines.

[1]DEHN SE, Neumarkt i. d. Oberpfalz, Germany
[2]BENDER GmbH & Co. KG, Grünberg, Germany

The main part of IEC 61400-24 deals with:

- the lightning environment of wind turbines (lightning current parameters): reference to lightning protection standard IEC 62305-1 [2]
- evaluation of the lightning effect, adapted risk analysis for wind turbines
- lightning protection of sub-components and associated test methods for: rotor blades, nacelle and structural components (hub, tower, etc.), mechanical power train (bearings, sliding contacts, etc.), low-voltage electrical system and electronic systems (surge protection)
- earthing of wind turbines and wind farms
- safety of people; safety precautions in the event of a lightning strike
- documentation of the lightning protection system
- inspection of the lightning protection system.

With respect to protection measures for the lightning effect, reference is made to a complete lightning protection system based on the IEC 62305 series of norms for lightning protection:

- According to IEC 62305-2 [2], the overall risk of lightning damage is composed of the frequency of a lightning strike, the probability of damage and the loss factor.
- With regard to material damage and risk of fatality in the event of direct lightning strikes to a wind turbine, IEC 61400-24 refers to IEC 62305-3 [2] in certain parts of the turbine in order to reduce the extent of damage by setting up the lightning protection system (LPS).
- For protection in structures with electrical and electronic systems, especially if high demands are placed on their reliability of function and supply, protection of these systems against conducted and radiated interference must also be ensured. Interference of this kind is caused by lightning electromagnetic impulses (LEMP) in the event of direct and indirect lightning strikes. This requirement can be satisfied by a LEMP protection system as per IEC 62305-4 [2]. In order to achieve a continuous and functioning surge protection concept, the energy coordination between the arrester types according to IEC 62305-4 must also be ensured.

DNV, an international classification society and service provider in the fields of technical consulting, engineering services, certification, and risk management, is often used for the certification of wind turbines. The guidelines of DNV [3] also refer to IEC 61400-24 and the IEC 62305 series of standards.

#### 1.1.1.1 Damage and causes

The sources of damage or causes for the malfunction or even destruction of electrical and electronic systems and components of wind turbines are highly diverse and range from direct and indirect lightning interference to surges caused by switching operations, earth faults, and short circuits or tripping of fuses (SEMP: Switching Electromagnetic Pulse) (Figure 1.1). With regard to lightning

Direct lightning strikes (LEMP)
• Galvanic coupling
• Inductive / capacitive coupling



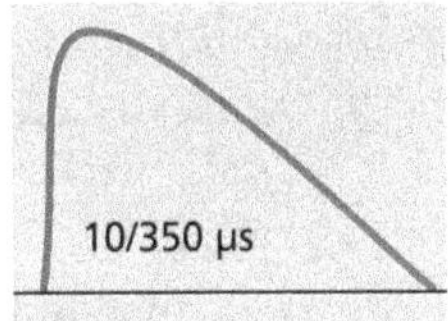

Indirect lightning strikes
• Conducted partial lightning currents
• Inductive / capacitive coupling



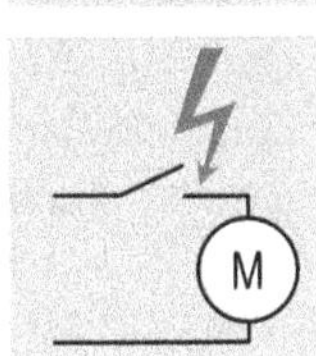



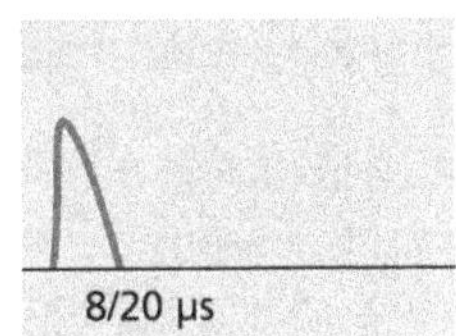





*Figure 1.1 Causes of overvoltages*

strikes, these can be divided into four groups depending on the point of strike according to IEC 62305-2:

- direct lightning strike to a wind turbine
- lightning strike near a wind turbine
- direct lightning strike to an incoming supply line
- lightning strike near an incoming supply line.

Conducted interference impulses can be transmitted through both the medium-voltage and low-voltage sides.

In the case of copper-based communication interfaces, there is an additional coupling path. A direct lightning strike into the respective conductor system would be a conceivable threat or even a nearby lightning strike close to the respective conductor system.

The annual number of cloud-to-earth flashes for a given region can be determined with the isokeraunic level. In Europe, an average number of one to three cloud-to-earth flashes per km$^2$ per year applies to coastal and low-mountain-range landscapes. When planning lightning protection measures, not only cloud-to-earth flashes but also earth-to-cloud flashes, so-called upward leaders, must be considered for objects at exposed locations with a height of more than 60 m. This results in higher values than in the aforementioned definition. Furthermore, earth-to-cloud flashes emanating from high, exposed objects have a high charge, which is of great importance, especially for protective measures on rotor blades and the design of lightning current arresters. IEC 61400-24 contains a risk analysis adapted for wind turbines to assess the lightning effect on these installations.

#### 1.1.1.2 Lightning protection zone concept

Due to their exposed location and height, wind turbines are often subjected to direct lightning strikes. In order to prevent damage and the resultant downtimes from lightning, an integrated lightning protection concept as outlined in IEC 61400-24 is imperative. The IEC 61400-24 recommends protecting all subcomponents of the wind turbine in accordance with lightning protection level (LPL) I unless a risk analysis demonstrates that a lower LPL is sufficient. Dividing a wind turbine into so-called lightning protection zones forms the basis for devising a protection concept.

The lightning protection zone concept (Figure 1.2) is a structuring measure. Within a wind turbine, measures can thus be identified and specified to create a defined electromagnetic compatibility (EMC) environment. This EMC environment depends on the immunity of the electrical equipment used. Conducted and field-based disturbance variables at interfaces can thus be reduced to agreed values. For this reason, the wind turbine is subdivided into protection zones (LPZ). The rolling sphere method is used to determine the lightning protection zone LPZ $0_A$, namely the parts of a wind turbine which may be subjected to direct lightning strikes, and LPZ $0_B$, namely, the parts of a wind turbine which are protected from direct lightning strikes by external air-termination systems or air-termination systems integrated in parts of a wind turbine (e.g., in the rotor blade) (Figure 1.3). According to the IEC 61400-24 standard, the rolling sphere method must not be used for the rotor blade itself. For this reason, the design of the air-termination system should be tested according to subsection 8.2.3 of the IEC 61400-24 standard (Figure 1.4).

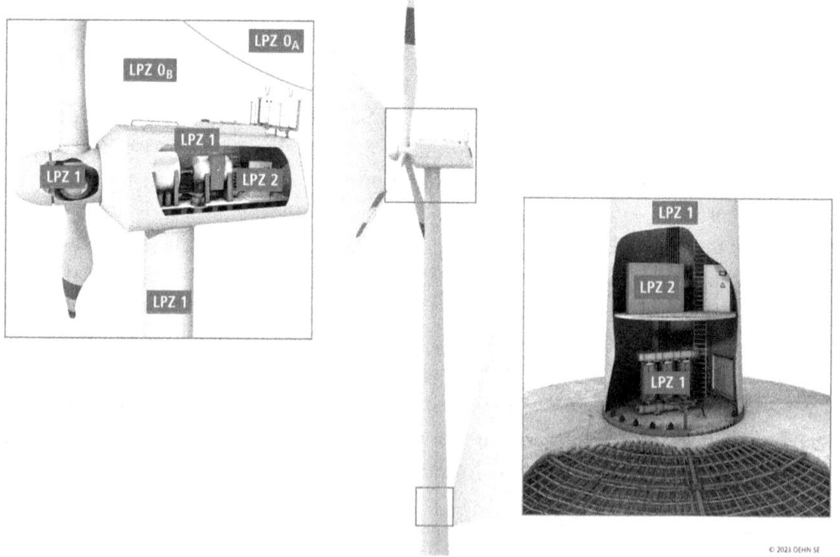


*Figure 1.2 Lightning protection zones of a wind turbine*

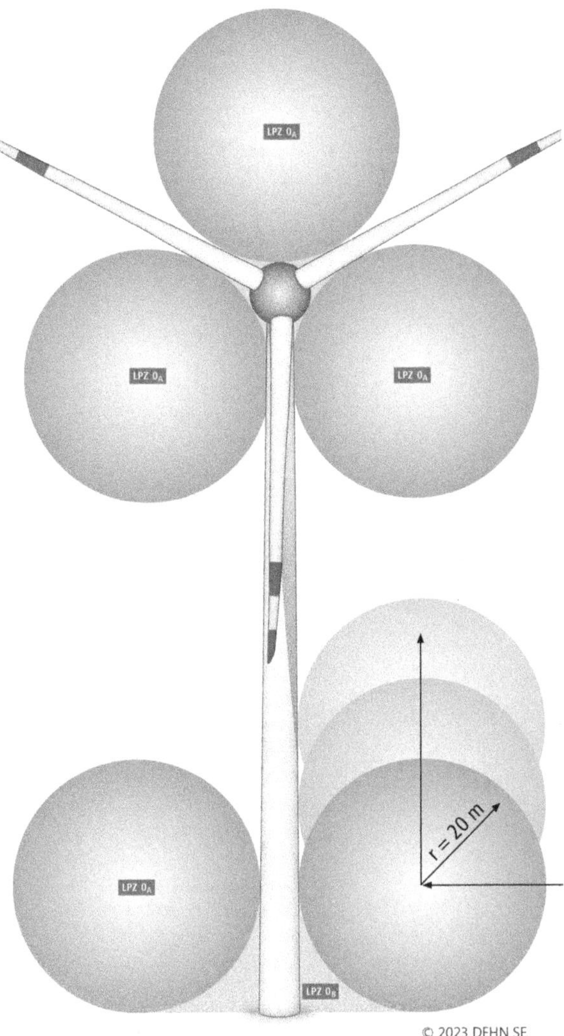


*Figure 1.3 Rolling sphere method*

However, it is decisive that the lightning parameters that are injected into LPZ $0_A$ from the outside are reduced by suitable shielding measures and surge protective devices at all zone boundaries so that the electrical and electronic devices and systems inside a wind turbine are not interfered with.

### 1.1.1.3 Shielding measures

The nacelle should be designed as a closed metal shield. Thus, a volume with an electromagnetic field that is considerably lower than the field outside the wind turbine is generated in the nacelle. In accordance with IEC 61400-24, a tubular

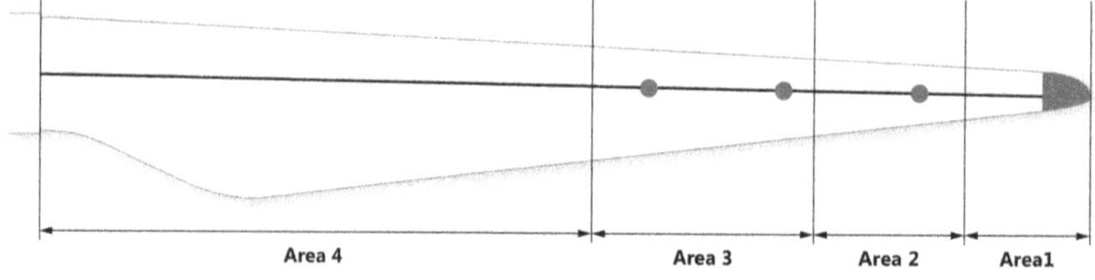


| Rotor blade area | | Typical area dimensions [m] | | | Current level |
|---|---|---|---|---|---|
| Area | Length [m] or % of blade length | 40 m rotor blade | 60 m rotor blade | 80 m rotor blade | I [kA] |
| 1 | 5% | 2 | 3 | 4 | 200 |
| 2 | 10% | 4 | 6 | 8 | 150 |
| 3 | 15% | 6 | 9 | 12 | 100 |
| 4 | Rest of blade | 28 | 42 | 56 | 10 |

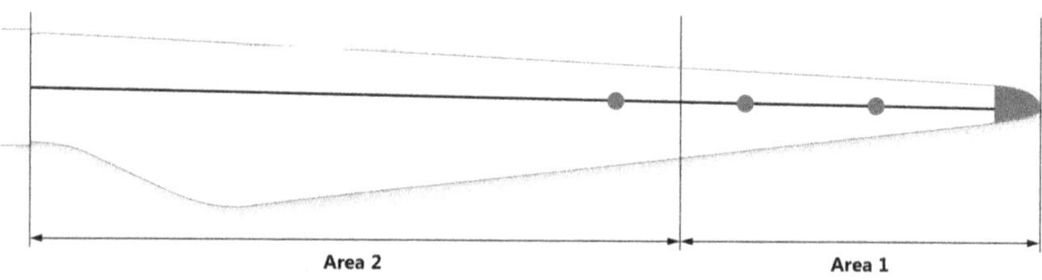


| Rotor blade area | | Typical area dimensions [m] | | | Current level (10/350 µs current components) |
|---|---|---|---|---|---|
| Area | Length [m] or % of blade length | 40 m rotor blade | 60 m rotor blade | 80 m rotor blade | I [kA] |
| 1 | 40% | 16 | 24 | 32 | 200 |
| 2 | 60% | 24 | 36 | 48 | 150, 100, 10, as defined for the specific rotor blade |



*Figure 1.4 Classification of rotor blades according to IEC 61400-24*

steel tower, which is frequently used for large wind turbines, can be regarded as an almost perfect Faraday cage for electromagnetic shielding. In the case of concrete hybrid towers, the function of the galvanic cage must be ensured by reinforcing steel as well as earthing and electrical connection of the individual components. The switchgear and control cabinets in the nacelle and, if any, in the operations building should also be made of metal. The connecting cables should feature an external shield capable of carrying lightning currents. Shielded cables are only resistant to EMC interference if the shields are connected to the equipotential bonding system on both ends. The contacting of the shields must be EMC-compliant with all-round-contacting terminals.

Magnetic shielding and cable routing should be performed as per section 4 of IEC 62305-4. For this reason, the general guidelines for an EMC-compatible installation practice according to IEC TR 61000-5-2 [4] should be observed. Shielding measures are, e.g.,

- installation of a metal braid on GRP-coated nacelles
- metal tower
- metal switchgear cabinet
- metal control cabinets
- lightning current carrying, shielded connecting cables (metal cable duct, shielded pipe, or the like)
- cable shielding

#### 1.1.1.4 External lightning protection measures

The function of an external lightning protection system (LPS) is to intercept direct lightning strikes including lightning strikes to the tower of a wind turbine and to discharge the lightning current from the point of strike to the ground. It also serves to distribute the lightning current in the ground without causing thermal or mechanical damage or dangerous sparking which may lead to fire or explosion and put people at risk.

External lightning protection measures are:

- air-termination and down-conductor systems in the rotor blades
- air-termination systems for protecting nacelle superstructures, the nacelle itself and the hub
- using the tower as an air-termination system and down conductor
- earth-termination system consisting of a foundation earth electrode and a ring earth electrode

The rolling sphere method can be used to determine potential points of strike for a wind turbine (except for the rotor blades).

Class of LPS I is recommended for wind turbines. Therefore, a rolling sphere with a radius $r = 20$ m is rolled over the wind turbine to determine the points of strike. Air-termination systems are required where the sphere touches the wind turbine (potential points of strike).

The nacelle construction should be integrated into the lightning protection system to ensure that lightning strikes to the nacelle hit either natural metal parts that are capable of withstanding this stress or an air-termination system designed for this purpose (Figure 1.5). GRP-coated nacelles or the like should be fitted with an air-termination system and down conductors forming a cage around the nacelle (metal braid). The air-termination system including the bare conductors in this cage should be capable of withstanding lightning strikes according to the relevant lightning protection level. Other conductors in the Faraday cage should be designed to withstand the amount of lightning current to which they may be subjected. The IEC 61400-24 standard requires that air-termination systems for protecting measuring equipment, etc. mounted outside the nacelle be designed in compliance with the general requirements of IEC 62305-3 and that down conductors be connected to the cage described above.

*Figure 1.5 Example of an air-termination system for the weather station and the aircraft warning light*

With high-voltage-resistant, insulated down conductors, lightning currents can be safely routed past sensitive components. This reduces the burden on electrical and mechanical systems.

Natural components made of conductive materials which are permanently installed in/on a wind turbine and remain unchanged (e.g., lightning protection system of the rotor blades, bearings, mainframes, hybrid tower) may be integrated in the LPS. If wind turbines consist of a metal construction, it can be assumed that they fulfil the requirements for an external lightning protection system of class of LPS I according to IEC 62305.

This requires that the lightning strike be safely intercepted by the lightning protection system of the rotor blades so that it can be discharged to the earth-termination system via the natural components such as bearings, mainframes, tower, and/or bypass systems (e.g., open spark gaps, carbon brushes).

Rotor blades, the nacelle and its attachments, the rotor hub and the tower of the wind turbine can be struck by lightning. If all these wind turbine components are capable of safely intercepting the maximum anticipated lightning impulse current of 200 kA and discharging it to the earth-termination system, they can be used as natural components of the air-termination system of the wind turbine's external lightning protection system.

Metallic receptors, which represent a defined point of strike for flashes, are frequently attached to the tip and the surface of the GRP blade to protect the rotor blades from lightning strikes. A down conductor is routed from the receptors to the blade root. In case of a lightning strike, it can be assumed that lightning hits the receptors and then travels through the down conductor inside the blade via the nacelle and the tower to the earth-termination system.

#### 1.1.1.5 Laboratory tests as per IEC 61400-24

IEC 61400-24 describes two basic system-level immunity test methods for wind turbines:

- When performing impulse current tests under operating conditions, impulse currents or partial lightning currents are injected into the individual lines of a control system while mains voltage is present. Thus, the equipment to be protected including all lightning current and surge arresters is subjected to an impulse current test.
- The second test method simulates the electromagnetic effects of the LEMP. To this end, the full lightning current is injected into the structure which discharges the lightning current, and the behavior of the electrical system is analyzed by simulating the cabling under operating conditions as realistically as possible. The lightning current steepness is a decisive test parameter.

Engineering and testing service examples:

- lightning current tests for bearings and gearboxes of the mechanical drive string
- high-current tests for the receptors and down conductors of rotor blades
- system-level immunity tests for important control systems such as pitch systems, wind sensors or aircraft warning lights (in accordance with the recommendation of IEC 61400-24)
- testing of customer-specific connection units.

#### 1.1.1.6 Earth-termination system

The earth-termination system of a wind turbine must perform several functions such as personal protection, EMC protection and lightning protection. Specifications regarding materials, their shapes and minimum cross-sections can be found in IEC 62305-3 Table 7.

An effective earth-termination system is essential to distribute lightning currents and save the wind turbine from destruction. Moreover, the earth-termination system must protect people and animals against electric shock. In case of a lightning strike, the earth-termination system must discharge high lightning currents to the ground and distribute them in the ground without causing dangerous thermal and/or electrodynamic effects.

The tasks of a wind turbine's earthing system are:

- protective earthing to safely connect electrical installations to the earth's potential and, in the event of an electrical fault, to protect human life and property.
- functional earthing with the task of ensuring that the operation of electrical and electronic equipment is as reliable and faultless as possible.

- lightning protection earthing with the task of safely carrying the lightning current from the down conductors and discharging it into the ground.

When it comes to lightning protection, a single, common earthing system for the wind turbine for all purposes (e.g., medium-voltage systems, low-voltage supply, lightning protection, electromagnetic compatibility, telecommunication and control systems) is advantageous in order to manage the tasks listed.

For this purpose, the foundations of the wind turbines made of reinforced concrete should primarily be used as foundation earth electrodes. They provide a low earth resistance and are an excellent basis for equipotential bonding.

Due to the installation of a medium-voltage transformer in the wind turbine, its earthing system must comply with IEC 61936-1 [5]. As in the lightning protection standard IEC 62305-3, IEC 61936-1 also describes all types of earth electrodes, whereby the foundation earth electrode is referred to as the most effective earth electrode.

According to IEC 61936-1, earthing systems must be designed to meet four requirements:

- Mechanical strength and corrosion resistance must be ensured.
- The maximum fault current (usually calculated) must be controlled in terms of the thermal impact.
- Equipment and resources must be protected from damage.
- The safety of people must be ensured with respect to the voltages that arise in earthing systems during peak earth fault currents.

Note: Applicable national standards describe how to design an earth-termination system to prevent high touch and step voltages caused by short circuits in high or medium-voltage systems. With regard to the safety of people, the IEC 61400-24 standard refers to IEC / TS 60479-1 [6] and IEC TR 60479-4 [7].

A site plan of the earthing network should be devised, containing the materials, the location of the earth electrodes, their branching points and installation depths.

Prior to commissioning, an inspection report must be produced demonstrating that all the requirements of the standards to be complied with have been fulfilled. The earth resistance required by the lightning protection standard IEC 62305-3 is $<10\ \Omega$.

#### *1.1.1.6.1 Arrangement of earth electrodes*

The IEC 62305-3 standard describes two basic types of earth electrode arrangements for wind turbines:

- **Type A:** According to the informative Annex I of IEC 61400-24, this arrangement must not be used for wind turbines themselves, but for adjoining buildings of wind turbines (e.g., buildings containing measuring equipment or office sheds of a wind farm). Type A earth electrode arrangements consist of horizontal or vertical earth electrodes connected to the building by at least two down conductors.

- **Type B:** Type B earth electrodes for wind turbines are recommended by informative Annex I of IEC 61400-24. These consist of either an external ring earth electrode, which is laid in the ground with at least 80% of its length in direct contact with earth, or a foundation earth electrode. Ring earth electrodes and metal parts in the foundation must be connected to the tower construction.

In any case, the reinforcement of the tower foundation should be integrated into the earth-termination system of a wind turbine. To ensure an earth-termination system ranging over as large an area as possible, the earth-termination system of the tower base and the operations building should be connected by means of a meshed earth electrode network. Corrosion-resistant ring earth electrodes (made of stainless steel (V4A), e.g., material no. 1.4571) with potential control prevent excessive step voltages in case of a lightning strike and must be installed around the tower base to ensure that no-one comes to harm (Figure 1.6).

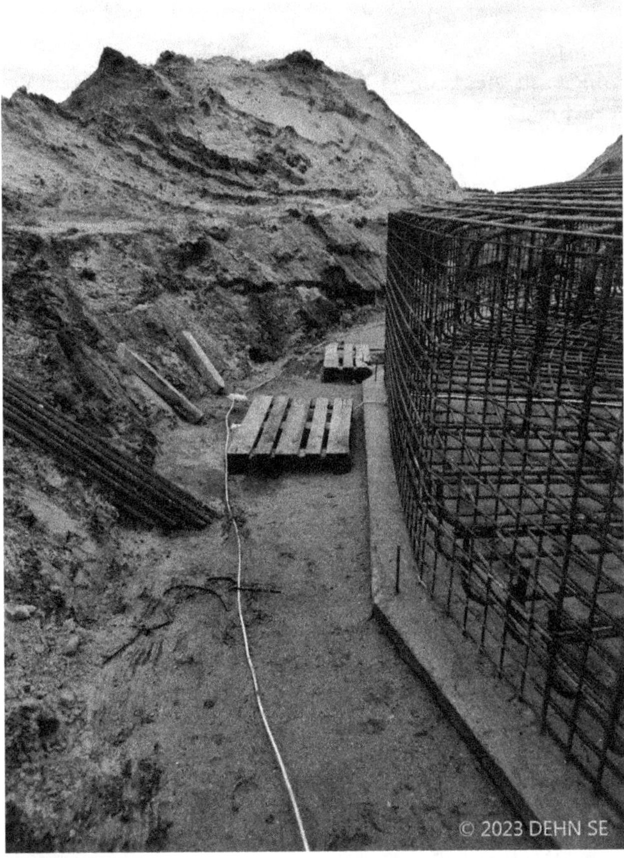


*Figure 1.6 Wind turbine reinforcement with a ring earth electrode that has been installed in direct contact with the earth*

The design of an earthing system for a concrete tower is laid out in the standard IEC 61400-24. It describes the continuous, necessary connection of the reinforcement in concrete towers of wind turbines (Figure 1.7). This can be used as a lightning conductor if it is connected with 2–4 parallel and vertical connections. In addition, these vertical connections must be connected every 20 m to a horizontal ring connection with a conductor capable of carrying lightning currents. Only in this way can adequate lightning current distribution be achieved, thus reducing the magnetic field inside the tower.

Since the tower is also used for equipotential bonding, fixed earthing terminals must be installed on the reinforcement at favorable points inside the tower (Figure 1.7) to ensure termination points for electrical equipment as well as conductive structures, such as ladders or lifts. Due to it being a tall structure, direct lightning strikes into the tower are to be expected and must be taken into account in the design. If an external lightning protection concept is used, it must be connected to the tower's reinforcement and implemented in accordance with IEC 62305-3 Lightning Protection – Part 3: Physical damage to structures and life hazard.

#### *1.1.1.6.2 Foundation earth electrode*

The foundation earth electrode is part of the electrical installation and fulfills essential safety functions. For this reason, it must be installed by or under the

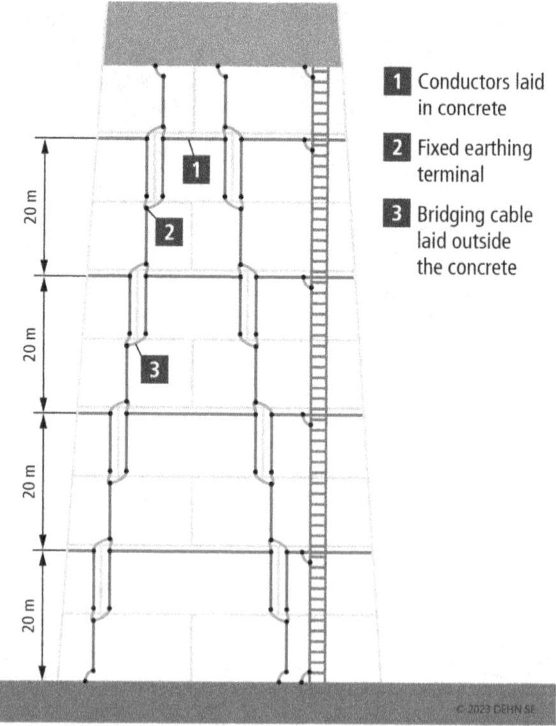


*Figure 1.7 Integration of the reinforcement of the wind turbine*

supervision of an electrician. The metals used for earth electrodes must comply with the materials listed in Table 7 of IEC 62305-3. The corrosion behavior of metal in the ground must always be taken into account.

The foundation earth electrode must be designed as a closed loop and arranged in the tower's foundation in accordance with IEC 61400-24.

Galvanized steel (round or strip steel) has proven to be a suitable material for foundation earth electrodes:

- round steel of at least 10 mm diameter
- strip steel with minimum dimensions of 30 mm × 3.5 mm

It must be observed that this material must be covered with a concrete layer of at least 5 cm (corrosion protection). The foundation earth electrode must be connected to the main earthing busbar in the wind turbine. Corrosion-resistant connections must be established via fixed earthing terminals or terminal lugs made of stainless steel (V4A). A ring earth electrode made of stainless steel (V4A) must also be installed in the ground (Figure 1.6).

In the case of reinforced foundations, such as those used for wind turbines, the round or strip steel is placed on the lower reinforcement layer. It is to be connected to the reinforcement in a secure, electrically conductive manner at intervals of 2 m. Welded, clamped, or press-fitted connections are permissible. If the concrete is compacted with a machine (e.g., a vibrator), no wedge connectors may be used.

An installation plan is required for designing the foundation earth electrode. Photographs, diagrams, and measurement reports are used for documentation.

##### *1.1.1.6.3 Foundation earth electrode as part of the lightning protection system*

If the foundation earth electrode is used as part of the lightning protection system, connecting parts must be used according to IEC 62561-1 [8]. For equipotential bonding in lightning protection systems and for EMC purposes, round or strip steel must be installed in the foundation and connected to the reinforcement and the equipotential bonding bar (Figure 1.8).

In the event of a lightning strike, there must be no flashover from the foundation through the insulation to the earthing system. This is achieved with a maximum ring earth electrode mesh size of 10 m × 10 m as per IEC 62305-3. The ring earth electrode and the terminal lugs must be corrosion-protected (high-alloy stainless steel V4A, material number 1.4571).

#### 1.1.1.7 Internal lightning protection measures

Internal lightning protection measures include:

- earthing and equipotential bonding measures
- spatial shielding and separation distance
- cable routing and cable shielding
- installation of coordinated surge protective devices

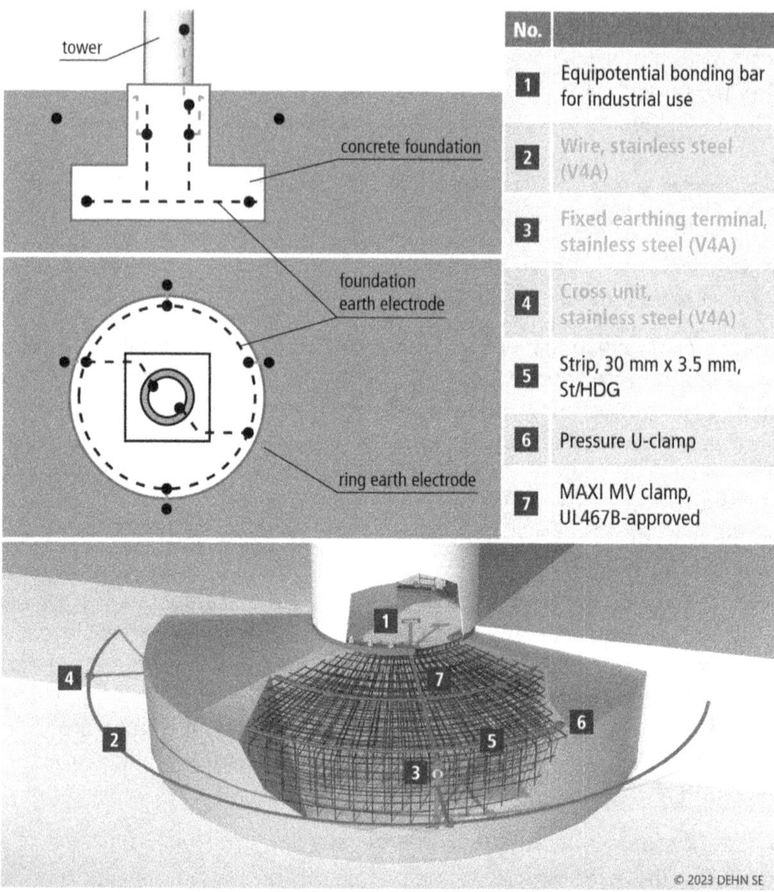


*Figure 1.8 Wind turbine earthing example*

In accordance with IEC 61400-24 surge protective devices capable of discharging high partial lightning currents without destruction must be installed at the transition from LPZ $0_A$ or LPZ $0_B$ to LPZ 1 (also referred to as lightning equipotential bonding). These surge protective devices are referred to as type 1 lightning current arresters and are tested with impulse currents of 10/350 μs waveform. At the boundary from LPZ 1 to LPZ 2 and higher, low-energy impulse currents caused by voltages induced on the system or surges generated in the system must be coped with. These protective devices are referred to as type 2 surge arresters and are tested with impulse currents of 8/20 μs waveform.

This affects both power supply and communication lines. An additional local equipotential bonding system where all cables and lines entering this boundary are integrated must be established for every further zone boundary within the volume to be protected. The function of type 2 and type 3 surge arresters is to further reduce the residual interference of the upstream protection stages and to limit the surges induced on the wind turbine or generated in the wind turbine.

#### 1.1.1.8 Selection of lightning current and surge arresters based on the voltage protection level ($U_p$) and the immunity of the equipment

To describe the required protection level ($U_p$) in a lightning protection zone (LPZ), the immunity levels of the equipment within an LPZ must be defined. These are specified for power supply connections in IEC 61000-6-2 [9]. For signalling or telecommunication connections, the ITU-T recommendations K.21 and K.20 must be followed. Manufacturers of electrical and electronic components or devices should be able to provide the required information on the immunity level according to the EMC standards. If this is not the case, the wind turbine manufacturer should perform tests to determine the immunity level. The specific immunity level of components in an LPZ directly defines the voltage protection level required at the LPZ boundaries. The immunity of a system must be proven, where applicable, with all those arresters installed and the equipment they are intended to protect.

##### *1.1.1.8.1 Protection of power supply systems*

The transformer of a wind turbine may be housed at different locations (in a separate distribution station, in the tower base, in the tower, in the nacelle). In case of large wind turbines, for example, the unshielded 20 kV cable in the tower base is routed to the medium-voltage switchgear installation consisting of a vacuum circuit breaker, mechanically locked selector switch disconnector, outgoing earthing switch and protective relay. The medium-voltage cables are routed from the medium-voltage switchgear installation in the tower of the wind turbine to the transformer which may be situated in the tower base or in the nacelle. The transformer feeds the control cabinet in the tower base, the switchgear cabinet in the nacelle and the pitch system in the hub by means of a TN-C system (L1, L2, L3, PEN conductor). The switchgear cabinet in the nacelle supplies the electrical equipment in the nacelle with an AC voltage of 230/400 V.

According to IEC 60364-4-44 [10], all electrical equipment installed in a wind turbine must have a specific rated impulse withstand voltage according to the nominal voltage of the wind turbine (see IEC 60664-1 (EN 60664-1 [11], Appendix F, Table F.1)). This means that the surge arresters to be installed must have at least the specified voltage protection level according to the nominal voltage of the wind turbine. Surge arresters used to protect the 400/690-V supply must have a minimum voltage protection level $U_p \leq 2.5$ kV, and those used to protect the 230/400-V supply must have a voltage protection level $U_p \leq 1.5$ kV to ensure protection of sensitive electrical/electronic equipment.

The MV-transformer infeed is protected by medium-voltage arresters. These must be adapted to the system configuration and voltage of the medium-voltage system.

Type 2 surge arresters should be used to protect the voltage supply of the control cabinet in the tower base, the switchgear cabinet in the nacelle and the pitch system in the hub with a 230/400 V TN-C system.

Coordinated single-pole lightning current arresters with a high follow current limitation for 400/690 V systems should be installed to protect the 400/690 V transformer, inverters, mains filters and the measuring equipment (Figure 1.9). It must be ensured at the frequency converter that the arresters are dimensioned for the maximum voltage peaks, which are higher than in case of pure sinusoidal voltages. In this context, surge arresters with a nominal voltage of 600 V and $U_{mov}$ = 750 V have proven their worth. The DEHNguard DG M WE 600 FM arresters can be installed at both sides of the converter (grid and machine side) and on the generator. Only if doubly-fed induction generators are used, an arrester combination with an increased electric strength must be used on the rotor side. Installing a powerful surge protective device (SPD) is recommended, which is designed for continuous voltages of up to 1,000 V and discharge currents up to 40 kA (8/20 μs), has a low protection level $U_p \leq 5$ kV and does not even trip under temporary voltage surges of up to 2,200 $V_{peak}$. There is also a suitable solution for larger voltage peaks of up to 3 kV (with a protection level of $\leq 10$ kV) in the form of a 3+1 Neptune circuit.

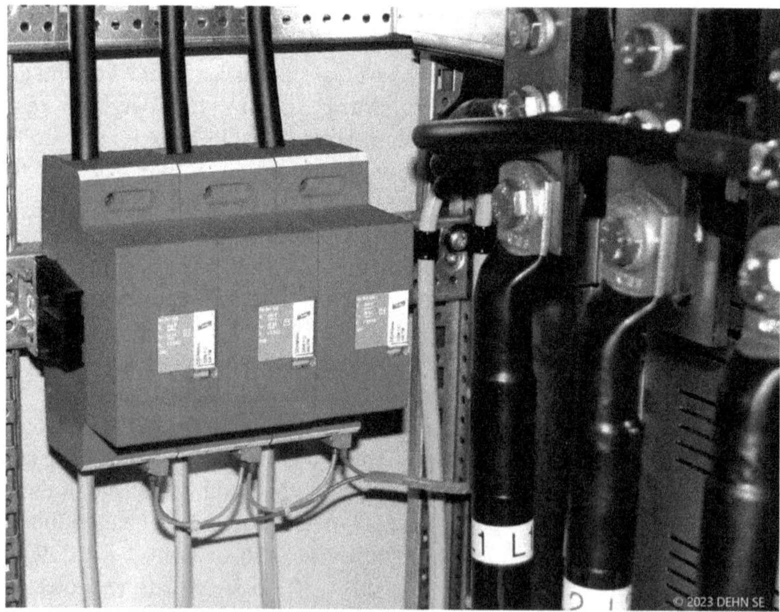


*Figure 1.9 Spark-gap-based lightning current arrester with integrated arrester backup fuse capable of withstanding lightning currents to protect the transformer on the low-voltage side*

#### 1.1.1.8.2 Surge arresters for information technology systems

Surge arresters for protecting electronic equipment in telecommunication and signalling networks against the indirect and direct effects of lightning strikes and other transients are described in IEC 61643-21 [12] and are installed at the zone boundaries in conformity with the lightning protection zone concept. Multi-stage arresters must be designed without blind spots. In other words, the different protection stages must be coordinated with one another. Otherwise, not all protection stages will be activated, thus causing faults in the surge protective device. Glass fiber cables are frequently used for routing information technology lines into a wind turbine and for connecting the control cabinets in the tower base to the nacelle. Shielded copper cables are used to connect the actuators and sensors to the control cabinets. Since interference by an electromagnetic environment is excluded, the glass fiber cables do not have to be protected by surge arresters unless they have a metal sheath which must be integrated in the equipotential bonding system either directly or by means of surge protective devices. In general, the following shielded signal lines connecting the actuators and sensors with the control cabinets must be protected by surge protective devices:

- signal lines of the weather station and aircraft warning light on the sensor mast
- signal lines routed between the nacelle and the pitch system in the hub
- signal lines for the pitch system
- signal lines to the inverter
- signal lines to the fire extinguishing system

#### 1.1.1.8.3 Signal lines of the weather station and aircraft warning light

The weather station and aircraft warning light on the sensor mast in LPZ $0_B$ should be protected by a type 1 lightning current arrester at the relevant zone transitions (LPZ $0_B \rightarrow 1$). Depending on the system, components from, for example, the DEHNventil series (low voltage) and/or BLITZDUCTOR family can be used here for extra low voltage/signal lines. Also to be considered is whether or not the wind measurement equipment is fitted with a heating system. In this case, additional energy-coordinated surge protective devices are required for protecting unearthed DC power supplies.

#### 1.1.1.8.4 Signal lines for the pitch system

A universal DEHNpatch surge arrester can be used if information between the nacelle and the pitch system is exchanged via Ethernet data lines. This arrester is designed for Industrial Ethernet and similar applications in structured cabling systems.

Whether the signal lines for the pitch system must be protected by surge protective devices depends on the sensors used which may have different parameters depending on the manufacturer (Figure 1.10).

#### 1.1.1.8.5 Condition monitoring of the surge arrester

IEC 61400-24 and the DNV certification guideline always recommend the monitoring of surge arresters.

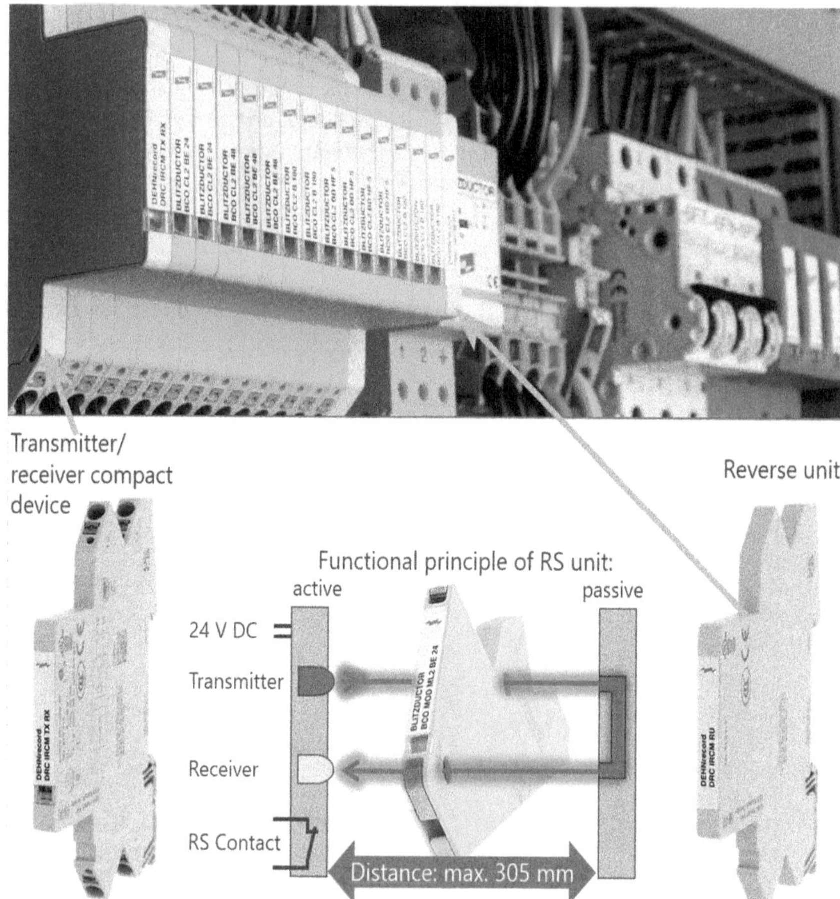


*Figure 1.10 Surge protection on the pitch system of a wind turbine, including condition monitoring*

In the case of lightning current or surge arresters for low voltage, monitoring can be implemented via an integrated remote signalling contact (floating changeover contact). This contact signals a disconnection of the protection module due to an overload.

Lightning current or surge arresters for information technology, such as the BLITZDUCTORconnect series with status indication, can be integrated into the condition monitoring system (Figure 1.10) with an optional remote signalling unit. The function of the surge arresters can thus be permanently monitored to guarantee maximum protection against lightning strikes and surges. Permanent availability and system reliability are made a priority.

#### *1.1.1.8.6 Lightning current measuring system*

Continuous system monitoring helps to operate the installations economically. Operating data and status changes can be detected using sensors. This sensor

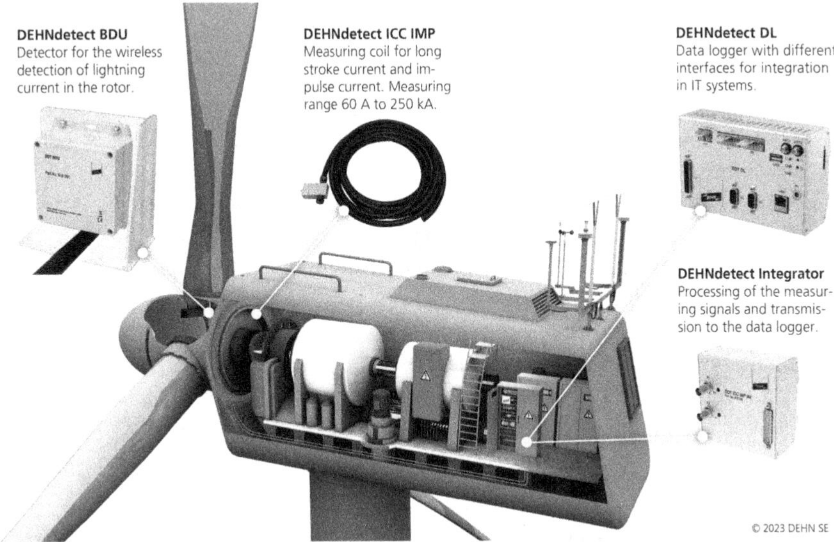
DEHNdetect BDU
Detector for the wireless detection of lightning current in the rotor.
DEHNdetect ICC IMP
Measuring coil for long stroke current and impulse current. Measuring range 60 A to 250 kA.
DEHNdetect DL
Data logger with different interfaces for integration in IT systems.
DEHNdetect Integrator
Processing of the measuring signals and transmission to the data logger.
© 2023 DEHN SE

*Figure 1.11 DEHNdetect lightning current measuring system*

information is necessary to be able to do preventive servicing and maintenance planning. For offshore installations in particular, it is important to always be abreast of the current status of the system, so that unnecessary site inspections can be avoided as much as possible. The DEHNdetect lightning current measuring system, which meets the strictest requirements of IEC 61400-24 Annex L, provides support here (Figure 1.11). With a measuring range from 60 A to 250 kA, the DEHNdetect lightning current measuring system not only detects impulse currents but also dangerous long stroke currents (ICC only events – Initial Continuous Current). This enables the operator to assess whether a lightning strike was harmless or whether it is necessary to shut down and test the turbine to rule out damage to the rotor blades or other components. Damage resulting from a lightning strike does not necessarily lead to the immediate failure of the turbine. That is why lightning events often go undetected. Especially in the case of upward flashes, initial long stroke current of no more than a few 100A flows and can be the main cause of melting, e.g., on the receptors of rotor blades. The resulting damage may be severe. To prevent damage to the rotor blades or other components, DEHNdetect registers the total current flowing through the system as well as individual partial currents in the rotor blades. In detail, **DEHNdetect** registers the peak current values in the system, the specific energy, the load and the rise time of the lightning current and long stroke current. The minimum trigger level is 60 A. This can be adjusted upwards by the operator.

In order to be informed about the lightning event in real time, the measured data can be integrated into the IT infrastructure of the wind turbine via existing interfaces using a Modbus TCP interface. The data can then be easily read out and managed via SCADA systems.

The wind turbine operators are thus always aware of any lightning events that occur and can use this information when planning maintenance and service activities.

There is also the option of transferring the data to a cloud and evaluating it via a Web application. Monitoring several systems or even entire wind farms therefore poses no problem.

The modular design of this measuring system is an advantage. It can be installed directly by the wind turbine manufacturer or retrofitted in existing systems. Installation is quick and easy and can be adjusted to the specifications of the individual turbine. Since no cable connections between the rotor blade and the hub or between the hub and the nacelle are necessary, DEHNdetect is a device that is particularly easy to retrofit. This means that all essential information about a lightning event is now also available for existing plants in order to optimize maintenance and service work and thus increase plant efficiency.

The system records:

- impulse current [kA]
- long stroke current [A]
- charge [C]
- specific energy [MJ/Ω]
- rise time [kA/μs]

DEHNdetect identifies the affected blade, and a notification appears in the SCADA system or online.

A direct hit from lightning or overvoltages due to indirect hits with all their consequences can severely damage a wind turbine (WT) and adversely affect its operation with lasting effect. This applies both to material damage at the outer shell and to damage to the installed electrical components. Therefore, the design for surge protection described in the preceding chapter is very important.

There are further factors which adversely affect the functioning of a wind turbine and may even result in its complete breakdown. Contrary to damage due to direct lightning hits or overvoltages, some damage to wind turbines due to environmental influences or stress caused by their operation will appear only with a time delay and unexpectedly.

#### 1.1.1.9 Environmental influences

The relevant standard for the building and operation of wind turbines is IEC 61400-1 [13] and describes all the possible environmental influences in scenarios:

With respect to environmental conditions, it distinguishes various influencing variables. Requirements for the foundation and the wind turbine's construction serve to counter wind influence, but wind does not directly affect the electrical installation. All other environmental conditions, however, do. They directly impact the electrical components and also the functioning of the overall electrical installation.

Environmental influences on an onshore wind turbine include the following factors:

- temperature, air humidity, air density,
- sun, rain, hail, snow, and ice,
- chemically active substances,
- mechanically active particles,
- salt content of the air,
- lightning,
- earthquakes.

In the case of offshore wind turbines, also sea conditions need to be considered, such as waves, sea currents, sea ice, marine growth, scouring, and seabed mobility. These, however, do not directly impair the functioning of the electrical installation, they rather influence the construction of the wind turbine and will therefore not be considered in more detail in this chapter.

Temperature changes (e.g., from day to night) often lead to the formation of condensate. As is well known, the resulting humidity poses a risk to all electrical components. The hazard due to extremely dry air constitutes no lesser risk as it can result in static electricity charge. Excessive salinity, however, can cause electrochemical processes and lead to corrosion. This may impair the conductivity of the contacts of circuit breakers.

#### 1.1.1.10 Rotation and vibration

Wind turbines always move, not only because the rotor blades and the connected gearbox turn, but also because the rotor blades and nacelle are continuously oriented toward the wind. These constant movements for optimum orientation toward the wind cause vibrations, and accordingly all the installed electrical components and cables must meet special requirements to avoid failures and downtimes.

For the afore-mentioned reasons, one requirement of Section 10.2 of IEC 61400-1 is that electrical installations be designed in such a manner that "hazards to people and animals as well as the risk of damage to the wind turbine and the outer electrical installation is reduced to a minimum during operation and maintenance." The standard also stipulates in general that all the installed electrical components shall be suitable for the environmental influences described above. When it comes to the planning and design of the electrical installation the standard becomes much more specific but nevertheless not less flexible.

## 1.2 The electrical installation as protection against electrical hazards

IEC 61400-1 requires that electrical installations be designed in accordance with the machinery directive 2006/42/EG [14], which corresponds to standard IEC 60204-1 [15]. As protection against electrical hazards and to achieve the related

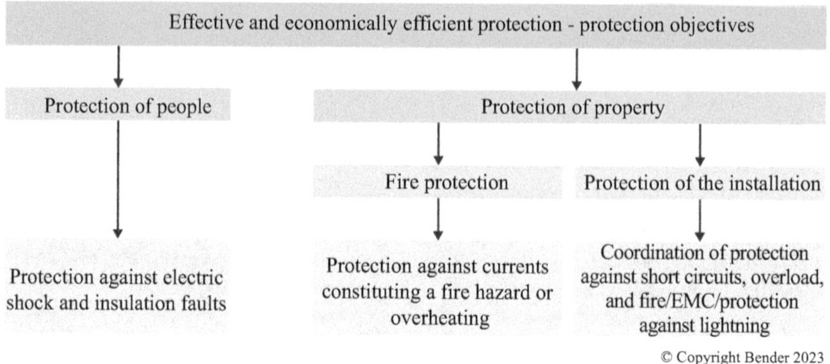


*Figure 1.12 Protection objectives for wind turbines*

protection objectives for wind turbines, the low-voltage directive 2014/35/EU [16] is applicable in connection with the harmonized standards but is not applicable in itself. This inevitably leads to the question which protection objectives must be aimed at and reached for wind turbines (Figure 1.12).

Insulation faults in control circuits of machinery or installations must not lead to unwanted starting up or termination of a process. Furthermore, drive systems may not create momentum of their own and in doing so endanger personnel. Therefore, control circuits must be incorporated into the safety plan for the electrical installation in the same manner as main circuits are considered.

IEC 60364 [17] contains all essential aspects to be considered when erecting electrical installations in general. Hence for the electrical design of wind turbines two principle approaches are employed:

- Wind turbines are always designed based on IEC 60204-1.
- Regarding the protection of people i.e. the protection against electric shock, the definitions and measures of IEC 60364-4-41 [18] must be employed. These are:
    - Basic protection, i.e., that enclosures can only be opened with the aid of a tool, that live parts are switched off before an enclosure is opened, protection by insulation, voltage remaining after 5 s < 60 V
    - Fault protection, i.e., the prevention of potentially dangerous situations: provisions for fault protection shall be made for live parts and exposed conductive parts in the event of an insulation fault.

When the situation escalates, especially the fault protection measures will result in the respective circuit being switched off and hence in failure or standstill of the installation. Of course such a stoppage is in contradiction to the aimed at availability.

Apart from the safety of people and the protection of the installation from damage, keeping the installation functioning and making operation and maintenance easy play a decisive role. Both the design engineers and the manufacturer, but also the future operator of a wind turbine must draw up a risk assessment and document it. Such a risk assessment should also include critical conditions during which operation can still continue.

To facilitate compliance with the safety requirements and to make fail-safe operation as well as the maintenance of the installation easier, IEC 60364-4-41 offers the following technical options:

- In TN and TT-systems (earthed systems) a residual-current monitor (RCM) in accordance with IEC 62020 can be used for preventive maintenance.
- In IT-systems (unearthed systems) an acoustic or optical signal must be permanently present for the duration of the first insulation fault. Switching the installation off for the first fault is not necessary.

The residual-current monitors, however, should satisfy the newer product standard IEC 62020-1 [19], which stipulates higher EMC requirements than its predecessor version IEC 62020.

IEC 61557-8 [20] is the applicable standard for the selection of insulation monitoring devices (IMDs). Here it is essential that the employed IMDs recognize both symmetrical and asymmetrical insulation faults.

As mentioned above, electrical safety refers not only to the main circuits but also to the control circuits. Among other requirements IEC 60204-1 therefore stipulates for functional reliability "(. . .) for insulation faults in control circuits that these may not lead to malfunction such as unintentional starting, potentially hazardous motions or prevent stopping of the machine. (. . .)." One reason for this requirement is that control circuits often contain electronic devices with a low signal level, and therefore do not trigger upstream safety units reliably in the event of a fault. Therefore insulation faults in control circuits must be avoided or signalled early on.

Whether control circuits function properly can be determined by necessary function checks. For these the following possibilities exist:

- automatic checking during operation,
- manual checking during inspections,
- checks during start-up,

and after defined periods, or as a combination depending on the local requirements. Starting from the assumption that wind turbines necessitate high availability, automatic monitoring by insulation monitoring devices in unearthed systems or residual-current monitors in earthed systems is recommended.

### *1.2.1 General types of power supply systems*

The possible distribution system types are distinguished by their design as either earthed or unearthed systems. Earthed means that the neutral conductor is connected to the protective earth (PE). In contrast to this, in unearthed systems, no active conductor has a conducting, low-resistance connection to protective earth.

In technical terms, the distribution system types are described as follows:

- TN-S system = French: terre neutre séparé, i.e., separate neutral earth (Figure 1.13)
- TT system = French: terre terre, i.e., direct earthing of one point (Figure 1.14)
- IT system = French: isolé terre, i.e., isolated earth (Figure 1.15)

To avoid making the following discussion unnecessarily complicated, here only the applications earthed and unearthed system in general shall be considered, since only these fundamental design differences are important for the system behavior in the event of a fault, for the protective measures required, and for automated monitoring during operation.

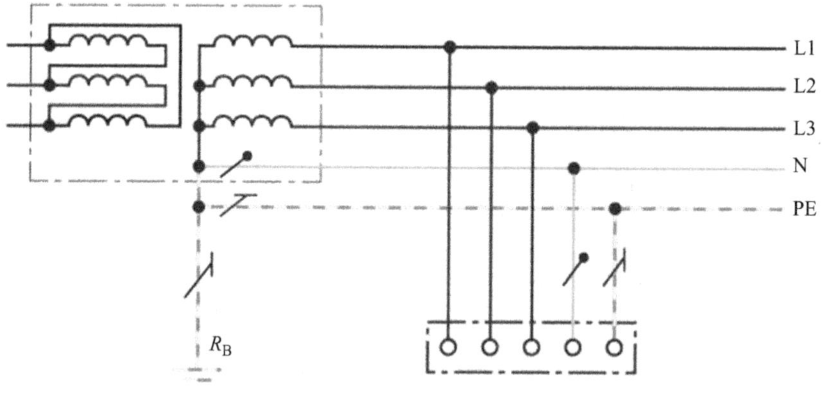


*Figure 1.13 Schematic diagram of a TN-S system*

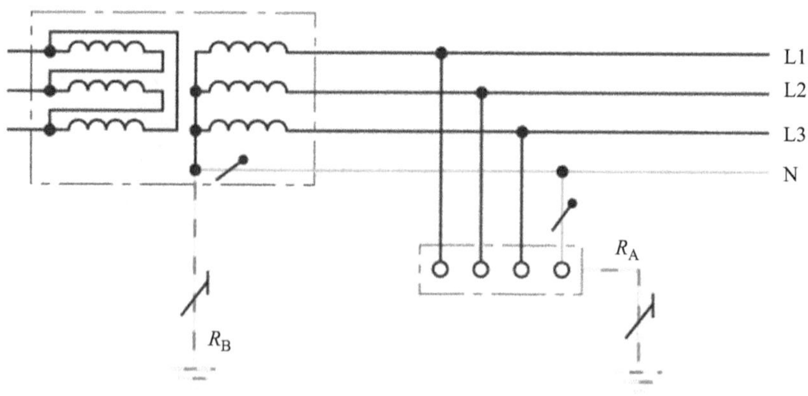


*Figure 1.14 Schematic diagram of a TT-system*

In an earthed system, already the first insulation fault results in a direct connection of an active conductor to the protective earth. This causes a high fault current to flow which is equal to the short-circuit current $I_k$. The protective device trips, and this inevitably leads to a standstill in the application (Figure 1.16).

A fault current is rarely a sudden event, rather it often develops over a certain time due to damage to the insulation or aging of electrical components.

With an unearthed system, the first fault does not seriously influence the functioning of the installation as the star point of the transformer is not connected to protective earth, and hence no fault loop is closed via PE, and no high fault current can flow. Only parasitic capacitances $C_e$ cause a slight current flow, which, however, is negligibly small (Figure 1.17).

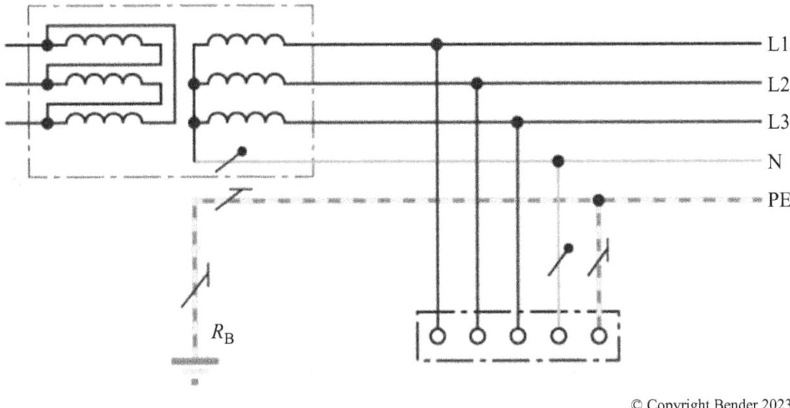


*Figure 1.15 Schematic diagram of an IT-system*

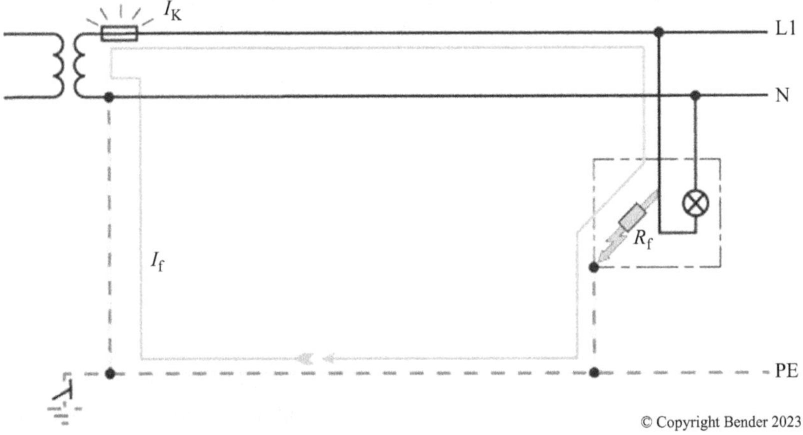


*Figure 1.16 Fault current in an earthed system*

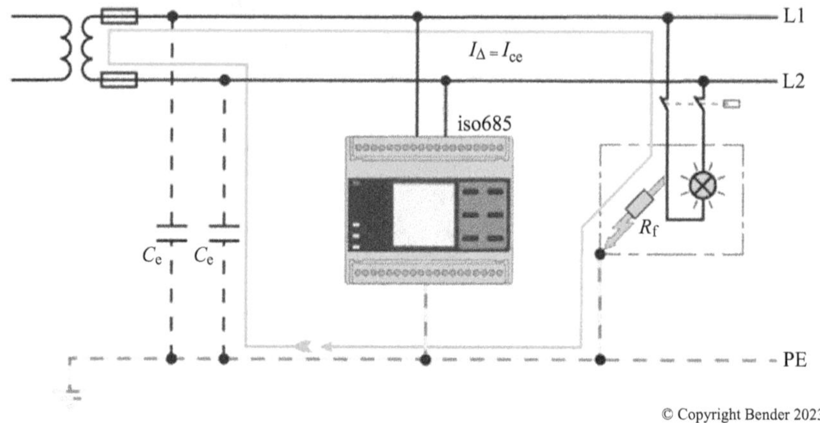


*Figure 1.17 Fault current $I_\Delta = I_{Ce}$ in an unearthed system*

For unearthed systems, standards prescribe the use of insulation monitoring devices that alert to the first fault with an acoustic and optical signal, since the application itself – in this case the wind turbine – may continue to be operated. As the fault is signalled, the operator has the opportunity to initiate countermeasures to correct the fault, while at the same time continuing to operate the installation. This maximizes availability and constitutes an enormous economic advantage. Nevertheless, the protection of people and the installation is ensured at any time.

### 1.2.2 Residual-current monitoring (RCM)

Modern wind turbines often need extensive monitoring via a huge variety of sensors so that the arising malfunctions can be recognized early on. This permits, e.g., analyzing the recorded data from residual-current sensors and drawing conclusions about gradual insulation changes. The general principle here is based on the vectorial addition of all currents in the active conductors, which ideally should be zero (Figure 1.18).

As shown in Figures 1.18 and 1.20, in every electrical installation there are parasitic system leakage capacitances $C_e$ as well as inductive couplings – both created by cables running in parallel – and these result in leakage currents $I_{Leakage}$ on the protective-earth conductor. Leakage currents are specific to the respective installation and rather uncritical for the functioning of the installation. As a rule, component manufacturers indicate the relevant values in the respective datasheets. When a wind turbine has been first put into service, the measured leakage currents should be compared to the datasheets of the installed components (e.g., frequency converters). When no anomalies are noted, the measured values should be recorded in the commissioning log so that reference values are available for future maintenance work. Based on the values measured for the leakage currents, threshold values for warnings and alarm messages can be defined in consultation with the future operator of the installation.

If during operation fault currents $I_{Fault}$ occur in addition to the leakage currents $I_{Leakage}$, the residual currents $I_{\Delta}$ increase, which is an indication of a deterioration of the insulation. When a first threshold value is reached, the RCM system signals a warning, which can lead to an inspection or maintenance work performed without switching off the wind turbine. In this manner, the operator has an information advantage, as it were, and can proactively counteract a possible outage of the electrical installation (Figure 1.19).

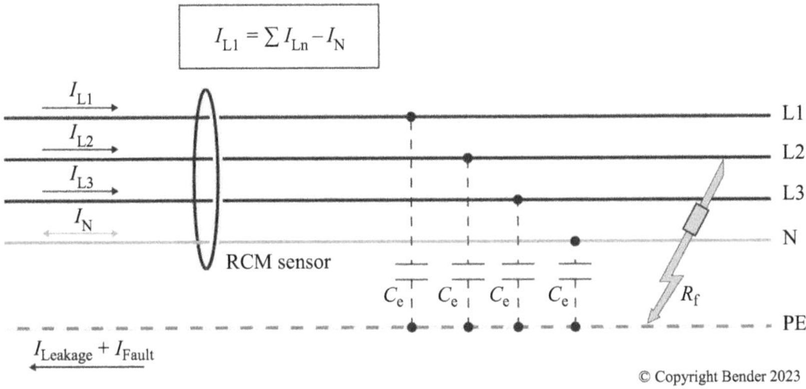


*Figure 1.18 Principle of residual-current measurement*

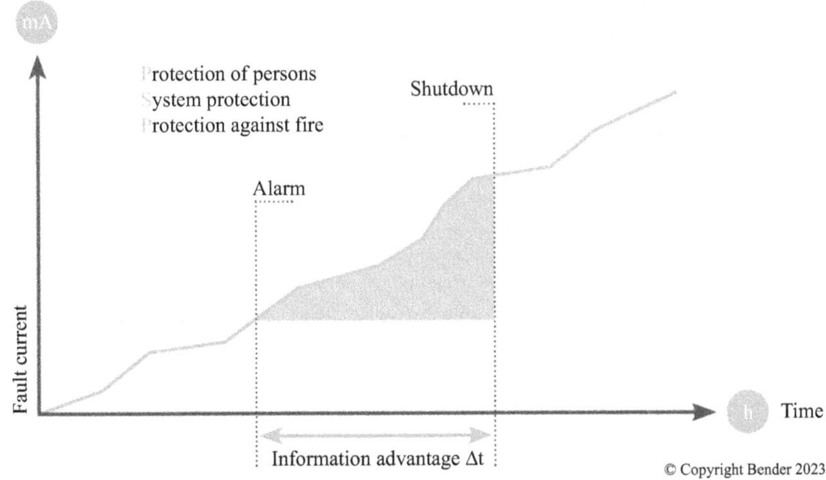


*Figure 1.19 Information advantage based on fault-current analysis*

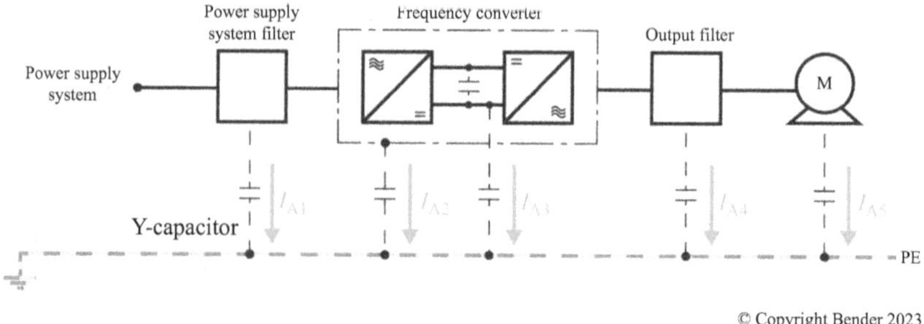


*Figure 1.20 Schematic diagram of a frequency converter with leakage currents*

The selection of the current transformer is decisive for reliably obtaining accurate measured values. For this, two transformer types are distinguished:

- RCM type A: detects AC fault currents and pulsed DC fault currents
- RCM type B: in addition to the type A functions, it detects also smooth DC fault currents

Dependent on the parts of the installation that are to be monitored, the transformers need to be selected already in the planning phase. In the past few years, the number of industrial applications using frequency converters has risen significantly. In theory and practice frequency converters can generate all kinds of leakage or fault currents. Therefore, the use of type B RCM sensors is recommended (Figure 1.20).

Residual-current monitoring (RCM) offers the following advantages to the operator:

- continuous monitoring of the electrical installation and the certainty that no insulation fault is present in the installation
- measurement and evaluation of residual, fault, and nominal currents of loads and installations
- recognition of gradual changes in the insulation supported by preventive maintenance
- EMC monitoring of earthed systems for "stray" currents and additional PEN-connections,
- measured-value transmission to a building management system (BMS)

### 1.2.3 Insulation monitoring (IMD – insulation monitoring device)

The parts of the wind turbine operated as an unearthed system must be monitored by an insulation monitoring device. These parts include for instance the control circuits. The IMDs measure and record the essential parameters such as the voltage in volts [V], the system leakage capacitance $C_e$ in microfarads [$\mu$F], and the insulation resistances $R_{iso}$ in [$\Omega$] (Figure 1.21).

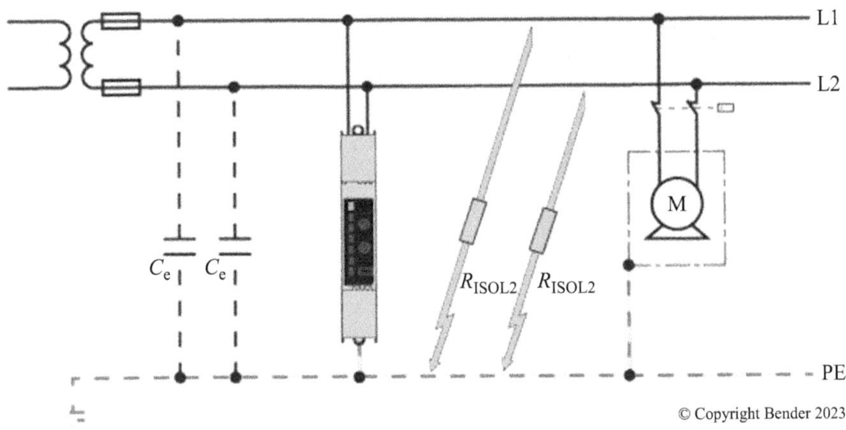


*Figure 1.21 System leakage capacitance $C_e$ and insulation resistance $R_f$*

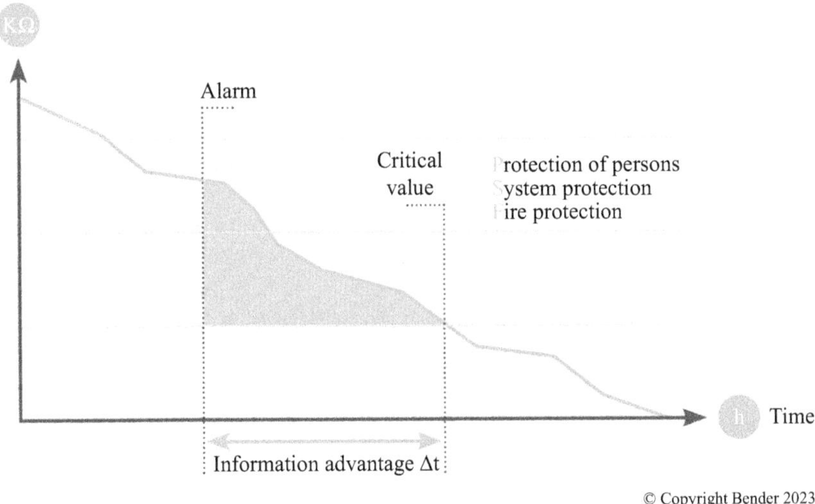


*Figure 1.22 Information advantage owing to insulation monitoring*

Newly built IT systems initially have very high insulation resistance, which deteriorates in the course of time due to outside influences, damage to or ageing of the insulation material. Figure 1.22 illustrates the change in insulation level.

As soon as the first threshold is reached, the insulation monitoring device emits an optical and acoustic signal. To complement this, information about the alert can be sent to a higher-level control system via a data interface (e.g., Modbus). The operator of the installation must eliminate the fault promptly as a second fault will

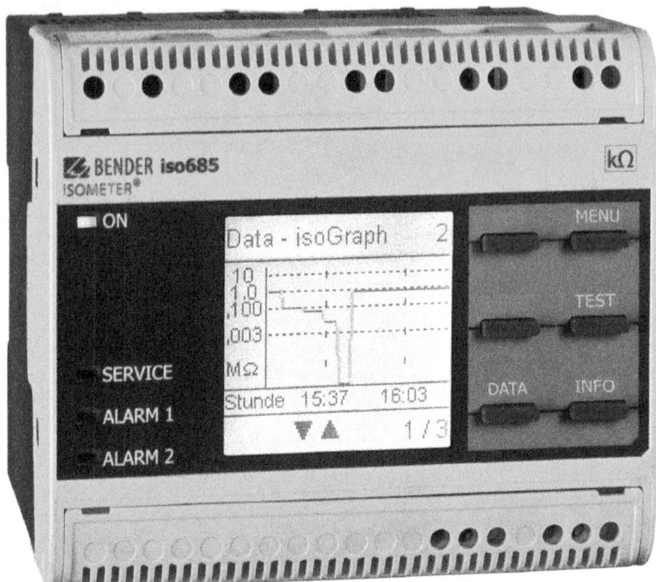


*Figure 1.23 Bender Isometer® type iso685 with graphic display*

result in a short-circuit current, and the installation is then switched off by the protective devices.

The system leakage capacitance $C_e$, which depends on the overall design of the installation and on the environmental conditions, plays a special role. Even with newly erected installations, it cannot be assumed that the system leakage capacitance will always have the same value. Even minimum variations in air humidity can markedly change the measured value. Therefore, insulation monitoring devices for which the theoretically forecast system leakage capacitance must be set to a fixed value have not proven helpful in practical application. In contrast to this approach, adaptive measuring procedures determine the actual system leakage capacitance with every measuring cycle and hence yield accurate measurement results.

Also helpful are graphic displays in the measuring devices which inform the maintenance staff directly about the current status of the parameters described above. Trends can be shown directly on location via graphics, which facilitates the evaluation of an installation's condition (Figure 1.23).

### *1.2.4 Protection of the power-supply system and the wind turbine*

Of course the generated energy shall be fed into the public power grid, and this means certain conditions must be fulfilled. The operators of the respective local public power grid have certain requirements for decentralized energy generation plants so that these will not adversely affect the grid when feeding power in. Regarding this, individual country-specific regulations are applicable for every network operator.

On the other hand, the public power grid must not interfere with the safe operation of the wind turbine. For this reason, there is the so-called network and system protection which separates the connection between the public power grid and the energy generation plant by actuating coupling switches when limit values are exceeded. If the voltage and frequency values are not within the threshold values stipulated by the respective standard, the relays of the network and system protection switch and separate the energy generation plant from the public power grid. The currently measured values for voltage [V] and frequency [Hz] are continuously shown on the display. Ideally, critical measured values that resulted in a switching of the relays are logged in the history memory with date and time stamp/real-time clock.

#### 1.2.4.1 The wind turbine as a whole

Energy generation using wind force follows a simple principle: the movement of the rotor blades first converts wind energy (kinetic energy) into mechanical energy. The generator inside the nacelle transforms the mechanical energy into electrical energy. An upstream gearbox can serve to reduce the generator size. A frequency converter converts the electrical energy generated by the generator to the desired voltage and frequency and feeds it into the power grid. For this reason, the functioning of all drive and control systems is so important, for only this can ensure effective and fail-safe operation.

Due to the rotary motion of the nacelle, the vertical cables in the tower that conduct the energy from the nacelle to the transformer station at the ground are slightly twisted and hence subject to permanent mechanical stress. Because of these mechanical movements, the insulation value $R$ [$\Omega$] may be impaired. Moreover, the cables are installed in the immediate vicinity of the metallic outer wall of the tower, which gives rise to the parasitic system leakage capacitances $C_e$ toward protective earth (PE) described above (Figure 1.24).

In the end, the wind turbine must be considered as an overall design that is not only comprised of a multitude of individual electrical installations. The parts of the overall installation are monitored by separate monitoring units, for only all electrical and mechanical components working together ensure fault-free operation. For the operator, it is indispensable that all the measured data are transmitted to a central process control system. Figure 1.25 shows how all this is connected. Several monitoring components ensure electrical safety in the respective subsystems (TN-S or IT systems). All monitoring components are interconnected via a bus system (e.g., Modbus), and all measured data are forwarded to the process control system via a central data interface (Bender COM465IP). Figure 1.14 also shows the network and system protection described above (VMD460NA) at the transformer output.

Regardless of whether an insulation fault occurs in an unearthed or earthed sub-system, quick localization and elimination of the fault is indispensable for the

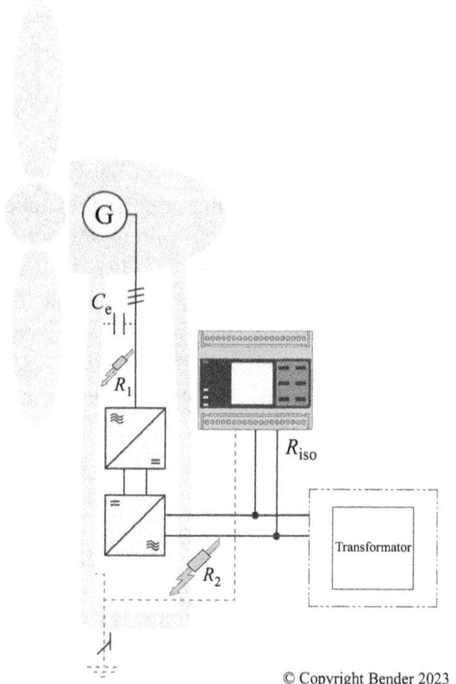


*Figure 1.24 Insulation monitoring of the vertical cables within the tower*

effective operation of a wind turbine. In an earthed TN-S system, residual current transformers (RCMs) identify the affected circuit. This is possible since several sensors can be installed within an earthed system. In contrast to this, in unearthed systems (IT systems) only one central insulation monitoring device can be installed, which complicates the location of a fault. Bender therefore has integrated a so-called locating current injector into its insulation monitoring devices which is started automatically in the event of a fault. Service technicians can also use mobile current probes by Bender or sensors integrated into the electrical installation to identify the affected circuit within the IT system and to thus eliminate the cause of a fault more quickly.

Which type of distribution system will be used exclusively or partially in subsystems differs from one manufacturer to the other and of course also depends on the costs involved. With wind turbines with an output in the megawatt range, the galvanic separation needed for the IT system is already located within the wind turbine or in its immediate vicinity in the form of a medium-voltage transformer. Hence there are no reasons against designing it as an unearthed (IT) system with appropriate insulation monitoring. On the one hand investment costs remain within reasonable limits, while on the other hand, the first insulation fault will not result in an outage of the wind turbine.

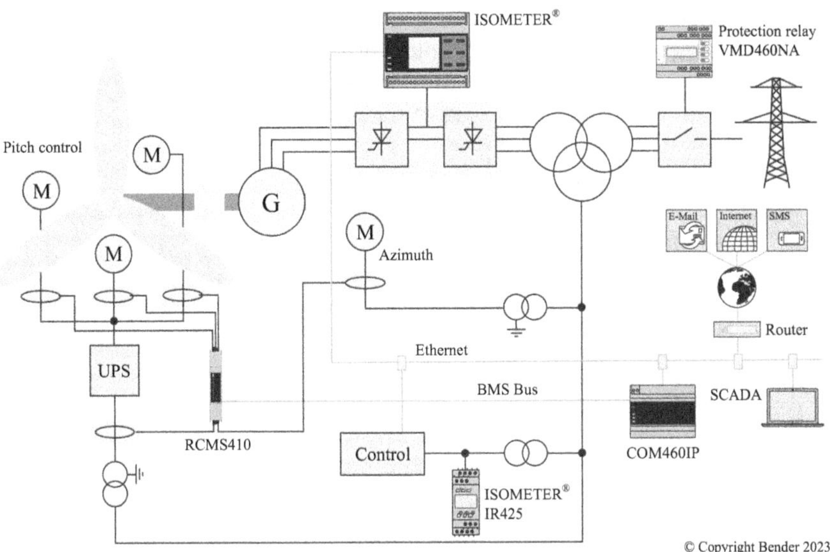


*Figure 1.25 Residual current and insulation monitoring incl. data exchange*

## 1.3 Conclusion

Damage to the power-electronic components of wind turbines, especially to converters, is one of the main cost drivers in the plant operation phase, both in terms of repair and spare parts costs and in terms of the resulting loss of yield. Even more serious than for onshore wind farms is the impact on system availability and thus on yield losses in the case of offshore wind farms. With these facilities, limited accessibility due to wind and waves leads to significantly longer downtimes on average until repairs are carried out. Therefore, in addition to a good earthing system, lightning and surge protection of wind turbines and monitoring of lightning events, another essential element for electrical safety is the rapid detection of insulation faults. An undetected insulation fault is synonymous with unexpected system downtimes or increases fire hazards, and unscheduled service calls triggered in this way cause high time and cost expenditure. Quickly pinpointing and eliminating insulation faults is therefore imperative.

At a first glance, the recommended measures described above for functioning surge protection as well as for high operational availability seem to be contradictory, because the arrester for possible lightning events is designed as functional earthing, whereas the IT system should not have any connection to potential earth. That both measures still work together harmoniously becomes obvious when one considers their function in closer detail. Modern surge protective devices consist of a combination of varistors and gas discharge tubes between the N and PE conductors (GDT). Though a GDT connects the N and PE conductors, it provides a very high insulation resistance in the giga ohm range when it is not tripped. The system leakage capacitance is $< 1$ pF (picofarad). Hence neither the leakage currents nor the insulation resistance in an earthed or unearthed system are affected by the two recommended measures.

Both when the electrical components and when the surge protection is selected, the expected voltage must be taken into account. Especially in unearthed systems, the first insulation fault can lead to so-called residual voltages, which then result in the following effect: With the first insulation fault, voltages between the conductors L1-3 and PE can rise from 230 to 400V, which however must not trip the surge protective device. The specific application should therefore be designed in consultation with the manufacturer of the respective surge protective device.

To operate wind turbines with high availability in the future, protective measures, continuous monitoring, and predictive maintenance are essential. Only in this manner can the costs of construction and operation be amortized over the average 20-year service life of wind turbines.

## References

[1] IEC 61400-24. 2019: Wind energy generation systems – Part 24: Lightning protection

[2] IEC 62305 Parts 1 to 4: IEC 62305-1:2010-12: Protection against lightning – Part 1: General principles IEC 62305-2:2010-12: Protection against lightning – Part 2: Risk management. IEC 62305-3:2010-12: Protection against lightning – Part 3: Physical damage to structures and life hazard IEC 62305-4:2010-12: Protection against lightning – Part 4: Electrical and electronic systems within structures

[3] Germanischer Lloyd Guideline for the Certification of Wind Turbines 2010 and DNVGL-ST-0076 "Design of Electrical Installations for Wind Turbines" from 2021.

[4] IEC TR 61000-5-2. 1997: Electromagnetic compatibility (EMC) – Part 5: Installation and mitigation guidelines – Section 2: Earthing and cabling

[5] IEC 61936-1: 2021: Power installations exceeding 1 kV AC and 1.5 kV DC – Part 1: AC

[6] IEC/TS 60479-1: 2005: Effects of current on human beings and livestock – Part 1: General aspects

[7] IEC TR 60479-4: 2020: Effects of current on human beings and livestock – Part 4: Effects of lightning strokes

[8] IEC 62561-1: 2017: Lightning protection system components (LPSC) – Part 1: Requirements for connection components

[9] IEC 61000-6-2: 2016 Electromagnetic compatibility (EMC) – Part 6-2: Generic standards – Immunity standard for industrial environments

[10] IEC 60364-4-44: 2007 Low-voltage electrical installations – Part 4-44: Protection for safety – Protection against voltage disturbances and electromagnetic disturbances

[11] IEC 60664-1:2020: Insulation coordination for equipment within low-voltage supply systems – Part1: Principles, requirements and tests
[12] IEC 61643-21:2012: Low voltage surge protective devices – Part 21: Surge protective devices connected to telecommunications and signalling networks – Performance requirements and testing methods
[13] IEC – International Electrotechnical Commission: "Wind energy generation systems – Part 1: Design requirements" (2019) – IEC 61400-1
[14] RICHTLINIE 2006/42/EG DES EUROPÄISCHEN PARLAMENTS UND DES RATES vom 17. Mai 2006 (Maschinenrichtlinie)
[15] IEC – International Electrotechnical Commission: "Safety of machinery – Electrical equipment of machines – Part 1: General requirements" – IEC 60204-1
[16] RICHTLINIE 2014/35/EU DES EUROPÄISCHEN PARLAMENTS UND DES RATES vom 26. Februar 2014 (Niederspannungsrichlinie)
[17] IEC – International Electrotechnical Commission: "Low-voltage electrical installations – Part 1: Fundamental principles, assessment of general characteristics, definitions" (2005) – IEC 60364-1
[18] IEC – International Electrotechnical Commission: "Low-voltage electrical installations – Part 4-41: Protection for safety – Protection against electric shock" Edition 5.1 (2017) IEC 60364-4-41
[19] IEC – International Electrotechnical Commission: "Electrical accessories – Residual current monitors (RCMs) – Part 1: RCMs for household and similar uses" (2021) IEC 62020-1
[20] IEC – International Electrotechnical Commission: "Electrical safety in low voltage distribution systems up to 1 000 V AC and 1 500 V DC – Equipment for testing, measuring or monitoring of protective measures – Part 8: Insulation monitoring devices for IT systems" Edition 3.0 (2014)

# Chapter 2
# Generator design for geared turbines

Henri Arnold[1]

## 2.1 Introduction and basics

### 2.1.1 Short history of electrical machines

Direct current (DC) machines, that is, motors and generators, had been known since the sixtieth of the eighteenth century. They were quickly established in industrial applications and their theory was fairly developed at that time. With the discovery of the dynamo-electric principle by Werner von Siemens in 1866, also self-excitation was possible and the generation of electricity mainly had been solved. Nevertheless, their transportation over large distances jet remained unsolved.

Alternating current (AC) proposed to be the solution for that. The help of transformers increased the voltage. Higher voltage levels allowed lower currents and lower transportation losses. At least since 1882, also poly-phase systems (Nikola Tesla) were under investigation. Within a very short time, the induction motor (Galileo Ferraris, 1886) and the synchronous motor (Nicolai Tesla, 1887) were introduced for two-phase AC systems. Giesbert Kapp thoroughly described the theory of transformers in 1888, although transformers itself had been known earlier.

The history of three-phase AC systems finally started in 1888, when Michail Dolivo-Dobrovolski first described the fundamentals of the chained three-phase system (denoted as "three-phase current"), and consequently the principle of the three-phase induction motor. The work from Dolivo-Dobrovolski more or less combined many of the ideas, which were already known at that time. His honor was to see the big advantage of the three-phase system. It was far cheaper and more efficient than the two-phase or poly-phase systems before.

Already three years later, in 1891, the first long-distance electricity transmission over 175 km was ready for demonstration. Electrical energy was sent from a waterpower generator in Lauffen (fed from the river Neckar) to the International Electrotechnical Exhibition in Frankfurt am Main. For this purpose, a three-phase AC synchronous generator provided 200 kVA at 55 V and 40 Hz. A transmission line with 20 kV and three-phase transformers demonstrated the energy transmission. Between May and October in 1891 over 1000 light bulbs, an artificial

[1]Electrical Engineer & Technical Consultant, Pirna, Germany

waterfall, electric horse race tracks, an electric mine locomotive, and an electric ship on the river Main amused and astonished over 1.2 million people and massively inspired engineers from all over the world. This marked the beginning of the modern electricity era.

In the following years, many scientific publications led to a very quick understanding of the conversation principle and the theory of three-phase AC motors and generators. The industrialization came within a very short time, especially the mass-to-power ratio of the machinery went down dramatically, e.g., from 88 kg/kW in 1891 to 25 kg/kW in 1910 [1]. A hundred years ago, at the beginning of the nineteenth century, the generation, distribution, and conversion of electrical energy had been solved sufficiently.

During the last century, almost any transmission line on the distribution level was a three-phase system providing alternating current (AC) with 50 or 60 Hz. The vast electrically excited synchronous machines in our conventional power plants provided a major part of our electrical energy. They were best suited for that task, as they provided highest efficiency, stable operation, good controllability, and reactive power support. Other generator concepts, e.g., asynchronous machines, were almost unknown. Mainly in small water power stations or as sliding frequency converters they had their use. On the other hand, induction machines, especially the squirrel-cage induction machine, became the workhorse for almost any industrial drive application as they where cheap and robust. Their need for reactive power was supported by power factor compensation units and synchronous machines.

Two things changed this picture: the invention and fast improvement of power electronics with the beginning of the seventies of the last century and the decentralization of our energy production due to the change to alternative energy resources from the early ninetieth onwards.

Power electronics came into practical sight at the late sixties of the last century, although patents for transistors reach back until 1925. In 1957 the first thyristor had been introduced by General Electric, and MOSFETs by Bell Labs in 1959. It then took another ten years, until the first frequency converter series was introduced by Danfoss (1968). Speed control, reactive power control and energy-efficient partial load operation was now possible for almost any motor and generator concept. Also new concepts, mainly the synchronous reluctance and permanent magnet machines were possible.

Today we see a mixture of different concepts for the generation and consumption of electrical energy. Transformers help to adapt the voltage level. Gearboxes help to adapt speed and torque. Frequency inverters adapt partial load, speed, torque, and reactive power. And they help to start up drive systems. It always depends on the boundary conditions, which concept is used. Not always an inverter is needed, neither for generator applications nor for motors. Not always gearboxes are needed. This is true, especially for wind power applications.

In this chapter, we want to focus on drive trains with gearboxes. Gearless drive trains will be introduced in Chapter 3.

### 2.1.2 Introduction to geared concepts

We have seen many discussions about gearbox-related failures in wind turbines during the last decades. The fluctuating load from wind was far above the values known from experience with industrial applications. Furthermore, gearboxes were told to be noisy, inefficient, expensive, and heavy. These problems, especially in the early days of wind power development, made geared wind turbine solutions somehow suspicious. Many publications therefore offered gearless turbines as a solution.

Nevertheless, the law of growth for electrical machines teaches us that there is a strong correlation between the volume of the stator bore and the torque capability of an electrical machine. Higher speeds lower the torque for a given power and make the generator size smaller. The use of a gearbox in general makes the generator smaller, less heavy, and cheaper.

Many of the technical issues with gearboxes have been solved by today. The load conditions for example are well understood, the gearbox design had been adapted to that. Geared drive trains can be compared to gearless solutions with respect to robustness, efficiency, and noise. Geared generator concepts are a good choice due to their advantage in size, weight, and cost. A big advantage of high-speed concepts for manufacturer is the low integration depth of the single components and quite clear interfaces between them (Figure 2.1). Gearbox, coupling, break, and generator can be described and specified independently. For dissolved and partly integrated drive trains this offers to choose from a set of different suppliers.

*Figure 2.1 Partly integrated drive train with high-speed generator, picture taken from [2]*

High-speed concepts try to bring down the size and mass of the generator as much as technically possible. One technical limitation is the transmission ratio of the gearbox. With two planetary stages and one spur stage, ratios of 1:100 are possible.

Medium-speed generator concepts try to level out the benefits and disadvantages from the high-speed concepts. Often they use only planetary gear stages with a lower total transmission ratio (~1:40). The generator then needs a higher pole number, typically at least eight. The size of the gearbox roughly stays the same, as the only difference is the absence of the last (high-speed) gear stage. Visible sign for such gearboxes is a concentric output shaft instead of a shaft with laterally offset. This, and the bigger dimensions from the output shaft and generator, typically led to integrated generator concepts. Gearbox and generator merge into each other. The big advantages are cost, size, weight, and efficiency efforts. On the other hand, there are also hard engineering issues, e.g., the integration of a break, serviceability, and overall a purchase dilemma: components need a very precise description of their interfaces and exact boundaries for responsibilities between component suppliers. Nevertheless, all geared concepts, high or medium speed, do have their own advantages and will find their markets and niches. Provided, they are technically mature.

This chapter tries to introduce to the basic principles of the major generator types for gear supported wind power applications, including their advantages, operational behavior, and simulation models. A systematic comparison of those concepts in respect of a ranking is not within the scope, as there are further facts to be taken into account – not only the generator.

### *2.1.3 Why do we need speed variability?*

Wind power is a very fluctuating power source. The mechanical power fed into the turbine strongly depends on the wind speed, wind gusts, turbulences, and shading effects. It is not possible, or rather not recommended, to maintain a very stiff rotational speed. In this respect, it is necessary to distinguish between small fluctuations and the mean speed value fitting to the offered wind.

Different speed set points enable the wind turbine to harvest a maximum of wind power (compare Figure 2.2). The optimum load torque is proportional to the rotational speed [3].

$$M_{\text{opt}}(n) = k\, n^2 \tag{2.1}$$

This means there is a cubic dependency for power from speed, remembering the general relation

$$P = \omega\, M = 2\pi\, n\, M \tag{2.2}$$

between mechanical power $P$, speed $n$, and torque $M$. However, turbines with fixed speed set point are possible and may have advantages, mainly, in respect of simplicity. At least the necessary speed range can be limited without a big loss of energy. As an example, typical values for large-scale double-fed asynchronous machines (DFIG) are 65…130% of their synchronous speed. There is no need for lower values; the energy in this speed range is neglectable.

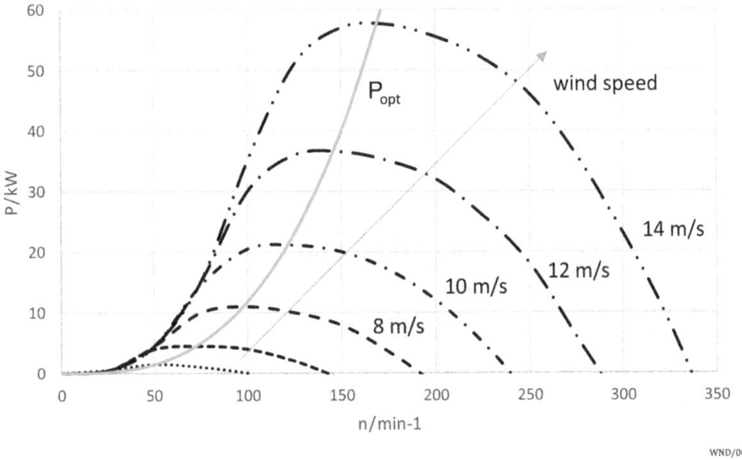


*Figure 2.2 Example curve of power optima for different wind speeds in a small wind turbine*

Regardless of the possibility to adapt the set point, all wind turbines must be able to change their speed slightly around the speed set point. This allows the turbine to store excessive energy into rotational energy from the rotor inertia in the case of wind gusts.

The squirrel-cage induction generator (IG) is the only generator concept that shows some sort of small speed variability only due to its asynchronous behavior. Provided that the rotor of this machine comes with a rather high rotor resistance, and consequently a rather high nominal slip value, this generator is able to react on wind gusts with acceleration or deceleration. This damps high torque input impulses. However, the generator speed still stays near the synchronous speed. An adaption to different set points is not possible without changing the frequency of the feeding AC voltage. Two-speed generators (with changeable pole pair configurations) may be used, but this is limited to small turbines only (in the range of some kW). Furthermore, current grid regulations may be hurt by this generator type.

All other generator concepts need power electronics in the form of a variable frequency drive (VFD) to translate variable generator frequency output into grid-conform constant values or a variable speed gearbox. An example is, the hydraulic Voith WinDrive that enables fixed generator speed at variable turbine speed.

One general problem of all wind turbines is the very low rotational speed of the main hub. This fact is mainly related to the maximum blade tip speed (typical values $\sim$ 80m/s) and the strict correlation between the square of the rotor diameter and the rated turbine power. The mechanical power can be calculated [3] according to

$$P_{\text{kin}} = \frac{1}{2}\, C_{\text{p}}\, \rho\, A\, v^3 \qquad (2.3)$$

In this equation, $P_{\text{kin}}$ is the kinetic power taken off from the wind in W. $C_{\text{p}}$ is a dimensionless performance coefficient, $\rho$ is the air density (near sea level

$\rho \approx 1.18\ \text{kg/m}^3$), $A$ is the swept area in m$^2$, and $v$ is the wind speed in m/s. Typical values for the performance coefficient start from 0.25 for smaller turbines and scratch 80% of the theoretical limit of 0.593 (referred as Betz limit [4]) for large turbines in the MW-range. For a performance coefficient of 1.0 we would have to stop the air completely, which is physically impossible.

For a given wind speed the swept rotor area is proportional to the power. In this way, the rotor diameter is proportional to the square root of the power. Consequently, the rotor speed goes down as the power goes up. With the assumption of a constant tip speed the possible power take off is reciprocal proportional to $n^2$.

### 2.1.4 The advantage of mid and high-speed generator concepts

The fundamental duty of a wind turbine is to turn the fluctuating kinetic (that means mechanical) power offered by the wind into a uniform, stable, and controllable electric power delivery. The mechanical power supplied to a rotating shaft can be expressed as

$$P_{\text{m}} = \omega_{\text{m}} T_{\text{m}} = 2\pi\, n_{\text{m}}\, T_{\text{m}} \tag{2.4}$$

In (2.4) the rotational speed $\omega_{\text{m}}$ and also the torque $T$ may be translated by the help of a gearbox. Between the hub and the generator shaft we can define the transmission ratio $i$ with

$$T_{\text{m}} = \text{i} \cdot T_{\text{hub}} \tag{2.5}$$

$$n_{\text{m}} = \frac{1}{\text{i}} \cdot n_{\text{hub}} \tag{2.6}$$

A gearbox ratio of up to $i = 1:100$ is achievable, typically with two planetary stages and one spur gear stage, which is typically situated at the high-speed output shaft. Such high ratios are more complicated in design and especially the fast spinning output shaft was subjected to failures on bearings and tooth flanks. This led to the development of mid-speed and gearless concepts for high-power wind turbines (e.g., in Table 2.1) that do not need those high-speed spur gear stages.

In geared wind power applications two major machine types are in use: induction generators and synchronous machines. Both type are available in

*Table 2.1 Example rotor characteristics of different high-power wind turbines*

| Turbine | Rated power | Rotor diameter | Rated speed | Gear ratio | Remarks |
|---|---|---|---|---|---|
| GE Haliade-X | 14,000 kW | 220 m | 7 rpm | – | Gearless |
| MHI Vestas V164-8.8 | 8800 kW | 164 m | 12.1 rpm | 1:41 | Mid speed |
| Senvion 6.2M126 | 6150 kW | 126 m | 12.1 rpm | 1:97 | High speed |
| Senvion 3.4M104 | 3400 kW | 104 m | 13.8 rpm | 1:87 | High speed |
| GE 1.5s | 1500 kW | 70.5 m | 22.2 rpm | 1:81 | High speed |

*Table 2.2 High-speed wind generator concepts*

| | IG-DOL | IG-VFD | DFIG | SPSG | PMSG |
|---|---|---|---|---|---|
| Rotor type: | squirrel cage rotor with enlarged rotor resistance | squirrel cage optimized for VFD | three-phase wound rotor with slip rings | electrically excited salient pole field winding | buried or surface-mounted permanent magnets |
| Excitation: | stator from grid | stator from VFD | rotor from VFD | rotor from brushless excitation system | permanent magnets and stator from VFD |
| Speed range: | 90...95% $n_s$ | 0...130% $n_s$ | 65...130% $n_s$ | 0...130% $n_s$ | 0...130% $n_s$ |
| Reactive power: | not capable | from VFD | limited capability | full capability from stator | VFD |
| Mid speed capability: | bad power factor | bigger inverter | high rotor currents | good | excellent |

different configurations, compare Table 2.2. Each of those concepts have advantages and disadvantages in respect of costs, weight and size, reliability, noise, grid compliance, and efficiency.

All generator types have one thing in common: a rotating electromagnetic field and a multi-phase (mostly three phases) armature winding in the stator that reacts with this field. DC machines are not used for large-scale wind power applications. This means, all generator types differ only in their rotor design, their principle of excitation, and the connection to the grid.

Typical high-speed concepts have pole numbers of 2p=4 or 2p=6 and nominal stator frequency of 50 or 60 Hz, if directly coupled to the grid (IG, DFIG, SPSG). This leads to nominal synchronous speeds between 1000 and 1800 rpm. For all VFD driven concepts (IG-VFD, SPSG, and PMSG), there is no need to use nominal frequencies of 50 or 60 Hz. This means, the pole pair number and nominal frequency can be chosen to fit efficiency and cost requirements. Typical values are in the range of 2p = 4–8. Higher pole numbers lead are used in mid-speed concepts.

### *2.1.5 Dimensioning and law of growth*

For comparison of generator topologies it is necessary to find general and simple analytical equations to calculate some sort of quantification numbers. Cost and weight do have a strong linearity in between them. Although very excessive instrumentation, high costs for magnet material or additional equipment, as, integrated rectifiers or measurement equipment will increase the cost above normal limits. As the first indication, between 10 and 15 €/kg will give quite good price indication. It is clear that this is finally also a question of market situation, supply

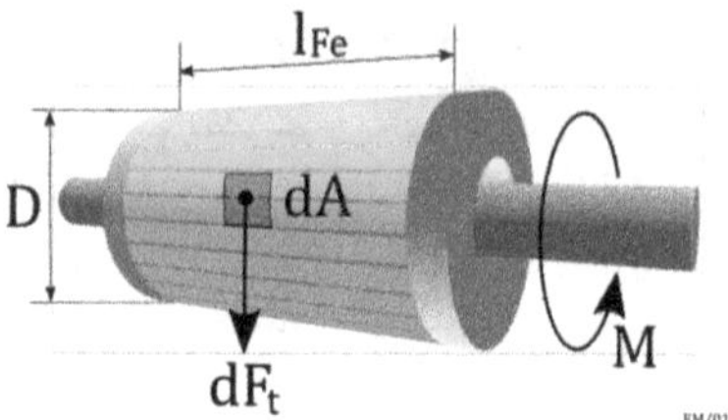


*Figure 2.3 Tangential force density definition in the air gap of an electric machine*

chains, and production place. Generator weight for high-speed concepts ranges from 5 t for small 1.5 MW generators to almost 20 t for 7 MVA class.

The relation between weight and power capability is more technically driven. One point has to be stated quite clearly: not the power but the torque demand is driving the size of a rotating electrical machine. The torque of an electric machine can be calculated from the tangential force produced by the magnetic field in the air gap. The quotient from tangential force and surface defines the tangential stress $\sigma_{\tan}$, or better: the tangential force density, in the air gap (compare Figure 2.3)

$$\sigma_t(r,x) = \frac{\mathrm{d}\,F_t(r,x)}{\mathrm{d}A} = B_\delta(r,x)\cdot A(r,x) \tag{2.7}$$

The product of air gap flux density $B_\delta$ and ampere turns $A$ is in correlation to Lorenz's force definition $F = I\cdot(l\times B)$ for conductors in the magnetic field. Integration over the entire air gap bore, that means over the surface from the rotor, we calculate the total tangential force with

$$F_\mathrm{t} = l_\mathrm{fe}\int_0^{\pi D} B_\delta(r)\cdot A(r)\mathrm{d}r \tag{2.8}$$

Taking the diameter $D$ from the air gap into account, we get from the mean tangential force density to the torque of the machine with

$$M = \frac{\pi}{2}\,D^2\,l_\mathrm{fe}\,\bar{\sigma}_{\tan} \tag{2.9}$$

At this point it is remarkable that neither the mean air gap flux density nor the ampere turns can be increased without big efforts. The ferromagnetic material in conjunction with the magnetization current (IG, SPSG) or the amount of magnet material (PMSG) limit the mean flux density in the air gap. The cooling conditions, and with this the possible current densities in the stator winding, limit the ampere turns. This means that also the mean force density in the air gap cannot be increased considerably without high technical efforts. The size of the machine, that is, the volume of the stator bore, is directly proportional to the torque requirement. Equation (2.9) is essentially for dimensioning of electrical machines.

Typical values of the tangential force density for electrical machines in the range above 1 MW are within $\bar{\sigma}_{\mathrm{tan}} = 25 \ldots 40$ kPa (1 kPa = 0.001 N/mm$^2$). The bigger the machine and the better the cooling, the higher will be the mean force density. As an example, water jacket-cooled high torque machines can reach up to 200 kPa in traction applications.

Sometimes it is more useful to compare the power instead of the torque with the volume of the stator bore. This is useful especially if an electrical machine is designed for the delivery of reactive power, as wind power generators typically do. This leads to Esson's utilization number $C$, which is originally based on the inner apparent power of an electrical machine. For induction generators, a common definition uses real power output and synchronous speed for the definition. The relation to tangential stress if given by

$$C_{\mathrm{P}} = \frac{P_{\mathrm{n}}}{D^2\, l_{\mathrm{fe}}\, n_0} = \cos\varphi_{\mathrm{n}} \frac{U_1}{E_1} \pi^2\, \bar{\sigma}_{\mathrm{tan}} \tag{2.10}$$

For synchronous generators, the utilization number is based on nominal apparent power and nominal speed. The relation to the tangential stress is given by

$$C_{\mathrm{S}} = \frac{S_{\mathrm{n}}}{D^2\, l_{\mathrm{fe}}\, n_{\mathrm{n}}} = \frac{U_1}{E_1} \pi^2\, \bar{\sigma}_{\mathrm{tan}} \tag{2.11}$$

Typical values are $C_{\mathrm{P}} = 4...7$ kW min/m$^3$ for induction generators and $C_{\mathrm{S}} = 5...7$ kVA min/m$^3$ for synchronous generators (Table 2.3).
The "relative length" can be used to compare machine's slimness.

$$\lambda = \frac{l_{\mathrm{fe}}}{\tau_{\mathrm{P}}} = \frac{2p\, l_{\mathrm{fe}}}{\pi D} \tag{2.12}$$

Typical values for high and mid speed wind power generators are $\lambda = 1.1 \ldots 2.8$.

### *2.1.6 Stator connection*

All the machine types that we cover in this chapter do have a very similar and typical stator design. They all have three-phase AC windings in their stator. The

*Table 2.3 Examples for tangential force density for high power wind turbine generators of different geared drive train concepts*

| Generator type | Nominal power | Stator bore | Nominal torque | Tan force density | Esson's number |
|---|---|---|---|---|---|
| Air cooled IG | 3750 kW | 0.32 m$^3$ | 25 kNm | 39 kPa | 6.4 kWmin/m$^3$ |
| Water cooled mid speed IG | 3500 kW | 1.00 m$^3$ | 64 kNm | 32 kPa | 5.2 kWmin/m$^3$ |
| Air cooled DFIG | 1500 kW | 0.16 m$^3$ | 8.4 kNm | 26 kPa | 4.3 kWmin/m$^3$ |
| Water cooled DFIG | 6500 kW | 0.64 m$^3$ | 53 k Nm | 41 kPa | 6.8 kWmin/m$^3$ |
| Air cooled SPSG, pf 0.8 | 2160 kVA | 0.23 m$^3$ | 11 kNm | 24 kPa | 4.9 kVAmin/m$^3$ |
| Air cooled PMSG | 2780 kVA | 0.32 m$^3$ | 15 kNm | 24 kPa | 4.2 kVAmin/m$^3$ |

only difference is in their rotor construction and consequently the working principle. Therefore, we want to explain the stator connection and the creation of a rotating magnetic field in the following paragraphs for all AC generator types together.

Wind power generators typically come with distributed winding systems made of form wound coils. Only small machines (up to one MW) are equipped with windings made of round wire, so-called random or insertion windings. In both cases, it is possible to describe the distributed winding system using a concentrated winding model, as shown in Figure 2.4 for the example of a common three-phase winding. The spatial angle between the coils from such winding system is always

$$\alpha_{ph} = \frac{360^\circ}{N_{ph}} \tag{2.13}$$

In (2.13), $N_{ph}$ is the number of phases. In the following we always want to assume that $N_{ph} = 3$. We get the typical phase angle of $\alpha_{ph} = 120^\circ$. Furthermore, we agree to use the index "1" for all stator values and small letters "a," "b," and "c" for the phase numbering inside of the winding.

The feeding voltage system, necessary for smooth operation of such three-phase windings, has to be a symmetrical three-phase voltage system with positive phase order. Unsymmetrical feeding leads to harmonic components in the power flow and torque, additional losses and noises. Therefore, only very small deviations from symmetry are allowed. A symmetrical feeding system is defined in (2.14). Additionally, Figure 2.8 shows as an example the voltages of such a three-phase system for $\varphi_U = 0$ on the left diagram. The peak value of this sinusoidal voltage can be expressed with the corresponding root-mean-square (RMS) value $U$ multiplied by the factor of $\sqrt{2}$. Note that the line voltages in contradiction to the phases

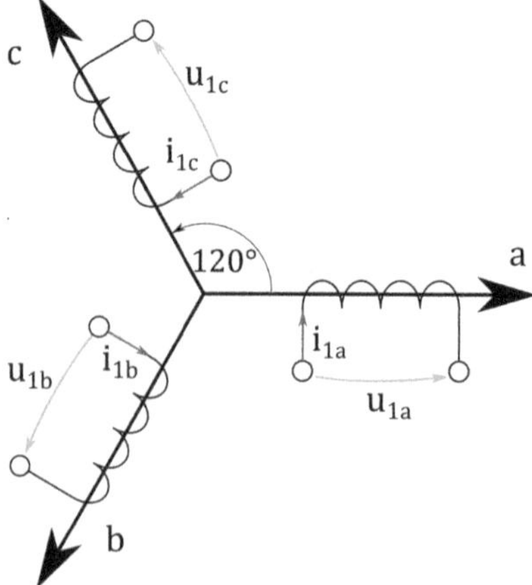


*Figure 2.4 Concentrated windings for a symmetrical three-phase system*

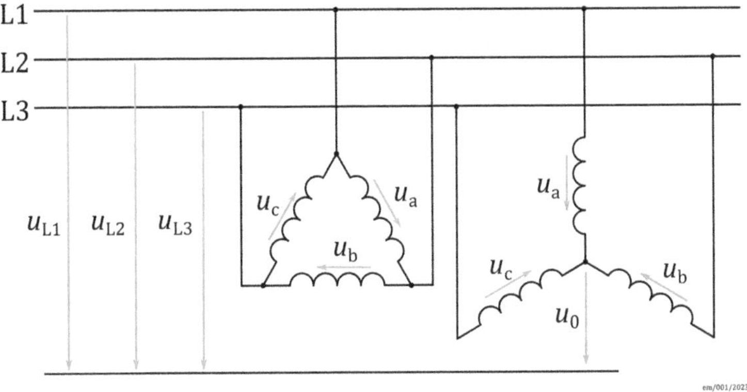


*Figure 2.5 Voltage definitions on delta (left) and star (right) connection of a three-phase-winding*

are numbered with L1, L2, and L3.

$$\begin{aligned} u_{L1} &= \sqrt{2}U\cos(\omega t + \varphi_U) \\ u_{L2} &= \sqrt{2}U\cos(\omega t + \varphi_U - 120^\circ) \\ u_{L3} &= \sqrt{2}U\cos(\omega t + \varphi_U - 240^\circ) \end{aligned} \tag{2.14}$$

The three phases from the machine are connected to the three-phase AC network either using delta or star connection. We have to distinguish between line voltage, line-to-line, and phase voltage. Their definition is according to Figure 2.5. The following equations apply for the phase voltages in a *delta connection*.

$$\begin{aligned} u_{1a} &= u_{L1} - u_{L2} \\ u_{1b} &= u_{L2} - u_{L3} \\ u_{1c} &= u_{L3} - u_{L1} \end{aligned} \tag{2.15}$$

For *star connection* the phase voltages are

$$\begin{aligned} u_{1a} &= u_{L1} - u_0 \\ u_{1b} &= u_{L2} - u_0 \\ u_{1c} &= u_{L3} - u_0 \end{aligned} \tag{2.16}$$

The star point voltage $u_0$ typically is unknown, but for symmetrical windings, as an approximation, the following equation can be used:

$$u_0 \approx u_{com} = \frac{1}{3}(u_{L1} + u_{L2} + u_{L3}) \tag{2.17}$$

Taking into account that the sum of all line voltages shall be zero, this leads to

$$\begin{aligned} u_{1a} &\approx u_{L1} \\ u_{1b} &\approx u_{L2} \\ u_{1c} &\approx u_{L3} \end{aligned} \tag{2.18}$$

Bear in mind that equivalent circuit diagrams always provide phase values, regardless of star or delta connection. Some inverters and simulation programs do not support machine data for delta connected machines, they suppose star connection. In this case, all resistances, reactances, and impedances have to be divided by the factor 3. As an example, Table 2.7 contains the equivalent circuit parameter of a 4200 kVA induction generator in delta connection.

Using complex phasors (compare next paragraph) we get by combining (2.29) and (2.15)

$$\begin{aligned} \underline{U}_{1a(\Delta)} &= \sqrt{3}U\, e^{j\varphi_U}\, e^{j\pi/6} \\ \underline{U}_{1b(\Delta)} &= \sqrt{3}U\, e^{j\varphi_U}\, e^{j\pi/6}\, \underline{a}^{-1} \\ \underline{U}_{1c(\Delta)} &= \sqrt{3}\, U e^{j\varphi_U}\, e^{j\pi/6}\, \underline{a}^{-2} \end{aligned} \tag{2.19}$$

For star connection we get

$$\begin{aligned} \underline{U}_{1a(Y)} &= U\, e^{j\varphi_U} \\ \underline{U}_{1b(Y)} &= U\, e^{j\varphi_U}\, \underline{a}^{-1} \\ \underline{U}_{1c(Y)} &= U\, e^{j\varphi_U}\, \underline{a}^{-2} \end{aligned} \tag{2.20}$$

Two things are remarkable: There is a difference in amplitudes of square root of 3 between star and delta connection, and there is also a phase difference of $\pi/6$ (30°). The base value $U$ describes the rms line-to-ground voltage from the feeding voltage system. Take note that the rms line-to-line voltage is used as characteristic parameter, not line-ground voltage.

### *2.1.7 Rotating magnetic fields*

Each phase winding in a three-phase machine produces a periodic (periodic along the air gap, not in time!) magnetic field in the air gap. Running along the air gap as an observer, we will notice a flux density wave, which has highest positive values in the direction of the exciting coil and a more or less sinusoidal shape (Figure 2.6 illustrated this). We neglect the harmonic content of this field, only the

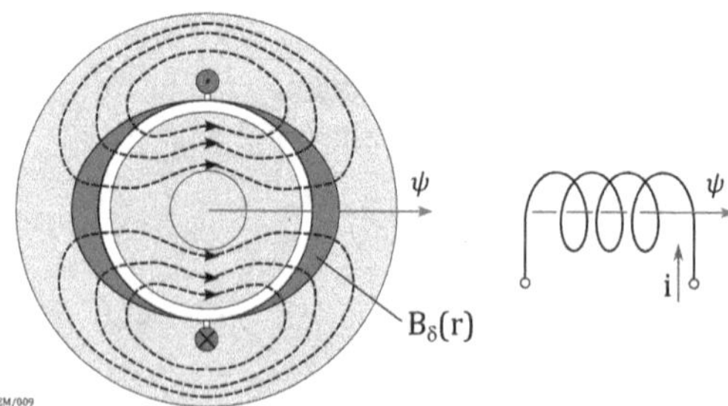


*Figure 2.6 Spatial orientation and distribution of the resulting magnetic field from one single winding in a two-pole machine, picture based on [5]*

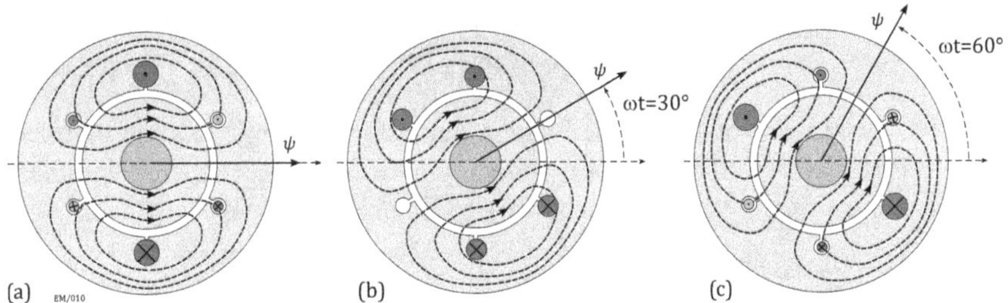


*Figure 2.7 Resulting magnetic field from a three-phase winding in a two-pole stator. The size of the current symbols correspond to the current value (red=phase a, green=phase b, blue=phase c).*

fundamental wave is of interest. The amplitude of this fundamental spatial wave directly depends on the instantaneous value of the exciting current and the inductance parameters of the winding. The maximum of this flux wave is oriented toward the center of the exciting coil.

Using more coils with different spatial orientation allows us to create magnetic fields in the machine in different directions. Each of these fields will have a sinusoidal shape. All instantaneous values sum up to one resulting magnetic field, which then will also have sinusoidal shape – but with its maximum according to the instantaneous values of the single fields.

In order to create any possible direction, as it is needed for a rotating magnetic field, a minimum of two coils with some angle in between them is needed, typically an angle of 90° is used then. An early example is the first regulated AC motor with a 90° two-phase rotor winding from Michael von Dolivo-Dobrowolsky, dated back to 1893 [6].

However, nearly all AC machine's stators come with a three-phase AC winding system and the typical 120° in between the phases. Figure 2.7 shows the creation of the magnetic field for three different current angles for such a stator.

The speed of the rotating field, the synchronous speed, equals the frequency of the three-phase alternating current (50/60 Hz). For a higher pole number the field rotates at a lower speed. Consider the difference between "pole pair number," which is $p$, and "pole number," meaning the total number of poles $2p$. Table 2.4 lists typical synchronous speeds.

$$n_{\mathrm{s}} = \frac{f_1}{p} \tag{2.21}$$

### 2.1.8 Complex phasors for time harmonic signals

All equations in the following part of this chapter use small letters for time varying signals. For example

$$i(t) = \sqrt{2}\, I \cos(\omega t + \varphi_{\mathrm{I}}) \tag{2.22}$$

This equation represents the phase current over time in a single coil. Capital letters shall be used for constant RMS or peak (time independent) values, e.g., the RMS value $I$ in (2.22).

If the mentioned coil is supplied with a stationary sinusoidal current that has a constant RMS value $I$ and the phase angle $\varphi_{\mathrm{I}}$ (as it is the case in many problems under investigation), it is advantageous to use Euler's identity

$$e^{\mathrm{j}x} = \cos x + \mathrm{j} \sin x \tag{2.23}$$

to represent this signal from (2.22) in an easier way. We may then use complex phasors for their representation. Stationary sinusoidal signals, e.g. (2.22), can then be rewritten by

$$i(t) = \mathrm{Re}\left\{\sqrt{2}\, I\, e^{\mathrm{j}\varphi_{\mathrm{I}}}\, e^{\mathrm{j}\omega t}\right\} = \mathrm{Re}\left\{\sqrt{2}\, \underline{I}\, e^{\mathrm{j}\omega t}\right\} \tag{2.24}$$

In (2.24) the phase angle and the amplitude (both represented by $\underline{I}$) of the signal is handled separately from frequency and time. An underlined capital letter denotes the complex phasor.

Notice that we use the RMS value $\underline{I}$ instead of the peak value for the complex phasor, because these values are also used to describe nominal conditions. Furthermore, the calculation of power values from RMS phasors is slightly easier.

The main advantage of using complex phasors instead of equations with trigonometric functions is that any fundamental mathematical operation, e.g., addition, subtraction, and especially integration and differentiation, can be done with the complex phasor directly. This makes calculation much easier.

If we want to calculate power values, that is, multiplication of voltage and current, we may also use the complex phasors rather than the sinus signals itself:

$$\underline{S} = P + \mathrm{j}\, Q = \underline{U} \cdot \underline{I}^* \tag{2.25}$$

In (2.25) $\underline{I}^*$ is the conjugate complex value of $\underline{I}$. Consider that $\underline{S}$ contains both real power and reactive power. We can write

$$\underline{S} = U\, I \cos(\varphi) + \mathrm{j}\, U\, I \sin(\varphi) \tag{2.26}$$

The magnitude of $\underline{S}$ is called apparent power.

$$\mathrm{S} = |\underline{\mathrm{S}}| = \mathrm{U\, I} \tag{2.27}$$

The phase angle

$$\varphi = \varphi_{\mathrm{U}} - \varphi_{\mathrm{I}} \tag{2.28}$$

between current and voltage phasor defines the power factor. Note that the term $\cos(|\varphi|)$ is typically used to define the power factor. If $\varphi_{\mathrm{U}} > \varphi_{\mathrm{I}}$, which means voltage is leading, we say "leading power factor," otherwise we call it "lagging power factor."

## 2.1.9 Complex space vectors (CSV)

For the calculation with time varying sinusoidal sizes in steady state the use of complex phasors may be advantageous. Comparing with (2.24) we get for the complex voltage phasors of our voltage system

$$\underline{U}_{L1} = U\, e^{j\varphi_U}$$
$$\underline{U}_{L2} = U\, e^{j\varphi_U}\, e^{-j2\pi/3} \tag{2.29}$$
$$\underline{U}_{L3} = U\, e^{j\varphi_U}\, e^{-j4\pi/3}$$

In (2.29) we see the rotational operator $e^{j\alpha_{ph}}$ (Figure 2.9). For a three phase system this means

$$e^{j2\pi/3} = \underline{a} = -\frac{1}{2} + j\frac{\sqrt{3}}{2} \tag{2.30}$$

$$e^{j4\pi/3} = \underline{a}^2 = -\frac{1}{2} - j\frac{\sqrt{3}}{2} \tag{2.31}$$

In Figure 2.8 on the right sight the complex representation for those three phasors is given for the example of $\varphi_U = 0$. Note, for symmetrical systems the sum of all three line voltages sums up to zero. This also applies for complex phasors, if geometrically summed up. In this case they form an equilateral triangle (right side of Figure 2.8).

Remembering (2.13), the $N_{ph}$ phase windings from the stator are equally distributed over the entire circumference of the machine. Their magnetic fields overlay each other. The resulting field from all phase windings together gives, again, a sinusoidal shaped field. The amplitude and spatial orientation of this resulting field can be described using a complex space vector in the complex number plane. The same description can be used for any other value, i.e., also the phase currents and the phase voltages can be summed up taking their instantaneous phase values and coil orientation into account.

If we define the real axis from the complex number plane to be aligned with the coil from phase a, we can use the rotational operators (2.30) and (2.31) to calculate the current space vector from the real instantaneous phase current values

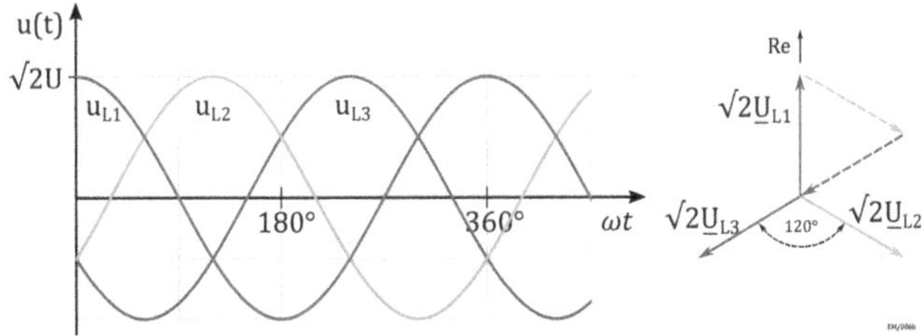


*Figure 2.8 Sinusoidal voltages in a symmetrical three-phase system with $\varphi_U = 0$ (left) and corresponding complex phasors (right)*

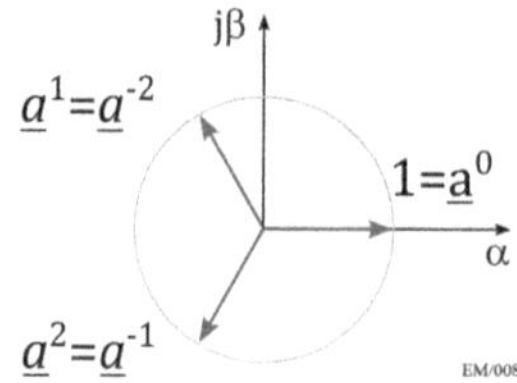


*Figure 2.9 Rotational operator for three-phase systems*

for a three-phase system by:

$$\underline{i}_1 = \frac{2}{3}\left(i_{1\mathrm{a}} + \underline{a}\, i_{1\mathrm{b}} + \underline{a}^2\, i_{1\mathrm{c}}\right) \tag{2.32}$$

Equation (2.32) becomes more simple if $i_{1\mathrm{a}} + i_{1\mathrm{b}} + i_{1\mathrm{c}} = 0$ is taken into account. This holds true for any star winding where the star point is not connected to ground. Then we can rewrite

$$\underline{i}_1 = i_{1\mathrm{a}} + \mathrm{j}\frac{1}{\sqrt{3}}\left(i_{1\mathrm{b}} + i_{1\mathrm{c}}\right) \tag{2.33}$$

The recalculation of the real phase values is possible with

$$\begin{aligned} i_{1\mathrm{a}} &= \mathrm{Re}\left\{\underline{i}_1\right\} \\ i_{1\mathrm{b}} &= \mathrm{Re}\left\{\underline{i}_1 \underline{a}^{-1}\right\} \\ i_{1\mathrm{c}} &= \mathrm{Re}\left\{\underline{i}_1\, \underline{a}^{-2}\right\} \end{aligned} \tag{2.34}$$

We have to consider that for the definition of complex space vectors no agreement to the shape of the real-time signals must be made. It is not necessary to have sinusoidal time signals. The only agreement is that the common effect of all phase values together is producing a sinusoidal field in circumferential direction in the air gap. The resulting complex space vector is time depending but describing a sinusoidal spatial distribution.

This is a big difference to the definition of complex phasors. They are used for the description of stationary sinusoidal time signals. In order to distinguish to complex phasors (denoted with underlined capital letters to show they are complex and time invariant), we use underlined small letters for complex space vectors. Remember that we already defined small letters to be used for time-varying signals.

However, for the special case that the phase values from a multi-phase winding do have stationary sinusoidal shape we can use the complex phasor $\underline{I}_1$ to describe their function (as done for the voltage in (2.14) and (2.29):

$$\begin{aligned} i_{1a} &= \mathrm{Re}\left\{\sqrt{2}\,\underline{I}_{1a}\,e^{j\omega t}\right\} = \mathrm{Re}\left\{\sqrt{2}\,\underline{I}_{1}\,e^{j\omega t}\right\} \\ i_{1b} &= \mathrm{Re}\left\{\sqrt{2}\,\underline{I}_{1b}\,e^{j\omega t}\right\} = \mathrm{Re}\left\{\sqrt{2}\,\underline{I}_{1}\,\underline{a}^{-1}\,e^{j\omega t}\right\} \\ i_{1c} &= \mathrm{Re}\left\{\sqrt{2}\,\underline{I}_{1c}\,e^{j\omega t}\right\} = \mathrm{Re}\left\{\sqrt{2}\,\underline{I}_{1}\,\underline{a}^{-2}\,e^{j\omega t}\right\} \end{aligned} \tag{2.35}$$

We find the following correlation between complex space vector $\underline{i}_1$ and complex phasor $\underline{I}_1 = \underline{I}_{1a}$

$$\underline{i}_1 = \sqrt{2}\,\underline{I}_1\,e^{j\omega t} \tag{2.36}$$

Using complex space vectors for voltage and current, the instantaneous power can be calculated with

$$\underline{s} = p + j\,q = \frac{3}{2}\,\underline{u} \cdot \underline{i}^* \tag{2.37}$$

Compare with (2.25), where the power was calculated using complex phasors for one single coil in stationary operation. The equation looks quite similar, except for the factor of 3/2, which takes the conversion from rms to peak values and the three phase system into account.

The sign of $p$ and $q$ leads to an information about the power flow from and to the coil. As we apply the consumer reference-arrow system, a positive power shall be power that flows into the system. If voltage and current are taken from the stator winding, a positive power according to (2.37) means that the machine consumes electrical power. It runs in motor mode. Negative power means, it runs in generator mode. The same applies for the reactive power: positive values "consume" reactive power as motor or choke does (lagging power factor). Negative values are equivalent for a leading power factor.

## 2.2 Induction generator (IG)

### *2.2.1 Basics*

The induction machine with squirrel cage rotor is the most robust machine type available. The stator compares to all other three-phase AC machine types, there is no difference. The main parts of an induction motor are shown in Figure 2.10. Major parts are the stator core including conventional three-phase AC winding, the terminal, bearings, fan, and the drive shaft.

The major difference to all other three-phase AC machines is the rotor design. It is very simple; it only consists of an iron stack and a simple rotor winding made of rotor bars. These bars are connected by the help of end rings, creating the well-known short circuit rotor, which typically is called a squirrel-cage rotor. It needs no electrical connection to moving parts and no additional power source for the rotor circuit.

Low-power induction motors come with the typical aluminum cast rotor cage (compare Figure 2.10). In this case, any suitable form of rotor bar is possible. The

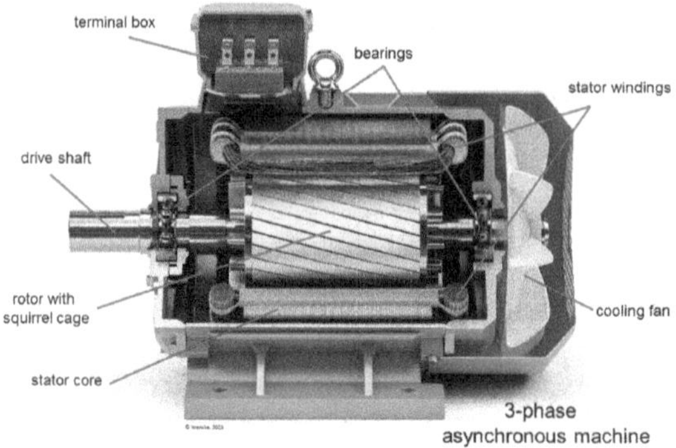


*Figure 2.10 Basic design elements of a squirrel cage induction generator (IG)*

*Figure 2.11 Typical squirrel cage rotor of a high-power induction machine. © VEM Sachsenwerk GmbH (part of VEM Group).*

rings are directly cast together with the bars. The production is very cheap and the construction extremely robust.

However, for induction machines with frame sizes above 500 mm this production method is not suitable. Now solid rotor bars, made of copper, brass, or bronze are used instead. They are short-circuited by end rings, mainly made of copper, sometimes of special alloys, e.g., Copper–Chromium–Zirconium (CuCrZr). In Figure 2.11 a typical example of such a rotor can be seen: an iron core holds solid rotor bars that are short-circuited by solid copper rings. Important to note is that there is always some bar overhang between the iron core and the ring. This is needed to give place for the end plates (also visible in the picture) and the welding process.

On the other hand, the induction machine is the only machine concept that always consumes reactive power from the feeding line in order to build up the magnetic flux inside of the stator. Induction machines with short circuit rotor never maintain a power factor of one or even deliver reactive power to the feeding grid. Also, the relatively low efficiency at full speed seems to be some disadvantages on the first sight. This might not hold true in any case, as the influence of additional components, e.g., frequency inverters need to be taken into account.

One of the advantages from the squirrel cage induction machine is that there is no strict relation between mains frequency and synchronous speed, described by (2.21). The real speed of the induction machine depends on the load torque, the synchronous speed strictly applies for no-load condition only. This soft behavior makes torque peaks less harmful for the drive train. A typical way to describe the relation between load torque and slip, defined by

$$s = \frac{n_s - n}{n_s} \tag{2.38}$$

is to use the well-known formulation according to Kloss, given by

$$M(s) = \frac{2\,M_b}{\frac{s_b}{s} + \frac{s}{s_b}} \tag{2.39}$$

Kloss's equation holds true exactly for a handful of definitions and simplifications. Major simplifications are:

- Neglecting stator resistance
- Constant saturation of the iron core
- No skin effect in the rotor bars
- Neglecting iron losses
- Neglecting mechanical losses
- Neglecting additional losses

Later in this chapter we will see that these assumptions may not be true. The equation for torque calculation will become more complicated. However, for operation in the near of the synchronous speed the obtained results have proven to be precise enough. For any operational point far away from the synchronous speed, (2.39) may not be used. Especially the breakdown torque $M_b$ and the pull-out torque are not covered correctly.

Figure 2.12 shows a typical torque over speed characteristic for a machine with breakdown torque two times nominal torque using (2.39). The dotted line is the even easier approximation

$$M(s) \approx \frac{2\,M_b}{s_b} s \tag{2.40}$$

which can be used near the synchronous speed without a big error.

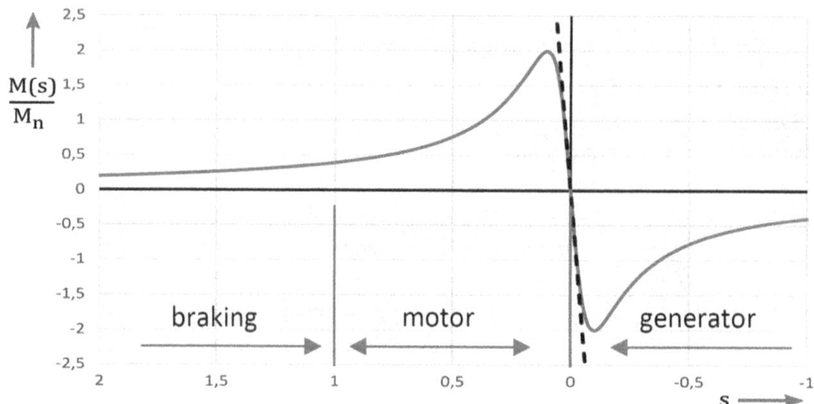


*Figure 2.12 Torque over slip and typical operating areas for an induction machine with short-circuited rotor fed by AC voltage with constant frequency*

*Table 2.4 Synchronous speed for common pole pair numbers at 50 and 60 Hz*

| **Pole number** | **4** | **6** | **8** | **10** | **16** |
|---|---|---|---|---|---|
| @50 Hz | 1500 rpm | 1000 rpm | 750 rpm | 600 rpm | 375 rpm |
| @60 Hz | 1800 rpm | 1200 rpm | 900 rpm | 720 rpm | 450 rpm |

The application as DOL connected generator is limited to small power units up to approximately 1500 kW. However, fed from an inverter, the induction machine perfectly fits as variable speed wind power generator. The operational behavior is different. Therefore both generator types have to be presented separately.

DOL machines run with direct connection to the feeding mains. This means, there is always a stator frequency with 50 or 60 Hz and a stiff voltage level that compares to the main transformer output. Low voltage grids shall manage the average values of the mains frequency for 95% of the time in the range of $\pm 1\%$ and the voltage average in the range of $\pm 10\%$ of their nominal values [7]. This means, the machine will keep a more or less constant excitation value and its speed only varies in the limits given by the maximum slip. A certain set point value of speed is not possible.

The only possibility to change to different speeds is the use of two-speed machines, for example with pol-pair number two and three ($2p = 4/6$). Table 2.4 lists possible synchronous speeds. However, this option leads to a high technical effort, as special switches are needed to change the stator connection according to speed requirement. Therefore, only very small wind turbines use this feature. For high-power wind turbines this is not in use.

Asynchronous machines have an especially strong relation between air gap and magnetizing current. Normally a very big part of the magnetic voltage will drop over the air gap. Compared to other machine types, this is a serious issue as the magnetizing current is fed into the machine from the stator side. The result is a very poor power factor. Therefore, additional losses may heat up the machine. This means, asynchronous machines need an air gap as small as anyhow possible. As a rule of thumb, the following approximation may be used [8]:

$$\frac{\delta}{\text{mm}} \approx 0.25\sqrt[4]{P_{\text{m}}/\text{kW}} \tag{2.41}$$

On the other hand, technical issues (tolerances in production) as well as high values of rotor tooth tip losses demand for higher air gap values. To find an optimal compromise is a serious engineering issue.

### *2.2.2 Grid connected IG in stationary operation (DOL)*

The cheapest way to convert wind power into electrical power is the use of an asynchronous induction generator with squirrel-cage rotor, directly connected on-line to the grid (DOL). Most of these applications use a four-pole or a six-pole machine, resulting in 1500 or 1000 rpm of synchronous speed at 50 Hz. Therefore, a gearbox is needed to convert the speed. The generator must have a higher rotor resistance in order to increase the nominal slip. Otherwise, the impact from load peaks would directly lead to higher stress for the drive train and also for the grid.

The two biggest disadvantages of the IG with direct grid connection are that they always need reactive power from the grid and that they run at a rather fixed speed. Typical power factors are 0.85–0.91 lagging at full load. There is no possibility to correct the power factor but to use capacitor banks or to mix them with other machine concepts in a wind park. Figure 2.13 shows the main principle of this extremely simple electrical drive train concept.

All investigations of the static behavior of induction machines with direct grid connection (DOL) are based on the following assumptions:

- The stator winding is assumed to produce a pure sinusoidal magnetic field in the air gap

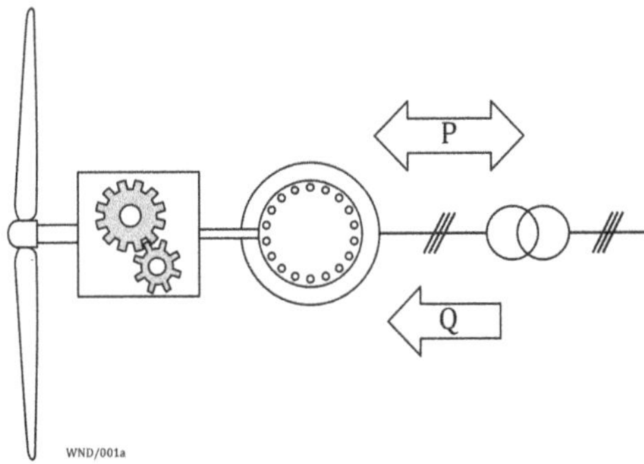


*Figure 2.13 Induction generator drive train concept using direct grid connection*

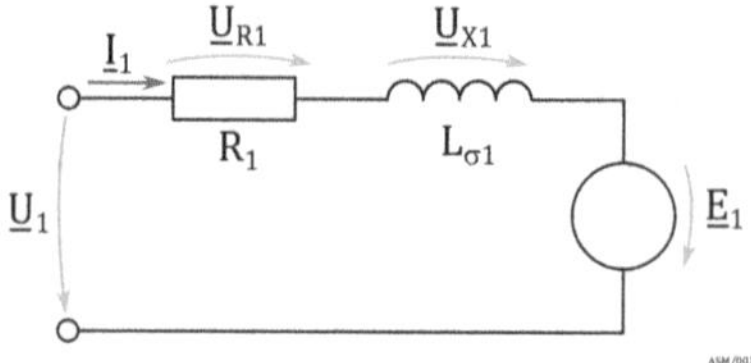


*Figure 2.14 Equivalent circuit diagram of the stator winding*

- The air gap between stator and rotor is constant over the entire circumference
- Neglecting saturation in the iron core, that is, constant inductances of the stator and rotor winding
- Neglecting iron losses
- Neglecting current displacement effects in stator and rotor winding

Furthermore, the machine shall be fed from a stiff voltage source with a constant grid frequency $f_1$ and a constant phase voltage with an RMS value $U_1$.

One phase of the stator is then represented by an equivalent circuit diagram as shown in Figure 2.14. The representation with complex phasors is used for this diagram. Take (2.36) into account to convert into stationary complex space vectors.

Looking at the armature equivalent diagram on Figure 2.14 and applying Kirchhoff's voltage law (Kirchhoff's second law) we find the following equation:

$$\underline{U}_1 = R_1\,\underline{I}_1 + \mathrm{j}\omega_1\,L_{\sigma 1}\,\underline{I}_1 - \underline{E}_1 \tag{2.42}$$

The resistance $R_1$ is the total resistance from one stator phase. It includes all winding parts from that phase, including winding overhang and connections inside of the machine. This means we use the value that is measured at the terminal box.

The leakage inductance $L_{\sigma 1}$ covers all leakage flux parts from the armature phase winding, including

- Slot leakage inductance,
- end winding leakage inductance, and
- air gap leakage inductance (zigzag leakage flux).

We have to consider the difference between inductance and reactance. For any static operation at fixed frequencies, we can use the reactance instead of inductance in the equations. At a given frequency $f_1$ we get for the stator reactance

$$X_{\sigma 1} = \omega_1\,L_{\sigma 1} = 2\pi f_1\,L_{\sigma 1} \tag{2.43}$$

The absolute value of the induced voltage $E_1$ can be calculated using

$$E_1 = c_1\,\omega_1\,\Phi_\mu \tag{2.44}$$

where $c_1$ is a machine depending constant (the stator winding factor), $\omega_1 = 2\pi f_1$ the electrical frequency and $\Phi_\mu$ the main flux inside of the machine. For machines in the high power range is $E_1 \gg R_1 I_1$ and also $E_1 \gg X_{\sigma 1} I_1$. Therefore, the rough approximation $U_1 \approx E_1$ can be used. This leads to the fundamental correlation

$$\Phi_\mu \sim \frac{U_1}{f_1} \tag{2.45}$$

From (2.45), we have to keep $U_1$ and $f_1$ in the same relation to each other in order to maintain a constant flux in the machine. For DOL-connected machines, this is nearly fulfilled because the voltage and frequency are given by the grid. The machine will run at nominal flux or at least close to it. Deviations mainly come from grid tolerances and armature reaction effects. Typical tolerances lead to a flux variation from ±10%, mainly coming from voltage tolerances.

In any electrical machine, the main flux is subjected to saturation due to saturation of the iron core. In order to get a maximum of flux from a minimum of excitation current, the core material is used above the point of highest permeability. That means, we operate the machine at flux density values at roughly 1.0...1.2 T, exactly in the region of the saturation "knee." Figure 2.15 shows as an example the magnetizing curve of the very common core material M400-50A. As a result the machine behavior is quite sensitive to main flux variations. A DOL-driven machine will show different behavior at 90% undervoltage condition or at 110% overvoltage condition.

The equivalent circuit diagram in Figure 2.16 from the rotor circuit looks very similar to that from the stator circuit in Figure 2.14. The only difference is that rotor bars are short-circuited by the two rings and therefor the rotor voltage is $U_2 = 0$. Applying Kirchhoff's second law again, we read from the equivalent circuit

$$0 = R_2 \underline{I}_2 + j\,\omega_2 L_{\sigma 2} \underline{I}_2 - \underline{E}_2 \tag{2.46}$$

The rotor resistance $R_2$ represents the resistance from the bars including technically necessary bar overhang and the proportional part from the two short circuit rings. Bar and ring resistances are transformed into an equivalent three-phase rotor resistance.

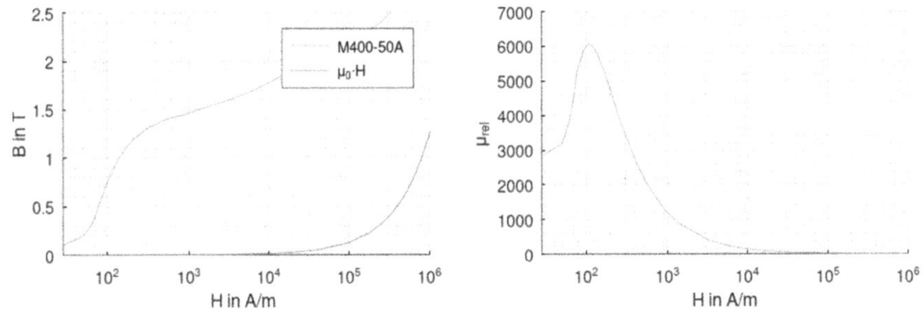


*Figure 2.15 Saturation curve of typical core material M400-50A*

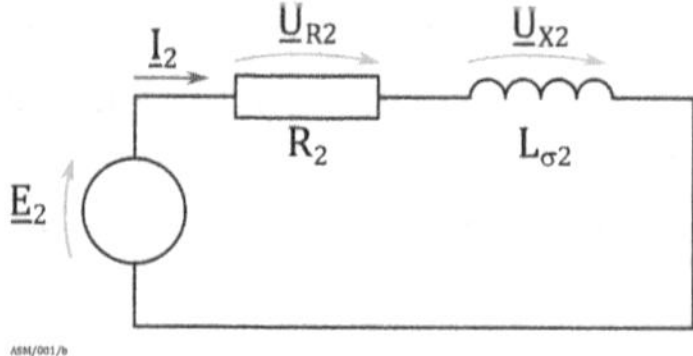


*Figure 2.16 Equivalent circuit diagram of the rotor circuit*

Again, the rotor leakage inductance $L_{\sigma 2}$ contains slot leakage, air gap leakage and the leakage flux from the bar overhang and rings. In stationary operation, all sizes in the rotor circuit are sinusoidal values with the rotor frequency

$$\omega_2 = s\,\omega_1 \tag{2.47}$$

The rotor leakage reactance usually is based on $\omega_1$, therefore, it is

$$X_{\sigma 2} = \omega_1 L_{\sigma 2} \tag{2.48}$$

The induced voltage in the rotor circuit is accordingly

$$E_2 = c_2\,\omega_2\,\Phi_\mu = c_2\,s\,\omega_1\,\Phi_\mu \tag{2.49}$$

Again, $c_2$ is a constant value, reflecting the rotor winding or rotor cage. It is remarkable in (2.49) that there will be no voltage in the rotor circuit, as long as the slip of the machine is zero. For $s = 1$, that is, for machine at standstill, (2.49) becomes

$$E_{20} = c_2\,\omega_1\,\Phi_\mu \tag{2.50}$$

This value is called the locked rotor voltage. Using the locked rotor voltage, (2.49) becomes easier

$$\mathrm{E_2 = s \cdot E_{20}} \tag{2.51}$$

The torque can be calculated by multiplying the absolute values from the main flux with that part of the rotor current, which is oriented with the induced rotor voltage $E_2$ (compare Figure 2.17). The multiplication with 3 and $p$ is obvious, as we do have three phases and $p$ pole pairs.

$$M = 3p\,c_2\,\Phi_\mu\,I_2 \cos\varphi_2 \tag{2.52}$$

For the absolute value of the rotor current we find

$$I_2 = \frac{E_2}{\sqrt{R_2^2 + \omega_2^2 L_{\sigma 2}^2}} \tag{2.53}$$

The angle between $E_2$ and $I_2$ can be expressed by

$$\cos\varphi_2 = \frac{U_{R2}}{E_2} = \frac{R_2}{\sqrt{R_2^2 + \omega_2^2 L_{\sigma 2}^2}} \tag{2.54}$$

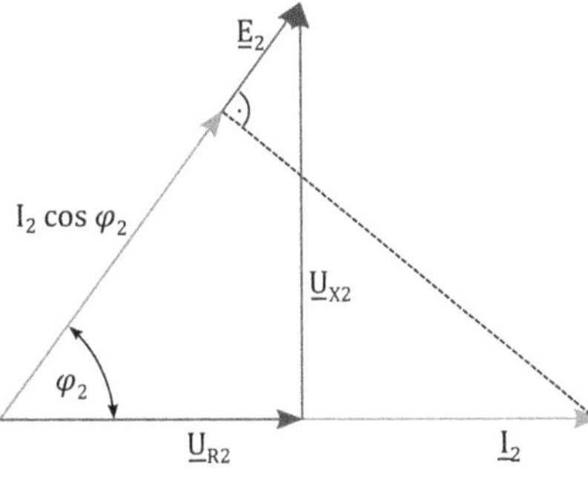


*Figure 2.17 Phasor diagram for the rotor of an induction motor*

Inserting (2.53) and (2.54) into (2.52) we get a representation of Kloss's equation.

$$M = \frac{2}{\frac{R_2}{s\,\omega_1 L_{\sigma 2}} + \frac{s\,\omega_1 L_{\sigma 2}}{R_2}} \cdot \frac{3p}{2L_{\sigma 2}} c_2^2 \Phi_\mu^2 \tag{2.55}$$

Compare with (2.40) to find the accordance with the following substitutions for the breakdown torque

$$M_b = \frac{3p}{2L_{\sigma 2}} c_2^2 \Phi_\mu^2 \tag{2.56}$$

and for the breakdown slip

$$s_b = \frac{R_2}{\omega_1 L_{\sigma 2}} \tag{2.57}$$

From (2.56) we see that the breakdown torque is dependent from the square of the main flux of the machine. Remembering (2.44) and using again the approximation $U_1 \approx E_1$ we see that the torque capability especially depends on the square of the stator voltage divided by stator frequency.

$$M_b \sim \left(\frac{U_1}{f_1}\right)^2 \tag{2.58}$$

Note that the breakdown torque is the maximum load value that is achievable in DOL-driven applications. This also applies for inverter-driven machines running in field weakening area. Induction machines running with VDF below the voltage limit do not show a breakdown torque.

The induced voltage $E_1$ in the armature circuit and also $E_2$ in the rotor circuit are both linked to the main flux $\Phi_\mu$, only using different constants $c_1$ and $c_1$ (winding turns) and working at different frequencies. We can combine both circuits by the help of a suitable transformation ratio $K$. For this purpose, it is helpful to scale the rotor values in such a way that the induced stator voltage $E_1$

equals the scaled locked rotor voltage $K \cdot E_{20}$. We say, $E_{20}$, the voltage from the locked rotor winding, refers to the primary winding. This transformation procedure is well-known from transformer theory and can be adapted for induction machines. For induction motors the effective stator winding turns $(w\xi_1)_1$ and for the short circuit squirrel-cage winding $N_2/(2 \cdot 3)$ must be used for the transformation.

$$K = \frac{c_1}{c_2} = \frac{(w\xi_1)_1}{N_2/(2 \cdot 3)} \tag{2.59}$$

All further rotor sizes, current, reactances and resistances, have to be treated accordingly. Table 2.5 lists all transformation rules. For the rotor (2.46), we get now

$$0 = \frac{R_2'}{s} \underline{I}_2' + \mathrm{j}\, X_{\sigma 2}' \underline{I}_2' - \underline{E}_1 \tag{2.60}$$

with the inner voltage

$$\underline{E}_1 = -j\omega_1 c_1 \underline{\phi}_\mu = -\mathrm{j}\omega_1 \underline{\psi}_\mu \tag{2.61}$$

For the main flux linkage we use

$$\underline{\psi}_\mu = L_\mu \left( \underline{I}_1 + \underline{I}_2' \right) = L_\mu \underline{I}_\mu \tag{2.62}$$

For a given mains frequency and voltage, a constant main inductivity value may be used (compare Figure 2.21), as there is not too much change by means of grid tolerances.

With this, it is now advantageous to put both equivalent circuit diagrams from Figures 2.14 and 2.17 into one. Additionally, we add an extra resistance in parallel to the main reactance.

From this ECD we read the stator equation

$$\mathrm{j}\, \omega_1 \underline{\psi}_1 = \underline{U}_1 - R_1 \underline{I}_1 \tag{2.63}$$

The rotor circuit finally can be expressed in the same way as the stator using

$$\mathrm{j}\, \omega_2 \underline{\psi}_2' = -R_2' \underline{I}_2' \tag{2.64}$$

*Table 2.5 Reference system for all rotor sizes*

| **Quantity** | **Transformation** | **Quantity** | **Transformation** |
|---|---|---|---|
| Voltage | $U_2' = K\, U_2$ | Current | $I_2' = K^{-1}\, I_2$ |
| Reactance | $X_{\sigma 2}' = K^2\, X_{\sigma 2}$ | Resistance | $R_2' = K^2\, R_2$ |
| Flux linkage | $\psi_2' = K\, \psi_2$ | Inductance | $L_{\mu 2}' = K^2\, L_{\mu 2} = L_\mu$ |

In the case of a squirrel cage induction machine the rotor circuit is short circuited, and therefore $U_2 = 0$.

The resulting flux linkages from stator and rotor are

$$\underline{\psi}_1 = L_{\sigma 1}\,\underline{I}_1 + \underline{\psi}_\mu = L_1\,\underline{I}_1 + L_\mu\,\underline{I}_2^{'} \tag{2.65}$$

$$\underline{\psi}_2^{'} = L_{\sigma 2}^{'}\,\underline{I}_2^{'} + \underline{\psi}_\mu = L_\mu\,\underline{I}_1 + L_2^{'}\,\underline{I}_2^{'} \tag{2.66}$$

The stator and rotor inductance are

$$\begin{aligned} L_1 &= L_{\sigma 1} + L_\mu \\ L_2^{'} &= L_{\sigma 2}^{'} + L_\mu \end{aligned} \tag{2.67}$$

The relation between stator and rotor frequency is

$$\omega_2 = s \cdot \omega_1 \tag{2.68}$$

The stator electric power is

$$\underline{P}_1 = 3\underline{U}_1\,\underline{I}_1^* \tag{2.69}$$

The torque now can be calculated using

$$\begin{aligned} M &= 3p\,\psi_\mu\,I_2^{'}\cos\varphi_2 \\ &= 3p\,\mathrm{Im}\left\{\underline{\psi}_\mu \cdot \mathrm{conj}\left(\underline{I}_2^{'}\right)\right\} \end{aligned} \tag{2.70}$$

The common equivalent circuit diagram take stator and rotor copper losses into account. The fictitious iron loss resistance in Figure 2.18 represents the stator iron losses. They go up with the square root of the flux density and they depend on the stator frequency. The quadratic dependency from flux is fully covered by the placement in parallel to the main inductance. Consideration of the frequency dependency is not needed for DOL machines, as the mains frequency varies only in small tolerances. In order to calculate suitable efficiency data, the following additional elements have to be taken into account. The iron losses can be calculated from the ECD with

$$P_{\mathrm{Lfe}} = 3\frac{E_1^2}{R_{\mathrm{fe}}} \tag{2.71}$$

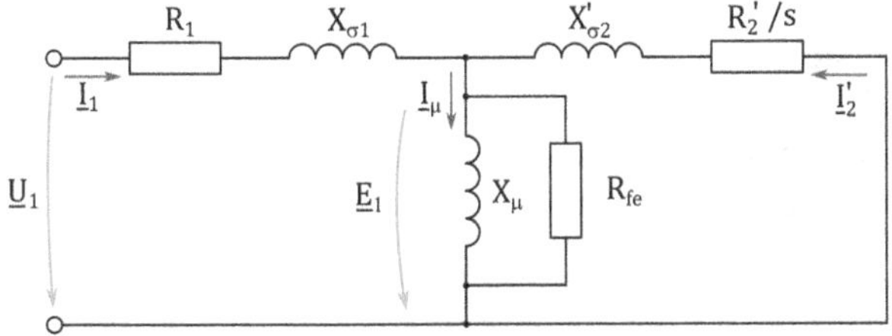


*Figure 2.18 Equivalent circuit diagram (ECD) of an induction machine with squirrel-cage rotor for static operation*

#### 2.2.2.1 Skin effect in the stator winding

For low voltage machines (below 1000 V) with high power the number of winding turns in the stator winding is very small. Typical winding configurations do have only 2 or 3 turns per coil. Even at the relatively low frequency of 50 Hz skin effect may occur. That means, that the AC losses $P_{AC}$ in the winding are higher than the DC losses calculated with $P_{\mathrm{DC}} = 3 \cdot R \cdot I^2$. The skin effect can be taken into account by using a correction factor $k_R = (P_{\mathrm{AC}} + P_{\mathrm{DC}})/P_{\mathrm{DC}}$. Typical values for $k_R$ are between $k_R = 1 \ldots 1.2$. The winding losses are for the stator then are

$$P_{\mathrm{Lw1}} = 3k_{\mathrm{R1}}R_1I_1^2 \tag{2.72}$$

Skin effect can be reduced with the following actions:

- Using delta connection leads to 1.73 more winding turns and less skin effect, but most of the high-power IG for low voltage come with delta connection anyway.
- Dividing the conductors into smaller conductors reduces the skin effect, but this leads to a smaller slot fill factor.
- Less stator slots lead to higher number of turns per coil but leads also to higher harmonic content in the air gap.

For machines with random winding the calculation of the AC losses is difficult and typically the additional skin effect losses are interpreted as load-dependent additional losses. For form windings analytical solutions for skin effect are available and calculation is quite easy. However, for high voltage machines (above 1 kV) their influence is very small at 50 Hz and, again, losses will not be treated separately.

For the rotor the frequency at normal operation condition is very small. Skin effect is then neglectable. The rotor winding losses can be calculated very simple with

$$P_{\mathrm{Lw2}} = 3R_2^{'}I_2^{'}2 \tag{2.73}$$

#### 2.2.2.2 Additional losses

Additional losses sum up load-dependent losses such as tooth tip losses, harmonic losses in the rotor bars, axial currents flowing in the core of skewed machines, and, if not treated separately, skin effect. They show a quadratic dependency from the current and therefor a fictitious additional resistance can be integrated in series to the stator resistance. If there are no other information available, IEC 60034-2-1 recommends to calculate the additional losses with

$$\begin{aligned} P_{\mathrm{LL,n}} &= P_{\mathrm{m,n}} \cdot k_{\mathrm{LL}} \\ &= P_{\mathrm{m,n}} \cdot \left[2{,}5\% - 0{,}5\% \cdot \log\left(\frac{P_{1n}}{1kW}\right)\right] \end{aligned} \tag{2.74}$$

*Table 2.6 Example ECD for a 4-pole induction generator with squirrel cage rotor for wind power applications without inverter. Nominal conditions are: 690 V, 50 Hz, 1650 kVA, delta-connection, 95 °C winding temperatures.*

| Parameter | Value | Parameter | Value |
|---|---|---|---|
| $R_1$ | 11 mΩ | $R_2^{'}$ | 14 mΩ |
| $X_{\sigma 1}$ | 200 mΩ | $X_{\sigma 2}^{'}$ | 100 mΩ |
| $X_{\mu}$ | 5.8 Ω | $R_{fe}$ | 132 Ω |

Equation (2.64) applies for machines in the range of up to 10 MW. For other load conditions than nominal load, the value of the additional losses is

$$P_{LL} = P_{LLn} \cdot \frac{I_1^2 - I_{\mu}^2}{I_{1n}^2 - I_{\mu n}^2} \tag{2.75}$$

The total losses are

$$P_L = P_{LL} + P_{Lwf} + P_{Lfe} + P_{Lw1} + P_{Lw2} \tag{2.76}$$

The efficiency of the induction generator can be calculated with

$$\eta = \frac{P_1}{P_m} = \frac{P_1}{P_1 + P_L} \tag{2.77}$$

Table 2.6 lists the ECD data for an induction generator used in wind power applications for rather high wind speeds and for areas with rough environment. Remarkable is the slightly higher rotor resistance (made of an CuZn alloy) and the smaller rotor leakage inductance.

### 2.2.3 *Speed variable IG in stationary operation using VFD*

We learned in the previous paragraph that there are two major disadvantages with the IG, if they are connected to a stiff voltage source. They consume reactive power and they run at rather fixed speed. Using a VFD resolves both issues – on behalf of higher cost and more complexity. Figure 2.19 shows the principle design of an electrical drive train with IG using a VDF.

Reactive power now is exchanged between the grid and VFD in both directions. Full reactive power support is possible and only limited by the current capability from the line side converter (LSC) and the DC link capacity. A typical power factor range is from 0.9 lagging to 0.9 leading. It is important to note that reactive power support capability has no impact on the generator, and it is only a feature of the inverter.

The possible generator speed now ranges literally from zero up to approximately 140% of the nominal speed. Nominal speed shall be that speed, where the generator reaches nominal power at nominal voltage and nominal frequency. More

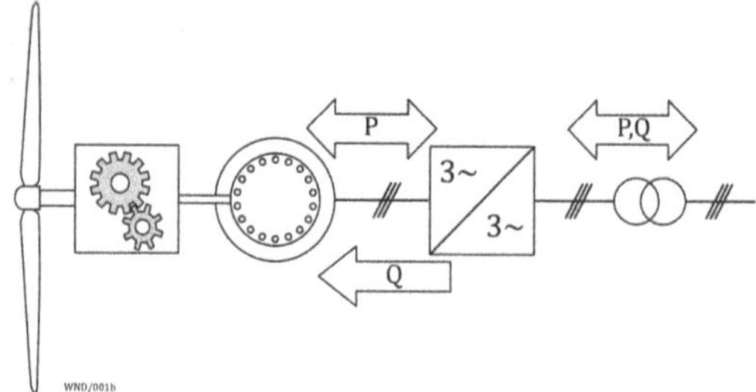


*Figure 2.19 Induction generator drive train concept using VFD for speed and reactive power control*

than 140% will lead to bigger machine dimensions. The generator speed is independent from the mains frequency but dependent from the maximum fundamental frequency of the machine side converter. Typical maximum values for large-scale inverters are 120 Hz (in some cases up to 200 Hz for newer developments with higher switching frequency) due to switching losses in the semiconductors. Typical pole numbers for the induction generator are 4 or 6 for high-speed concepts. Choosing pole number 2 leads to a more complicated electromagnetic design, because then flux will pass perpendicular through the shaft and over half the machine in the stator yoke. On the other hand, generators with higher pole numbers than 8 need more magnetizing current and lead to more expensive machine side converter.

The engineering task is now, to choose the gearbox ratio, generator pole number and inverter parameter (max frequency) to harvest a maximum of power. Inverter frequency and generator pole number determine the generator nominal speed and possible speed range, considering (2.21).

For inverter-driven machines, the flux can be maintained near nominal value within a speed range of approximately 10%–95% of the nominal speed. Below this limit, the influence from the armature resistance becomes decisive. We have to increase the voltage by a certain extend above (2.45) to take the extra voltage drop into account. At around nominal speed the voltage reaches the maximum machine side converter (MSC) voltage. At higher speed values the voltage must stay at that level, we now have less flux in the machine than nominal value. This is called the flux weakening area.

As we have seen in (2.56), with limited flux the possible torque is limited to the breakdown torque, with squared dependency from the possible torque to the flux. Therefore the speed range above nominal speed is limited, if we want to maintain a certain load. As the speed goes above nominal speed, flux in the machine goes down and so does the maximum torque and power. A safety margin is needed between generator torque and breakdown torque. Figure 2.20 shows a typical torque-speed range as an example. We clearly see that the overload capability at nominal speed decides over the size of the flux weakening range.

A further degree of freedom is the possibility to choose the nominal speed point different to the point where maximum voltage from the MSC will be reached.

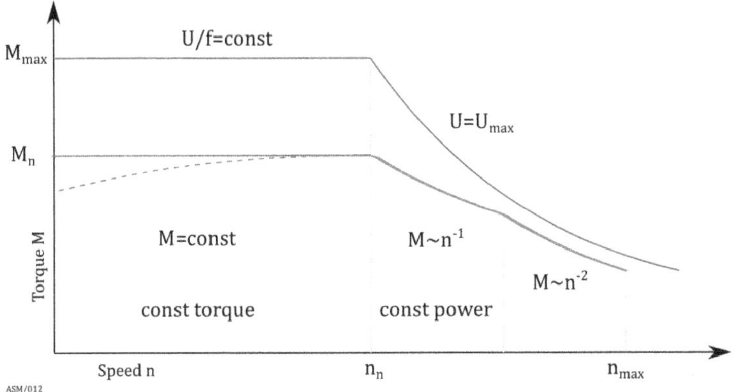


*Figure 2.20 Torque-speed range of an induction machine with VFD*

This slightly changes the torque over speed curve. Setting the u/f-corner point higher than nominal speed enlarges the breakdown torque at the end of the constant power range, but leads to higher currents at lower speeds. Typically this option is not used for wind power generators and in order to keep things easy we want to assume that the nominal speed marks both, the constant torque corner point and the beginning of the flux weakening area.

Another advantage of using a VFD is that it decouples the generator from the grid voltage. The magnetizing state of the generator now can be chosen freely in accordance to the best efficiency. That means operating points with low power will not necessarily need full flux, what enables lower iron losses (remembering $P_{LFe} \sim B^2$). This so-called "Max-Efficiency" strategy is in contradiction to the other widespread control scheme, called MTPA (max torque per ampere) or MPPA (max power per ampere) that could be used to minimize the size of the VFD. All strategies apply for the base speed range only, as in flux weakening area we need the flux for high torque values.

At this point, it is necessary to mention that the maximum output voltage from the MSC depends on the DC link voltage. To maintain a constant DC link voltage is the task of the LSC. The LSC power output is following a closed loop control for the DC link voltage. A higher power output lowers this voltage (as the DC link capacitors will drain), a lower power output lifts this voltage (as the capacitors will be charged from the MSC again). Typical 690 V IGBT inverters can maintain a DC link voltage between 1100 and 1200 V, depending on the voltage withstand capability of the semiconductors. For the MSC this means a maximum output voltage of approximately 840 V RMS line-line voltage to the generator. Additional voltage peaks from the switching itself will also stress the generator. This is much more than typical 690 V insulation systems have to bear in DOL applications! On the other hand, there is the big advantage of lower current, resulting in lower costs for inverter, switchgears and cables.

The drive train concept according to Figure 2.19 is also suitable for mid speed concepts where the gearbox and generator are partly integrated into each other. Typical pole numbers then are 8, 12, or 16.

Investigations of steady-state points for inverter-driven generators can be done in the same way as we have seen in the previous chapter for DOL generators. The equivalent circuit diagram stays untouched (compare Figure 2.18), but we now need to adapt the parameter according to the working point. The reason is that we have now much more flexibility in the working points. They are spread over a huge speed range, use different stator frequencies, and lead to very different saturation states in the iron core. Additional to the DOL application, the following effects must now be considered:

- Saturation of the main field, $L_\mu = f(I_\mu)$
- Frequency-dependent iron losses in the stator, $R_{fe} = f(f_1)$
- Skin effect in the stator winding, $R_1 = f(f_1), L_{\sigma 1} = f(f_1)$
- Speed-dependent friction losses, $P_{Lwf} = f(n)$
- Additional harmonic losses from the inverter in all parts of the machine.

Saturation effects in the stator and rotor leakage inductances, still can be neglected, because the currents in inverter-fed applications do not lead to currents far above their nominal values. These effects only arise at situations with high slip values and in short circuit situations – which we do not have in controlled operation. Table 2.7 lists ECD parameter for a 4200kVA IG.

In the following, we want to go slightly deeper into these points. The main flux linkage $\psi_\mu$ and so also the main inductivity $L_\mu$ are subjected to saturation. For proper calculation of static operation points, the identity $L_\mu = f(I_\mu)$ must be taken into account. Figure 2.21 shows as an example the main inductivity values for a medium voltage IG with 4 MVA. The nominal magnetizing current is 160 A. It is clear that each deviation from that leads to lower values for the main inductance. Deviations will be noticed mainly in situations, where the stator voltage cannot maintain the nominal flux any more (flux weakening area) or if the flux value in the machine will be changed in order to drive the generator into a more efficient working point (Max-Efficiency control strategy). In either case, (2.62) has to be

*Table 2.7 Example equivalent circuit data for a for-pole induction generator with squirrel cage rotor for wind power applications. Nominal conditions are: 730 V, 50 Hz, 4200 kVA, delta-connection, 95 °C winding temperatures.*

| Parameter | Value | Parameter | Value |
|---|---|---|---|
| $R_1$ | 1.8 mΩ | $R_2'$ | 1.1 mΩ |
| $X_{\sigma 1}$ | 47 mΩ | $X_{\sigma 2}'$ | 29 mΩ |
| $X_\mu$ | 1.3 Ω | $R_{fe}$ | 62 Ω |

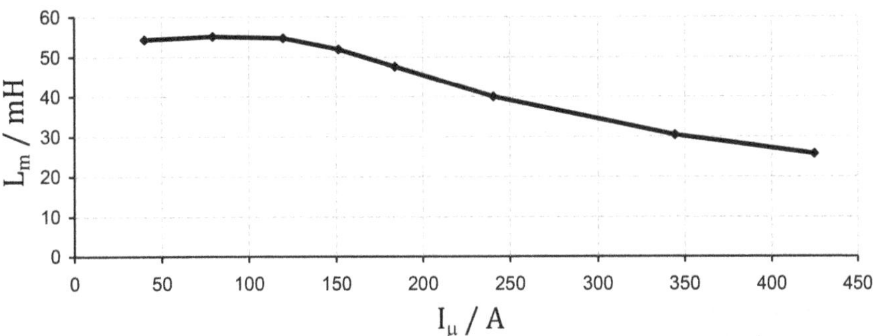


*Figure 2.21 Typical magnetizing current versus main inductance of a 4000 kVA medium voltage IG, nominal magnetizing current is 160 A*

adapted to

$$\underline{\psi}_{\mu} = f\left(\underline{I}_{\mu}\right) \tag{2.78}$$

The implementation of (2.78) leads to a non-linear problem and a numerical solutions needs iteration, e.g., by the regula falsi.

Connected to the main flux saturation is also a change in the iron losses. But wisely enough, we positioned a resistor in parallel to the main inductance to represent the iron losses. The voltage drop over the main inductance always represents the flux correctly and so the iron losses will automatically be adapted to different flux values and its dependency to $B^2$. The only change will be noticed due to different stator frequencies. The main parts from the iron losses are the hysteresis losses (linear depending on frequency) and the eddy current losses (approximately square dependency from frequency).

$$P_{\mathrm{Lfe1}} = k_{\mathrm{hys}} \cdot f \cdot B^2 + k_{\mathrm{eddy}} \cdot f^2 \cdot B^2 \tag{2.79}$$

Remembering (2.71) we find the following adaption for the iron loss resistance

$$R_{\mathrm{fe1}}(f_1) = \frac{R_{\mathrm{fe1,n}} \cdot f_1/f_{1\mathrm{n}}}{k_{\mathrm{hys}} + k_{\mathrm{eddy}} \cdot f_1/f_{1\mathrm{n}}} \tag{2.80}$$

For low stator frequencies (up to 50 Hz) the influence from the eddy loss part is neglectable. Then the iron loss resistance is linear depending on the stator frequency.

The stator frequency also effects the losses in the stator winding and the leakage inductance value from the stator by skin effect. Especially for low voltage generators the number of winding turns in the stator winding is very low and the phase currents are quite high. In those cases, massive copper cross-sections are needed, including a higher number of parallel conductors. In this case, skin effect has to be taken into account by the help of AC correction factors $k_{r1}$ and $k_{x1}$.

$$R_1(f_1,\theta_1) = k_{r1}(f_1,\theta_1)\cdot[1+\alpha\,(\Theta_1 - 20^\circ C)]\cdot R_{1,20} \tag{2.81}$$

$$X_{\sigma 1}(f_1,\theta_1) = k_{x1}(f_1,\theta_1)\cdot X_{\sigma 10} \tag{2.82}$$

Closed analytical solutions are available for form windings [8]. Some care is needed, as the skin effect also depends on the conductivity. Therefore, the winding temperature will be needed as second input.

Friction losses are clearly speed dependent. A rather simple but suitable suggestion is to use the exponential approach with an exponent between $k_{wf} = 2\ldots3$.

$$P_{\mathrm{Lwf}} = P_{\mathrm{Lwf,n}}\cdot\left(\frac{n}{n_{\mathrm{n}}}\right)^{k_{\mathrm{wf}}} \tag{2.83}$$

The additional losses can be taken into account in the same way as we did in the previous paragraph, but now adapted for different speeds according to the suggestion in [9]. Equation (2.75) now becomes

$$P_{\mathrm{LL}} = P_{\mathrm{LLn}}\cdot\left[c_{\mathrm{LL}}\,\frac{n}{n_{\mathrm{n}}} + (1-C_{\mathrm{LL}})\left(\frac{n}{n_{\mathrm{n}}}\right)^2\right]\frac{I_1^2 - I_\mu^2}{I_{1\mathrm{n}}^2 - I_{\mu\mathrm{n}}^2} \tag{2.84}$$

*Figure 2.22 Air-cooled IG prototype with 3750 kW for wind power application on the test bench. The inverter for system testing is in the background. Losses can be measured using accurate power analyzer and torque meters. © VEM Sachsenwerk GmbH (part of VEM Group).*

The most uncertain part of all losses is the amount of additional losses induced by the harmonics from the inverter, so-called high-frequency losses $P_{\mathrm{Lhf}}$. These harmonics induce additional losses in any part of the magnetic and electric circuit. The calculation of these losses is rather complicated as they do not rely only on the machine but on the entire system. Therefore it is advantageous to measure those losses in system tests. As a rule of thumb, 15% of the losses from the nominal working point can be used as rough estimation without measurement [10]. Those losses are independent from the working point and can simply be added [9] as long as the pulse pattern, DC link voltage and switching frequency from the inverter does not change.

Last but not least, the losses $P_{\mathrm{Linv}}$ from the full-scale inverter itself must be added as well. Typical values of switching losses in the semiconductors and filters of active front-end inverters are 3...4% of their full load capability. The total amount of losses then is

$$P_{\mathrm{L}} = P_{\mathrm{LL}} + P_{\mathrm{Lwf}} + P_{\mathrm{Lfe}} + P_{\mathrm{Lw1}} + P_{\mathrm{Lw2}} + P_{\mathrm{Lhf}} + P_{\mathrm{Linv}} \tag{2.85}$$

An easy way to measure the losses in the entire torque-speed range is described in DIN EN 60034-2-3 including an interpolation method based on only seven load points to be measured. Figure 2.23 shows an example of such a loss map for a smaller induction machine. Unfortunately, this is only suitable and described for the fundamental speed range and not suitable for field weakening area, and only for motors.

Finally, the impact of fast voltage switching from the inverter needs to be taken into account. Two major effects are of importance: The first point is that our generators need to have a stronger insulation system, as there will be higher stress in the form of voltage peaks from the IGBT of the inverter. The inverter can either be placed in the nacelle directly beside the generator or it can be placed in the base of the tower. In the first case, there is only a neglectable cable length in between them, just a couple of meters. In the second case, we can have above 100 m, resulting in

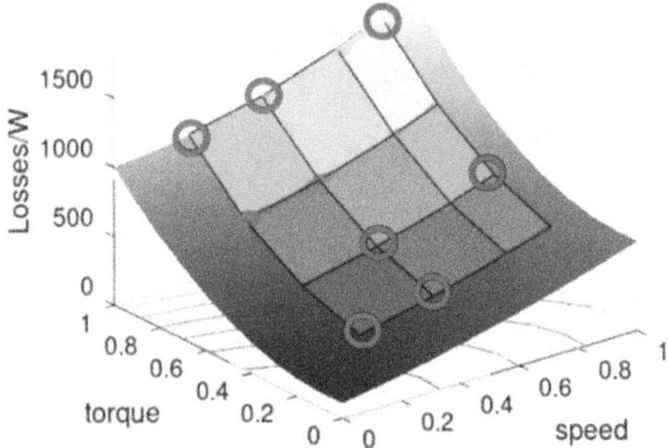


*Figure 2.23 Interpolated loss map and measurement points for a small 48kW induction machine according to DIN EN 60034-2-3 [9]*

very high voltage peaks at the generator terminal. Reflection effects in the cables may lead to voltage peaks of twice the DC link voltage and to voltage rise times of up to 10 kV/μs. Refer to [11], for a rough estimation of these values.

The second effect is that all inverters charge the winding system with a common mode voltage. Each time, when all output lines from the inverter switch to high DC link potential or to low DC link potential together (this is needed if we want to have all line-line-voltages to be zero), the entire winding will be charged with this potential, too. We charge and discharge the winding, which now behaves as a capacity, with several kHz. This can lead to unwanted electric charging also in other parts of the generator (e.g., the shaft and the rotor) an may lead to disruptive electric currents flowing through the bearings. Server damages and bearing failures are the consequence. To avoid this, special bearings can be used with ceramic balls or with a special insulation between outer ring and bearing seat. Additionally, the stator housing must be properly grounded. Also, the rotor may by grounded by a set of grounding brushes.

### 2.2.4 Dynamic operation

For stationary investigations, that is, the calculation of static load points, we took common losses into account and we used complex phasors for calculation. For dynamic calculations this is different. Now the use of complex instantaneous space vectors is more advantageous, and also more common. Additionally, in a lot of cases it is not necessary to take losses into consideration. Especially, iron losses, load depending additional losses, and mechanical losses can be neglected. The accuracy of the calculated solutions will not be effected. The goal is to investigate the behavior of the machine during a short period. Typical examples for such calculation tasks are:

- Sudden stator short circuit (two-phase or three-phase)
- Load jumps
- Voltage dips or peaks
- Oscillating loads
- Open circuits
- Synchronization problems

For the calculation of all those problems, a system of differential equations is needed. The stator differential equation in stator reference frame is

$$\frac{\mathrm{d}\,\underline{\psi}_1}{\mathrm{d}t} = \underline{u}_1 - R_1\,\underline{i}_1 \tag{2.86}$$

The rotor differential equation in rotor reference frame is accordingly with $\underline{u}_2 = 0$ for squirrel cage machines

$$\frac{\mathrm{d}\underline{\psi}_2}{\mathrm{d}t} = 0 - R_2\,\underline{i}_2 \tag{2.87}$$

Both equations are only valid in their own reference system. The stator reference system rests still with the stator, and the rotor reference system is fixed to the rotor and moves with the rotational speed and the corresponding angle of the rotor. Several steps are needed to set up a system of differential and algebraic equations.

The first step is to apply a transformation ratio as has been explained in Section 2.2.2. Again, an apostrophe indicates a rotor value that is referring to the stator winding. Compare Table 2.5 for more information. The rotor differential equation is then

$$\frac{\mathrm{d}\,\underline{\psi}_2'}{\mathrm{d}t} = 0 - R_2'\,\underline{i}_2' \tag{2.88}$$

The second step is to put all equations into one coordinate system. Table 2.8 lists common coordinate systems for use in rotating electrical machines. To demonstrate this, Figure 2.24 shows the stator current in stationary reference frame S and in the rotating universal reference frame K, rotated by the frame angle $\vartheta_K$. It is

$$\begin{aligned} \underline{i}_1^{(\mathrm{K})} &= i_{1\mathrm{x}} + \mathrm{j}\, i_{1\mathrm{y}} \\ &= i_{1\alpha}\cos\vartheta_\mathrm{K} + i_{1\beta}\sin\vartheta_\mathrm{K} + \mathrm{j}\,\left(i_{1\beta}\cos\vartheta_\mathrm{K} - i_{1\alpha}\sin\vartheta_\mathrm{K}\right) \end{aligned} \tag{2.89}$$

The general transformation is then

$$\underline{i}_1^{(\mathrm{K})} = \underline{i}_1^{(\mathrm{S})} e^{-\mathrm{j}\vartheta_\mathrm{K}}$$

*Table 2.8 Typical coordinate system orientation in rotating electrical machines*

| **Reference** | **Vector labeling** | **Typical application case** |
|---|---|---|
| Stationary reference frame | $\underline{g}^{(\mathrm{S})} = g_\alpha + \mathrm{j}\, g_\beta$ | Run-up, stator short circuits, reconnections, etc. |
| Rotor reference frame | $\underline{g}^{(\mathrm{R})} = g_\mathrm{d} + \mathrm{j}\, g_\mathrm{q}$ | Rotor-related faults, e.g., contact loss in the slip ring, rotor short circuits, faults in synchronous machines |
| Universal reference frame | $\underline{g}^{(\mathrm{K})} = g_x + \mathrm{j}\, g_\mathrm{y}$ | Closed loop control (FOC) in flux-oriented reference frame |

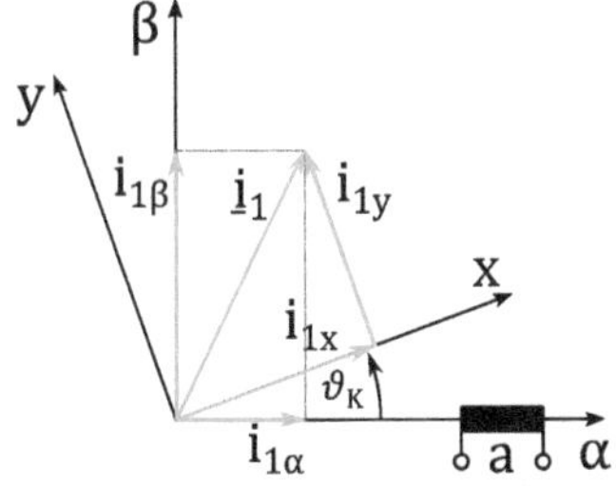


*Figure 2.24 Stator current space vector represented in stationary α–β reference frame S and in universal* x–y *reference frame K.*

By applying the rotation operator $e^{-\mathrm{j}\vartheta_K}$ to the stator and the rotor differential equation, both equations can be transformed and written in one reference frame together. Take care that you use the chain rule of differentiation.

$$\frac{\mathrm{d}\,\underline{\psi}_1}{\mathrm{d}t} = \underline{u}_1 - R_1\,\underline{i}_1 - \mathrm{j}\,\omega_\mathrm{K}\,\underline{\psi}_1 \tag{2.90}$$

$$\frac{\mathrm{d}\,\underline{\psi}_2'}{\mathrm{d}t} = 0 - R_2'\,\underline{i}_2' - \mathrm{j}\,(\omega_\mathrm{K} - \omega_\mathrm{m})\underline{\psi}_2' \tag{2.91}$$

with the rotational speed of the reference frame

$$\frac{\mathrm{d}\,\vartheta_\mathrm{K}}{\mathrm{d}t} = \omega_\mathrm{K} \tag{2.92}$$

With $\omega_\mathrm{K} = 0$ these equations are valid for the stationary stator reference frame. With $\omega_K = \omega_\mathrm{m}$ the equations are valid for the rotor reference frame.

The flux linkage equations that connect the stator and the rotor system are independent from the coordinate system. They are

$$\underline{\psi}_1 = L_1\,\underline{i}_1 + L_\mu\,\underline{i}_2' \tag{2.93}$$

and

$$\underline{\psi}_2' = L_\mu\,\underline{i}_1 + L_2'\,\underline{i}_2' \tag{2.94}$$

The electromagnetic torque is

$$m = \frac{3}{2}p\left(\underline{\psi}_1 \times \underline{i}_1\right) \tag{2.95}$$

As we agreed at the beginning, for dynamical investigations losses can be neglected. This means the internal electromagnetic torque equals also the mechanical torque at the shaft and $m_\mathrm{m} = m$. We can use the very simple approach of a one-mass-oscillator for the mechanical system, although more sophisticated approaches, with more oscillating masses, easily can be implemented.

$$J_\mathrm{m}\frac{\mathrm{d}\,\omega_\mathrm{m}}{\mathrm{d}t} = m - m_\mathrm{load} \tag{2.96}$$

Now all needed equations are available. The parameter can be taken from equivalent circuit data sheets. For the simulation, any integration method can be used. The most common and well proven is the Runge-Kutta method, which is shipped with most modern computation programming languages as MATLAB or Python out of the shelf. To make the calculation easier and numerically more stable, it is advantageous to use per-unit values instead of physical values. Per-unit values have been normalized with the help of suitable base values. The basic goal is, to remove the physical quantity and to normalize to a value near to the value one in nominal condition. This avoids the computation of very small with very large values on limited numeric resolution.

The base for the reference system are the nominal peak values from phase voltage and phase current, the nominal apparent power and the nominal electrical frequency from the stator. This leads to the definition of all other reference values. Table 2.9 lists typical values.

Care is needed, as the reference system is based on nominal apparent power, not real power. A generator, running at nominal conditions, will have an apparent power of 1 and a real power in the size of the nominal power factor. For the torque, as its reference value is defined based on nominal apparent stator power rather than on nominal mechanical power, the ratio between both torque values is for motors:

$$\frac{M_n}{M_r} = \frac{P_n}{P_r} = \frac{\eta_n \cos\varphi_n}{1 - s_n} \tag{2.97}$$

and for generators

$$\frac{M_n}{M_r} = \frac{P_n}{P_r} = \frac{\cos\varphi_n}{\eta_n(1 - s_n)} \tag{2.98}$$

Equivalent circuit diagrams typically give data in real values, that is, the parameters from the T-ECD are in Ohm. In some cases they are given also in their normalized version in percent of the base resistance value according to Table 2.9.

*Table 2.9 List of reference values for per-unit system definition*

| Physical | Reference definition | Physical value | Reference definition |
|---|---|---|---|
| Voltage | $U_r = \sqrt{2}\,U_{1n}$ | Angular frequency | $\omega_r = 2\pi f_{1n}$ |
| Current | $I_r = \sqrt{2}\,I_{1n}$ | Time | $t_r = 1/\omega_r$ |
| Resistance, reactance | $R_r = X_r = U_{1n}/I_{1n}$ | Flux linkage | $\psi_r = U_r/\omega_r$ |
| Frequency | $f_r = f_{1n}$ | Power | $P_r = 3U_{1n}I_{1n}$ |
| Speed | $n_r = f_{1n}/p$ | Torque | $M_r = pP_r/\omega_r$ |

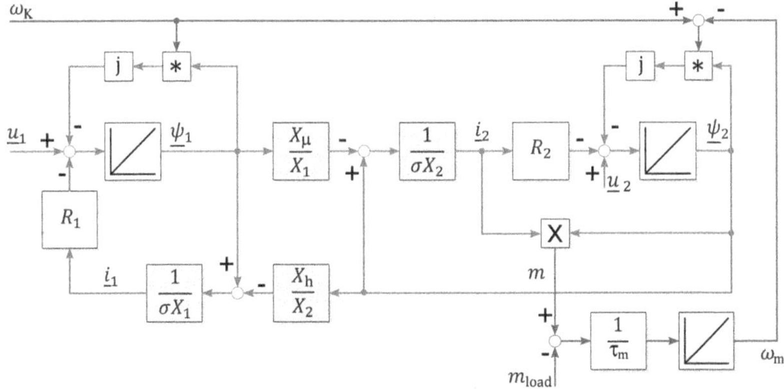


*Figure 2.25 A simple signal flow chart for IG using complex space vectors in universal reference frame*

## 2.3 Double-fed induction generator (DFIG)

### 2.3.1 Basics

The history of the double-fed induction machine started in the year 1893 as Dolivo-Dobrowolsky patented [6] his idea for "Regulation of alternating-current motors" – an induction machine with wound rotor and slip rings for the connection of external resistors (Figure 2.26). This was not jet a double-fed machine, but the fundamentals (theory and machine design) were available.

Six years later, the "double-field motor" was the next great idea, 1899 patented from M. Kloss [12], and the first real double-fed induction motor was born. Stator and rotor were fed from two different three-phase AC voltage sources. Even if this was 50 Hz in both cases, a speed different to that according to (2.38) and (2.21) was possible.

Some publications from the following years indicate that there was some scientific interest in those machines, but a real technical application could not be found. The two main disadvantages in motor applications were the difficult motor starting due to low torque and torque oscillations during operation. At that time, there was no possibility to control the voltage sources to overcome both problems.

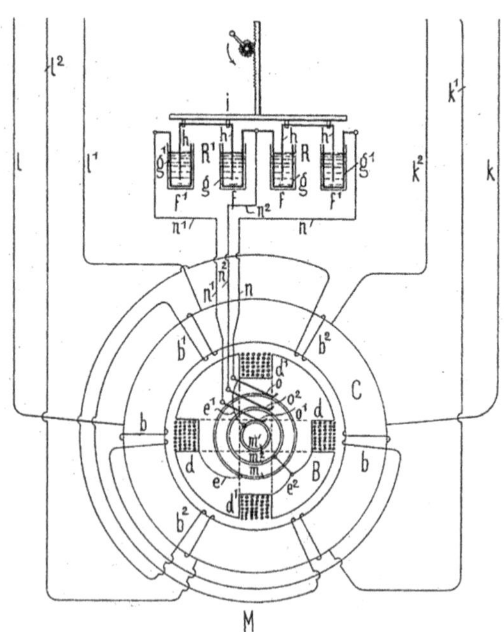


*Figure 2.26 The first slip ring induction motor with external resistors according to US Patent No. 503,038 "Regulation of alternating-current motors," August 1893*

The first technical applications were later the use in different cascaded drive configurations, first with a secondary electrical machine in the background, later with thyristor based inverters as so-called sub synchronous converter cascade. The stator was connected to a stiff-feeding three-face AC network. The rotor was connected to the same network via a thyristor-based current source inverter to control the amplitude of the rotor current, what controlled the torque of the motor. The difference between real stator power and mechanical power at the shaft, the slip energy, was fed back into the mains. This technology gained some popularity in the eighties, as speed control could be maintained even for drives with several MW of drive power. The thyristor technology was well proven, low cost, and without further requirements to the insulation or grounding system of the motor. However, the speed range was limited to sub synchronous operation (50%–97% of synchronous speed). The motor's reactive power consumption and the inverter's commutation currents were heavy loads for the feeding grid and typically compensation by filters and capacitors was needed. For the use in power generation, this concept was not feasible at all.

A breakthrough was the use of four-quadrant voltage source inverters (IGBT, IGCT) instead of current source thyristor technology. Now a controlled AC voltage source for the rotor feeding was available, not only by amplitude but also by its angle. By this, not only the torque was controllable, but also the flux inside of the machine. This made the AC induction machine comparable to synchronous machines: the excitation in the machine could be chosen such that a certain amount of reactive power was delivered to the grid.

The general principle for double-fed wind power generators can be seen in Figure 2.27. The inverter, and also the stator of the generator, can consume or deliver real power and also imaginary power in both directions.

A typical rotor voltage of this configuration is a maximum of 740 V line-line as typical low-voltage inverter technologies allow a DC link voltage of 1100 V. The wound rotor must have an insulation system to withstand this voltage including the resulting voltage peaks from the inverter.

Typical stator voltages are in the low-voltage area are 575, 690, or 950 V. These voltage levels can be used up to the 5 MW class. In an electrical machine, the

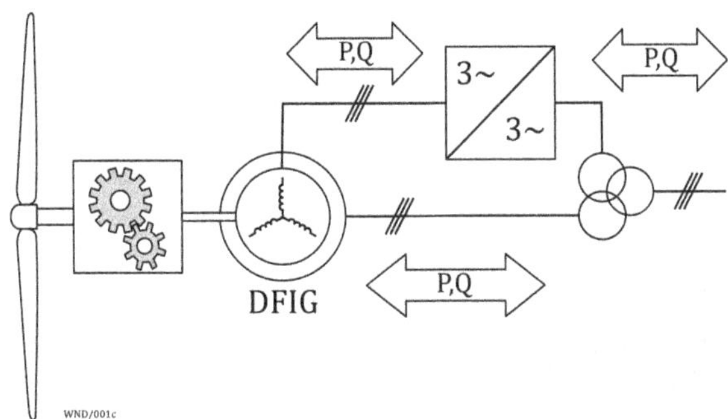


*Figure 2.27 Double-fed induction generator concept*

number of winding turns goes down as the power goes up. Around 5 MW there is a technical limit, as the number of turns in the coils reaches two. Above this power, a higher voltage level is needed. Typical values here are 3, 6.6, or 10 kV. Any of these voltages still is smaller than the typical voltage level from park internal distribution of 20 kV or 30 kV. Therefore, any wind turbine needs at least one transformer to the low-voltage from the rotor inverter or a three-system transformer to connect both to the distribution line (see Figure 2.27).

A very sensitive part of DFIG is the slip ring, which is needed to exchange power between the wound rotor and the inverter. Currents of up to 2400 A must be handled in case a 6.5 MW DFIG. This leads to very big slip ring compartments with many carbon brushes in parallel. Although slip ring motors have been known by one century already, the design of such a slip ring apparatus was not clear straight away. Some major differences to industrial drive applications are very obvious, but had to be learned first:

- There is no need to handle big starting currents, as wind power generators are accelerated by the wind.
- In contradiction to industry drives, wind power generators run in partial load quite a bit of their operational time.
- Wind power generators may run idle for some time, without any rotor current.
- The voltage harmonics and $dU/dt$ values from the inverter feeding adds additional requirements to the insulating parts.
- Service intervals above 12 months.

Those differences lead to a new design, where the main concern was a longer service span. Longer brushes (100 mm) with constant force coiled clips and wear indicator came. Metal-graphite brushes with higher metal content and different additives lowered the electrical losses. Slotted stainless-steel slip rings lowered the slip ring temperature. Much higher current densities of 20–25 A/cm$^2$ lowered the number of brushes and the mechanical losses. Very excessive cooling, air filtering and insulator design lowered the risk of an electrical breakthrough between phases or phases to ground. Figure 2.28 shows as an example the slip ring compartment of a typical 2 MW double-fed wind power generator.

### 2.3.2 DFIG in stationary operation

For any calculation of stationary operation points the equations from the previous chapter may be used, but they must be adapted in respect to the following points:

- The stator is connected to a stiff grid, leading to the same requirements as for DOL connected IG. However, the saturation state now depends on the excitation from the rotor rather than on the voltage level of the mains. This leads to very different saturation values for the main inductance, if the demand for reactive power changes.
- There is a rotor voltage to be taken into account. The rotor is not short circuited.
- The rotor frequency is not limited to small values, a frequency up to 20 Hz is typical. This leads to skin effect in the rotor conductors.

*Figure 2.28* *Typical slip ring compartment of a 2-MW DFIG,* $E_{20}$ = *1770V,* $I_2$ = *520A, with four metal-graphite brushes 20 × 40 × 100 mm³ per slip ring on slotted stainless steel slip rings*

- There are additional iron losses in the rotor iron core.
- There are additional harmonic losses from the inverter.
- There are additional losses from the slip rings and brushes.

The stator voltage (2.63) stays untouched. The rotor voltage equation (2.64) needs to be extended by the rotor voltage

$$\mathrm{j}\,\omega_2\,\underline{\psi}_2' = \underline{U}_2' - R_2'\,\underline{I}_2' \tag{2.99}$$

The transformation ratio, now defined by the effective winding turns from stator and rotor winding

$$K = \frac{(w\xi_1)_1}{(w\xi_1)_2} = \frac{E_{1\mathrm{n}}}{E_{20\mathrm{n}}} \approx \frac{U_{1\mathrm{n}}}{U_{20\mathrm{n}}} \tag{2.100}$$

must be taken into account in this equation. All rotor sizes can be transformed using Table 2.5. For squirrel-cage induction machines there is almost no interest in the physical dimension of the rotor current, voltage, or flux. Therefore, the transformation ratio K was not of any interest. For slip ring motors the knowledge of current and voltage is essential. Therefore the transformation ratio according to (2.100) must be delivered in data sheets. It can be measured with open rotor circuits in the locked-rotor test, if unknown.

The flux linkage equations (2.65) and (2.66) are valid also for slip ring machines. The same applies for slip and torque, which are still valid without

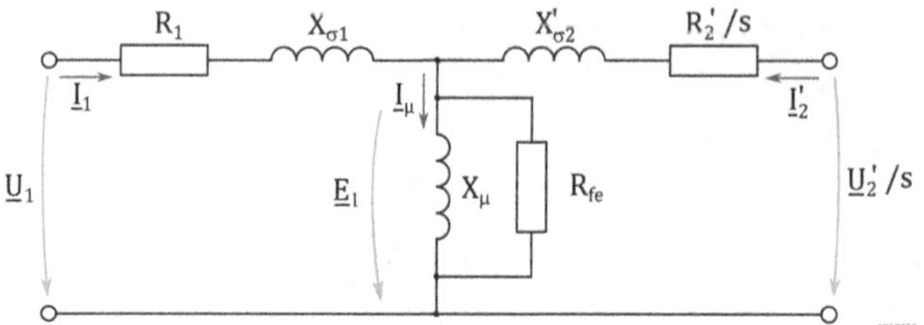


*Figure 2.29 Simple equivalent circuit diagram for a double-fed induction generator in stationary operation*

change. Special care is needed, as DFIG are very sensitive to saturation of the main flux, that is, Equation (2.78) and Figure 2.21 must be regarded. The saturation curve must be provided by the manufacturer or measured in suitable no-load tests. The ECD for a DFIG can be seen in Figure 2.29. Using (2.47) and (2.66) we get for the rotor voltage

$$\underline{U}_2' = \left(R_2' + \mathrm{j}s\,\omega_1 L_{\sigma 2}'\right)\underline{I}_2' + \mathrm{j}\,s\,\omega_1\,\underline{\psi}_\mu \tag{2.101}$$

The locked-rotor test is done with open rotor circuit ($I_2 = 0$) and with slip $s = 1$. This leads to the already mentioned locked-rotor voltage $U_{20}' = K \cdot U_{20} = E_1$. The load reaction from the rotor current is neglectable, as rotor resistance and rotor leakage inductance are small values. Consequently, the rotor voltage approximately depends linear on the slip. For typical low-voltage inverters used for DFIG applications with up to 740 V output and a speed range 70–130% of synchronous speed, a locked-rotor voltage between 1600 and 2100V can be chosen. The smaller the speed range – the higher the locked-rotor voltage – and the lower the rotor current. A small speed range makes the inverter small for a given turbine power! The upper speed is limited to thermal issues as well. Therefore often the stationary speed range ends at approximately 120% (Figure 2.30).

The complex (active and reactive) power for the rotor circuit is

$$\underline{S}_2 = 3\underline{U}_2'\,\mathrm{conj}\left(\underline{I}_2'\right) \tag{2.102}$$

Using (2.101) we find for the active rotor power

$$P_2 = 3R_2' I_2' 2 - 3s\,\omega_1 \mathrm{Im}\left\{\underline{\psi}_\mu\,\mathrm{conj}\left(\underline{I}_2'\right)\right\} \tag{2.103}$$

which is after some reforming

$$P_2 = P_{\mathrm{Lw2}} - s\,P_\delta \tag{2.104}$$

From this equation we see that the rotor power flow in super synchronous operation (above synchronous speed) always points in the same direction as the air gap power.

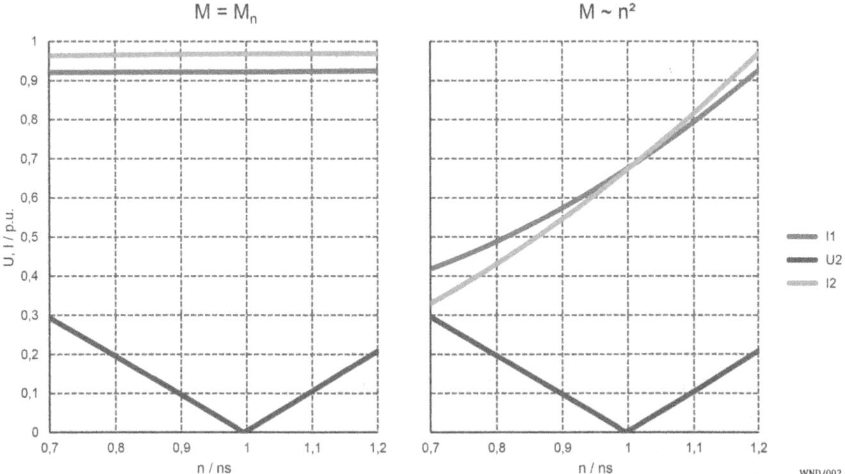


*Figure 2.30 Typical speed range of a double-fed induction machine at different load conditions (left: constant torque, right: torque rises with power of two)*

The double-fed asynchronous machine has additional loss components compared to the squirrel-cage induction machine, which shall be discussed in the following.

First of all, there is a considerable frequency in the rotor and consequently *rotor iron losses* must be taken into account. This can be done by interpreting the iron loss resistance in Figure 2.29 as a representation of stator and rotor iron losses together. It is then for stator and rotor, respectively,

$$P_{\text{Lfe}} = 3\frac{E_1^2}{R_{\text{fe}}} = P_{\text{Lfe1}} + P_{\text{Lfe2}} = 3\frac{E_1^2}{R_{\text{fe1}}} + 3\frac{E_1^2}{R'_{\text{fe2}}/s} \tag{2.105}$$

The resulting iron loss resistance is

$$R_{\text{fe}} = \frac{R_{\text{fe1}}R'_{\text{fe2}}}{s\,R_{\text{fe1}} + R'_{\text{fe2}}} \tag{2.106}$$

The iron loss resistance for the stator depends on the stator frequency, the rotor iron loss resistance depends on the rotor frequency. Equation (2.80) can be adapted to determine the rotor iron resistance.

$$R_{\text{fe2}}(f_2) = \frac{R_{\text{fe2,n}} \cdot f_2/f_{2\text{n}}}{k_{\text{hys}} + k_{\text{eddy}} \cdot f_2/f_{2\text{n}}} \tag{2.107}$$

The iron losses from the stator can be measured in a classical no-load test. The iron losses from the rotor can be determined in an additional test with blocked rotor and open rotor circuits after withdrawing the stator iron losses.

Due to *skin effect in the rotor winding*, the winding losses have an additional AC component. High-power DFIG have wave-windings made of preformed copper rods with quite big cross sections. Even at rotor frequencies of 10...20 Hz, massive skin effects, resulting in 20%...50% higher copper losses, are possible. The rotor resistance must be adapted then according to (2.81) to take these losses into account. AC wining losses cannot be measured with low effort.

$$P_{\mathrm{Lw2}} = 3k_{R2}\, R_2\, I_2^2 \tag{2.108}$$

The calculation of the AC loss factor $k_{R2}$ is possible with closed analytical equations and can be found in the literature [5,8].

The *additional losses due to the slip rings* and the brushes cannot be measured. They have to be calculated and two parts need to be taken into account: mechanical friction between brush and slip ring and an electrical voltage drop over the contact point between brush and ring. Their calculation is

$$P_{\mathrm{Lbr}} = 3U_{\mathrm{br}}I_2 + N_{\mathrm{br}}\, \mu_{\mathrm{br}}\pi D_{\mathrm{sr}} n\, F_{\mathrm{br}} \tag{2.109}$$

For modern metal-graphite brushes with a rather high metal content, the voltage drop is typically $U_{\mathrm{br}} = 0.2\ldots0.4\ V$. The voltage drop depends on the brush pressure. A higher force $F_{\mathrm{br}}$ to press the brush against the slip ring lowers the voltage drop but enlarges the mechanical losses. It is remarkable that the number of brushes does not change the electrical brush losses. The mechanical brush losses depend on friction coefficient, number of brushes, machine speed, and the brush force. The material combination of slip ring and brush determines the friction coefficient $\mu_{\mathrm{br}}$. Typical values are between $\mu_{\mathrm{br}} = 0.07-0.15$. The entire losses for the slip ring apparatus of a 7-MVA generator can reach 10 kW.

Windage and friction losses $P_{\mathrm{Lwf}}$, stator winding losses $P_{\mathrm{Lw1}}$, and load depending additional losses $P_{\mathrm{LL}}$ do not differ from those in induction machines with squirrel-cage rotors. Also, the handling of high-frequency losses does not differ. The total amount of losses in a DFIG drive system is then

$$P_{\mathrm{L}} = P_{\mathrm{Lw1}} + P_{\mathrm{Lw2}} + P_{\mathrm{Lfe}} + P_{\mathrm{Lwf}} + P_{\mathrm{Lbr}} + P_{\mathrm{LL}} + P_{\mathrm{Lhf}} + P_{\mathrm{Linv}} \tag{2.110}$$

All those losses can be sorted to be related to the stator, to the rotor, to the inverter, or to be mechanical for thermal evaluation.

Although there are more loss components to be taken into account and even if some of them are really in addition to the classical losses inside a normal IG driven by a full-size inverter, both concept still can compete. One big advantage of the concept of DFIG is that the inverter losses, and also the size of the inverter itself, are much smaller than in normal induction and synchronous machines with full-size converters. Especially in the speed range slightly above the synchronous speed, the power flow is quite optimal. To demonstrate this, the power flow shall be investigated a little further. According to (2.104), the active rotor power depends on the air gap power and the slip

$$s \cdot P_\delta = P_{L2} - P_2 = s \cdot M \cdot \frac{\omega_1}{p} \tag{2.111}$$

That means: in super synchronous (above synchronous speed, $s < 0$) and generator operation ($P_\delta < 0$), the rotor power is negative too. That means the rotor delivers active power. Below synchronous speed ($s > 0$), a positive power indicates consumption of active power. For the mechanical (shaft) power we get the equation

$$P_m + P_{Lwf} = (1 - s)\, P_\delta = 2\pi\, n \cdot M \tag{2.112}$$

Care must be taken, as the electromagnetic torque $M$ is not necessarily equal to the shaft torque $M_m$ due to mechanical losses as windage and friction losses $P_{Lwf}$. The total electrical system output power is the sum of stator and inverter output power.

$$P_{el} = P_1 + P_{inv} \tag{2.113}$$

Figure 2.31 shows the two possible power flows in a DFIG. Two things are obvious from that: in super synchronous operation, one part of the power flow does not cross the air gap. This leads to lower losses in the machine. Secondly, the inverter power bears only a proportion of the total power, consequently causing less losses. Also quite clearly we see that in super synchronous operation, the generator can transform more energy compared to an IG with a squirrel cage for a given air gap power $P_\delta$.

The determination of losses and efficiency values can be done by measuring or calculation of all single loss components or by testing the entire system efficiency including all needed components. As an example, in Figure 2.32 a test configuration is shown to system-test the 7 MVA double-fed system from the Senvion 6M turbine in the test bay from VEM Sachsenwerk GmbH in Dresden. The test included inverter cabinet, control cabinets, transformer, cooling system, and even a sloped pedestal to simulate the mounting position inside of the nacelle. A

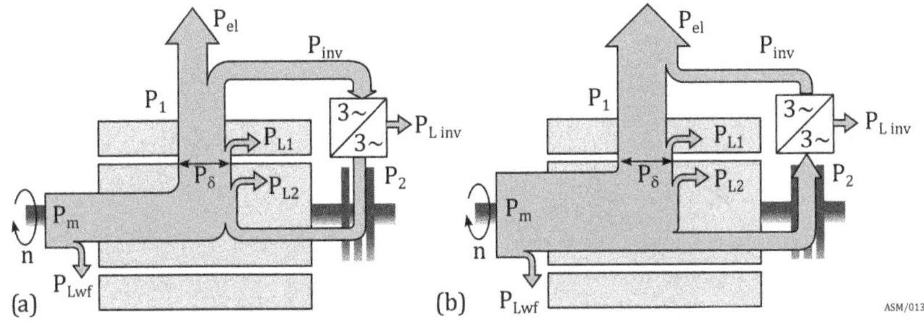


*Figure 2.31 Power flow in a double-fed induction machine in (a) sub synchronous operation and (b) super synchronous operation*

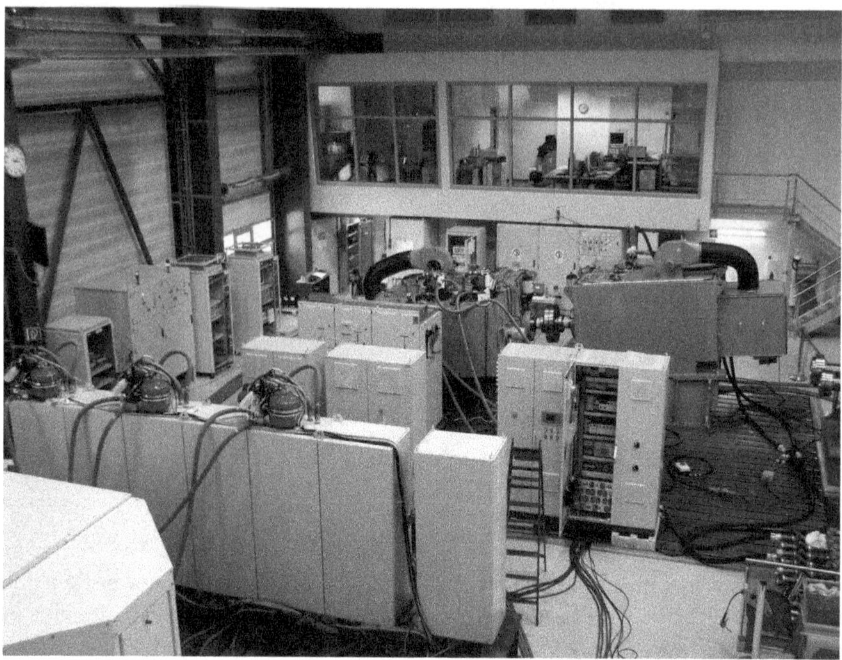

*Figure 2.32 System test of a complete 7-MVA DFIG system including all original components connected to it. © VEM Sachsenwerk GmbH (part of VEM Group).*

technically equal generator served as load machine. The grid was simulated via 20 kV test bay power line and the original three-system transformers 20/6,6/0,69 kV.

### 2.3.3 Dynamic operation

The dynamic behavior of induction machines has been introduced already. Again, for the purpose of simplicity, we neglect the iron, brush and additional losses in the equations. We only need to take the rotor voltage into account and Equations (2.90) and (2.91) become

$$\frac{\mathrm{d}\,\underline{\psi}_1}{\mathrm{d}t} = \underline{u}_1 - R_1\,\underline{i}_1 - \mathrm{j}\,\omega_\mathrm{K}\,\underline{\psi}_1 \tag{2.114}$$

$$\frac{\mathrm{d}\,\underline{\psi}_2'}{\mathrm{d}t} = \underline{u}_2' - R_2'\,\underline{i}_2' - \mathrm{j}\,(\omega_\mathrm{K} - \omega_\mathrm{m})\underline{\psi}_2' \tag{2.115}$$

The flux linkage equations are

$$\underline{\psi}_1 = L_1\,\underline{i}_1 + L_\mu\,\underline{i}_2' \tag{2.116}$$

$$\underline{\psi}_2' = L_\mu\,\underline{i}_1 + L_2'\,\underline{i}_2' \tag{2.117}$$

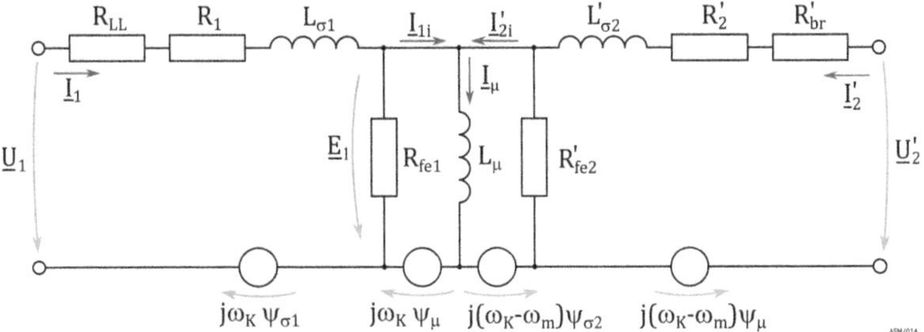


*Figure 2.33 Universal equivalent circuit diagram for IG, considering additional losses for dynamic operation*

The electromagnetic torque is

$$m = \frac{3}{2} p \left( \underline{\psi}_1 \times \underline{i}_1 \right) \tag{2.118}$$

Remember the transformation for rotor sizes according to Table 2.5. The equations are valid for a universal reference system, rotating with the angular speed $\omega_K$. For simulation tasks with DFIG all three reference systems according to Table 2.8 are needed:

- stator related faults as short-circuits and voltage dips are preferably simulated in stator reference frame
- rotor related faults, e.g., slip ring failures or crow-bar events may be simulated in rotor reference frame
- dynamic events in field-oriented control may be simulated in a universal reference frame oriented toward the stator flux space vector.

For dynamic simulations, the signal flow chart from Figure 2.25 can be used. Again, care must be taken, as all rotor sizes are transformed to the stator winding and the transformation ratio of Table 2.5 must be regarded!

For the case, that even losses are needed in dynamic simulations, Figure 2.33 shows an adapted equivalent circuit diagram that implemented all loss components, except windage and friction losses in one equivalent circuit diagram. This diagram can be used for stationary and dynamic investigations and it is valid for any reference frame. It also can be adapted for squirrel cage induction machines by setting the rotor voltage to zero.

## 2.4 Salient pole synchronous generator (SPSG)

### *2.4.1 Basics*

During the last century, almost any electrical power has been generated using electrically excited synchronous machines. A deep understanding for this generator design is obvious, leading to most mature machine designs with the highest efficiency values and highest reliability. Machines for nominal power of more than 1000 MW are possible. There are many important advantages in using those machine types for generating electricity. One of the most important among them is their ability for automatically stabilizing the grid in case of unwanted failure events, e.g., its low voltage and high voltage ride-through capability.

However, there is also one big disadvantage: the frequency of the output alternating voltage and the generator speed do have a strong correlation according to (2.21). Not the slightest deviation from this relation is possible; the speed must fit the stator frequency. Moreover, any load change on the mechanical side will automatically lead to a load fluctuation at the electrical side. The result is a high mechanical torque for the gearbox, shaft and coupling in the case of load peaks. The synchronous machine, directly coupled to a grid with fixed electrical frequency, is unsuitable for wind power applications. There is the strong need to adapt the speed, either mechanically by the help of a hydraulic gearbox, as the Voith WinDrive, or electrically by a frequency converter between generator and grid. Although the mentioned WinDrive was a good choice from technical point of view, it stayed quite exotic. Therefore, we want to focus on a system as it is depicted in Figure 2.34. This system uses either a full size converter to transform frequency, voltage and reactive power between generator and grid. Also possible is the use of a B6 diode bridge between DC link and stator winding. The DC link voltage is then a result of the excitation of the machine on the one side and the power delivery to the grid on the other side. Even generators with integrated B6 bridge are possible, feeding DC voltage directly from the generator to the DC link from the inverter (e.g., the Eno 82 was using this concept).

A gearbox with a fixed transmission ratio is used for speed adaption between the hub and generator. The rotor field winding is fed by an external voltage source to adapt the magnetic field inside of the machine according to speed and current wind power supply.

Synchronous machines can be divided into two major groups:

- non-salient pole (cylindrical) synchronous machines
- salient pole synchronous machines

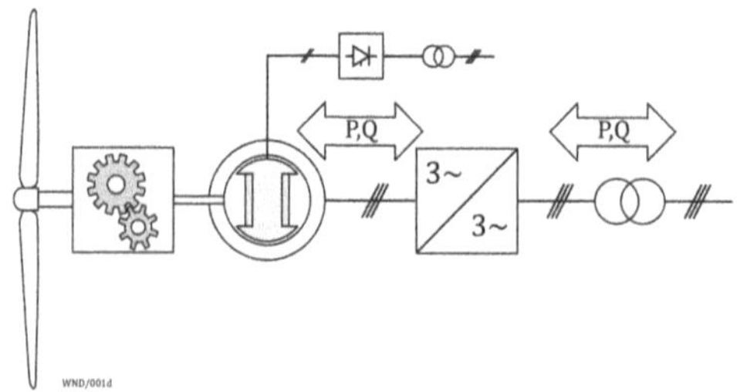


*Figure 2.34 Salient pole synchronous generator*

Non-salient pole synchronous machines have a cylindrically shaped rotor, which is made either of solid iron or of laminations. The excitation field is created by a distributed rotor field winding: conductors from the field winding are placed in slots within the cylindrical rotor. The position of the slots in circumferential direction is not equidistant in order to create a good sinusoidal shaped excitation field in the air gap. As the conductors are fitted into the rotor slots, a good mechanical strength can be reached. It is difficult to include additional damper windings in this machine type, some machine type do not have them. Non-salient synchronous machines can be designed with quite low rotor inertia. This leads to the main applications: high torque motors where high controllability is needed (e.g., rolling mills) or high-speed generators (e.g., 3000 rpm) for high power at fixed speed. Those generators then have very long stacks and a small rotor diameter (turbo generators). For wind power application, this is not suitable.

Salient pole synchronous machines do have concentrated rotor windings, wound on separated poles. The poles are made of solid or laminated iron. The pole shoes can be manufactured separately and screwed to the rotor hub after the excitation winding has been wound onto them. This enables the manufacturing of large pole numbers. With the shape of the pole roof, a sinusoidal magnetic field in the air gap shall be formed. Therefore, the outer surface from salient pole synchronous machines is never cylindrical. Each pole typically bears additional damper bars near the outer surface or, in case of solid poles, the pole itself acts as damper winding. Figure 2.34 shows the cross section of a salient pole synchronous motor.

The rotor construction is less robust and has more inertia as those from non-salient synchronous machines. The main reason is that the rotor winding, wound with several layers onto the poles, is made of pure copper with quite low mechanical strength. The pole roof must support the high mass of the copper winding, limiting the speed to low values (<3000 rpm). On the other hand, the excitation winding can deliver more ampere-turns, effective damping can be achieved and the additional use of the reluctance effect ($L_d \gg L_q$) can increase the torque density. This all makes them suitable for applications with variable speed and load peaks, but limited maximum speed. Very

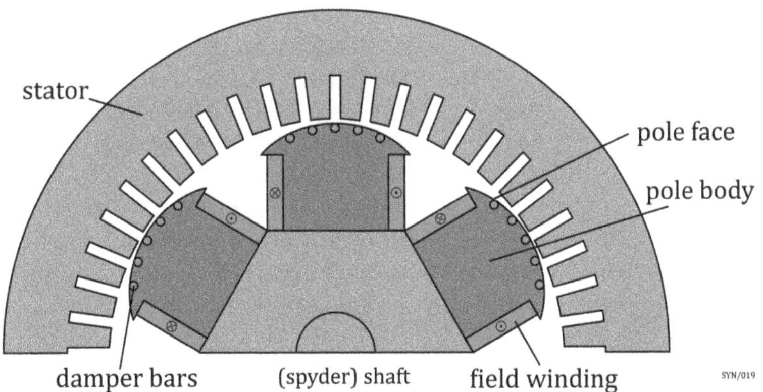


*Figure 2.35 Salient-pole synchronous generator*

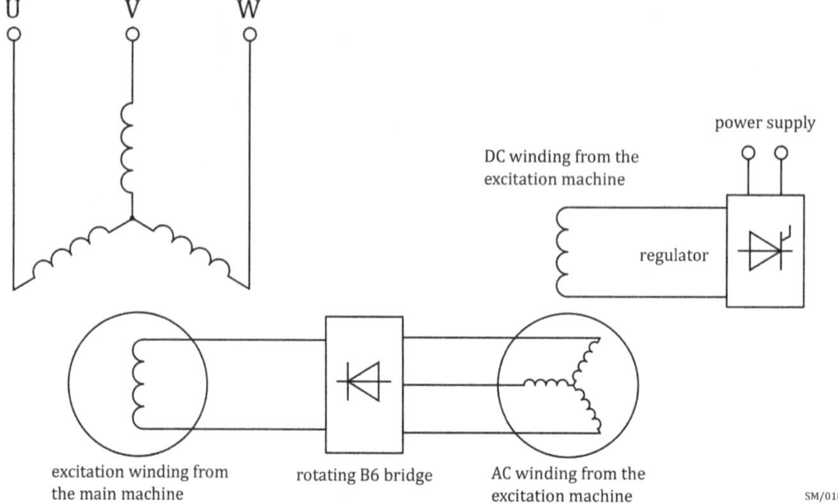


*Figure 2.36 Brushless excitation system for synchronous generators*

typical for this machine type is the application as diesel generator (6–8 poles), shaft generator (16 poles), water power generator (very high pole number), and wind power generator. In wind power applications, generators do have a pole number of 4 or 6 in high-speed geared applications. Geared medium speed applications do have 12 or 16 poles and gearless multipole machines may have pole numbers of up to 200.

A very important part of electrically excited synchronous machines is the excitation system. A certain DC current has to be fed into the excitation winding of the rotor, electrical energy must be transported to the rotating part of the machine. This can be done using two slip rings with carbon brushes, which are subjected to unwanted wear and maintenance. Very often brushless excitation systems are used instead. Due to the fact that a generator does not need magnetic flux at standstill, synchronous excitation systems can be used. Figure 2.36 shows the principle of such a system. Frankly speaking, a second small synchronous generator with a three-phase AC winding on the rotor shaft and pole windings on the stator side will be used to transport the energy into the rotor. A direct current feeds an excitation winding from that auxiliary machine. A voltage will be induced in a three-phase AC winding on the rotor. The AC will be rectified by a rotating B6 bridge and fed into the excitation winding from the main machine. The DC current fed to the poles of the auxiliary machine is proportional to the internal direct current to the main excitation winding. There is no need for slip rings or any other part subjected to wear. The energy for the excitation comes from a separate output from the inverter.

## 2.4.2 SPSG in stationary operation

The stator, and also the stator winding, of an SPSG compares exactly to those from all the other three-phase AC machines from this chapter. The same equivalent circuit diagram according Figure 2.14 can be used. This also means we can use the same stator voltage equation again.

$$\underline{U}_1 = R_1\,\underline{I}_1 + \mathrm{j}\omega_1\,L_{\sigma 1}\,\underline{I}_1 - \underline{E}_1 \tag{2.119}$$

As the rotor construction is different to that from the induction machine, we need also a new understanding and new set of equations for the rotor and the main flux. Figure 2.35 shows a typical cross section of an SPSG.

We see that the rotor is far away from being symmetrical as it was the case for the previous machine types. Therefore, there is a strong dependency of rotor and main field parameters to the rotor angle. The field winding is defined aligned with the direction of the rotor poles. In addition, the machines main inductance differs considerably, when magnetized in the direction of the rotor pole or the pole gap. It is obvious that the magnetic flux can pass very easily along the poles, but it will not pass easily perpendicular to them. The axis aligned with the poles is called the direct magnetic axis or *d*-axis. The perpendicular orientation is called the quadrature magnetic axis or q-axis. Stator current and stator voltage now can be represented in the *d*–*q*-reference system, strictly connected to the rotor (*d* is aligned with the rotor pole) (Figure 2.38).

The voltage from the field winding is oriented perpendicular to the rotor pole. We get for the stator voltage equation

$$\underline{U}_1 = R_1\,\underline{I}_1 + \mathrm{j}X_d\,\underline{I}_d + \mathrm{j}X_q\,\underline{I}_q + \underline{U}_p \tag{2.120}$$

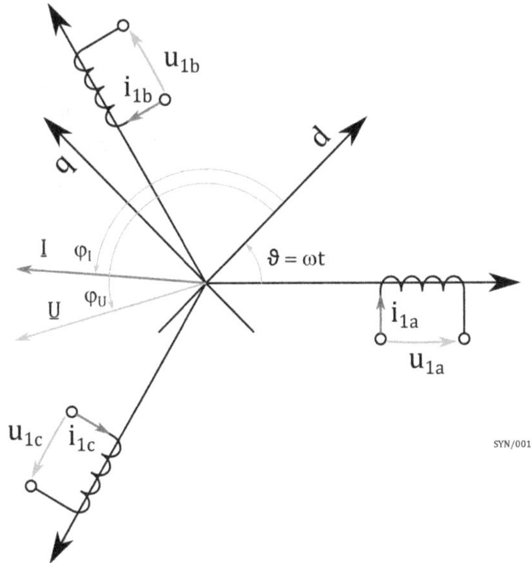


*Figure 2.37 Stator voltage and stator current represented in rotor (d–q) reference frame*

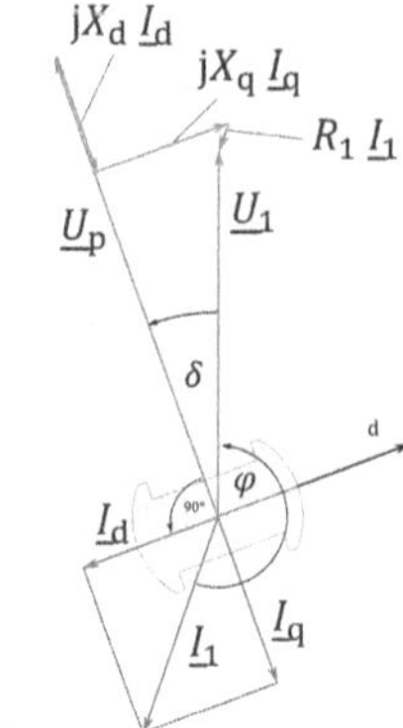


*Figure 2.38 Phasor diagram for salient pole synchronous machines in generator operation*

The stator reactance can be expressed by pole form factors C and the main reactance. Typically, for salient pole synchronous machines is $X_d/X_q \approx 0.5...0.8$.

$$
\begin{aligned}
X_{\mathrm{d}} &= C_{\mathrm{ad}} \cdot X_{\mu} + X_{\sigma 1} \\
X_{\mathrm{q}} &= C_{\mathrm{aq}} \cdot X_{\mu} + X_{\sigma 1}
\end{aligned} \tag{2.121}
$$

Furthermore is

$$
\underline{I}_1 = \underline{I}_{\mathrm{d}} + \underline{I}_{\mathrm{q}} \tag{2.122}
$$

In Figure 2.38, the phasor diagram shows an example for generator operation.

One significant condition for stationary operation of synchronous machines is the absence of rotating fields in the rotor. All sizes in rotor-aligned coordinates do not change, they are strictly constant. For the damper windings and the field winding this means that no voltage can be induced into them. Therefore, at least the damper windings can be neglected completely for stationary processes.

The stationary behavior from SPSG on first sight equals those of DC machines. The excitation current directly leads, according to Ampere's law, to a magnetic field in the machine. Neglecting the saturation of the iron core, the flux and the excitation current are proportional to each other. However, due to saturation the flux from the pole winding will be smaller.

$$
\underline{U}_{\mathrm{p}} = \mathrm{j}\,\omega(w\xi_1)_1\,\Phi_{\mathrm{p}} = f(I_{\mathrm{f}}) \tag{2.123}
$$

For no-load condition, if the stator current is zero, the pole voltage $\underline{U}_p$ exactly equals the stator voltage $\underline{U}_1$ and can be measured at the stator terminals. Figure 2.39 shows such a curve as an example.

If the machine is loaded with a torque, an electrical power is needed and a current will arise in the armature winding. As the armature reaction field, resulting from the stator current, has a significant phase angle to the rotor field, the resulting magnetic field will change its orientation. Then the stator voltage

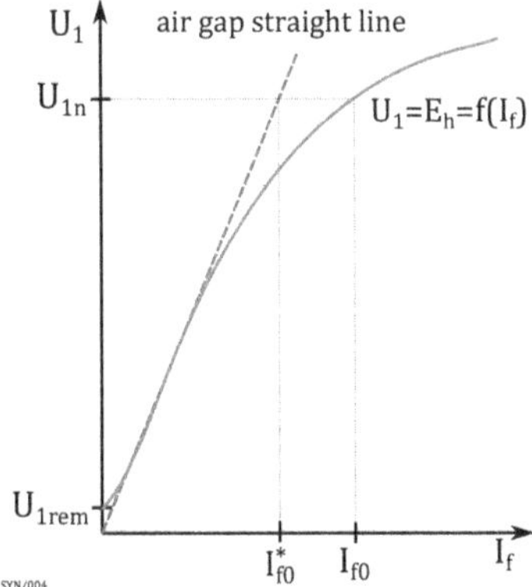


*Figure 2.39 No-load curve of a salient pole synchronous generator*

and the voltage from the excitation winding do not fit. Besides a change in the amplitude, they show a phase angle that relates to the load. This angle $\delta = \varphi_{UP} - \varphi_U$ is called the load angle (Figure 2.37). The air gap torque for SPSG can be calculated with the help of this angle

$$M = -\frac{3p}{\omega}\frac{U_1 U_\mathrm{p}}{X_\mathrm{d}}\sin\delta - \frac{3p}{2\omega}U_1^2\left(\frac{1}{X_\mathrm{q}} - \frac{1}{X_\mathrm{d}}\right)\sin 2\delta \tag{2.124}$$

In this equation, it is remarkable that even without any voltage from the pole winding a torque can be achieved, as long as there is any difference between $X_\mathrm{d}$ and $X_\mathrm{q}$. We call this part of the torque reluctance torque or reaction torque. It adds to the torque produced by the armature current. The minus sign in the equation above indicates that the voltage from the rotor pole winding lags behind the stator voltage at positive torque (=motor) operation.

From (2.124) and Figure 2.40, we can derive one more information: at a given stator voltage and excitation current there is a maximum torque that we cannot exceed. We can enlarge the machine torque by a higher excitation current and we can control it by the angle and amplitude of the stator voltage. The load angle must stay within

$$\delta_\mathrm{b} = \pm\mathrm{acos}\left(\frac{U_\mathrm{p}}{4U_1}\frac{X_\mathrm{q}}{X_\mathrm{q} - X_\mathrm{d}} + \sqrt{\left(\frac{U_\mathrm{p}}{4U_1}\frac{X_\mathrm{q}}{X_\mathrm{q} - X_\mathrm{d}}\right)^2 + \frac{1}{2}}\right) \tag{2.125}$$

in order to get a stable operation at a certain voltage ratio $U_p/U_1$.

If additional reactive power is needed, the amplitude from the pole voltage must be changed. But at this point we have to take into account that wind power generators in SPSG design may be mostly connected to a B6 rectifying bridge, which has a typical power factor of 0.96 lagging. That means, the generator only has to supply the very small commutating reactive power from that B6-bridge. This is a big difference to grid connected synchronous machines. The excitation only must stabilize the output voltage from the generator, in order to achieve a constant DC link voltage in the full-size converter.

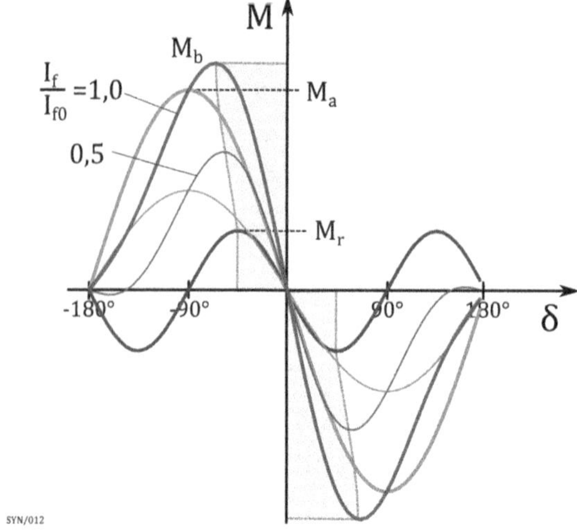


*Figure 2.40 Armature torque $M_a$, reluctance torque $M_r$ and resulting electromagnetic torque of the SPSM overload angle at different excitation currents*

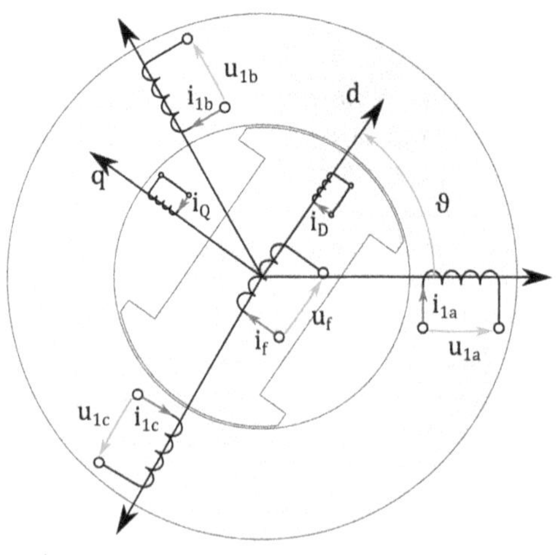


*Figure 2.41 Salient-pole synchronous machine, schematic presentation*

## 2.4.3 Dynamic operation

For dynamic operation, also the damper winding must be taken into account. Figure 2.41 shows the typical schematic representation of a salient pole synchronous machine, for simplicity with the lowest possible pole number (2p=2). Remember, typical SPSG do have higher pole numbers than two. The figure includes a representation of the field winding and the damper winding in the rotor. The figure also clearly shows an unsymmetrical rotor construction. Whereas the field winding is strictly oriented along the rotor pole, the damper winding is thought to be separated into two windings, one strictly oriented along and the other part perpendicular to the pole. The stator bears the same three-phase winding as all the machine types within this chapter.

For simplicity, we also transform the stator winding into a two-phase representation, always one component oriented with the rotor pole, the other component-oriented perpendicular to it. This transformation is known as Park's transformation or d/q-transformation.

$$i_{\mathrm{d}} = \frac{2}{3}\left[i_{\mathrm{a}}\cos(\vartheta) + i_{\mathrm{b}}\cos\left(\vartheta - \frac{2\pi}{3}\right) + i_{\mathrm{c}}\cos\left(\vartheta - \frac{4\pi}{3}\right)\right] \tag{2.126}$$

$$i_{\mathrm{q}} = \frac{2}{3}\left[-i_{\mathrm{a}}\sin(\vartheta) - i_{\mathrm{b}}\sin\left(\vartheta - \frac{2\pi}{3}\right) - i_{\mathrm{c}}\sin\left(\vartheta - \frac{4\pi}{3}\right)\right] \tag{2.127}$$

The stator voltage equations are

$$u_{\mathrm{d}} = R_1 i_{\mathrm{d}} + \frac{\mathrm{d}\psi_{\mathrm{d}}}{\mathrm{d}t} - \omega_{\mathrm{m}}\psi_{\mathrm{q}} \tag{2.128}$$

$$u_{\mathrm{q}} = R_1 i_{\mathrm{q}} + \frac{\mathrm{d}\psi_{\mathrm{q}}}{\mathrm{d}t} + \omega_{\mathrm{m}}\psi_{\mathrm{d}} \tag{2.129}$$

In this equation, $\omega_{\mathrm{m}}$ is the rotor speed resulting from the rotor angle in electrical degree. Mechanical speed and rotational angle are

$$\omega_{\mathrm{m}} = 2\pi\, p\, n = \frac{\mathrm{d}\vartheta}{\mathrm{d}t} \tag{2.130}$$

Equations (2.128) and (2.129) still compare to (2.90) and (2.91), now taking the rotor angle for transformation into account. From the same quantities, the torque can be calculated with

$$m = \frac{3p}{2}\left(\psi_{\mathrm{d}} i_{\mathrm{q}} - \psi_{\mathrm{q}} i_{\mathrm{d}}\right) \tag{2.131}$$

The rotor equations are slightly more difficult. From the first sight, they appear very easy because they can be written directly in the rotor reference frame. We get for the field winding

$$u_{\mathrm{f}} = R_{\mathrm{f}} i_{\mathrm{f}} + \frac{\mathrm{d}\psi_{\mathrm{f}}}{\mathrm{d}t} \tag{2.132}$$

For the two components of the damper winding we get

$$0 = R_{\mathrm{D}} i_{\mathrm{D}} + \frac{\mathrm{d}\psi_{\mathrm{D}}}{\mathrm{d}t} \tag{2.133}$$

$$0 = R_{\mathrm{Q}} i_{\mathrm{Q}} + \frac{\mathrm{d}\psi_{Q}}{\mathrm{d}t} \tag{2.134}$$

The flux linkage equations related to the stator and in rotor reference frame are

$$\psi_{\mathrm{d}} = L_{\mathrm{d}}\, i_{\mathrm{d}} + L_{\mathrm{aD}}\, i_{\mathrm{D}} + L_{\mathrm{af}}\, i_{\mathrm{f}} \tag{2.135}$$

$$\psi_{\mathrm{q}} = L_{\mathrm{q}}\, i_{\mathrm{q}} + L_{\mathrm{aQ}}\, i_{\mathrm{Q}} \tag{2.136}$$

For the rotor flux linkages we get the following dependencies

$$\psi_{\mathrm{f}} = \frac{3}{2} L_{\mathrm{fa}} i_d + L_{\mathrm{fD}}\, i_{\mathrm{D}} + L_{\mathrm{ff}}\, i_{\mathrm{f}} \tag{2.137}$$

$$\psi_{\mathrm{D}} = \frac{3}{2} L_{\mathrm{Da}} i_d + L_{\mathrm{DD}}\, i_{\mathrm{D}} + L_{\mathrm{Df}}\, i_{\mathrm{f}} \tag{2.138}$$

$$\psi_{\mathrm{Q}} = \frac{3}{2} L_{\mathrm{Qa}} i_q + L_{\mathrm{QQ}}\, i_{\mathrm{Q}} \tag{2.139}$$

For the mutual inductances between two windings we agree that the first index is always the affected winding and the second index is the source winding. The inductances mentioned above can be calculated from main inductance, leakage inductances and mutual inductances from all windings if all equations are scaled to the same main inductance. Therefore as next step, the rotor equations need to be translated by the help of a suitable transformation ratio to fit to the stator quantities. We have done this for the induction generator as well. The problem is that we do not have one but three different transformation ratios, as we have three different windings in the rotor. Table 2.10 lists all three ratios, and also the transformation results of all rotor quantities. With this transformation, the rotor equations (2.132) and (2.133) can be transformed. Again, we will use an apostrophe for signing transformed quantities.

The inductances can be rewritten to

$$\begin{array}{lll} L_{\mathrm{d}} = L_{\mu\mathrm{d}} + L_{\sigma 1} & L'_{\mathrm{DD}} = L_{\mu\mathrm{d}} + L'_{\sigma 2} + L'_{\sigma\mathrm{D}} & L'_{\mathrm{fD}} = L_{\mu\mathrm{d}} + L'_{\sigma 2} \\ L_{\mathrm{q}} = L_{\mu\mathrm{q}} + L_{\sigma 1} & L'_{\mathrm{QQ}} = L_{\mu\mathrm{q}} + L'_{\sigma\mathrm{Q}} & L'_{\mathrm{Df}} = L'_{\mathrm{fD}} \\ & L'_{\mathrm{ff}} = L_{\mu\mathrm{d}} + L'_{\sigma 2} + L'_{\sigma f} & \end{array}$$

*Table 2.10 Transformation ratios and transformation of all rotor quantities*

| $K_{\mathrm{f}} = \frac{L_{\mathrm{hd}}}{L_{\mathrm{af}}}$ | $K_{\mathrm{D}} = \frac{L_{\mathrm{hd}}}{L_{\mathrm{aD}}}$ | $K_{\mathrm{Q}} = \frac{L_{\mathrm{hd}}}{L_{\mathrm{aQ}}}$ |
|---|---|---|
| $i'_{\mathrm{f}} = \frac{1}{ü_{\mathrm{f}}} i_{\mathrm{f}}$ | $i'_{\mathrm{D}} = \frac{1}{ü_{\mathrm{D}}} i_{\mathrm{D}}$ | $i'_{\mathrm{Q}} = \frac{1}{ü_{\mathrm{Q}}} i_{\mathrm{Q}}$ |
| $u'_{\mathrm{f}} = \frac{2}{3} ü_{\mathrm{f}}\, u_{\mathrm{f}}$ | | |
| $\psi'_{\mathrm{f}} = \frac{2}{3} ü_{\mathrm{f}}\, \psi_{\mathrm{f}}$ | $\psi'_{\mathrm{D}} = \frac{2}{3} ü_{\mathrm{D}}\, \psi_{\mathrm{D}}$ | $\psi'_{\mathrm{Q}} = \frac{2}{3} ü_{\mathrm{Q}} \psi_{\mathrm{Q}}$ |
| $L'_{\mathrm{ff}} = \frac{2}{3} ü_{\mathrm{f}}^2\, L_{\mathrm{ff}}$ | $L'_{\mathrm{DD}} = \frac{2}{3} ü_{\mathrm{D}}^2\, L_{\mathrm{DD}}$ | $L'_{\mathrm{QQ}} = \frac{2}{3} ü_{\mathrm{Q}}^2\, L_{\mathrm{QQ}}$ |
| $R'_{\mathrm{f}} = \frac{2}{3} ü_{\mathrm{f}}^2\, R_{\mathrm{f}}$ | $R'_{\mathrm{D}} = \frac{2}{3} ü_{\mathrm{D}}^2\, R_{\mathrm{D}}$ | $R'_{\mathrm{Q}} = \frac{2}{3} ü_{\mathrm{Q}}^2\, R_{\mathrm{Q}}$ |
| $L'_{\mathrm{fD}} = L'_{\mathrm{Df}} = \frac{2}{3} ü_{\mathrm{f}} ü_{\mathrm{D}}\, L_{\mathrm{fD}}$ | | |

Here is

- $L_{\sigma 1}$ the leakage inductance from the stator winding,
- $L_{\mu d}$ and $L_{\mu q}$ the main inductance in *d*- and *q*-axis,
- $L'_{\sigma 2}$ common part of the leakage inductance for damper and field winding,
- $L'_{\sigma D}$ and $L'_{\sigma Q}$ the leakage inductance from the damper winding in *d*- and *q*-axis, and
- $L'_{fD}$ the leakage inductance from the field winding.

These values must be provided by the manufacturer of the machine in an ECD or in parameter files.

We can write now the resulting system of equations for dynamic process calculation.

$$\frac{d}{dt}\begin{bmatrix}\psi_d\\ \psi'_f\\ \psi_D\\ \psi_q\\ \psi'_Q\end{bmatrix} = \begin{bmatrix}u_d\\ u'_f\\ 0\\ u_q\\ 0\end{bmatrix} - \begin{bmatrix}R_1 & 0 & 0 & 0 & 0\\ 0 & R'_f & 0 & 0 & 0\\ 0 & 0 & R'_D & 0 & 0\\ 0 & 0 & 0 & R_1 & 0\\ 0 & 0 & 0 & 0 & R'_Q\end{bmatrix}\cdot\begin{bmatrix}i_d\\ i'_f\\ i'_D\\ i_q\\ i'_Q\end{bmatrix} + \begin{bmatrix}0 & 0 & 0 & +\omega_R & 0\\ 0 & 0 & 0 & 0 & 0\\ 0 & 0 & 0 & 0 & 0\\ -\omega_R & 0 & 0 & 0 & 0\\ 0 & 0 & 0 & 0 & 0\end{bmatrix}\cdot\begin{bmatrix}\psi_d\\ \psi'_f\\ \psi_D\\ \psi_q\\ \psi'_Q\end{bmatrix}$$

$$\begin{bmatrix}\psi_d\\ \psi'_f\\ \psi_D\\ \psi_q\\ \psi'_Q\end{bmatrix} = \begin{bmatrix}L_{hd}+L_{\sigma 1} & L_{hd} & L_{hd} & 0 & 0\\ L_{hd} & L_{hd}+L'_{\sigma 2}+L'_{\sigma f} & L_{hd}+L'_{\sigma 2} & 0 & 0\\ L_{hd} & L_{hd}+L'_{\sigma 2} & L_{hd}+L'_{\sigma 2}+L'_{\sigma D} & 0 & 0\\ 0 & 0 & 0 & L_{hq}+L_{\sigma 1} & L_{hq}\\ 0 & 0 & 0 & L_{hq} & L_{hq}+L'_{\sigma Q}\end{bmatrix}\cdot\begin{bmatrix}i_d\\ i'_f\\ i'_D\\ i_q\\ i'_Q\end{bmatrix}$$

The last step, same as for induction machines, is to use a suitable per-unit-system to normalize all physical quantities. The same stator quantities can be used for this purpose: the amplitude from the stator phase voltage and stator phase current and the apparent power.

Figures 2.42 and 2.43 show the corresponding ECD for a salient pole synchronous machine.

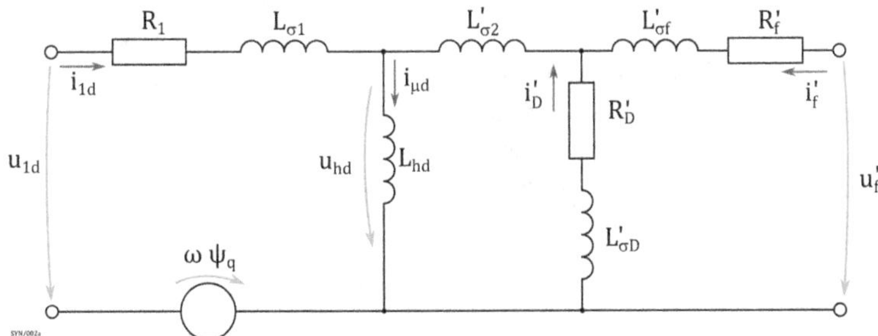


*Figure 2.42 Direct axis ECD of an SPSG for dynamic processes*

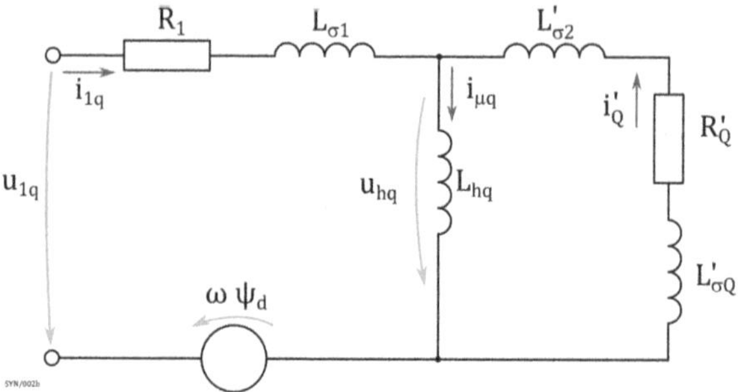


*Figure 2.43 Quadrature axis ECD of an SPSG for dynamic processes*

## 2.5 Permanent magnet synchronous generator (PMSG)

### 2.5.1 Basics

To use permanent magnets in synchronous machines is an interesting alternative to the use of excitation windings. No current is needed for the excitation of the machine. This also means that no ohmic losses (Joule losses) due to an excitation current will heat the rotor. With modern rare earth magnets, employing samarium–cobalt or neodymium, very high flux densities and a low rotor inertia can be achieved. This promises lightweight, robust and efficient generators.

However, also disadvantages must be taken into account:

- The excitation is constant. The only way to influence the excitation are suitable stator currents. This limits the speed range and makes the stator current, and possibly the MSC bigger.
- The generator always delivers voltage, as soon as the rotor is rotating. This may be dangerous and unwanted.
- 95% of the rare-earth elements are produced in China [13], leading to unwanted market dependencies.
- The prices for rare-earth magnets changed quite heavy during the last years [14].
- Eddy-current losses in rare-earth magnets can be significant [15], heating up the magnets. This leads to lower flux densities and lowers efficiency.
- Magnet material may be subjected to high flux densities or high temperatures, e.g., during short circuit events, which may de-magnetize the magnet material.
- Rotor assembling and disassembling (e.g., in case repair of bearing failures is needed) is much more difficult due to strong forces between stator and rotor. Special tooling is needed for this purpose.

Those disadvantages must be balanced with a higher efficiency, less complexity, and less rotor inertia. Especially for high-speed generator concepts, the disadvantages may overweight the advantages. Therefore, PM generators mainly can be found in gearless configurations and especially in the Chinese market.

However, in any case a full-scale and bi-directional inverter is mandatory between generator and grid. The necessary size of the machine side converter (MSC) depends on maximum of the generator current. For high speed, the magnetic field from the permanent magnets induces high voltages in the stator winding. In order to get the terminal voltage down, an additional direct axis current must be pushed into the generator in order to generate a field acting against the field from the magnets.

The q-axis current still is used to generate torque and active power respectively. The MSC must be bigger than for electrically excited SG, where only the q-axis current was needed to generator active power. We remember that we could use the excitation current to control the field in the machine.

The line-side converter (LSC) must take over all active and reactive power. This converter will be in the same size as for SPSG. There is no difference.

The entire drive train configuration can be seen in Figure 2.44.

For synchronous generators with permanent magnets two main configurations are possible:

- The magnets are mounted on top of the circular rotor surface (surface mounted PM).
- The magnets are buried deep into the rotor (interior mounted PM) forming paths and barriers for the magnetic flux.

There is some kind of analogy to non-salient and salient pole synchronous machines. Non-salient pole synchronous machines have the excitation winding mounted near the surface of a cylindrical rotor; salient pole synchronous machines have the excitation winding in the inner part of the rotor.

With surface-mounted magnets (Figure 2.48), the rotor shape is almost cylindrical and the magnetic resistance is equal in any orientation of the rotor. There will be only a small difference between *d*- and *q*-axis inductances. For surface-mounted PMSG applies

$$L_{\mathrm{d}} \approx L_{\mathrm{q}} \qquad (2.140)$$

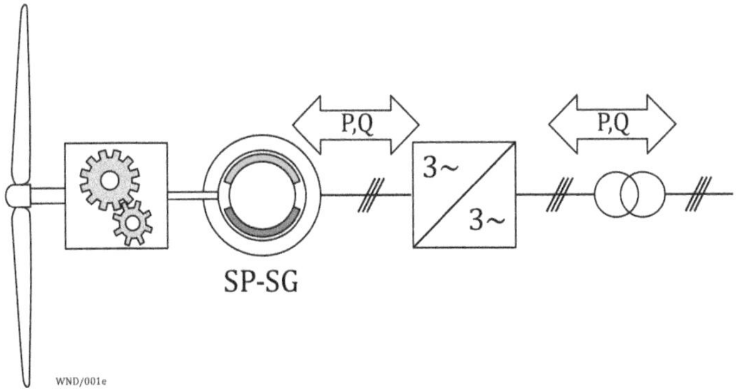

*Figure 2.44 Permanent magnet generator drive train concept*

No reluctance torque will be noticeable in those machines. The flux linkage equations are not different between direct and quadrature axes, except for the magnetic field from the magnets.

Equal to non-salient pole synchronous machines, a sinusoidal-shaped magnetic field in the air gap can be designed with the position and the high of the surface-mounted magnets. Magnets with different magnetization level and direction can be used alternatively, to achieve a sinusoidal flux density.

Preferably, we use only one magnet dimension. Then all magnets have the same height. Typical for this configuration is that the magnet is glued to the outer rotor surface. The construction is mechanically reinforced by the help of glass fiber, carbon fiber, or stainless steel bandages. The magnets in this construction are situated right next to the air gap. Any harmonic content in the air gap flux from stator slotting or from inverter voltage harmonics lead to eddy currents in the magnet material. Therefore, magnets must be segmented in axial or tangential direction.

Figure 2.45 shows an example for a surface-mounted magnet configuration using different magnet dimensions. We must consider that the use of magnets of different height or different magnetization-leads to extra cost. Typical for this configuration is that the magnets will be mounted in pockets situated near the rotor surface. The outer rotor surface then can be shaped additionally to support a sinusoidal field. For this configuration, no bandage is possible. Only the rotor pocket mechanical support the magnets. This magnet configuration has slightly better magnet loss behavior. The field magnets are not directly subjected to field harmonics. The rotor pockets must be designed as a compromise between high mechanical strength and very low magnetic conductivity. Otherwise, they short-circuit the magnetic field from the magnets.

Mounting the magnets inside of the rotor offers additional possibilities. If we arrange the magnets in V- or U-configuration, wider magnets can be used and a flux concentration can be achieved. Furthermore, two or three layers of magnets are possible with magnetic conductive parts between the layers. Magnets have much lower magnetic conductivity compared to laminated steel. In this way we can design magnetic conductive paths in d-axis and magnetic barriers in q-axis. This

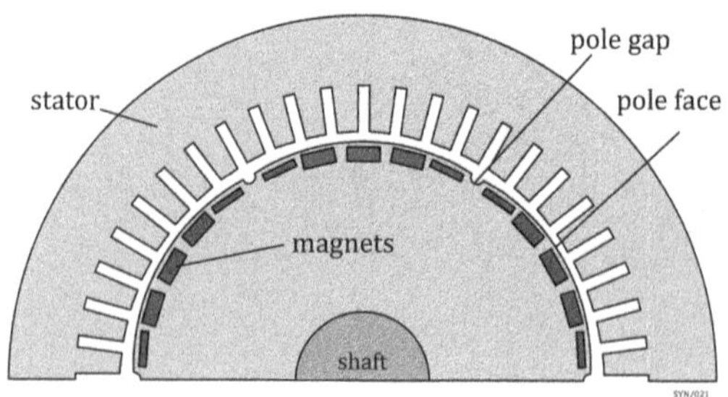


*Figure 2.45 Example of surface-mounted magnets using pockets near the rotor surface*

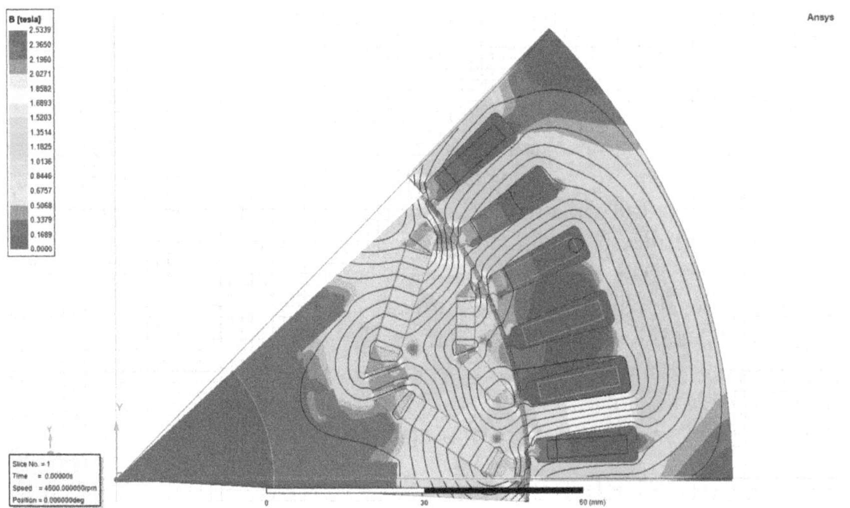


*Figure 2.46 Example of an interior PM synchronous machine with two magnet layers in V-arrangement in a typical high torque operating point*

leads to a considerable difference in *d*- and *q*-axis, resulting in a quite high reluctance torque (Figure 2.46).

Interior permanent magnet machines use the arrangement of the magnets and the shape of rotor to form a sinusoidal magnetic field in the air gap, comparable to the pole roof from SPSG. As the magnets are fixed inside of magnet pockets, a robust mechanical behavior and an easy assembling process can be assumed. However, for high-speed applications it is crucial to design the flux barriers and the flux paths with an eye on the mechanical strength. The remaining sheet material between the magnets must bear the entire centrifugal force from the magnets and the outer rotor material. Quite often, a compromise between electrical and mechanical behavior must be found. This makes the development somehow more complicated as for surface mounted magnets. Another point that we must mention here is that in contradiction to salient pole SG an interior permanent magnet generator has a good magnetic conductivity perpendicular to the pole and a bad conductivity with the pole. Therefore for interior PMSG applies

$$L_{\mathrm{d}} \leq L_{\mathrm{q}} \tag{2.141}$$

## 2.5.2 PMSG in stationary operation

The drive train concept has been introduced in Figure 2.44 already. The three-phase stator winding from the PMSG is connected to the low voltage inverter, typically in the 690 V class. The maximum stator voltage is then approximately 740 V. The possibility to use only a diode bridge between inverter and stator winding is not possible, because the field from the magnets is constant, resulting in a stator voltage being linear depending on the speed. The flux inside of the machine must be manipulated with the stator current. Current amplitude and phase angle can be controlled using either a field-oriented control (FOC) and space vector modulation (SVM) or a direct torque control (DTC) scheme. Open loop control is also possible, but untypical. The control scheme puts the stator current in the correct amplitude and angle in respect to the rotor position and torque (or power respectively) demand. For all control algorithms, it is important to know the actual rotor angle. This angle can be measured with position sensors (incremental encoders, variable reluctance resolvers) or estimated (sensorless control). This is an additional effort for PM machines in comparison to electrically excited machines.

All sizes, that is, current, voltage, and flux linkage, have to be transformed into rotor reference frame using the same equations as for the salient pole SM, Equations (2.126) and (2.127). We get then the same stator voltage equation, but now decomposed into d- and q-components:

$$U_{\mathrm{d}} = R_1 I_{\mathrm{d}} - \omega_1 \, \psi_{\mathrm{q}} \tag{2.142}$$

$$U_{\mathrm{q}} = R_1 I_{\mathrm{q}} + \omega_1 \, \psi_{\mathrm{d}} \tag{2.143}$$

The flux linkage in the stator winding is dependent from stator current and magnet flux

$$\psi_{\mathrm{d}} = L_{\mathrm{d}} \, I_{\mathrm{d}} + \psi_{\mathrm{p}} \tag{2.144}$$

$$\psi_{\mathrm{q}} = L_{\mathrm{q}} \, I_{\mathrm{q}} \tag{2.145}$$

If we put those equations together, we get the stator voltage space vector in rotor reference frame definition.

$$\underline{U}_1 = R_1 \, \underline{I}_1 + \mathrm{j}\omega_1 \psi_1 \tag{2.146}$$

with

$$\underline{\psi}_1 = \psi_{\mathrm{d}} + \mathrm{j} \, \psi_{\mathrm{q}} \tag{2.147}$$

and

$$\underline{I}_1 = I_{\mathrm{d}} + \mathrm{j} \, I_{\mathrm{q}} \tag{2.148}$$

We see that the stator frequency $\omega_1$ enlarges the stator voltage, but with a negative direct axis current, we can control the stator voltage as we act against the flux from the permanent magnets. We also see that the q-current, responsible for the torque, leads to an armature reaction. The torque can be calculated using:

$$M = \frac{3}{2} p \left( \psi_{\mathrm{d}} I_{\mathrm{q}} - \psi_{\mathrm{q}} I_{\mathrm{d}} \right) \tag{2.149}$$

This equation can be re-written to

$$M = \frac{3}{2} p \Big( \left( L_{\mathrm{d}} - L_{\mathrm{q}} \right) I_{\mathrm{d}} + \psi_{\mathrm{p}}^{'} \Big) I_{\mathrm{q}} \tag{2.150}$$

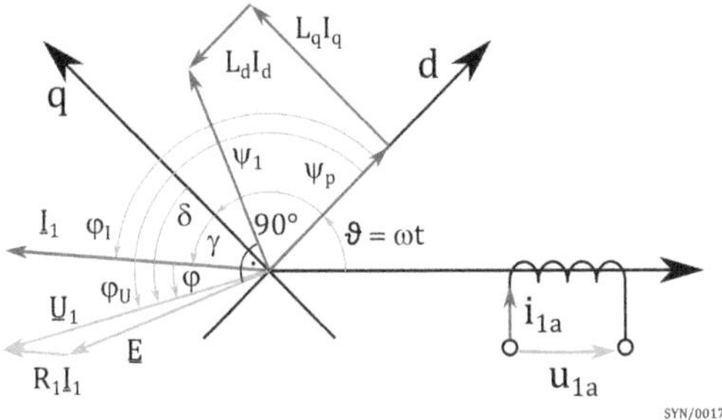


Figure 2.47 Definition of voltage and current phase angles in rotor coordinates

The q-current and the magnet flux linkage produce the main part of the electromagnetic torque. Therefore, the angle between the q-axis and stator current phasor is called the phase advance, which gives information about which proportion of the stator current value is transfer to torque and which to flux. Refer to Figure 2.47 to get an impression. In this figure, also the angle between stator voltage and $q$-axis can be seen. This angle is still called the "load angle" $\delta$, as the load, that is, the $q$-current, moves the stator voltage phasor away from the q-axis as it was described for the SPSG already.

From (2.150) it is also very obvious that a difference between $L_d$ and $L_q$ leads to an extra part of torque, comparable to the reluctance torque from SPSG. This torque only is produced by the field from the d-current, not by the magnetic field from the magnets. The magnets only interact with $I_q$.

For stationary operation, the damping effects in the rotor of the machine can be neglected. Any damping currents in the magnets or in conductive parts of the rotor are zero. This leads to the rotor voltage equations

$$0 = R_D I_D \tag{2.151}$$

$$0 = R_Q I_Q \tag{2.152}$$

This does not mean that in stationary operation all rotor losses are zero. We only consider the fundamental current and neglect harmonic components. Those harmonic components lead to flux variations that induce iron losses in the rotor core and eddy current losses in the magnets.

### 2.5.3 Dynamic operation

Even if there is no explicit damper winding in the rotor, the magnets themselves or parasitic circuits in the iron may bear small damping currents in the case of flux variation in the rotor. Therefore, in dynamic operation damper windings must be considered. Figure 2.48 shows a very simple model of a synchronous machine with surface mounted magnets including a theoretical representation of two damper windings along and perpendicular to the magnet pole.

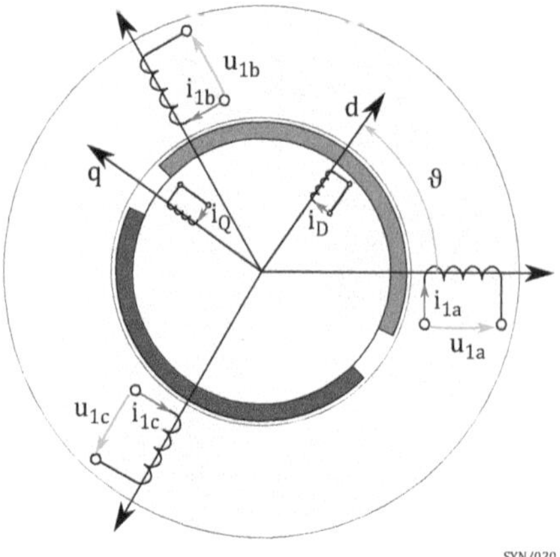


*Figure 2.48 PM synchronous machine with surface mounted magnets and damper winding representation in d and q axes*

The stator differential equation is exactly the same as for the synchronous machine with salient pole rotor. We apply the same Park transformation to transform from the three-phase system into *d–q*-reference frame. The following equations apply in rotor reference frame without any special marking.

$$u_{\mathrm{d}} = R_1 i_{\mathrm{d}} + \frac{\mathrm{d}\psi_{\mathrm{d}}}{\mathrm{d}t} - \omega_{\mathrm{m}}\,\psi_{\mathrm{q}} \tag{2.153}$$

$$u_{\mathrm{q}} = R_1 i_{\mathrm{q}} + \frac{\mathrm{d}\psi_{\mathrm{q}}}{\mathrm{d}t} + \omega_{\mathrm{m}}\,\psi_{\mathrm{d}} \tag{2.154}$$

In these equations $\omega_{\mathrm{m}}$ is the speed of the shaft calculated from electrical degrees, which is transformed by the pole pair number from the rotational speed. Compare (2.130).

The rotor differential equations, that is, the voltage equations for the two fictitious damper windings, can be written directly in rotor coordinates, as we align all equations with the rotor pole. The only thing needed is to transform the rotor sizes using Table 2.10 with reference to the stator winding. We get

$$0 = R'_{\mathrm{D}}\, i'_{\mathrm{D}} + \frac{\mathrm{d}\,\psi'_{\mathrm{D}}}{\mathrm{d}t} \tag{2.155}$$

$$0 = R'_{\mathrm{Q}}\, i'_{\mathrm{Q}} + \frac{\mathrm{d}\,\psi'_{\mathrm{Q}}}{\mathrm{d}t} \tag{2.156}$$

The flux linkage equations look equal to those of the salient pole SG. The only difference is that we do not have a field winding. For this purpose $\psi'_{\mathrm{p}}$ is in the equations, representing the flux from the magnets and we do not find a field current or field voltage.

$$\psi_{\mathrm{d}} = L_{\mathrm{d}}\, i_{\mathrm{d}} + L_{\mu\mathrm{d}}\, i'_{\mathrm{D}} + \psi'_{\mathrm{p}} \tag{2.157}$$

$$\psi_q = L_q\, i_q + L_{\mu q}\, i_Q^{'} \tag{2.158}$$

$$\psi_D^{'} = L_{\mu d}\, i_d + L_D^{'}\, i_D^{'} + \psi_p^{'} \tag{2.159}$$

$$\psi_Q^{'} = L_{\mu q}\, i_q + L_Q^{'}\, i_Q^{'} \tag{2.160}$$

The electromagnetic torque can be calculated with the common equation, which we have seen a couple of times now.

$$m = \frac{3}{2} p \left(\psi_d i_q - \psi_q i_d\right) \tag{2.161}$$

Last but not least we can calculate the speed if we know the load torque.

$$\frac{d\,\omega}{dt} = \frac{p}{J_m}\left(m - m_{load}\right) \tag{2.162}$$

## List of indices

| Index | Description | Index | Description |
|---|---|---|---|
| 1 | stator | $m$ | mechanical (related to shaft) |
| 2 | rotor | $n$ | nominal value |
| b | breakdown | $\sigma$ | leakage |
| load | load | $\delta$ | air gap |
| K | universal reference frame | $\mu$ | magnetizing |
| L | loss | $\alpha$ | real part in the stationary reference frame |
| Lfe | iron loss | $\beta$ | imaginary part in the stationary reference frame |
| Lhf | high-frequency harmonic losses | $x$ | real part in the universal reference frame |
| Lw | winding loss | $y$ | imaginary part in universal reference frame |
| Lwf | windage and friction loss | $d$ | real part in the rotor reference frame |
| Lbr | brush loss | $q$ | imaginary part in the rotor reference frame |
| LL | Load-dependent additional loss | $r$ | rated value (base for per-unit) |
| Linv | inverter losses | $s$ | synchronous |
| ph | phase | $t$ | tangential |

## List of symbols

| Symbol | Unit | Description |
|---|---|---|
| $f$ | Hz | frequency |
| $i$ | – | Gear ratio |
| $j$ | – | Imaginary unit $\sqrt{-1}$ |
| $k$ | Nm min$^2$ | Ratio between optimum hub torque and squared hub speed |
| $m$ | Nm | dynamic (electromagnetic) torque |

*(Continues)*

(*Continued*)

| Symbol | Unit | Description |
|---|---|---|
| $n$ | $\text{min}^{-1}$ | Rotational speed |
| $p$ | – | Pole pair number |
| $p$ | kW | instantaneous active power |
| $q$ | kVar | instantaneous reactive power |
| $s$ | % | Slip |
| $s$ | kVA | instantaneous apparent power |
| $t$ | s | Time |
| $J$ | kg $\text{m}^2$ | Total inertia |
| $I$ | A | Current |
| $K$ | – | Transformation ratio for rotor sizes |
| $M$ | Nm | Torque |
| $N$ | – | Number (integer value) |
| $P$ | kW | Power (active power) |
| $Q$ | kVar | Reactive Power |
| $S$ | kVA | Apparent Power |
| $U$ | V | Voltage |
| $\sigma$ | – | total leakage coefficient |
| $\omega$ | $\text{s}^{-1}$ | rotational frequency |
| $\psi$ | Wb | flux linkage |
| $\vartheta_{\text{m}}$ | ° | Mechanical rotor angle |

## References


[1] P. Schubert, Zur Genese von Theorie und Berechnung elektromagnetischer Energiewandler als Leitdisziplin bei der Hausbildung der Elektrotechnik als technikwissenschaftliche Disziplin, Dresden: TU Dresden, 1984.

[2] Bundesverband WindEnergie e.V. [Online]. Available: https://www.wind-energie.de/fileadmin/_processed_/7/f/csm_teilintegrierte-bauweise_295da44c9e.jpg. [Accessed 15 June 2023].

[3] G. L. Johnson, *Wind Energy Systems*, Engelwood Cliffs, NJ: Prentice-Hall, 1985.

[4] A. Betz, *Introduction to the Theory of Flow Machines*, Oxford: Pergamon Pres, 1966.

[5] A. Binder, *Elektrische Maschinen und Antriebe*, Berlin: Springer-Verlag, 2011.

[6] M. v. Dolivo-Dobrowolsky, "Regulation of alternating-current motors". United States Patent 503,038, 8 August 1893.

[7] *DIN EN 50160:2020-11*, Beuth Verlag, 2020.

[8] B. Ponick, G. Müller and K. Vogt, *Berechnung elektrischer Maschinen*, Weinheim: Wiley-VCH Verlag, 2008.

[9] *DIN EN 60034-2-3: 2021-07, Drehende elektrische Maschinen, Teil 2-3: Besondere Verfahren zur Bestimmung der Verluste und des Wirkungsgrades von umrichtergespeisten Wechselstrommaschinen*, VDE Verlag, 2021.

[10] *DIN VDE 0530-25 (VDE 0530-25): 2018-12, Drehende elektrische Maschinen, Teil 25 – Wechselstrommaschinen zur Verwendung in Antriebssystemen – Anwendungsleitfaden*, VDE Verlag, 2018.

[11] "IEC/TS 61800-8:2010 "Adjustable speed electrical power drive systems, Part 8: Specification of voltage on the power interface," IEC, 2010.

[12] M. Kloss, "Schaltungsweise für Drehstrommotoren zur Erziehlung zweier verschiedener Geschwindigkeiten". Deutschland Patent DE000000109986A, 20 Juni 1899.

[13] P. C. Dent, "Rare earth elements and permanent magners," *Journal of Applied Physics* 111, 2012.

[14] J. Goss, M. Popescu and D. Staton, "A comparison of an interior permanent magnet and copper rotor induction motor in a hybrid electric vehicle application," in *International Electric Machines & Drives Conference*, Chicago, IL, USA, 2013.

[15] H. Toda, Z. P. Xia, J. B. Wang, *et al.*, "Rotor eddy-current loss in permanent magnet brushless machines," *IEEE Transactions on Magnetics*, pp. 2104–2106, 2014.

# Chapter 3
# Generator design for direct-drive turbines

*Tobias Muik[1] and Stephan Jöckel[1]*

## 3.1 Introduction

Direct-drive systems that rotate at the same speed as the driven/driving device and avoiding any gear transmission are a usual choice when aspects like robustness, longevity, and reliability play a major role. In wind energy, they had been introduced in the early 1990s and could gain a global market share of approximately one third of the annual installed capacity.

The following chapter about direct-drive generator systems for wind turbine applications deals with the main aspects which determine the design of such generators, focusing on solutions with permanent-magnet excitation. Even though a considerable amount of current-excited direct-drive generators has been installed in the last 30 years – especially by the German pioneering company *ENERCON* (see chapter "Drivetrain concepts and developments" of Volume 1) – their latest generation favors permanent magnets.

After an insight into typical design considerations and the basic physics behind these generators, the chapter describes optimization strategies to achieve a minimum cost of energy (LCOE) of wind turbines with direct-drive generators. One section tries to clear the view on the "future of the direct drive in wind energy" by simplified scaling of existing solutions to rotor diameters of 300 m and power ratings of more than 20 MW that may appear on the market in the near future.

Touching on mechanical and cooling aspects, vibro-acoustics and production, the clear focus of this chapter lies on the electro-magnetic design. Starting with an overview of possible stator winding systems and their benefits as well as drawbacks, the chapter continues with different rotor configurations for permanent magnet excitation. A comparison between different rotor structures is given. The focus is set on one of the promising rotor structures – the V-shape configuration – which is further explained in detail and the associated optimization potentials are pointed out.

[1]Wind-direct GmbH, Germany

The chapter concludes with a short summary and an outlook into the possible future of direct-drive generator systems for wind turbine applications.

## 3.2 Basic aspects for designing direct-drive generators for wind turbines

In the past 45 years since the beginning of commercial wind energy in the late 1970s, wind turbines have dramatically grown in size and power rating and could achieve a tremendous reduction by factor 5 in the levelized cost of energy (LCOE) to current levels below 0.04 €/kWh, making it one of the cheapest sources of electricity besides solar PV.

The main driving factor for designing wind turbines as well as their components such as generators is to further reduce the LCOE, which can be generally minimized by maximizing annual energy production (efficiency, especially in partial load) and minimizing the overall cost. Even though permanent magnet materials are still so expensive that they can contribute to as much as one-quarter of the total generator cost (strongly depending on design), the superior efficiency of permanent-magnet generators (PMG) over all other generator types leads to the situation that it has become the preferable choice in wind energy, especially in direct-drive applications.

The second important aspect is to minimize the overall cost, which consists of the following parts:

- Material and production cost (CAPEX)
- Cost for transport and installation
- Operating expenses such as service and repair cost (OPEX)

Whereas the first aspect – minimization of CAPEX – needs no further explanation, the second aspects becomes one of the dominant factors for designing wind turbine components, since the dimensions and masses have grown tremendously. The situation may be slightly different for offshore applications, but especially for onshore wind turbines the question how to transport and install the device has become one of the most important ones.

One of the main aspects for minimizing the OPEX part is to maximize reliability of all major components (at minimum cost) and the possibility to repair certain faults and exchange certain parts without the necessity of large cranes or installation vessels.

The following sections will try to focus on the basic aspects of designing direct-drive generators for wind turbines, resulting from the need to further minimize the levelized cost of energy. It will concentrate on permanent-magnet generators since this has become the preferred choice in wind energy applications.

### *3.2.1 Basic design requirements*

The main aspects of this section have been presented in [1] for a 3 MW turbine. This section updates certain details and tries to list the most important items of the specification of an imaginary 8 MW onshore wind turbine with a rotor diameter of 190 m and a direct-drive generator system.

#### 3.2.1.1 Main dimensioning wind turbine data

The main specification of the wind turbine under development is essential for the design and must correspond to the requirements of the turbine manufacturer (client) in relation to the generator manufacturer (supplier). Table 3.1 shows example specifications for the considered turbine.

#### 3.2.1.2 Specifying the generator

In order to dimension any generator, e.g., the rated values for apparent power $S_N$, speed $n_N$ (taking into consideration a possible gear unit), voltage $U_N$ (rms value of phase-to-phase voltage), frequency $f_N$, power factor cos $\varphi_N$, have to be stipulated apart from the number of phases $m_s$. Table 3.2 shows exemplified specifications for a generator under development.

#### 3.2.1.3 Specifying the permanent-magnet material

Permanent-magnet materials of high magnetic energy density like Neodymium-Iron-Boron (NdFeB) allow for a simple, brushless rotor design of synchronous machines. Thus, not only excitation system and slip ring can be omitted, but also the excitation losses, resulting in considerably increased efficiency compared to conventional competitors. The disadvantage of non-adjustable excitation of the permanent magnets is of minor importance in variable-speed machines connected to frequency converters.

50 years ago, the economic power range for permanent-magnet motors had been seen below 5 kW because of moderate energy density and extremely high

*Table 3.1 Considered turbine specifications*

| Criteria | Figures | Remarks |
|---|---|---|
| Turbine rotor diameter | 190 m | |
| Rated power grid side | 8.0 MW | Transformer losses to be considered |
| Rated rotational speed | 8.00 rpm | Rated blade tip speed <80 m/s due to reasons of noise emission |
| Overspeed during normal operation | 9.20 rpm | Maximum operational overspeed of +15 % in very gusty winds |
| Maximum possible turbine speed | 10.00 rpm | Rough estimation: overspeed of 15% and safety margin of 10%, (loss of load when running at normal overspeed); detailed figure would be determined during the load calculations of the turbine |
| Peak torque requirements | | Maximum over-torque capacity of 10% may be sufficient because of the blade pitch regulation of the wind turbine |
| Ambient temperature range for operation | −15 to +20°C | Typical range for normal operation, maximum temperature affects the cooling system design |

*Table 3.2 Generator specifications*

| Criteria | Figures | Remarks |
|---|---|---|
| Transport and production dimensions | Maximum transport width <5.0 m | Small outer dimensions simplify production (e.g., size of the impregnation tank), transportation and integration into the turbine nacelle. |
| Generator mass (transport and lifting unit) | <80 t | Low mass is also important; components are mounted on top of tower in different lifts; each lift should not exceed a certain limit in order to use commonly available cranes. |
| Voltage level | 800 V rated generator voltage (Y connection) | Typical voltage level of wind generators |
| Number of phases | 3 or 6 | Common phase numbers |
| Stator winding | Form-wound coils with insulation class F | Usual choice for IGBT rectifiers, round wire winding may be difficult to insulate |
| Cooling method | Forced air cooling using filtered outside air | Typical choice for onshore generators |
| Protection class | IP23–IP54 | Typical range of values in wind energy |

cost. However, recently the cost of permanent-magnet materials (especially NeFeB) went down dramatically: Whereas Samarium-Cobalt did cost about 500 €/kg in early 1990, less than 60 €/kg have to be paid for magnets based on Neodymium-Iron-Boron presently (with minima of below 30 €/kg around 2015), with some heavy fluctuations in recent years. However, both cost as well as availability of permanent magnet materials are currently under heavy discussion, since China still dominates this market to over 90%. The production cost of direct-drive permanent-magnet electrical machines is still dominated by the magnet cost. Thus, minimizing the necessary magnet mass remains one of the main goals of the generator design, also considering strategic aspects as raw material availability and environmental questions of mining the raw materials.

Besides remanent flux density $B_r$ and coercive field strength $H_{cB}$, the maximum of remanence times coercivity $(B\,H)_{max}$, the so-called *BH* product or energy product, is characterizing a certain magnet material. Figure 3.1 presents historical development of permanent magnets concerning their *BH* product.

In the past 40 years, the achievable energy density of NdFeB magnets could be increased largely. By applying special production technology, permanent magnet suppliers could almost double the energy density from 200 kJ/m$^3$ (magnet quality "30" for 30 MGOe) values of more than 400 kJ/m$^3$ (magnet quality "54"), Figure 3.2. However, the highest magnet qualities typically require the addition of heavy rare earth materials such as Dysprosium (Dy), Terbium (Tb) or Holmium (Ho) so that the cost may increase over-proportionally. For direct-drive wind power applications, magnet qualities without or with a very small addition of heavy rare earth materials are usually chosen.

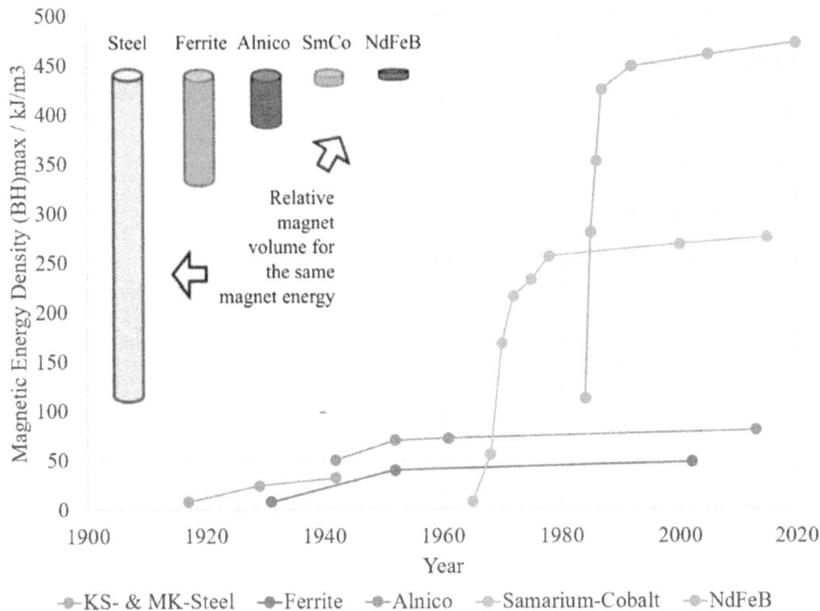


*Figure 3.1 Development of the energy density of permanent magnets within the last 100 years*

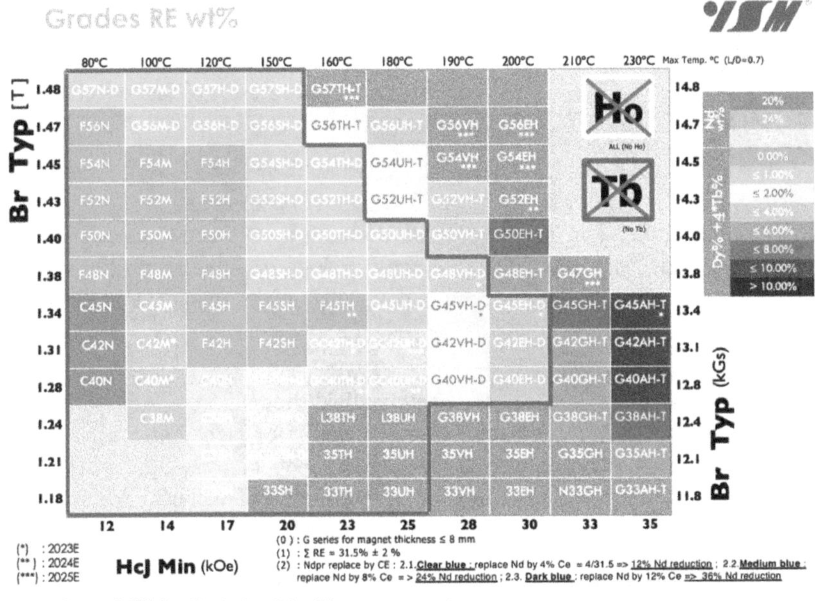

Grades RE wt%

YSM

| Br Typ [T] | 80°C | 100°C | 120°C | 150°C | 160°C | 180°C | 190°C | 200°C | 210°C | 230°C | Br Typ (kGs) |
|---|---|---|---|---|---|---|---|---|---|---|---|
| 1.48 | G57N-D | G57M-D | G57H-D | G57SH-D | G57TH-T *** | | | | | | 14.8 |
| 1.47 | F56N | G56M-D | G56H-D | G56SH-D | G56TH-T | G56UH-T | G56VH *** | G56EH *** | | | 14.7 |
| 1.45 | F54N | F54M | F54H | G54SH-D | G54TH-D | G54UH-T | G54VH *** | G54EH *** | | | 14.5 |
| 1.43 | F52N | F52M | F52H | G52SH-D | G52TH-D | G52UH-T | G52VH-T | G52EH ** | | | 14.3 |
| 1.40 | F50N | F50M | F50H | G50SH-D | G50TH-D | G50UH-D | G50VH-T | G50EH-T | | | 14.0 |
| 1.38 | F48N | F48M | F48H | G48SH-D | G48TH-D | G48UH-D | G48VH-D * | G48EH-T | G47GH *** | | 13.8 |
| 1.34 | C45N | C45M | F45H | F45SH | F45TH ** | G45UH-D | G45VH-D * | G45EH-D * | G45GH-T | G45AH-T * | 13.4 |
| 1.31 | C42N | C42M* | F42H | F42SH | [illegible] | [illegible] | G42VH-D | G42EH-D | G42GH-T | G42AH-T | 13.1 |
| 1.28 | C40N | C40M* | [illegible] | [illegible] | [illegible] | [illegible] | G40VH-D | G40EH-D | G40GH-T | G40AH-T | 12.8 |
| 1.24 | | C38M | [illegible] | | L38TH | L38UH | G38VH | G38EH | G38GH-T | G38AH-T | 12.4 |
| 1.21 | | | | [illegible] | 35TH | 35UH | 35VH | 35EH | G35GH | G35AH-T | 12.1 |
| 1.18 | | | | 33SH | 33TH | 33UH | 33VH | 33EH | N33GH | G33AH-T | 11.8 |
| HcJ Min (kOe) | 12 | 14 | 17 | 20 | 23 | 25 | 28 | 30 | 33 | 35 | |

Max Temp. °C (L/D=0.7)

Ho — ALL (No Ho)

Tb — (No Tb)

Nd wt%: 20%, 24%

Dy% + 4*Tb%: 0.00%, ≤ 1.00%, ≤ 2.00%, ≤ 4.00%, ≤ 6.00%, ≤ 8.00%, ≤ 10.00%, > 10.00%

(*) : 2023E
(**) : 2024E
(***) : 2025E

(0) : G series for magnet thickness ≤ 8 mm
(1) : Σ RE = 31.5% ± 2 %
(2) : Ndpr replace by CE : 2.1.**Clear blue** : replace Nd by 4% Ce = 4/31.5 => 12% Nd reduction ; 2.2.**Medium blue** : replace Nd by 8% Ce = > 24% Nd reduction ; 2.3. **Dark blue** : replace Nd by 12% Ce => 36% Nd reduction

Property of : **YSM** - Yantai Dongxing Magnetic Materials Inc — www.ysm-magnet.com

*Figure 3.2 Table of different permanent-magnet materials. © Yantai Dongxing Magnetic Materials Inc.*

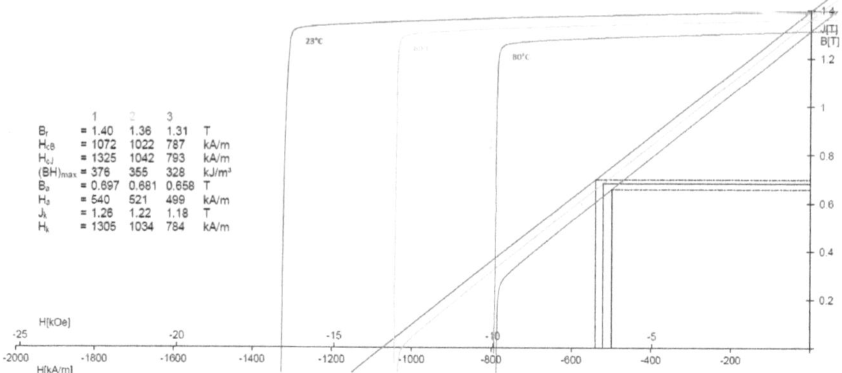


*Figure 3.3 Typical B–H characteristics of a permanent-magnet material based on NeFeB*

*Table 3.3 Magnet specifications*

| Criteria | Characteristics | Remarks |
|---|---|---|
| Magnet material | NeFeB | Lowest specific cost compared to any other magnet material in a range from 30 to 100 €/kg |
| Minimum coercivity | $H_C > 1000$ kA/m at 60 °C | This figure should be sufficient to withstand demagnetization during a short circuit at the maximal operating temperature of 60 °C |
| Average remanent flux density | $B_r > 1.42$ T at 20 °C | This rather high value can be used because of the low temperatures of the external rotor design, reducing the risk of irreversible demagnetization |

Permanent magnets can also be characterized by their hysteresis characteristic. In the second quadrant as shown in Figure 3.3 (where $H$ is negative and $B$ positive), this characteristic is called demagnetization curve, which is of special interest, since the magnet is operated there. Its linear part with the slope $\mu_r$ is the reversible load line, characterizing normal magnetic working points without irreversible demagnetization:

$$B_m = B_r + \mu_0 \mu_r \quad H_m \tag{3.1}$$

Besides remanent flux density and coercive field strength, the corrosion resistance of the magnet material in aggressive environments is of great importance in wind power. Regarding this, development of the magnet suppliers made great advantages in the past which is still ongoing. However, it is quite problematic to specify this in certain figures. The basic specifications may be summarized in Table 3.3.

#### 3.2.1.4 General frequency converter specifications

Apart from converting the frequency-variable electric power, the main functions of the rectifier are to influence the generator in order to control the power output and to protect and disconnect the generator from the grid in case of faults.

*Table 3.4 IGBT converter requirements*

| Criteria | Figures | Remarks |
|---|---|---|
| Maximum possible voltage generator side | | Exact figure depends on the generator overspeed at load loss, the generator power factor (thus on the applied rectifier!), inductances, and shape of the generator terminal voltage |
| Maximum required current generator side | $I_{max} = 1.1\ I_N$ | Maximum over-torque capacity of 10% may be sufficient because of pitch regulation |
| Power factor generator side for IGBT rectifier | $0.65 < \cos\varphi < 0.9$ inductive | Typical range, IGBT rectifier allows for flux orientated control; exact figure depends on the economical optimization for minimum LCOE |
| Power factor line side | $\cos\varphi > 0.9$ inductive | Usual specification |
| Maximum possible voltage line side | $U_{max} = 1.1\ U_N$ | Less critical than generator side due to line-side switch gear which can be switched off |
| Harmonic distortion grid side | | Latest IEC standards have to be met by connecting EMI filters between converter and transformer |
| Maximum voltage rise generator side | $dU/dt < 1$ kV/μs | Due to insulation and bearing stress, will be achieved by additional filters between converter and generator |
| Cable length between generator and rectifier | | Depends on the converter placement, whether inside the nacelle or at the tower bottom |
| Outer dimensions | | Depends on the converter placement, whether inside the nacelle, inside the tower at the tower bottom or outside in a container |

The power electronics technology for large-power AC drives during the last 50 years has been shifting from current source inverters (CSI) to voltage source inverters (VSI). Presently, the most used architecture is the VSI with six diodes and six IGBTs. The topology is driven by the power semiconductors IGBT, which are in a steady progress: The power losses are decreasing, they become more robust and the associated converter cost could be decreased tremendously. Refer to Chapter 4 of this book for more details.

Thirty years back, many generator systems for variable-speed wind turbines still featured converter systems with passive diode rectifiers. Due to the advantages of controlled IGBT rectifiers (mainly the possibility to feed the generator with reactive power), and the constant cost reductions of IGBTs, today's architectures are based on four-quadrant IGBT converters – especially when combined with permanent-magnet generators. Typical requirements of four-quadrant IGBT converters when combined with permanent-magnet generators are listed in Table 3.4.

### *3.2.2 Major aspects for minimizing the levelized cost of energy (LCoE)*

As summarized at the beginning of this chapter, the major challenge for any generator designer is to further minimize the energy cost. These costs can be influenced by numerous design variables – and this section tries to list and explain the most important of them.

#### 3.2.2.1 Generator efficiency

Even though any designer would be aiming to minimize the generator mass and associated production cost (influencing the capital expenses CAPEX), there is one more important aspect influencing the cost of energy: the annual energy production (AEP) which can be largely influenced by the generator efficiency. As has been shown in many publications, the main decisive factor in wind energy applications is a high partial-load efficiency: turbines would "pitch away" the surplus energy in high winds and so the rated efficiency is much less important.

However, experience in direct-drive generator design has revealed that super-effective cooling systems do typically not pay off: the reduction in CAPEX is (more than) eaten up by loss in AEP and the additional complexity of the cooling systems, for example, additional maintenance cost. This would explain why the vast majority of successful direct-drive generators (especially PMGs) does not rely on complicated liquid cooling – proven and robust air cooling is widely preferred.

The above statement differs from the findings in [1], where the best cooling concept yielded the lowest cost of energy. However, factors like installation cost and interest rates which lead to the fact that AEP is the decisive factor in the calculation of levelized cost of energy (LCOE) had been neglected in [1].

#### 3.2.2.2 Generator airgap force density

As, for instance, shown in [1], the following basic equation is applicable for the electromagnetic torque $M_\mathrm{e}$ of an electrical machine:

$$M_e = V_\delta \hat{A}_s \hat{B}_{\delta\mathrm{r}} \cos \psi \tag{3.2}$$

Thus, the torque depends on active volume $V_\delta$, stator electric loading $\hat{A}_s$, airgap flux density $\hat{B}_{\delta\mathrm{r}}$, and internal power factor cos $\psi$. For evaluating an electrical machine, indication of a tangential force density $\tau_{\mathrm{tan}}$ in the airgap may be very helpful. This force density $\tau_{\mathrm{tan}}$ results from division of total tangential force $F_{\mathrm{tan}}$ by lateral surface $A_\delta$ of stator bore:

$$\tau_{\mathrm{tan}} = \frac{F_{\mathrm{tan}}}{A_\delta} = \frac{2M_e}{d} \frac{1}{\pi\, d\, l} \tag{3.3}$$

By inserting (3.2) into (3.3), it follows that the force density only depends on the electromagnetic utilization of the machine:

$$\tau_{\mathrm{tan}} = \frac{1}{2} \hat{A}_s \hat{B}_{\delta\mathrm{r}} \cos \psi \tag{3.4}$$

Thus the tangential force density is determined by the electric loading of the stator $\hat{A}_s$, airgap flux density $\hat{B}_{\delta\mathrm{r}}$, and internal power factor cos $\psi$ (angle between stator electric loading and rotor field). All of the three parameters can be influenced by the generator designer:

- Electric loading of stator largely depends on the intended cooling system of the generator, but may also be chosen larger with increasing diameter (and consequently deeper slots): $\hat{A}_s \approx 40$ to 110 kA/m.
- Airgap flux density: With respect to eddy current and hysteresis losses in the stator core, the airgap flux density in high-speed electrical machines should not be too high; in low-speed generators herein described, however, the absolute upper limit should be aimed at: $\hat{B}_{\delta r} \approx 1.1$ T.
- The *internal* power factor $\cos\psi$ depends on the connected rectifier type and varies between 1.0 (for active IGBT rectifiers) and much lower values, if passive diode rectifiers are applied. However, the continuous dramatic technology improvements in the semi-conductor area lead to the fact that the vast majority of modern wind turbines uses controlled IGBT rectifiers today – see Chapter 4 of this book for details. The overall generator-converter system is usually optimized with respect to cost and efficiency so that minimum system LCOE can be achieved. Only "under-excited" direct-drive generator designs with an *external* generator side power factor in the range of $0.65 < \cos\varphi < 0.9$ are economically useful.

Using the latest high-energy permanent magnets, modern direct-drive generators for large wind turbines achieve typical airgap force density values of 70 or 80 kN/m$^2$ even with simple direct air cooling. Some designs even feature values of 100 kN/m$^2$ and above – however, the overall goal to achieve minimum system LCOE should always be considered:

- Does the specific design need complicated cooling to dissipate huge losses (does it sacrifice efficiency and annual energy production?)
- Does the specific design need a large amount of expensive permanent-magnet material (does it need excessive CAPEX, also considering the high volatility of magnet prices?)

### 3.2.2.3 Required generator diameter

#### *3.2.2.3.1 Mass of the active part*

The active part is the most expensive part of electrical machines, which applies particularly for the permanent magnets. For low-speed multi-pole generators, the mass of this active part $m_{act}$ is more or less proportional to their lateral airgap surface $A_\delta$

$$m_{act} \sim A_\delta = \pi\, d_\delta\, l \tag{3.5}$$

Looking at (3.5), the core length $l$ can be replaced by the rated torque $M_e$ and the tangential force density $\tau_{tan}$:

$$m_{act} \sim \pi\, d\, l = \pi\, d_\delta \frac{2M_e}{\pi\, d^2\, \tau_{tan}} = \frac{2M_e}{d\, \tau_{tan}} \tag{3.6}$$

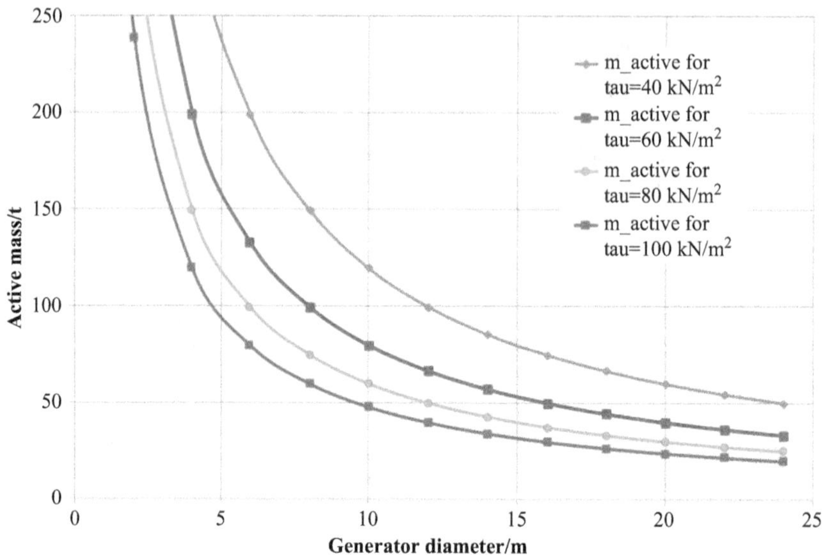


*Figure 3.4 8 MW generator: active mass vs. diameter for different force density values*

It is obvious that for given torque $M_e$ and tangential force density $\tau_{tan}$ the active mass is inversely proportional to the active diameter like shown in Figure 3.4. In order to minimize the active mass and associated cost, the largest possible active diameter should be used. However, it can be also clearly seen from Figure 3.4 that it seems not to be reasonable to go for diameters of more than approximately 12 m, if high tangential force densities can be achieved. Moreover, the economic optimum will again not be found for extremely large diameters, because additional structural, manufacturing and transportation costs will certainly overcompensate the gain in active mass reductions which is depicted in Figure 3.5.

#### *3.2.2.3.2 Mass and cost of the structural part*

The active masses reduce with increasing diameter – however, it is obvious that this will not be the case for the structural part! Whatever the exact values at low diameters, it is obvious that weight would be increasing with increasing diameter, since it becomes more and more challenging to ensure airgap stiffness as well as structural integrity at larger diameters. It is now the task of the generator designer to identify the diameter giving the lowest cost of energy.

A typical example is displayed in the following Figure 3.6 for a generator rating of 8 MW. It becomes obvious that an economical diameter would be well above the transportation limit of 5 m.

The exact amount of material (and even more the associated cost) largely depends on the design of the generator structure. Over the course of the last

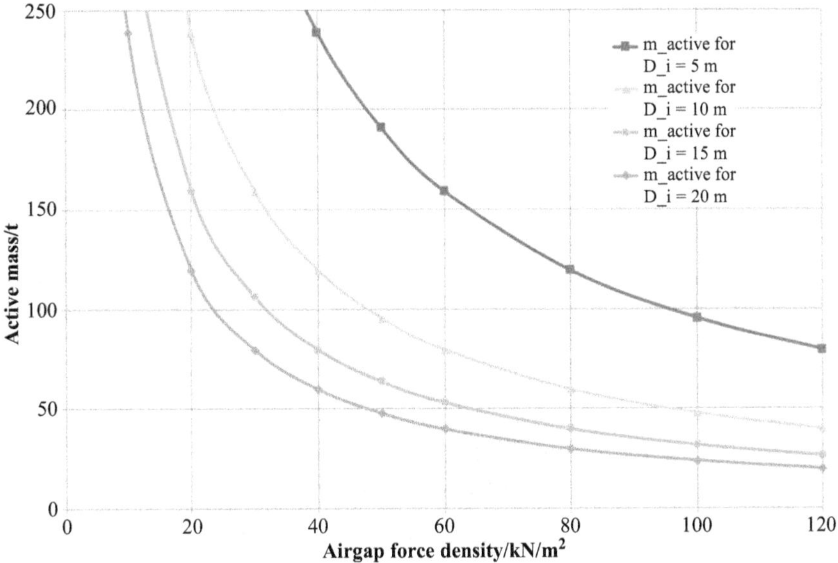


*Figure 3.5 8 MW generator: active mass vs. force density for different diameter values*

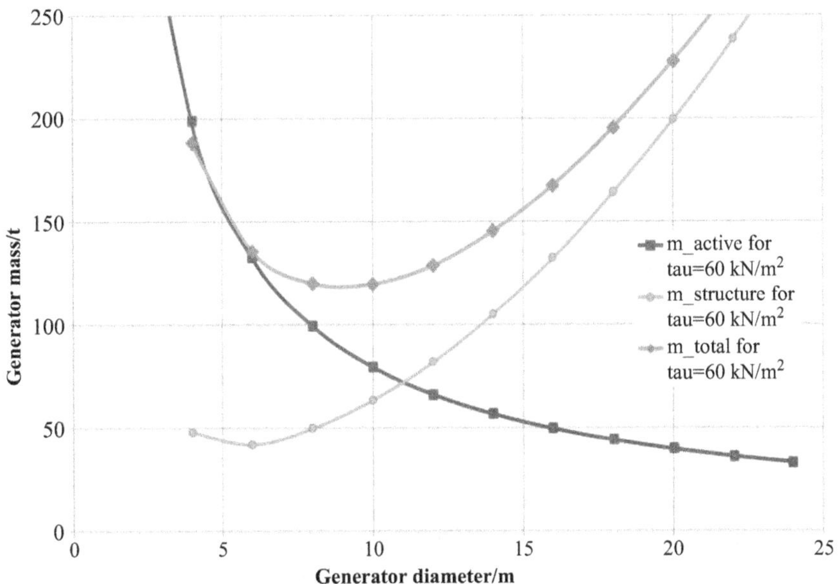


*Figure 3.6 8 MW generator: active, structural, and total generator mass vs. diameter (indicative)*

40 years, numerous solutions have been developed and successfully applied to the market:

- Compact designs with smaller diameter based on disc-shaped structures (onshore diameters typically less than approximately 5.0 m for simplified transportation in one piece)
- Compact disc-shaped designs with medium diameter based on self-supported laminated stacks (onshore diameters typically less than approximately 5.5 m for simplified transportation in one piece)
- Larger designs based on ring/support arm structures, with splitting options for higher ratings (and diameters), so that the transport dimensions can be still kept below 5 m width.

#### *3.2.2.3.3 Transportation limits*

In order to investigate directly driven generators under realistic conditions, the outer diameter has to be restricted to appropriate values. Two main limits of the loading gauges for road transportation can be stated:

- If the height of the transported good (plus low loader) remains under 4.3 m and its width under 3 m, it can be transported almost without any restrictions, yielding a maximum outer diameter of about 3.8 m and a total length of about 2.5 m, the generator being in upright position.
- Assuming the same height as before and a width of less than about 5.2 m, it can be transported on a special low loader escorted by police, yielding a maximum outer diameter of about 5 m and a total length of about 3.8 m, if the generator is transported in horizontal position.
- Any transported good having larger dimensions can only be shipped at exorbitant cost over road.

It becomes obvious that large-scale direct-drive generators with ratings of more than 6 MW will only become competitive by applying diameters of more than 5 m. In the onshore case, this leads to the requirement to split/segment the generator into pieces to avoid transport restrictions – be it for the width of more than five meters or excessive weight of more than 80 t.

#### *3.2.2.3.4 Outer rotor design for permanent-magnet generators*

For permanent-magnet machines, there exists another possibility to increase the active diameter for a limited outer diameter: the outer-rotor design. Compared to the thick stator part with its deep stator teeth, the rotor tube (with surface-mounted permanent magnets) is at least 100 mm less thick than the stator as illustrated in detail in Figure 3.7. Assuming a limited outer diameter, an outer rotor therefore leads to the larger airgap diameter of approximately 200 mm.

Concerning the demagnetization risk of the permanent-magnets, the outer-rotor concept is advantageous, too: The magnets are effectively cooled by the surrounding wind. This effect can be enhanced by fins welded onto the external surface of the rotor to increase its stiffness. This leads to an increased heat exchanging surface and thus a higher thermal heat flow can be realized, leading to low operational magnet

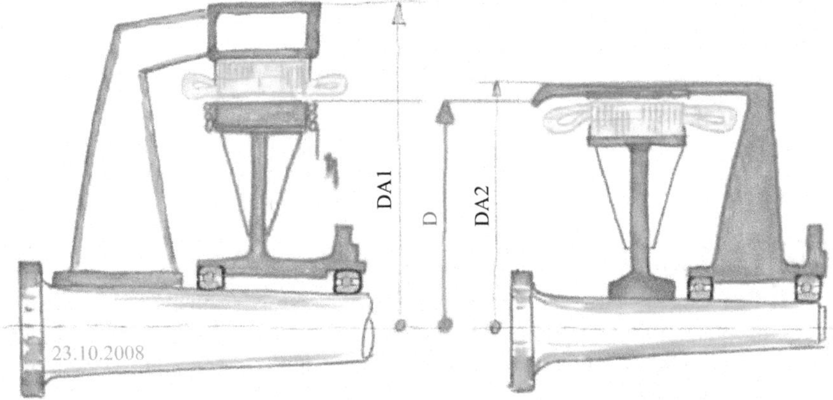


*Figure 3.7 Comparison of inner (left) and outer rotor: for identical airgap diameter D, the outer rotor becomes more compact with smaller outside diameter $D_{A2}$ [2]*

temperatures. In this way, cheap or more efficient magnets can be utilized, and magnet cost can be further decreased. Moreover, most gearless wind turbines are designed without any shaft and the hub can be easily flanged directly onto the outer rotor.

One prominent wind pioneer can be credited to have been the first to develop and introduce outer-rotor generators for wind turbine application: German emeritus professor *Friedrich Klinger* developed an innovative passive air-cooled 600 kW generator for the *Genesys* prototype that was installed already in 1997. The turbine, which also featured toothed belt pitch drives, had been developed in collaboration with student teams of the *Saarbrücken Hochschule für Technik und Wirtschaft des Saarlandes*. It became the basis for later *Vensys* and *Goldwind* designs of power ratings up to 4 MW which all featured air-cooled generators with outer rotor [2].

#### 3.2.2.4 Operating considerations – controlling PMSG

##### *3.2.2.4.1 Optimum generator current phase angle*

The choice of amplitude of the stator voltage and phase angle of the stator current and thus of the power factor at nominal load influence to a large extent construction volume and cost of the generator; however, it also determines the apparent power and thus the cost of the converter. Regarding (3.2) for the torque production, arising a question concerning the internal phase angle $\psi$ giving maximum torque seems to be superfluous, since the cosine is maximum for $\psi = 0°$ or $180°$, respectively. However, if saturation effects come into play, this question is no longer trivial.

Detailed investigations in the course of [1] had been performed to calculate the force density $\tau_{\tan}$ as a function of the internal phase angle $\psi$ for given geometry of the magnetic circuit of several PMG. Also experience has shown that even for surface-magnet PMGs the maximum force density is typically achieved at values slightly less (5°–10°) than what could be expected from linear analysis. Slight "flux weakening" operation obviously reduces the saturation and thus increases the torque production.

#### *3.2.2.4.2 Fault cases of PMSG – influences upon the converter voltage*

Overspeed of the wind turbine directly leads to an increase of the e.m.f. of the generator because of permanent-magnet excitation. This overvoltage may harm the power converter, i.e., the d.c.-link capacitor or the IGBT semiconductors. Since too high d.c. voltage may lead to destruction of very expensive parts in the converter the system must be designed so that too high d.c.-link voltage does never occur – especially if there are no circuit breakers between generator terminals and converters. This being such an important point to be considered, a summary of different failures will be carried out here taking into account aspects of converter control possibilities and the overspeed performance of pitch-controlled wind turbines.

**Overspeed performance of the wind turbine**

Normal overspeed due to variable speed operation: When the turbine operates at full load and rated speed, a normal transient overspeed of 10% will occur. This is because of wind gusts which will lead to an increase of turbine speed, because the electrical power is controlled to remain constant and the additional wind power is stored in the kinetic energy of rotation. Afterwards, the rotational energy store will be discharged by decelerating the turbine.

Overspeed due to loss of load: During acceleration, the pitch drives will pitch the blades in order to limit the speed. In normal operation with all three pitch drives operating, this normal overspeed is limited to 10%. However, at maximum speed a loss of load may occur, leading to an additional increase of speed up to 10%. Altogether, this yields a maximum turbine speed of 120% of the rated value, if all three pitch drives operate.

**Permitted overspeed/overvoltage of the converter**

Permitted overspeed with the power converter operating: If the IGBT rectifier is in operation, even at two or three times rated speed no overvoltage problem can occur, because the converter can protect itself by supplying the generator with reactive current (field weakening mode). The e.m.f. may be very high, but the terminal voltage remains on its rated value, because the terminal voltage is still actively controlled by the converter.

Permitted overspeed with the power converter NOT operating: In numerous cases, the converter will block its IGBT valves (i.e., they remain in the off-mode) in order to protect itself. This may happen because of:

- software or hardware faults in the driving control of rectifier and/or inverter units
- faults in the optical fibers between driving control and power modules
- any other fault in the control rack of the power converter
- release of the contactors between converter and transformer/grid, which may happen because of loss of grid voltage
- any other fault in the grid connection (frequency and/or voltage too high/too low).

The rated d.c. link voltage $U_d$ of low-voltage power converters using 1700 V modules is usually about 1150 V, the rated rms value of the generator voltage is assumed to be 800 V. If the converter blocks, the freewheeling diodes of the rectifier act as an uncontrolled diode rectifier being supplied by the generator e.m.f. If $E_p$ is the rms value of the (sinusoidally assumed) phase-to-phase voltage of the generator at no load, the d.c.-link voltage $U_d$ at maximum possible overspeed (loss of load, but all pitch drives operating) would rise up to

$$U_{\text{dmax}} = \sqrt{2}\, E_{\text{pmax}} = 1.20\,\sqrt{2}\, U_{s\,N} = 1350 \text{ V} \tag{3.7}$$

For exceptional cases it should be possible, however, to exceed the rated value of the d.c. link voltage $U_d$, because there will be safety margins for the converter parts. The utilized IGBT valves can block a maximum voltage of 1700 V, but for safety reasons this value should never be reached. More seriously will be the d.c. link capacitor. Capacitors can sustain certain voltages during their expected lifetime. However, high voltages times the number of cycles they occur may shorten the capacitor lifetime distinctly.

The matter is, however, less serious in case of PMSG with concentrated windings and inherently large inductance. Then the rated internal e.m.f. $E_p$ is usually about 10%–15% lower than the rated terminal voltage $U_s$! Keeping this in mind, the maximum d.c. link voltage $U_d$ would never exceed the rated value. However, this simplified consideration does not take into account that the d.c.-link voltage may increase further, due to the energy stored in the generator inductance. As the currents are decreasing, energy will be transferred from the generator inductance to the d.c.-link capacitance.

#### 3.2.2.5 Short paragraph on generator cooling design

The cooling system of an electrical machine is intended to dissipate operational losses from the machine in order to avoid operating temperatures possibly dangerous for the material used. The more efficient the cooling system and the higher the temperature limit of the materials, the higher losses and consequently electric loadings and flux densities can be chosen. In wind power, this is more important, since the efficiency at partial load decides the energy production, whereas the efficiency at rated load is less decisive.

##### *3.2.2.5.1 Admissible stator electric loading*

The following conclusions were earlier presented in [1]. As can be derived from (3.2), the maximum electric loading of the stator $A_s$ together with the airgap flux density of the rotor determines the generator volume required for transmission of a specified torque of an electrical machine. In many generators, the heat losses must flow through the lateral surface in order to be absorbed in the outer surface by the cooling medium. For this, the definition of a determined admissible heat flow density $\dot{q}$ per lateral surface of the generator is a physically sensible assumption:

$$\dot{q} = \frac{P_{\text{Cu}}}{\pi\, d\, l} \tag{3.8}$$

The ohmic power losses in the stator winding $P_{\mathrm{Cu}}$ – prevailing in low-speed machines by far – can be calculated according to

$$P_{\mathrm{Cu}} = R\, I_s\, I_s = \rho \frac{l}{A_{\mathrm{Cu}}} \left(A_s\, \pi\, d\right)^2 \tag{3.9}$$

which leads after conversion to the following equation:

$$\dot{q} = \frac{\rho\, \pi\, d}{A_{\mathrm{Cu}}} A_s^2 \tag{3.10}$$

The heat flow density $\dot{q}$ which can be dissipated via the lateral surface of the generator depends on the possible heat transfer coefficient $\alpha$ and the temperature difference $\Delta T$. The latter is fixed and is simply the difference between admissible stator winding temperature (depending on the utilized insulation class, usually class F) and ambient temperature, given in the specification:

$$\dot{q} = \alpha\, \Delta T \tag{3.11}$$

Equating both expressions for $\dot{q}$ and solving for the stator electric loading $A_s$ gives:

$$A_s = \sqrt{\frac{\alpha\, A_{\mathrm{Cu}}\, \Delta T}{\rho\, \pi\, d}} \tag{3.12}$$

Increasing the conductor cross-sectional area $A_{\mathrm{Cu}}$ will only be possible by applying a winding with a large slot fill factor, because the slot area itself will be limited by the required space for the teeth cross-sectional area needed for the magnetic circuit. A large slot fill factor may be possible with advanced insulation systems, however, there will be certain limits. Increasing the temperature difference $\Delta T$ may be possible by improving the insulation class, possibly from class F to H. This, however, would increase the winding and in particular the winding insulation cost.

Thus, the most effective way for maximizing the electric loading $A_s$ consists simply in maximizing the heat transfer coefficient $\alpha$ by applying a highly efficient cooling system! The following reference figures may be used for the initial layout of the low-speed generators:

- Natural cooling by free convection (IC 410): $\alpha = 20$ W/(m$^2$ K)
- Forced cooling by external fan (IC 411): $\alpha = 80$ W/(m$^2$ K)
- Water jacket cooling: $\alpha = 200$ W/(m$^2$ K)

###### *3.2.2.5.2 Electric parameters determining the temperature rise*

This section tries to clarify matters, whether electric loading or current density are the decisive parameters for the temperature rise of a given winding and cooling system. Starting again from (3.9) the ohmic power losses in the stator winding $P_{\mathrm{Cu}}$ can also be calculated according to

$$P_{\mathrm{Cu}} = R \cdot I_s \cdot I_s = \rho \frac{l}{A_{\mathrm{Cu}}} \cdot J_s\, A_{\mathrm{Cu}} \cdot A_s\, \pi\, D \tag{3.13}$$

which leads after conversion to the simple equation

$$\dot{q} = \rho\, A_s\, J_s \tag{3.14}$$

For the copper resistivity $\rho$ being constant at a certain temperature, it is obvious that electric loading $A_s$ times current density $J_s$ is the decisive factor, determining the temperature rise. This simple equation, however, neglects, that also other losses, such as ohmic losses in the winding overhang and rotor, hysteresis, and eddy current losses in the cores and additional losses, have to be dissipated via the lateral surface. By conversion, (3.14) indicates the maximum admissible current density in the stator windings.

The more efficient the cooling, the higher the heat flow density $\dot{q}$ through the stator outside and thus the stator current density $J_s$ can be realized. The following figures may be an initial help in designing permanent-magnet generators [1]:

- Natural cooling by free convection (IC 410): $J_s \approx 2.0$ A/mm$^{-2}$
- Forced cooling by an external fan (IC 411): $J_s \approx 3.5$ A/mm$^{-2}$
- Water jacket cooling: $J_s \approx 5.0$ A/mm$^{-2}$.

#### 3.2.2.6 Geometry of the magnetic circuit

##### *3.2.2.6.1 Airgap length*

The airgap length influences to an outstanding extent the magnetic resistance in an electrical machine, thus determining essentially the excitation required for a given flux. In order to minimize excitation losses, the lowest possible mechanic airgap should be aimed at. The definition of a certain airgap length yields consequences upon the mechanical structures which have to assure that contact between rotor and stator is excluded in all operating conditions.

Depending upon the construction of the supporting structure, the mechanical airgap should be chosen in such a way that a contact between the rotor and stator is excluded in all operating conditions. According to long-term experience [3] the airgap length in machines with large diameters should amount to at least one per thousand of the bore diameter because of mechanical reasons (manufacture tolerances, casing compliance, thermal expansion). Experiences with different supporting structures of generators with one-sided support of stator or rotor prove that the length of the machine should also be taken into account. Typical mechanical airgap values for large direct-drive generators would be in a range of 3–10 mm, depending largely on the generator diameter and stack length, but also on special technology such as adjusting possibilities.

One major advantage of any permanent-magnet machine should not be forgotten here: since the magnetic permeability of the magnet material is almost as low as air, the magnet thickness itself adds to the mechanical airgap when it comes to calculate the magnetically effective airgap! Let's look at some typical values: mechanical airgap of 5 mm, magnet thickness of 20 mm yields a magnetic airgap of 25 mm – being five times the value of the mechanical airgap alone! So all parasitic effects due to non-uniform airgap, like forces, vibrations, circulating currents, etc. will always be much lower for PMSG than for EESG.

#### 3.2.2.6.2 Optimum pole pitch

The pole pitch has an enormous influence on the generator inductance, which possesses no direct influence upon the torque production, but which may limit the power output, too. This is obvious for generators with diode rectifiers since the synchronous inductance determines the power angle for a given current, which again influences the internal phase angle. However, even generators with forced-commutated converters are influenced by their inductance, because the power angle may reach intolerable values, thus limiting the power instead of the thermal performance. Determining the minimum inductance for a given design will in any case minimize the necessary apparent power of the generator and thus cost and size of the rectifier to be installed.

The synchronous inductance consists of the magnetizing inductance (being proportional to the pole pitch) and the leakage inductance, which is inversely proportional to the pole pitch. Thus, for a given (magnetically effective) airgap length and stator slot depth there exists a pole pitch yielding minimum synchronous inductance. Typical pole pitch values for large permanent-magnet direct-drive generators would be in a range of 100–250 mm, increasing with generator rating and diameter.

For permanent-magnet excitation, the pole pitch has also to be chosen with respect to a possible demagnetization in case of short-circuit: the then obtained demagnetizing stator ampere-turns are directly proportional to the pole pitch, [1].

Section 3.4 of this chapter shows a "deep dive" into the optimization of the magnetic circuit, especially with respect to innovative rotor topologies.

## 3.2.3 Other important aspects for designing direct-drive generators

### 3.2.3.1 Vibro-acoustic performance and generator noise emission

One of the major aspects during the design of wind turbines is to predict the noise emission of the total turbine as accurately as possible. Besides the overall sounds pressure level, which must not exceed a certain level in a certain operation mode, the main requirement is the complete absence of any tonal sound omission.

Typically, drivetrain vibrations are the main source of tonal sound emission. In case of geared drivetrains, the gearbox itself with the teeth meshing can be one major cause of tonality. One of the usual remedies is to de-couple the complete drivetrain from the mainframe by damping elements. In the case of direct-drive wind turbines, the airgap forces of the generator can lead to an excitation of the surrounding mechanical structure. This effect can be further increased by changes of the local airgap width due to the finite accuracy and stiffness of the generator structure. The resulting structural vibrations are transferred through the drivetrain component and finally to the large surface structures like blades, tower, and nacelle housing, which act as a "loudspeaker membrane." The loudspeaker converts these oscillations into an airborne sound that radiates into the turbine's surrounding area.

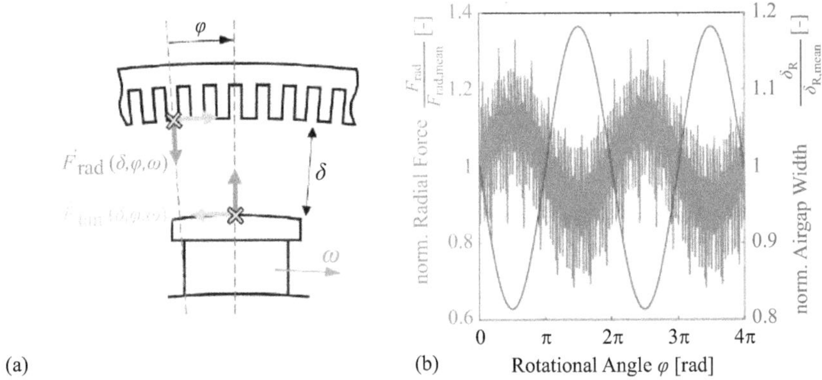


*Figure 3.8 Radial and tangential forces in the airgap of a generator [4]*

The first step to estimate the sound emission that is mainly driven by the electromagnetic forces in the large diameter generator, detailed calculations of the respective airgap forces have to be performed. Usually, the electromagnetic forces are calculated by means of finite element simulations of the electromagnetic field for all operating points and possible airgap widths that can occur during the turbine operation as illustrated in Figure 3.8 [4].

The second step to estimate the sound emission is to determine a suitable mechanical model for the generator structure, typically using a modal approach. The elastic deformation of the whole flexible body is obtained from the superposition of its modal behavior. The result of this approach is a system of differential equations with each equation describing the motion of its corresponding mode. This system of differential equations decouples when damping is neglected. This ensures numerical stability of the given system [5].

The third and final step is the calculation of the airborne sound. The characteristic wind turbine tonal sound emission is determined by the sound waves emitted by tower, blades, and nacelle housing as seen in Figure 3.9. In order to identify the discrete frequencies of the tones and to determine the corresponding sound pressure levels at first the time signals of the modal participation factors of the flexible bodies from the multi-body system have to be transformed into the frequency domain [5].

###### *3.2.3.1.1 Special considerations for PMSG with factional slot concentrated windings*

As will be presented in Section 3.4.3 in detail, a low number of stator coils and an extremely compact design of the winding overhang and thus lower stator winding resistance and lower losses compared to conventional distributed windings are major advantages of factional slot concentrated windings. However, besides these advantages there is also the disadvantage of much higher harmonic content of

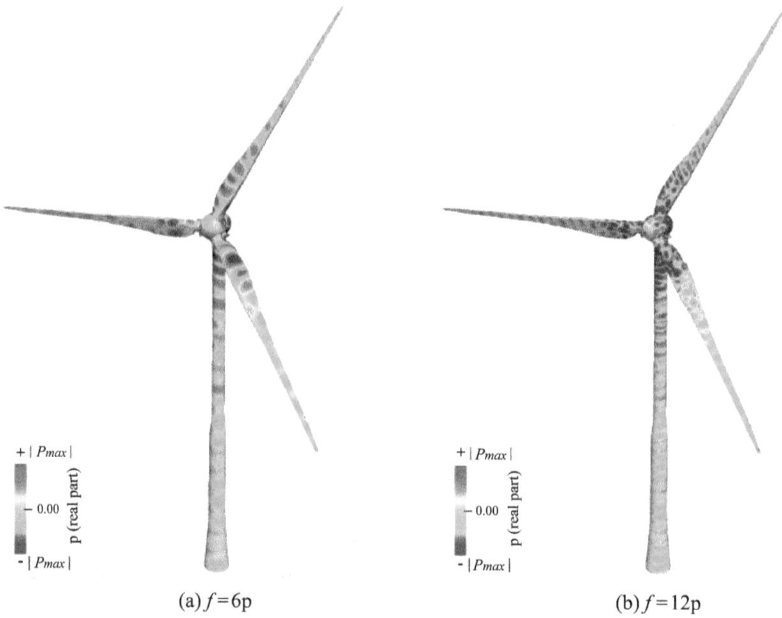


*Figure 3.9 Sound pressure distribution on the surface of a wind turbine at f = 6p (a) and f = 12p (b) [5]*

airgap field waves generated by these innovative winding types. Also, the *local* forces (typically the second harmonic) acting on teeth and poles will be much higher than for distributed windings – even if the overall torque is very smooth. Moreover, these strong local forces can usually not be reduced by traditional measures such as pole/slot skewing or arrow-shaped poles: otherwise large losses in the electro-magnetic performance would be the result.

All this calls for special attention toward the vibro-acoustic performance during the design phase of these PMSG:

- Proper analysis of the airgap harmonics for both stator and rotor for all possible operating points: selection of favorable pole/slot combinations and optimization of the overall electro-magnetic circuit to minimize possible excitations for critical frequencies.
- Modal analysis of the generator structure in an early design phase already: optimization of the mechanical design to avoid possible resonances in important operating points, so that critical frequencies are not further amplified during their way toward blades and tower. Furthermore, both stator and rotor structure should be stiff enough to handle the large local forces acting on teeth and poles around the circumference: the generator surface itself should not become an emitter of airborne sound, since these local vibrations can usually not be tackled by active vibration control.

- Design, implementation, and validation of active vibration reduction measures by the frequency converter: all global oscillations – both in tangential (torque) and radial (breathing) direction – can usually be dampened by the generator currents. By adding the required harmonics in the correct phase angles to the generator currents, a significant reduction of turbine tonality in most operating modes is usually achievable. Moreover, since the countermeasures may also depend on actual temperatures, production imperfections etc., suitable machine-learning algorithms can be applied to arrive at reasonable results for individual turbines.

Typically, the aspect of vibro-acoustic assessment and optimization is one of the major ones during the design and validation of direct-drive PMSG for wind turbines: it can easily eat up more resources than any other aspect – and sometimes much more resources than initially planned.

#### 3.2.3.2 Aspects of production and supply chain

Wind turbines types of a certain OEM are usually differentiated along different power platforms. The turbines of a certain rating are typically produced in numbers of several hundred units per year. A robust supply chain consisting of several suitable suppliers for the major items on the bill of materials (BOM) is one of the most important aspects for any wind turbine OEM.

Seen from a component buyer's perspective, a large direct-drive generator must be considered somehow differently from other main components such as bearings, castings or gearboxes. It has to be deeply integrated into the turbine's mechanical design (see Chapter 5 "Drivetrain concepts and developments" in Volume 1), so that independent suppliers could not succeed in designing a product that fits all customers. Moreover, in most of the cases, direct-drive generator systems could only be successfully established when being designed and produced in house by the respective OEM – just like rotor blades.

Production of any large direct-drive generator at a typical OEM is typically split into three main phases:

1. Production of large mechanical structures at suitable suppliers for heavy steel works, then shipment to OEM
2. Pre-assembly of the generator: laminations stacking, pressing, coil and magnet insertion and connection, winding impregnation and coating
3. Final assembly of the generator: "wedding" of stator and rotor, assembly of all secondary items, final testing

##### *3.2.3.2.1 Major issues of large-diameter, single-piece generators*

*Supply of mechanical structure*

Machining of mechanical structures becomes difficult with increasing diameter, only very few suppliers can machine large structures of more than 10 m diameter. Within Figure 3.10 a large carousel lathe up to 18 m diameter is depicted. However, single-piece structures with diameters of more than 10 m may be producible, but would remain unique components and not recommended for large-scale series production, since the required machining hours would be very expensive and the associated specific cost would be very high.

*Vacuum-pressure impregnation (VPI) devices*
The stator winding of most electrical machines is typically impregnated using a "global vacuum-pressure impregnation (VPI)," which means that the total stator is dipped into a resin bath under (first) vacuum and (second) pressure application. This removes any air inclusions in the insulation and yields the best possible resistance to external voltages stress.

In case of single-piece stators of diameters of 10 m and more, the application of global VPI becomes more and more challenging, because the investment and operating cost of such large equipment (approximately proportional to the required volume) become very high. This fact increases the final generator cost and reduces flexibility.

One possible alternative in case of the stator winding of large direct drives would be the move from "global VPI" to "local VPI," be it by stator segmentation or treatment of individual coils ("single bar or coil VPI"). For the latter case, the so called "resin rich insulation systems" are available as well. A "global VPI" impregnation facility can be seen in Figure 3.11.

*Figure 3.10 Huge vertical machining center. © Pietro Carnaghi.*

*Figure 3.11 VPI of an inner stator of the wind-direct wd2.8+121 direct-drive wind generator. © wind-direct GmbH.*

#### *3.2.3.2.2 Possible escape route: segmentation*

In 2009, *wind-direct GmbH* started the development of the direct-drive wind turbine *wd2.8+109* (Figure 3.12) with full segmentation on both the stator and rotor side (Figure 3.13). The basic design features and advantages seen at the time had been the following:

- 12 glued segments (30°) form basic elements, both in stator and rotor as depicted in Figure 3.14
- Low-cost casted components (3 discs with <5.0 m) as supporting structure
- Reliability through passive cooling: cooling fins stamped directly to the stator laminations
- More fault tolerant and robust design: two fully independent generator systems with full redundancy
- Easier to assemble, install, and repair
- Reduced life-cycle cost associated with assembly, transportation, installation, and O&M
- Fully segmented/modular: better scalable to higher MW ratings

The generator and the turbine were developed for an Indian customer, the generator segments were produced at a German supplier and the prototype could be installed,

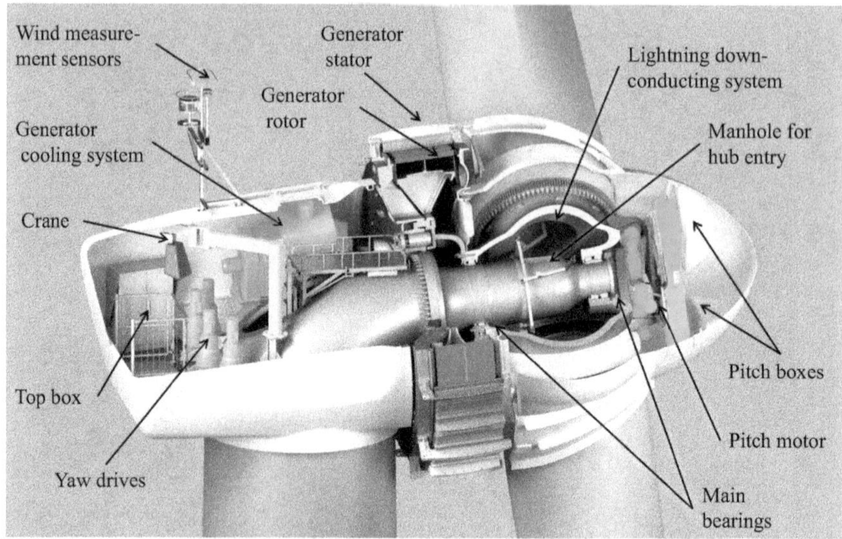


*Figure 3.12 wind-direct wd2.8+109 from 2014 – both rotor and stator consist of 12 individual segments [6] (© wind-direct GmbH)*

*Figure 3.13 wd2.8+109: assembling the stator segments around the rotor [6] (© wind-direct GmbH)*

tested, and certified in 2014 on a South Indian site. The machine had been probably the very first large-scale wind turbine without any mechanical brake: rotor locking is achieved by switching the stator of the PMG manually into short circuit which puts the turbine into a "creeping mode," then the rotor lock can be easily inserted.

However, the reality proved to be slightly different and more challenging than expected:

- Producing the segments with bonding technology ("Backlack") proved to be more difficult and costly than expected

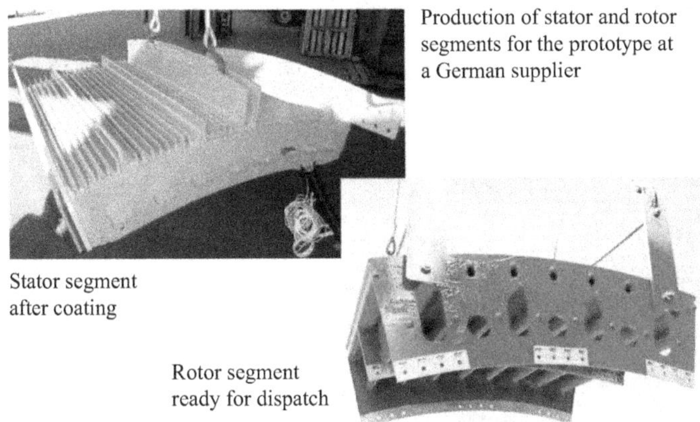

Production of stator and rotor segments for the prototype at a German supplier
Stator segment after coating
Rotor segment ready for dispatch


*Figure 3.14 wind-direct wd2.8+109 stator and rotor segment [6] (© wind-direct GmbH)*

- Production processes had not been specified in adequate detail, assembling the stator around the rotor had not been as simply as intended
- Overall, the technology of "bonded segments" had not been mature enough: insufficient bonding strength caused airgap imperfections and associated vibrations
- Airgap imperfections also caused imbalanced generator losses and non-symmetric heating up.

*Flender/Winergy's stator segments*

Another segmentation approach – commercially very successful – is the Flender/Winergy technology for the Siemens Gamesa direct-drive wind turbines. The following information is taken from a recent feature by Eize de Vries [7].

"Flender" – the company behind brand name Winergy – manufactures these stator segments at its lead factory in Subotica, Serbia (see Figure 3.15) established in 2003, currently Europe's largest for all types of wind turbine generators. Since product launch, Winergy has supplied Siemens Gamesa with nearly 50,000 stator segments – initially for 3 MW onshore turbines but gradually expanding to segments for the latest, most powerful, 14 MW offshore range.

Stator segments are modular building blocks that are factory pre-tested and pre-commissioned, ready to use in Siemens Gamesa's DD permanent magnet generators (PMGs). They are shipped to a generator facility such as the Cuxhaven plant in Germany, where they are mechanically mounted into full circles on matching support structures and then adjusted and electrically interconnected.

The generator rotor of all Siemens Gamesa turbines is still a single piece without any segmentation: since it is for offshore application, road transport of the assembled generator can be fully avoided and installation happens by large vessels with large lifting capacities.

*Figure 3.15 Large series stator segment production at Flender Factory in Subotica/Serbia [7] (© Flender GmbH)*

#### 3.2.3.3 Generator reliability, repair, and service aspects

Besides energy production and capital expenses, another main contributing factor in the LCOE equation is the regular cost for keeping the wind turbines in operation and their availability high. Besides regular maintenance, the failure rate of the main components largely influences the "service cost" and thus the final LCOE.

There are many publications regarding this reliability of certain main components – most of them confirm the picture seen in Figure 3.16, that drivetrain components such as generator, gearbox, and bearings largely influence wind turbine availability: may the failure rates be low, the associated downtimes (and associated AEP losses) are usually long.

Gearless wind turbines with direct-drive generators are typically considered to have an inherent higher reliability due to the absence of gears and high-speed bearings. It is, however, very difficult to get reliable data confirming this judgement, since the windings of large direct-drive generators can also fail prematurely due to:

- Environmental impacts of the cooling air, perfect encapsulation is typically impossible
- Electrical stress (peak voltages, voltage rise and spikes) due to operation by IGBT rectifiers
- Thermal stress due to extended operation at maximum temperatures
- Foreign bodies hitting the winding, often after service works inside the generator

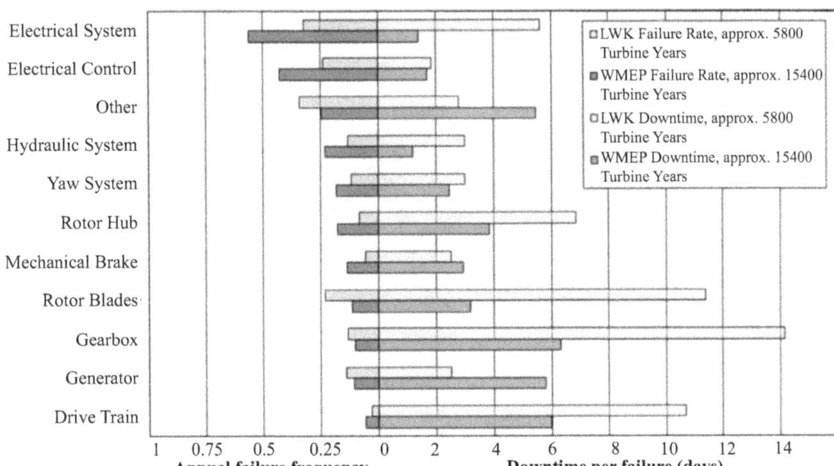


*Figure 3.16 Wind turbine reliability: annual failure rate and downtime per failure [8]*

There are two completely different strategies to cope with these challenges: either to invest in expensive insulation materials, encapsulated cooling concepts and fine filters (which will increase the CAPEX part), or develop repair concepts uptower, so that a replacement of the complete direct-drive generator (involving the rental cost of an expensive main crane and associated downtime) can be limited to absolute rare cases.

As already mentioned, *wind-direct GmbH* developed a direct-drive wind turbine (Figure 3.12) with fully segmented generator (both stator and rotor consisted of 12 segments each) and claimed that individual segments could be replaced uptower by an external winch. This turbine was installed as prototype in 2014, but no series production followed so that the repair concept could not be validated properly. An image of the intended strategy can be seen in Figure 3.17. If other segments than the 12 o'clock segment would have failed, it had been planned to lock the turbine rotor and then turn the stator in motoring mode until the faulty segment moved to the exchange location at 12 o'clock.

Other wind turbine OEMs using extremely compact direct-drive technology seem to be following the "high reliability approach," where the generators have to be brought down after every minor incident. In case of a proper design and proven insulation technology, this strategy may work out fine.

On the other hand, there are manufacturers which feature uptower winding repair works in case of minor winding faults. To this aim, the overall generator design must take all involved aspects into account:

- Mechanically, the generators have to provide proper internal access to the winding and adequate working areas so that the teams have a chance to accomplish the tasks in a safe manner

External winch (to be installed by 500 kg service crane)
- Exchange of single stator segments
- Exchange of single rotor segments

*Figure 3.17 Wind-direct wd2.8+109 – generator segment exchange concept [6] (© wind-direct GmbH)*

- Electrically, starting with a winding and insulation design that ensures "repairability" and a high probability of a certain lifetime after repair, and ending with adequate fault-current detection devices/measures to limit the damage of a short circuit to a limited space of the winding.

Both "maximized reliability approach" and "repair approach" seem to work, so that direct-drive generators seem to outperform gearbox drivetrains in terms of service cost still today. This may be one reason that gearless technology is the favored option for offshore wind energy, where the cost of downtimes is higher and service works are more difficult and more expensive than onshore.

## 3.3 Scaling laws of direct-drive generators

Section 5.3.5 in Volume 1 already touches the aspect "Scalability of designs and performance indicators." From Table 5.2 in volume 1, the "square-cube law" of wind turbine scaling can be derived: larger turbines should become less attractive, because torque increases faster than power. However, reality does not seem to care: wind turbines continue to grow in size, especially for offshore applications. Recent announcements indicate that 300 m rotor diameter and ratings of more than 20 MW will be achieved soon. Even for onshore applications – where transport and installation restrictions are much more rigid – there seems to be a continued trend toward larger dimensions. However, the following restrictions for onshore installations exist (see Section 3.2.2.3.3):

- Maximum total height 4.3 m (bridge height 4.5 m)
- Maximum transport width 5.0 m (avoiding expensive traffic police patrol)
- Maximum total weight 100 t (bridge carrying capacity, weight of load and truck!)

One question arising is whether direct-drive generators would be still competitive for such enormous torque values and the implied logistic challenges. This chapter tries to scale up existing designs – based on some general scaling laws for this technology. The initial comparison was first presented in [6] in 2014 for maximum rotor diameters of 180 m – here the basic rules have been applied again, extending the maximum rotor diameter to 300 m and the rated power to almost 22 MW.

### 3.3.1 Basic assumptions for scaling

Constant values:

- Constant specific power of 300 W/m$^2$ (typical for onshore wind turbines)
- Constant tip speed of 80 m/s (typical for onshore wind turbines)
- Constant tangential airgap force density of 75 kN/m$^2$ – typical value for surface magnet configuration with air cooling
- Constant steel sheet thickness of the supporting structure

Varied parameters:

- Generator diameter and stack length grow proportionally to the rotor diameter
- Airgap width increases with generator diameter, yielding decreasing airgap flux density with diameter
- Stator current loading increases with diameter to compensate reduced airgap flux density
- Stator tooth width decreases with reduced airgap flux density, yielding constant tooth flux density
- Pole pitch and stator slot depth increase with generator diameter

### 3.3.2 Scaled results for generator diameter

Figures 3.18 and 3.19 display the required generator diameter as a function of rotor diameter and rated power. The results are obviously linked closely to the given assumptions for scaling. It can be seen that the level of 10 m is crossed at a rotor diameter of 180 m and rated power of 7.5 MW. The largest variant requires a diameter of more than 16 m – yielding challenges for production, transport, and installation.

### 3.3.3 Scaled results for generator losses and efficiency

The calculated losses in the active part consist of losses in the stator winding (copper losses), losses in the laminated stacks (iron losses) and losses in the permanent magnets (magnet losses). It can be seen from Figure 3.20 that the vast

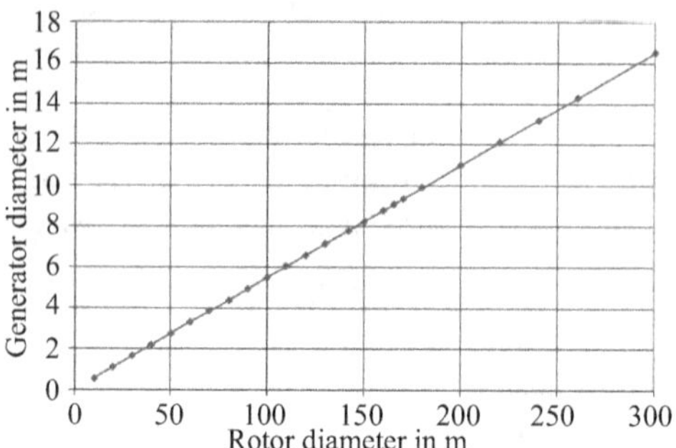


*Figure 3.18 Generator diameter vs. rotor diameter*

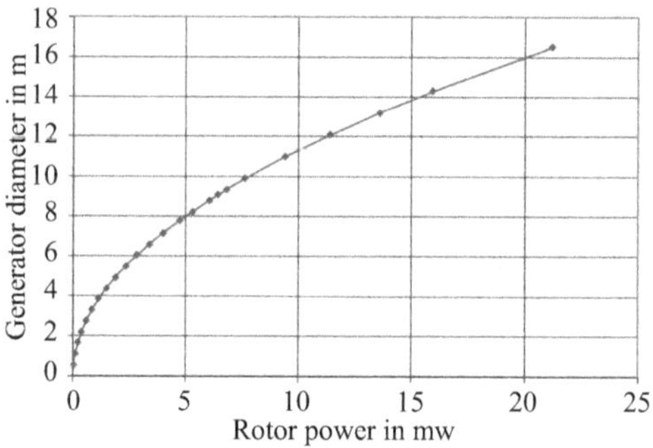


*Figure 3.19 Generator diameter vs. turbine power*

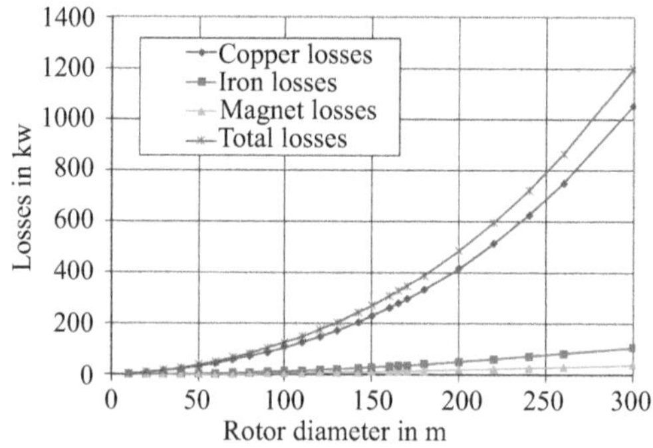


*Figure 3.20 Generator losses vs. rotor diameter*

majority (almost 90%) of losses is created in the stator winding – a typical result for direct-drive generators. The basic scaling result is that the copper losses in the stator winding increase with diameter squared, yielding total losses of more than one megawatt for the largest variant. Would this create any cooling problems for larger ratings?

If we look at the scaled rated efficiency values in Figure 3.21, we can see that this important parameter increases to approximately 95% for turbine diameters of 100 m, and then remains almost constant at this level: even the largest ratings should be able to achieve an attractive drivetrain efficiency and linked energy production.

But how about the generator cooling performance? A useful performance indicator for heat dissipation by surface cooling of any specific component is to divide the total losses by the available cooling area, yielding specific losses in kW/m$^2$. It can be observed that the specific losses per outside area drop up to 3 MW, then remain almost constant up to the largest ratings. As useful conclusion could be drawn, that a certain proven cooling principle should also work for the largest turbines as well.

### *3.3.4 Scaled results for generator masses*

Figure 3.22 displays the specific masses in kg/kW for winding copper, laminations and magnets as a function of the power rating. Despite the fact that the specific

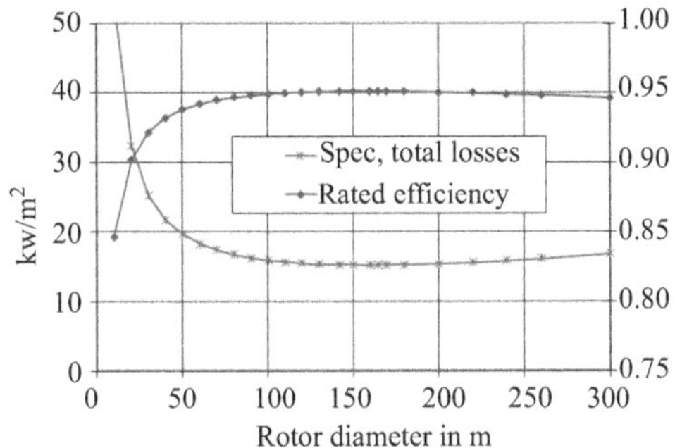


*Figure 3.21 Spec. losses and rated efficiency vs. diameter*

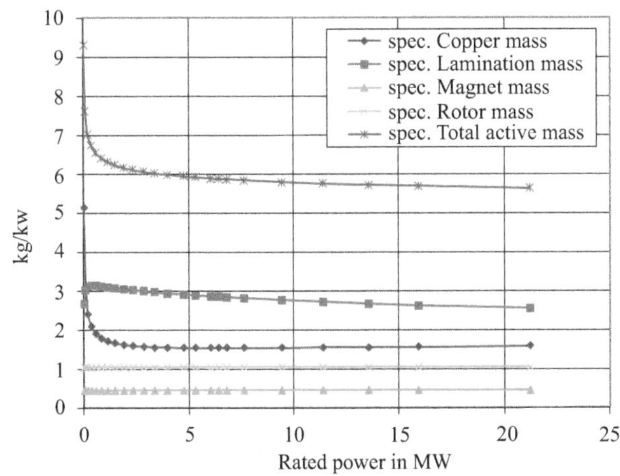


*Figure 3.22 Specific active masses vs. rated power*

torque in Nm/kW increases linearly with power because of constant blade tip speed, these specific active masses remain more or less constant, even slightly decreasing for the largest ratings. The specific structural masses show a somehow different behavior: they seem to be lowest for small power ratings, then increase up to 10 MW but also converge and stabilize at a certain constant level for largest ratings above 20 MW, see Figure 3.23.

Figure 3.24 displays the absolute masses in metric tons for winding copper, laminations and magnets as a function of the rotor diameter. These correlations indicate a proper quadratic increase of the active masses, reaching 120 t at the maximum diameter of 300 m. Similar applies to the structural and total generator mass vs. rotor diameter, see Figure 3.25. It can be concluded that the total mass of a direct-drive generator reaches levels of more than 400 t for largest diameters of 300 m.

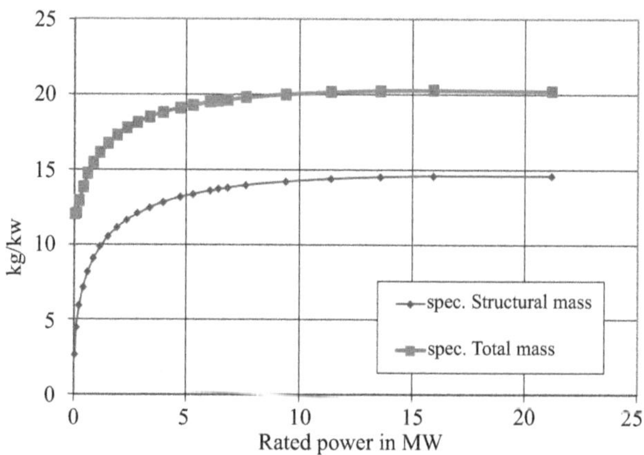


*Figure 3.23 Specific masses vs. rated power*

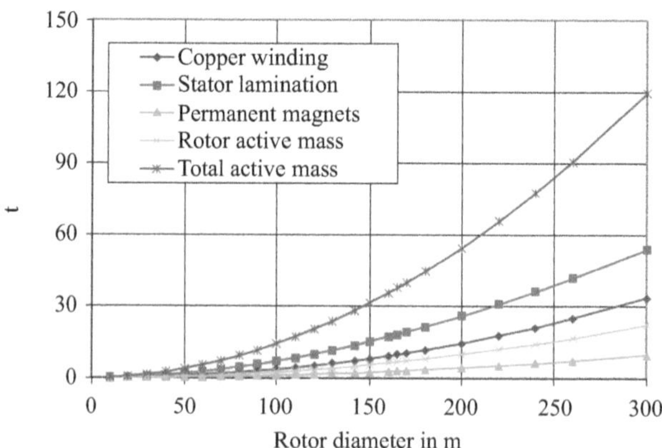


*Figure 3.24 Active generator mass vs. rotor diameter*

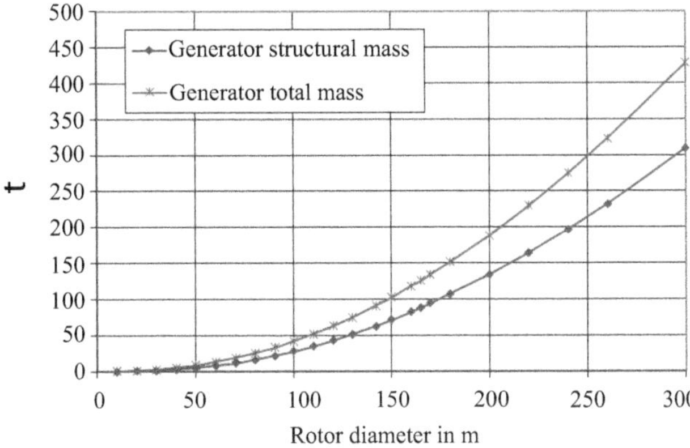


*Figure 3.25 Structural generator mass vs. rotor diameter*

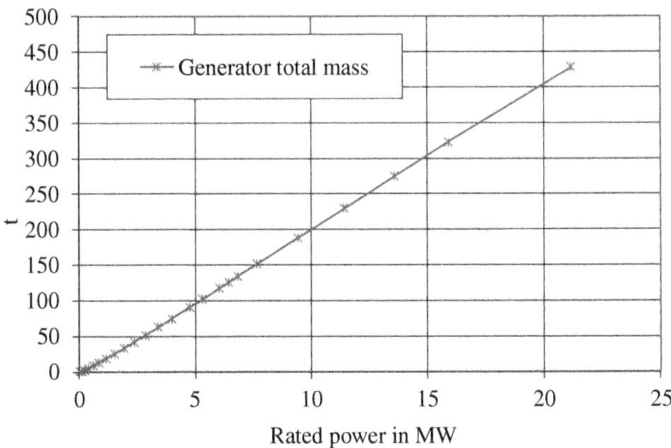


*Figure 3.26 Total generator mass vs. rated power*

Figure 3.26 displays the resulting total generator mass as function of rated power: it may be surprising to see that this is a rather linear graph instead of the assumed quadratic relation, for instance in the case of other turbine components.

If we convert this result into the well-known torque density (torque per generator mass) in Nm/kg and display this performance indicator vs. rated power, we get the graph in Figure 3.27. It shows that the torque density $T_d$ of direct-drive generators should increase with power, however with decreasing gradient: the relation may be somehow modeled by

$$T_d \sim C \cdot \sqrt{P} \tag{3.15}$$

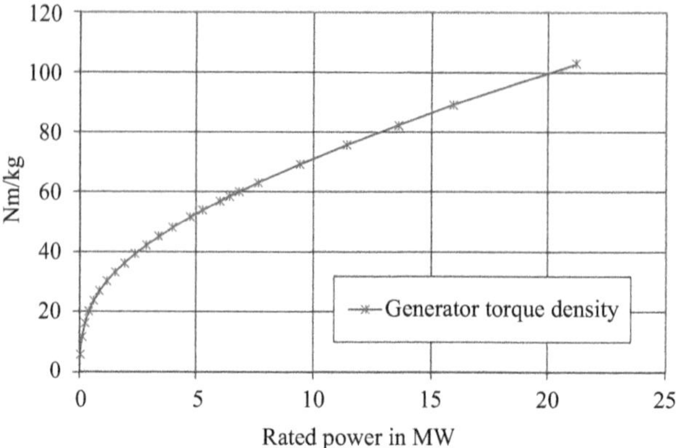


*Figure 3.27 Generator torque density vs. rated power*

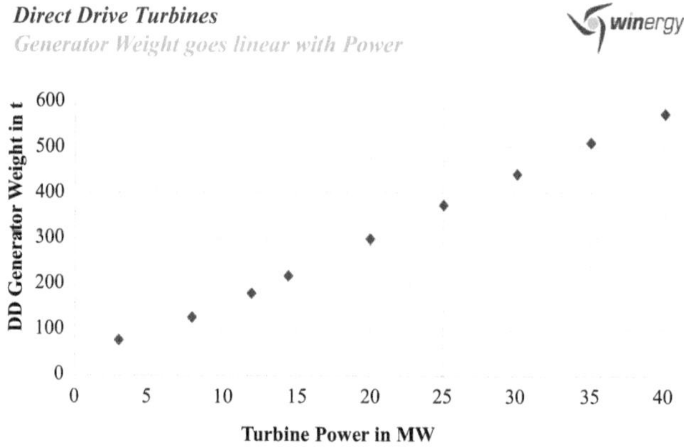


*Figure 3.28 Offshore scenario: total generator mass vs. rated power [9]*

These basic scaling results may be astonishing, but are (more or less) confirmed in a published scenario by the company Flender [9], see Figures 3.28 and 3.29. The first graph is taken directly from this publication, and the second graph showing the torque density is derived from the published numbers.

How to explain the higher torque density for larger generators? If we simplify the direct-drive generator as a cylinder of rather constant wall thickness, for which both diameter and length increase linearly with turbine rotor diameter. Since the torque of any electrical machine is proportional to its bore volume, whereas the weight is rather proportional to the surface of this almost hollow component, the displayed results make some sense.

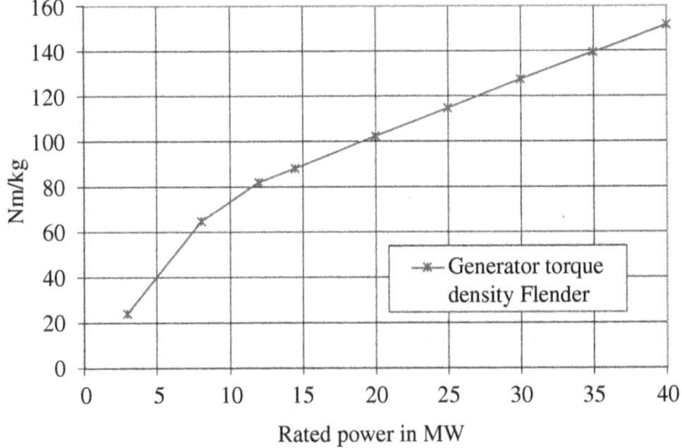


*Figure 3.29 Derived generator torque density vs. rated power [9]*

### 3.3.5 Conclusions from the scaling exercise: segmentation as the future of the direct drive?

The above section tried to throw some light on the question of whether direct-drive generators would be still competitive for ever increasing torque values of large wind turbines and the implied logistic challenges. By applying general scaling laws for this technology, the basic performance indicators for direct-drive generators have been derived, extending the maximum rotor diameter to 300 m and the rated power to almost 22 MW.

The length of the laminated stack has been increased linearly with rotor diameter as well, reaching a value of 1.4 m for the "final" variant of 22 MW. It is known from experience that also this value cannot be increased without limits, since challenges in production and cooling may become excessive: two meters may be some kind of "sonic wall" for this parameter.

It should have become clear that the most important design parameter is the generator diameter, which should be chosen to values of more than 15 m for powers higher than 20 MW. But how to produce and transport such giant components? It becomes obvious that – even for offshore applications – some form of segmentation of stator and rotor must be applied. Even though large carousel lathes are available up to diameters of 18 m (at excessively high hourly rates), the questions of winding impregnation, transport, and installation also remain to be addressed.

The total generator mass yields levels of more than 400 t – which is very difficult in offshore applications, but impossible for onshore wind energy. Separation into five or more parts may show an escape route to this issue, yielding individual component masses of less than 80 t. Even if modern installation cranes should be able to handle higher masses, transport is the bottleneck, as a total truck mass of 100 t is seen as "sonic wall" for standard bridges, as mentioned before.

However, all these limits are faced by wind turbines with geared drivetrains as well. Since it may be even less risky to split up generators than gearboxes, direct-drive concepts may regain market shares in the future, even for onshore wind turbines. A pre-requisite for this scenario will be proven segmentation concepts in large series production. Looking at the achievable torque density values – even considerably increasing for larger generators – this scenario may be well possible. An interesting outlook on the advantages of segmented concepts is given in the recent article about the Flender stator segment success story [7].

There we also find the statement: “A major challenge for all direct-drive generator concepts remains their high requirement for materials, especially structural and electric steel, copper and approximately ten times the amount of magnets compared with mid-speed geared generators.” Electrically excited direct drives are far less competitive compared to permanent-magnet design, especially because of their lower efficiencies and their higher masses. Thus the future of the direct-drive permanent-magnet technology should also remain dependent on the development of the raw material market, especially considering the rare earth part.

## 3.4 Stator winding topologies for direct-drive generators

For the stator winding there are various concepts available on how the stator is wound with a copper or aluminum winding which is carrying the stator currents. The stator of an electrical machine typically consists of a number of slots in the laminated stator iron in which the stator winding is inserted. The way of connecting the copper conductors in the slots together and the resulting shapes of the coils should be further explained here and a categorization of different stator winding types to be given subsequently in this chapter. Benefits and drawbacks will be mentioned to each approach and finally, the most promising stator winding setup for a direct-drive permanent-magnet synchronous generator for wind turbine applications will be identified and further explained in detail.

### *3.4.1 Definition of symmetrical $m_s$-phase stator windings*

A stator winding is defined to be symmetrical if the number of coils that are assigned to a phase is identical for all the phases that exist inside the stator winding. A stator winding consists in general of $m_s$-phases. For $m_s$ typically 3 as well as 6 is chosen. These types of windings are widely spread but also 9- or 12-phase machines are considered sometimes for various applications. For direct-drive generators in wind turbines, however, a 3- or 6-phase winding topology is typically chosen. The first value by which a stator winding is characterized is the so-called number of slots per pole and phase $q$ which can be determined using the following equation:

$$q = \frac{Q_s}{2 \cdot p \cdot m_s} = \frac{q_n}{q_d} \tag{3.16}$$

Whereas $Q_s$ is the total number of slots in the stator circumference, $p$ is the number of pole airs in the rotor, as well as the main exciting field from the stator winding. Finally, $m_s$ is the number of phases. Typically the number of slots per pole and phase is given as a reduced fraction with its nominator $q_n$ and its denominator $q_d$.

Sometimes the so-called number of coils per pole and phase $q_c$ is used as well which is defined as follows:

$$q_c = \frac{Q_c}{2 \cdot p \cdot m_s} = \frac{q_{cn}}{q_{cd}} \tag{3.17}$$

Wherein $Q_c$ is the number of coils inside the stator along the entire circumference. Having these expressions defined we can divide stator windings by its fist properties. This is to distinguish stator windings in so-called single layer as well as double layer windings. The following relations between single layer and double layer windings can be made:

$$q = \begin{cases} 2 \cdot q_c \text{ single layer winding} \\ q_c \text{ double layer winding} \end{cases} \tag{3.18}$$

The same relation is valid for the total number of slots

$$Q_s = \begin{cases} 2 \cdot Q_c \text{ single layer winding} \\ Q_c \text{ double layer winding} \end{cases} \tag{3.19}$$

The number of coils is equal to the number of slots in the stator for a double layer winding, whereas for a single layer winding the number of coils is equal to half the number of slots in the stator. For a single layer winding there is only one coil leg in each stator slot whereas for a double layer winding there are two coil legs inside a stator slot. The coil legs in case of a double layer winding can be arranged side by side or on top of each other in the stator slot.

The next important property of stator windings is to distinguish whether the coils inside the stator slots do have an overlap with adjacent coils or not. If, for example, every coil has its forward and return coil leg spaced by just one stator slot the type of stator winding is called non-overlapping winding. However, if there are multiple stator slots in between the forward and return conductor of a coil so that other coil legs can be put in the slots in between and all the coils along the stator circumference are nested somehow into each other the resulting stator winding is called an overlapping stator winding. This is also depicted in Figure 3.31.

In the following we are introducing a parameter $y_d$ which is basically an expression of the coil span or coil width in terms of stator slots along the circumference and the following expression is valid here:

$$y_d \begin{cases} = 1 \text{ non-overlapping} \\ > 1 \text{ overlapping} \end{cases} \tag{3.20}$$

In order to define the next property of a stator winding it is important to mention the next characteristic value $y_p$ which is defined as follows:

$$y_p = \frac{Q_s}{2 \cdot p} \tag{3.21}$$

Whereas $y_p$ is defined as the average coil span given in stator slots per pole pair. In particular, for the double layer windings there exists the option to pitch the windings and to obtain better properties in terms of current layer and winding factor which will be described later on. Pitching in this sense means defining a coil span which is typically lower than the average coil span $y_p$. Taking into account the average coil with $y_p$ the possible pitching can be determined as follows:

$$y_d = \lfloor y_p \rfloor \pm k \begin{cases} k \in \{1, 2, 3, \ldots\} \\ y_d \geq 1 \end{cases} \tag{3.22}$$

For $y_p = y_d$ the resulting winding is called non-pitched double layer winding. For $y_p \neq y_d$ a so-called pitched winding is actually obtained. The difference between pitched and non-pitched winding is in addition graphically depicted in Figure 3.32.

There are two more final aspects by which a stator winding can be further characterized and distinguished. This is by looking in detail into the defined equations to the calculation of number of slots per pole and phase $q$ as well as number of coils per pole and phase $q_c$. If the resulting denominator $q_d$ in the definition of slots per pole and phase $q$ is equal to one the winding is a so-called integer slot winding. This means in other words that the angular section of one pole pair on the rotor matches exactly to an integer number of slots on the stator. However, if the resulting denominator $q_d$ is greater than one the so-called winding is a fractional slot winding. This means that the number of slots per angular section of each pole pair on the stator side is just a fraction.

Finally, the last property for a classification of stator windings is given. This is the so-called property for concentrated winding as well as distributed winding. If the nominator $q_{cn}$ out of the equation which determines the number of coils per pole and phase is equal to one the stator winding is called concentrated winding. If the nominator $q_{cn}$ is greater than one the so-called stator winding is a distributed winding.

The entire overview of classification of stator windings can also be seen in Figure 3.33.

#### 3.4.1.1 Definition of a basic winding

As basic winding the smallest repetitive segment of a stator winding can be defined out of the entire circumference of the machine. The existing number of basic windings (bw) and therefore repetition along the machine circumference can be determined according to the greatest common divider (gcd) between the number of slots in the stator $Q_s$ and the number of pole pairs in the rotor $p$

$$\mathrm{bw} = \gcd(Q_s, p) \tag{3.23}$$

With the number of basic windings determined the number of slots and poles for each basic winding can be determined as follows:

$$Q_b = \frac{Q_s}{\mathrm{bw}} \tag{3.24}$$

$$p_b = \frac{p}{\mathrm{bw}} \tag{3.25}$$

Wherein $Q_b$ and $p_b$ are the number of stator slots and poles per basic winging respectively. If the greatest common divider is equal to one there is no repetition of several basic windings along the circumference.

#### 3.4.1.2 Examples of different winding types

As mentioned in the previous chapter various examples of different winding types are given here to get a better understanding and to distinguish between various winding types.

Figure 3.30 depicts the difference of a single and a double layer fractional slot non-overlapping concentrated winding. As usual for concentrated windings the coil legs are placed side by side into the slot in case of a double layer winding.

Figure 3.31 depicts two different types of single layer windings which are typically made of so-called diamond pulled coils. This coil shape is widely spread in so-called form coil stators.

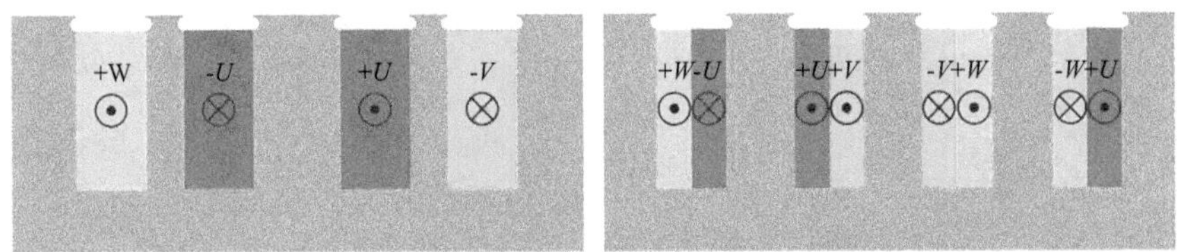


*Figure 3.30 Left: single layer fractional slot non-overlapping concentrated winding. Right: double layer fractional slot non-overlapping concentrated winding [10].*

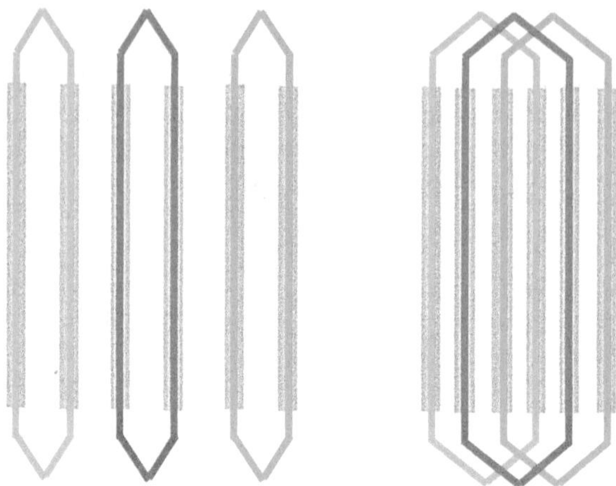

*Figure 3.31 Left: single layer non-overlapping winding. Right: single layer overlapping winding.*

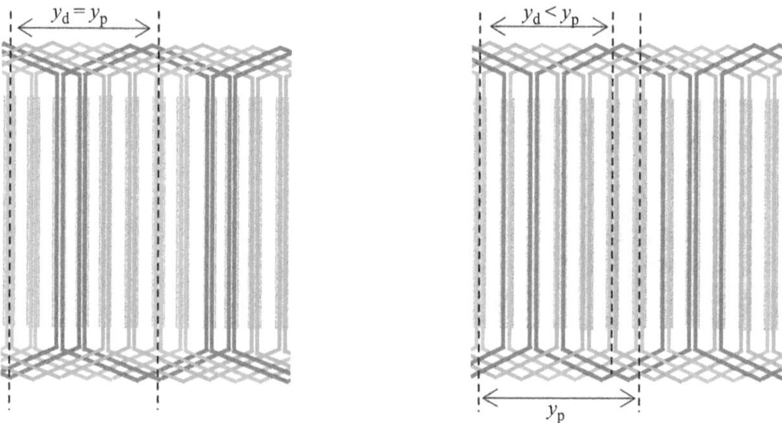


*Figure 3.32 Left: double layer non-pitched overlapping distributed winding ($q$ = 2). Right: double layer pitched overlapping distributed winding ($q$ = 2, $W/y_p$ = 5/6).*

Figure 3.32 graphically explains the method of pitching the coils. For this example, a stator winding with $q = 2$ is considered which is unpitched on the left picture but pitched by 5/6 on the right picture. The black dashed vertical lines indicate the pole pitch on the rotor as well as the implemented pitching on the stator winding.

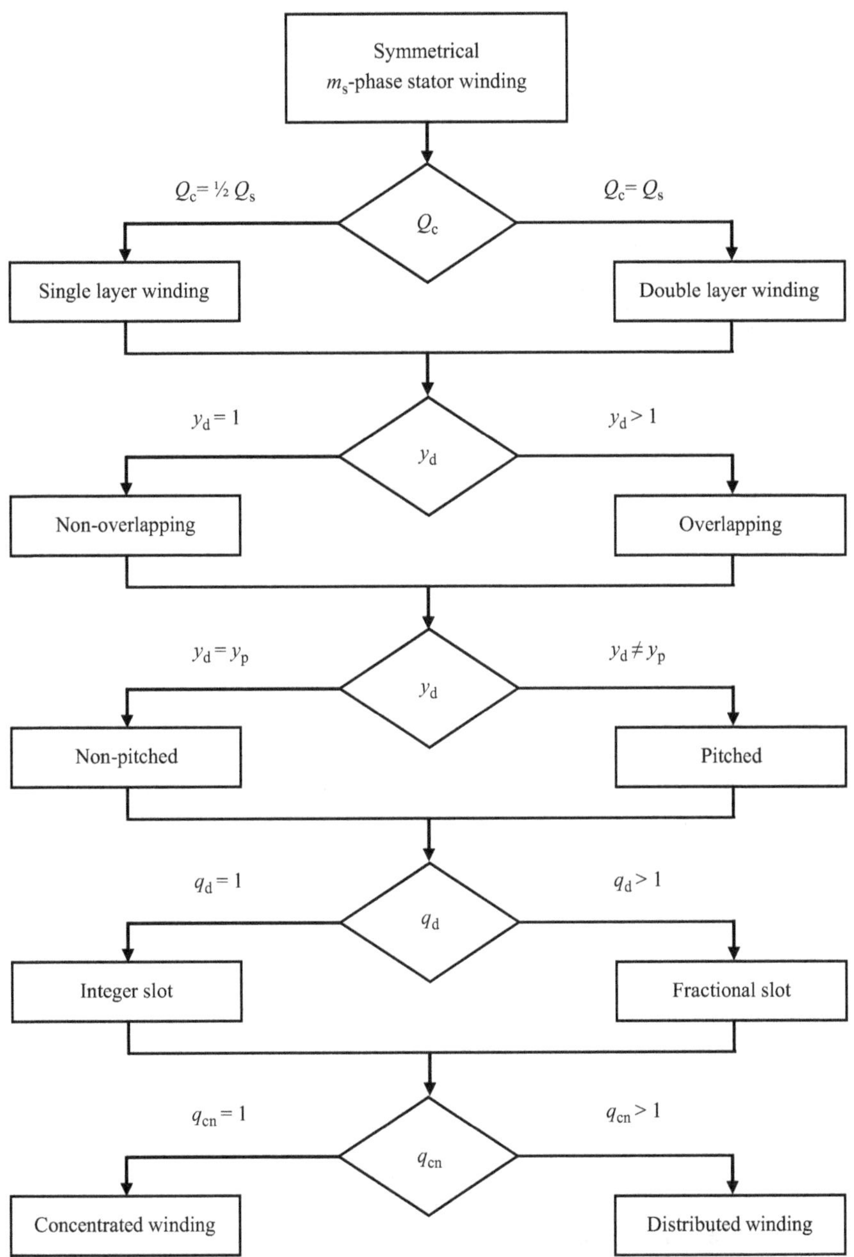


*Figure 3.33 Classification of symmetrical stator windings [10].*

## 3.4.2 Definition of the winding factor and current layer of a stator winding

After we have given an overview of symmetrical $m_s$-phase stator windings along with a classification in the previous chapter we would like to introduce two important properties of stator windings as well. These properties are the winding factor as well as the current layer of a stator winding.

### 3.4.2.1 The winding factor of a stator winding

One of the important properties of a stator winding is its resulting winding factor to its working harmonic $k_{w,p}$. The working harmonic or so-called fundamental harmonic is equal to the number of pole pairs $p$ on the rotor. The winding factor can be seen as the effectiveness of a stator winding in terms of converting the magnetic flux to induced voltage in the winding. Therefore the winding factor appears also in the equation of the induced voltage per phase:

$$U_{p,p} = \sqrt{2} \cdot \pi \cdot f_s \cdot k_{w,p} \cdot N_s \cdot \Theta_{p,1} \tag{3.26}$$

Whereas $f_s$ is the electrical frequency, $N_s$ is the number of turns per phase (which can consist of multiple single coils in series) and $\Theta_p$ is the magnet flux.

The winding factor is the product of the so-called pitching factor $k_{p,p}$ and the distribution factor $k_{d,p}$:

$$k_{w,p} = k_{p,p} \cdot k_{d,p} \tag{3.27}$$

If the rotor poles or the stator slots are in addition skewed the so-called skewing factor must be taken into account as well in the above-mentioned equation. But for simplicity is will be neglected here. The winding factor is a real number in the range of [0 1]. Due to the described pitching of a stator winding in the previous chapter and as illustrated in Figure 3.32 this number will get $<1$. Furthermore, if there are multiple coils connected in series within a stator winding within the section of one pole pair the so-called distribution factor $k_{d,p}$ needs to be taken into account as well. This distribution factor considers the geometrical addition of the voltage vectors for the induced voltage of individual coils. Therefore the winding factor results into unity only for an unpitched stator winding with coils connected in series where is no phase angle in between. This is the case for an unpitched winding with $q = 1$.

Other orders than the working harmonic of magnetic flux from the rotor side which are normally not avoidable shall be attenuated as much as possible by low winding factors of the corresponding stator winding in order to create an almost perfect sinusoidal voltage in the time domain in the stator winding. Ideally, the value has to be as high as possible for the working harmonic and zero for all other harmonics. However, this cannot be achieved by a real stator winding and in electrical machine design there is always a compromise between the maximum winding factor for the working harmonic and the lowest possible winding factors for harmonics other than the working harmonic needed.

#### 3.4.2.2 The current layer of a stator winding

The current layer of a stator winding is the spatial distribution of phase currents on the circumference of the stator. Since the phase currents vary in magnitude and time the resulting current layer is rotating inside the circumference of the stator in the direction of rotation of the rotor. If drawn in a spatial domain for a given time step the current layer results typically in a stepwise function of a certain shape. This is because the current is only flowing in stator slots and thus the steps in this spatial function are created whereas the current layer is constant over each stator tooth. The ideal current layer distribution would be a perfect sinusoidal function in the spatial distribution along the circumference. Due to the stator structure and the placement of coils is slots only this can be, however, not achieved in real world. Besides the sinusoidal fundamental current layer you need to accept that there are further sub as well as super harmonics in the corresponding spatial spectra. These are the so-called current layer harmonics.

### *3.4.3 Stator windings topologies with particular reference to low-speed permanent magnet synchronous generators*

The extremely compact design of the winding overhang and thus a lower stator winding resistance and lower losses compared to conventional distributed windings is one of the advantages of factional slot concentrated windings. The compact design of the winding overhang also reduces the overall axial length of the machine. Shorter axial length means less structural costs and a more compact machine design. Another benefit of fractional slot concentrated windings is that the number of coils inside the stator is much lower compared to distributed windings. So a lesser amount of coils has to be produced and inserted but on the other hand concentrated coils typically result in a more bulky and heavier design. Moreover, the segmentation of fractional slot concentrated windings is easily achievable since the winding is not distributed and therefore coils are not nested into each other. This makes the approach very attractive for modular designs and split generators.

However, besides these advantages, there is also a downside of fractional slot concentrated coils. The harmonic content of field waves generated by fractional slot concentrated windings is higher compared to conventional distributed windings and therefore special care must be taken while designing these kinds of machines. Otherwise, the machine design will result in too high losses, higher operating temperatures, and therefore, a too low efficiency. In general bulk electrical conductive materials have to be avoided and laminated (insulated) electrical steel can be used only for most of the active parts inside the machine. This is in particular the case for single layer concentrated windings. The generation of unwanted noise during machine operation because of these additional harmonics needs to be carefully studied in detail before a design can be released for manufacturing. The machine power factor of factional slot concentrated coils is in general slightly lower because of the additional harmonics which is resulting in an additional harmonic inductance and thus the apparent power demand is higher compared to conventional distributed windings.

However, all these downside points in particular the additional losses coming from additional harmonics can be minimized to some extend if the generator is in particular designed to low rotational speed like it is the case for direct-drive generators.

Moreover, permanent magnet generators result in a large air gap from the magnetic point of view since the magnet material itself behaves like air in the electromagnetic calculations. The typical relative permeability of magnet material is in the range of 1.05, so the magnet, which is exciting the magnetic field, behaves like air itself. This fact basically enables to use fractional slot concentrated tooth wound coils in the stator in combination with a permanent magnet rotor since a large air gap is dampening all the additional harmonics further down and thus the effect of parasitic harmonics is further minimized.

As mentioned before special care must be taken while designing generator concepts with permanent magnets on the rotor and fractional slot concentrated coils in the stator. However, if all the points are investigated carefully in detail during the design phase of such machines, the concept of fractional slot concentrated coils paired with a permanent magnet rotor can lead to very attractive solutions for direct-drive applications.

That's why we want to get more detail to the machine design of fractional slot concentrated coils in the stator paired with permanent magnets on the rotor side. As mentioned in the classification chart in Section 3.4.1.2 there are two options for fractional slot concentrated windings. These two options are:

- Single layer fractional slot concentrated coils
- Double layer fractional slot concentrated coils

In the following, we want to get more in detail to both of the above-mentioned types and highlight benefits as well as drawbacks.

Figure 3.34 depicts two different options on how the stator slots can be placed around the circumference for fractional slot concentrated coils. We do have the

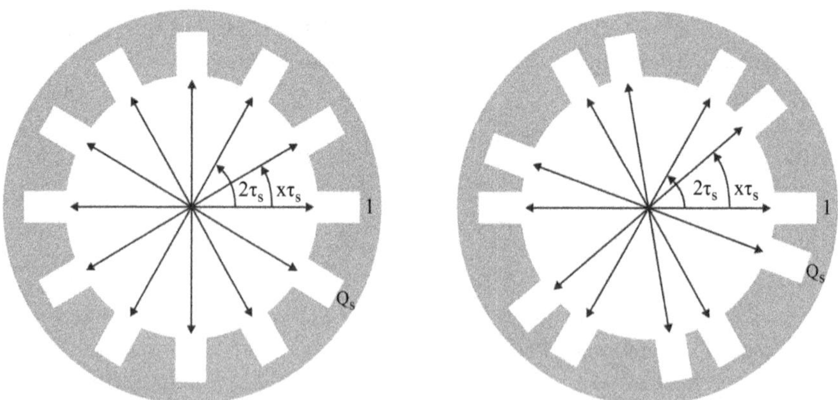


*Figure 3.34 Left: equal distribution of stator slots along the circumference ($x$ = 1.0, $\tau_s$ = slot pitch). Right: unequal distribution of stator slots along the circumference ($x > 1.0$) [10].*

option to distribute the stator slots in an equally or unequally spaced manner around the circumference. For double layer concentrated coils with only one coil geometry used there is only the option to use an equal distribution of stator slots. However, for single layer concentrated coils where only every second tooth is would we do have one additional degree of freedom to place the slots in an unequal way around the circumference. We will discuss this later to which extent this would be helpful to improve the machine design.

To depict further details and differences between single and double layer fractional slot concentrated windings we would like to introduce a few boundary conditions which should be the same for these two winding configurations. These boundary conditions are given in Table 3.5. In the following, a few examples of single layer as well as double layer fractional slot concentrated windings are given and the winding factors as well as the current harmonics to each example will be depicted and compared.

### 3.4.3.1 Single layer concentrated winding

In the following, a few examples of suitable single layer fractional slot concentrated windings are given. Since the number of slots per pole and phase mainly determines the pattern of stator current layer harmonics and winding factor one dedicated slot pole pair combination out of each group is depicted for an overview. In Table 3.6 a selection of suitable single layer concentrated windings is given. In general, there are many more variants possible. From these selected ones we think that these are the most promising ones for the design of direct-drive permanent-magnet generator systems with particular reference to low speed. The column with

*Table 3.5 Boundary conditions*

| | | |
|---|---|---|
| Generator stator air gap diameter | $d_{si}$ | 5000 mm |
| Number of phases inside the generator winding | $m_s$ | 3 |
| Range of amount of stator slots | $Q$ | 180–210 |
| Range of amount of pole pairs | $p$ | 70–100 |
| Fundamental winding factor | $k_{w,p}$ | $\geq 0.89$ |
| Number of slots per pole and phase | $q$ | 0.25–0.5 |
| Ratio tooth width/slot pitch in the stator | $b_d/\tau_q$ | 1/2 |

*Table 3.6 Overview of suitable single layer concentrated windings sorted by fundamental winding factor $k_{w,1}$ in descending order*

| ***Slots, poles, basic windings*** | | | | | ***Cogging torque*** | | ***Winding factor*** | | | | |
|---|---|---|---|---|---|---|---|---|---|---|---|
| $q$ | $Q$ | $p$ | $bw$ | $x$ | $\nu_{cog}$ | $n_{cog}$ | $k_{w,1}$ | $k_{w,5}$ | $k_{w,7}$ | $k_{w,11}$ | $k_{w,13}$ |
| 1/2 | 180 | 60 | 60 | 1.2 | 360 | 6 | 0.9511 | 0 | 0.5878 | 0.9511 | 0.5878 |
| 1/2 | 180 | 60 | 60 | 1.0 | 360 | 6 | 0.866 | 0.866 | 0.866 | 0.866 | 0.866 |
| 3/8 | 180 | 80 | 20 | 1.0 | 1440 | 9 | 0.9452 | 0.1398 | 0.0607 | 0.0607 | 0.1398 |
| 2/5 | 180 | 75 | 30 | 1.0 | 900 | 6 | 0.9659 | 0.2588 | 0.2588 | 0.9659 | 0.9659 |

the variable $x$ in Table 3.6 refers to the distribution factor of stator slots mentioned in Figure 3.34. The first two rows in Table 3.6 is basically the same stator winding design but the first is with unequal distribution of slots along the circumference and the second is with equally distributed slots along the circumference. This comparison is made in order to depict the impact on winding factor if this degree of freedom is taken into account. The two columns $\nu_{\text{cog}}$ and $n_{\text{cog}}$ is some other important information on the cogging torque of the machine. Cogging torque is the cogging between stator and rotor at no load. It is desired to have no cogging at all but due to the fact that the stator is made of iron tooth with slots for its winding in between this is unavoidable in reality. $\nu_{\text{cog}}$ is the number of cogging torque cycles along the entire circumference whereas $n_{\text{cog}}$ is the normalized order with respect to the number of pole pairs $p$ in the rotor. The number of cogging torque cycles can be determined by the least common multiple (lcm) of rotor poles and stator slots:

$$\nu_{cog} = lcm(2p, Q) \tag{3.28}$$

$$n_{cog} = \frac{\nu_{cog}}{p} \tag{3.29}$$

The higher the cycles of cogging along the circumference the lower the cogging torque amplitude and therefore the smoother the machine in operation. Therefore, a high number of cogging torque cycles is desired.

The corresponding current layer spectra to each mentioned variant in Table 3.6 are depicted in Figures 3.35–3.38. Each harmonic which shows up in a normalized

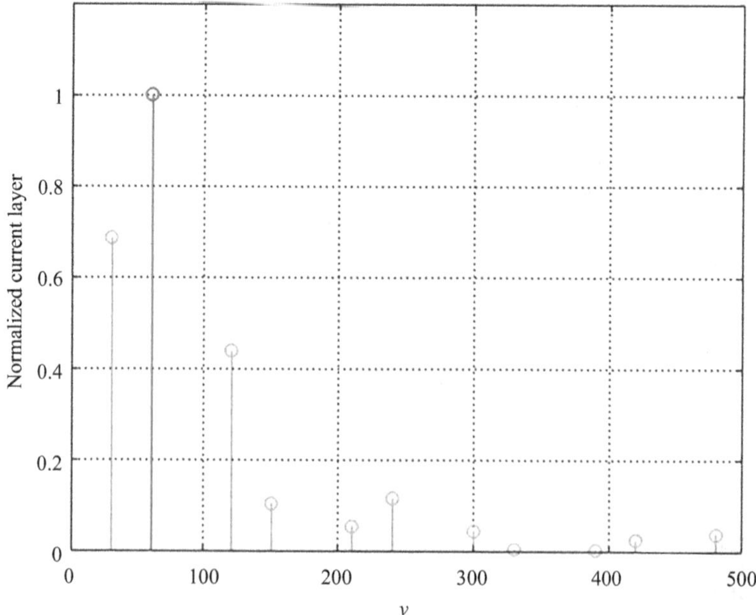


*Figure 3.35* $q = 1/2$, $Q = 180$, $p = 60$, $k_{w,1} = 0.866$, $x = 1.0$

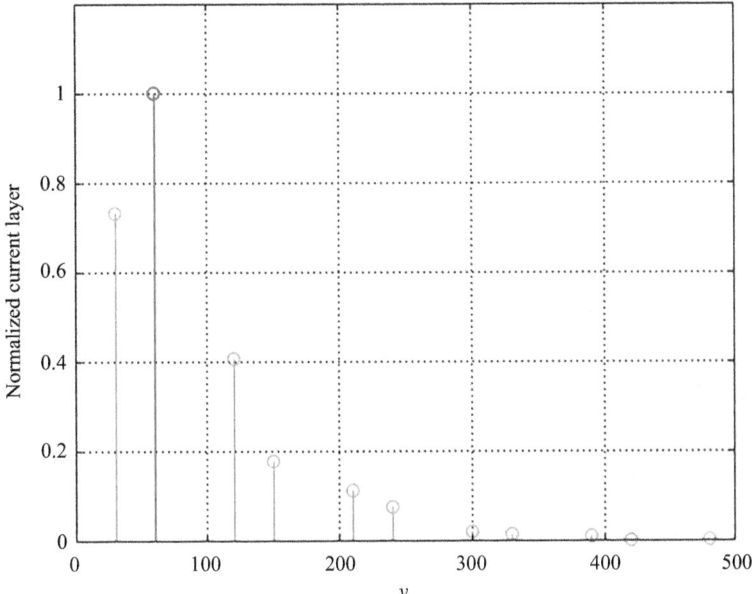


*Figure 3.36* $q = 1/2,\ Q = 180,\ p = 60,\ k_{w,1} = 0.9511,\ x = 1.2$

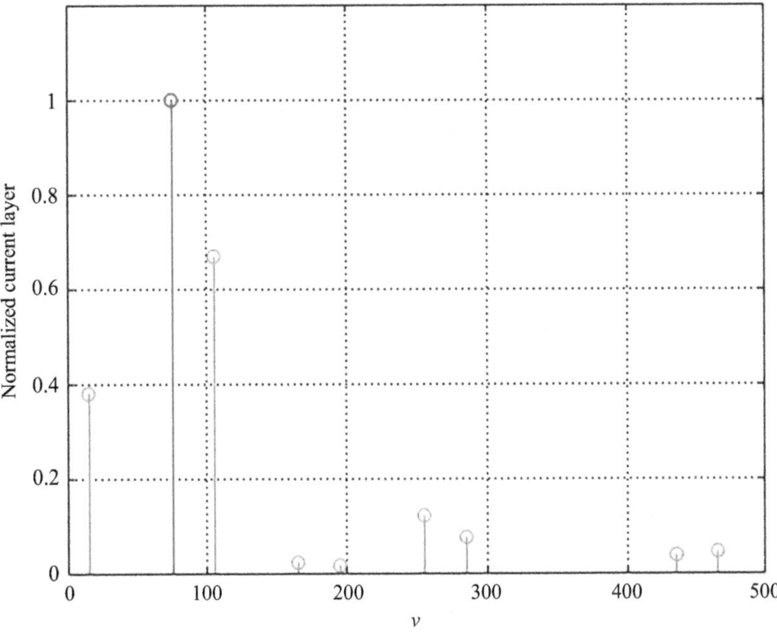


*Figure 3.37* $q = 2/5,\ Q = 180,\ p = 75,\ k_{w,1} = 0.9659,\ x = 1.0$

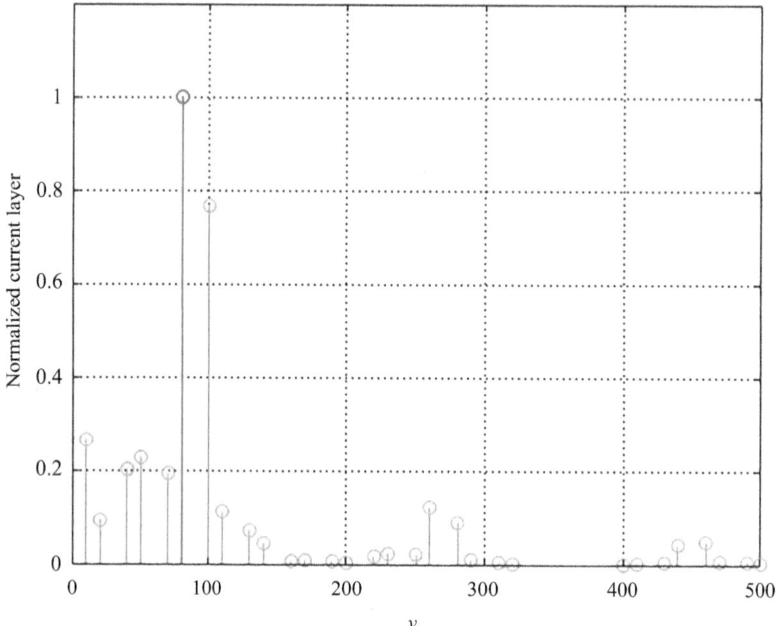


*Figure 3.38* $q = 3/8$, $Q = 180$, $p = 80$, $k_{w,1} = 0.9452$, $x = 1.0$

magnitude of less than 0.001 p.u. is neglected and the following plots have been truncated to depict all harmonics up to 500th order only for a better visibility of the dominating components. The working harmonic is depicted in red within each plot whereas the sub- and super harmonics are depicted in blue.

From these examples in Figures 3.35–3.38 it can be seen that for a single layer concentrated coil design there are two dominating current layer harmonics next to each other paired with some further non-negligible harmonic current layers in the spectrum. This is one of the inherent properties of fractional slot concentrated windings. Comparing Figures 3.35 and 3.36 you can see that for an almost identical current layer the winding factor for its working harmonic $k_{w,1}$ could be increased considerably, if an uneven distribution of stator slots is taken into account instead.

#### 3.4.3.2 Double layer concentrated winding

Within the previous chapter, we had a look at single layer fractional slot concentrated windings. Now we would like to give some further examples for double-layer fractional slot concentrated windings as well. Table 3.7 gives an overview of our selection. Of course also here many more variants and combinations do exist but from these examples we believe that these are the most promising for our application. The corresponding boundary conditions from Table 3.5 are also taken into account here as well.

*Table 3.7 Overview of suitable double layer concentrated windings sorted by fundamental winding factor $k_{w,1}$ in descending order*

| ***Slots, poles, basic windings*** | | | | ***Cogging torque*** | | ***Winding factor*** | | | | |
|---|---|---|---|---|---|---|---|---|---|---|
| $q$ | $Q$ | $p$ | $bw$ | $\nu_{cog}$ | $n_{cog}$ | $k_{w,1}$ | $k_{w,5}$ | $k_{w,7}$ | $k_{w,11}$ | $k_{w,13}$ |
| 5/14 | 210 | 98 | 12 | 2940 | 30 | 0.9514 | 0.1732 | 0.1111 | 0.0445 | 0.0213 |
| 4/11 | 192 | 88 | 16 | 2112 | 12 | 0.9495 | 0.1629 | 0.0959 | 0.0165 | 0.0165 |
| 3/8 | 180 | 80 | 20 | 1440 | 9 | 0.9452 | 0.1398 | 0.0607 | 0.0607 | 0.1398 |
| 5/13 | 180 | 78 | 6 | 2340 | 30 | 0.9358 | 0.1000 | 0.0156 | 0.0732 | 0.0934 |
| 2/5 | 180 | 75 | 30 | 900 | 6 | 0.9330 | 0.0670 | 0.0670 | 0.9330 | 0.9330 |
| 3/7 | 180 | 70 | 20 | 1260 | 9 | 0.9019 | 0.0378 | 0.1359 | 0.1359 | 0.0378 |

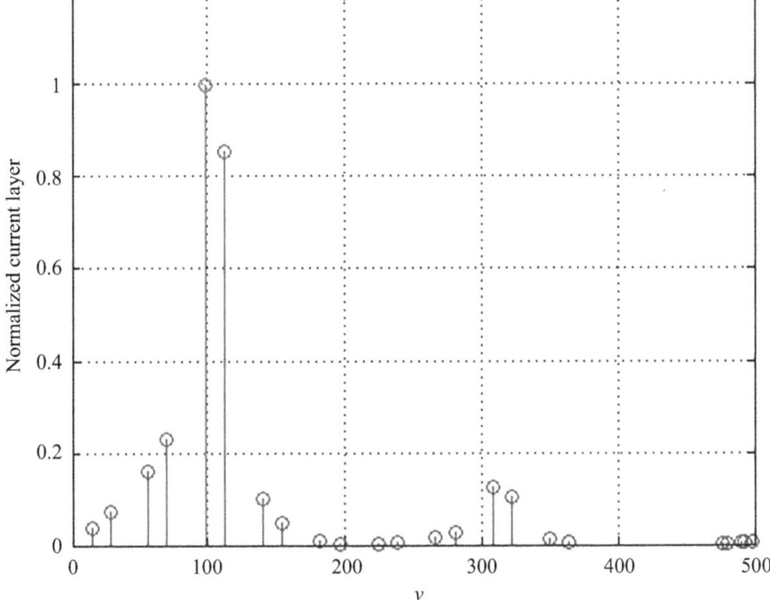


*Figure 3.39* $q = 5/14$, $Q = 210$, $p = 98$, $k_{w,1} = 0.9514$

Within Figures 3.39–3.44 the normalized corresponding current layer distribution is depicted for each proposed group of slots per pole and phase which has been presented in Table 3.7. Like for the single layer variants in the previous chapter two dominating components do occur in the spectrum along with other harmonics.

If the current layer spectra from the single layer concentrated winding in Figures 3.35–3.38 are compared with the spectra of the double layer winding in Figures 3.39–3.44 it can be seen that the normalized current layer for sub- as well as super harmonics is in general lower.

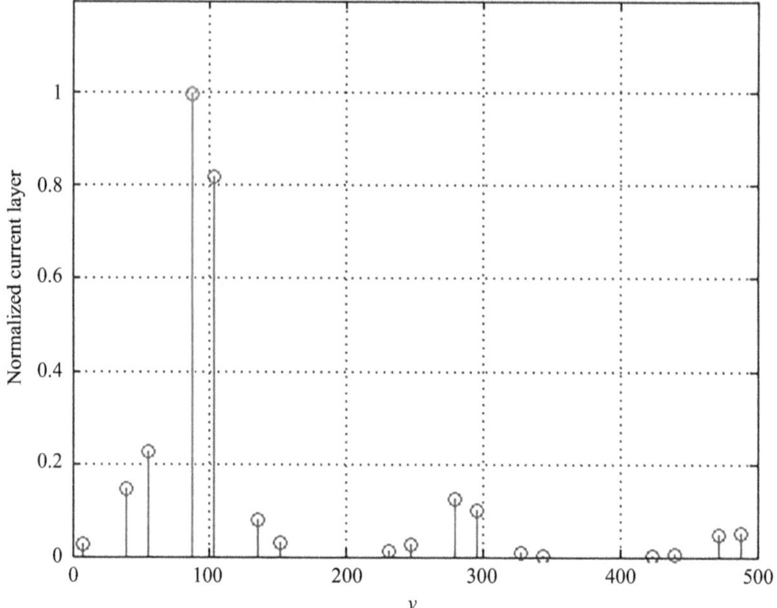


*Figure 3.40* $q = 4/11$, $Q = 192$, $p = 88$, $k_{w,1} = 0.9495$

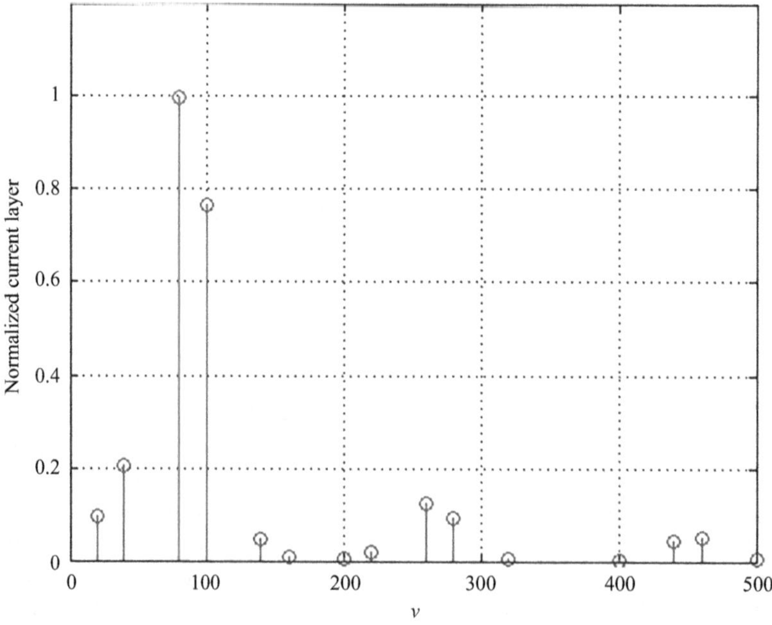


*Figure 3.41* $q = 3/8$, $Q = 180$, $p = 80$, $k_{w,1} = 0.9452$

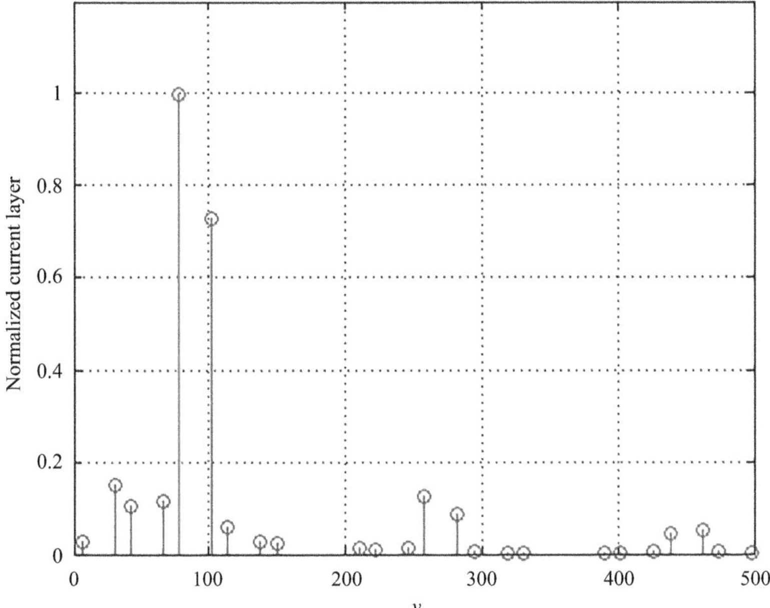


*Figure 3.42* $q = 5/13$, $Q = 180$, $p = 78$, $k_{w,1} = 0.9358$

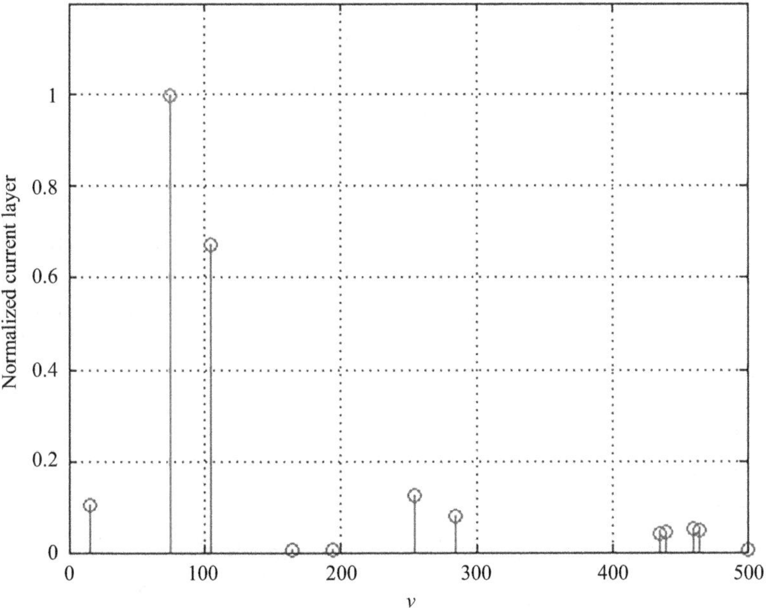


*Figure 3.43* $q = 2/5$, $Q = 180$, $p = 75$, $k_{w,1} = 0.9330$

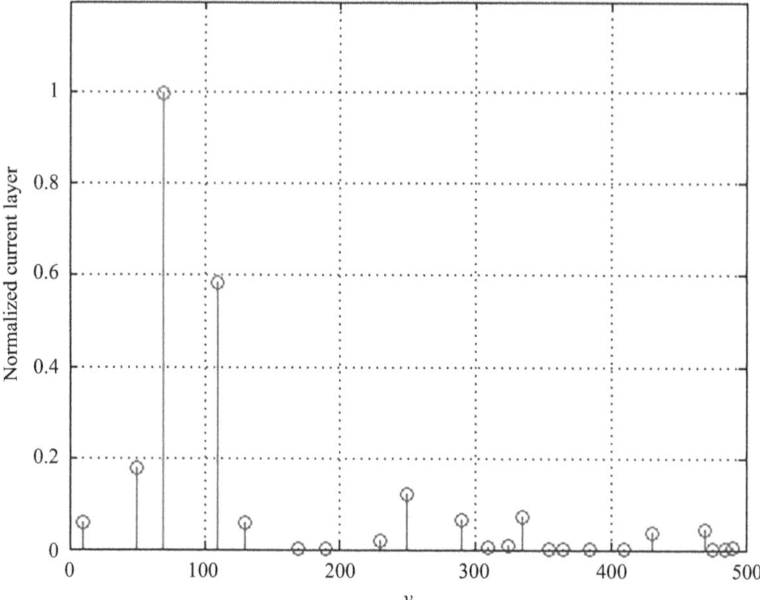


*Figure 3.44* $q = 3/7$, $Q = 180$, $p = 70$, $k_{w,1} = 0.9019$

### 3.4.3.3 Summary

As a summary, the following drawbacks and benefits of single layer and double layer concentrated windings shall be listed.

#### *3.4.3.3.1 Single layer concentrated coils*

*Benefits*

- Only 50% of coils compared to double layer
- Less winding connections in the overhang
- Degree of freedom to place the slots in uniform/non-uniform distribution in the stator to optimize the machine design
- Easy segmentation of stator windings is possible for a robust and modular machine design in case the machine needs to be split to solve transport issues

*Drawbacks*

- Higher harmonic content in the resulting current layer
- Higher losses, in particular in the rotor
- Lower harmonic orders in cogging torque and therefore higher cogging at no load
- Higher load-dependent torque ripple

- Placement of electrical conductive parts also a massive steel carrying structure in particular in the rotor not possible at all
- High fundamental winding factor on the cost of higher harmonic winding factors

##### *3.4.3.3.2 Double layer concentrated coils*

*Benefits*

- Lower harmonic content in particular lower magnitudes of harmonics in the resulting current layer
- Higher harmonic orders in cogging torque and therefore lower cogging at no load
- Smooth torque generation
- Conductive parts might be allowed in the rotor structure to some extent
- High fundamental winding factor with low harmonic winding factors possible

*Drawbacks*

- Double amount of coils needed compared to single layer concentrated coils
- More winding connections in the overhang
- Segmentation of stator windings possible but difficult to achieve a robust approach. The protection of coil legs is missing at segmentation boundaries and difficult to achieve.

### *3.4.4 Conclusion*

Our intention was always to design a modular and robust machine which can be segmented in stator as well as rotor (see Figure 3.14) to solve all the transport issues which are occurring with generators of this size. Our focus was therefore on single layer concentrated coils even if we knew before that there are some drawbacks. Along with a tailored design of an innovative generator carrying structure in particular on the rotor side which is minimizing part of the drawbacks we have started and designed a machine prototype.

## 3.5 Permanent magnet rotor topologies for direct drive generators

The main objective of this chapter is to depict and optimize the main generator characteristics with the chosen stator winding which has been presented in Section 3.4 as an overview and in particular mentioned in Section 3.4.4 but with a set of different magnet configurations in the rotor. To this aim, different permanent-magnet rotor configurations for the generator are investigated in detail. The most appropriate design should be pointed out along with drawbacks and benefits of each configuration. It is important that the generator and frequency converter are matching to each other. This total system consideration is offering various advantages to reduce the total system cost to a minimum.

The implemented spoke type magnet arrangement within the prototype design which is explained in chapter 3.5.2 is basically suitable but due to the following facts which have been experienced during prototype assembly and subsequent prototype testing, various other options to arrange the magnets are reconsidered further:

- A high nonlinear radial force dependency in terms of slight changes within air gap width leads to a considerable stress within structural parts and therefore structural deformations are a result.
- A non-negligible amount of leakage flux in the rotor yoke further amplifies the nonlinear behavior of the radial attracting force between the stator and rotor. Further to that the leakage flux results in a slightly mitigated magnet weight utilization in terms of torque production for a certain generator current. Magnet weight utilization in the following means magnet weight in terms of torque production for a certain phase current.
- A limited adaption possibility as it is required for a total system consideration.

Alternative magnet arrangements compared to the spoke magnet rotor should be suitable to easy replace the current electromagnetic active part within the rotor structure. As a boundary condition, the stator properties such as geometry as well as the winding schematic should be kept unchanged. This implies that the rotor pole count and the rotor pole surface toward the machine air gap, which was chosen at the beginning, remains also constant.

### *3.5.1 Boundary conditions*

For all subsequent magnet configurations, the following boundary conditions according to Table 3.8 are taken into account. These are the boundary conditions of the reference machine which has been built and successfully tested on the test bench as well as on top of the tower.

*Table 3.8 Boundary conditions*

| | | |
|---|---|---|
| Stator bore diameter | $d_{si}$ | 4900 mm |
| Mechanical air gap | $\delta$ | 4 mm |
| Number of pole pairs | $2p$ | 96 |
| Number of stator slots | $Q$ | 216 |
| Number of slots per pole and phase | $q$ | 3/8 |
| Winding topology | | Single layer concentrated coils |
| Rotational speed | $n$ | 13.2 rpm |
| Phase current | $I_s = I_q$, | 3200 A ($I_d = 0$ A) |
| Terminal voltage | $U_{LL}$ | 750 V |
| Magnet material | – | N38SH |
| Stator lamination material | – | M400-65A |
| Rotor lamination material | – | M1300-100A |

### *3.5.2 The spoke magnet rotor*

As an introduction, a short retrospect to the spoke magnet rotor design is given within this chapter. This configuration was chosen for the generator setup within the *wd2.8+109* turbine. Furthermore, this chapter is going to motivate, why an investigation of other permanent magnet rotor topologies such as surface magnets, simple interior magnets, and V-shape interior magnets in particular is reasonable.

#### 3.5.2.1 Rotor geometry and properties

The spoke-type magnet arrangement depicted in Figure 3.45 is the implemented rotor design within the prototype machine. An advantage of the spoke magnet design is the laminated discontinuous piecewise integrated rotor iron. This topology perfectly matches with a single layer concentrated winding approach within the stator because it attenuates in particular any parasitic sub-harmonics that are typical and always present within such a winding schematic. Furthermore, the eddy currents within the conductive magnet material are considerable small compared to any other competing design, if the same magnet dimensions and magnet segmentation is considered.

However, this topology has to have a very small machine air gap between the stator and rotor in the range of only a few millimeters. The larger the air gap, the larger will be the leakage flux content in the rotor back which is depicted within Figure 3.45 for 4 mm air gap. Large leakage flux content implies bad magnet utilization. Since the cost of magnet material is a dominating part of the total machine cost, it is difficult to obtain an economical solution with a spoke magnet design, if a certain machine air gap of several millimeters needs to be considered. Referring to the dimensions of a direct drive generator, air gaps below 5 mm are difficult to achieve at a reasonable cost for the carrying structure. In particular, small air gaps are difficult to achieve if the machine does not have a special reinforced housing and as provided conventionally, two bearing brackets.

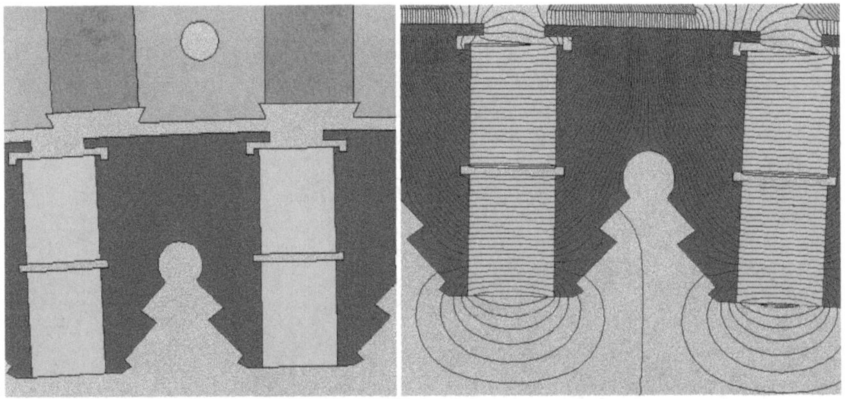

*Figure 3.45 Right: implemented spoke-type magnet arrangement. Left: stray field within the rotor yoke at no load.* $I_s = 0\ A$, $T_{mag} = 80\ °C$.

*Table 3.9 Spoke-type magnet arrangement properties*

| | | |
|---|---|---|
| Single magnet dimension | $b \times l \times h$ | 25 × 45 × 33 mm |
| Number of magnets in the rotor | – | 7680 pcs. |
| Total magnet mass | $m_{\text{mag}}$ | 2157 kg |
| Machine torque | $T_{el}$ | 2200 kNm |
| Machine air gap power | $P_d$ | 3041 kW |
| Power factor | $\cos(\varphi)$ | −0.72 |
| Short circuit peak torque | $Tsc$ | 2984 kNm |
| Short circuit peak torque ratio | – | 1.36 p.u. |
| Short circuit peak current | Isc | 10286 A |
| Short circuit peak current ratio | – | 3.21 p.u. |

This drawback has turned out more distinct during the assembly and testing of the prototype machine. Further to that, the points which have been mentioned at the beginning of Section 3.5 are also not desirable as well. That's why the search for other suitable rotor structures that accommodate permanent magnets in various shapes is considered in the following sections. The corresponding model properties of the implemented spoke-type magnet arrangement within the prototype are listed in Table 3.9.

### 3.5.3 Surface-mounted magnets

In the following section, a rotor with continuously laminated iron yoke along the circumference and surface-mounted magnets is taken into account as a benchmark because of its already known considerable small dependency of radial attraction force in terms of air gap width changes and thus less stress on structural parts.

#### 3.5.3.1 Rotor geometry and properties

Figure 3.46 depicts the geometry which has been considered for this approach. Herein the same magnet dimensions and therefore the same magnet weight is taken

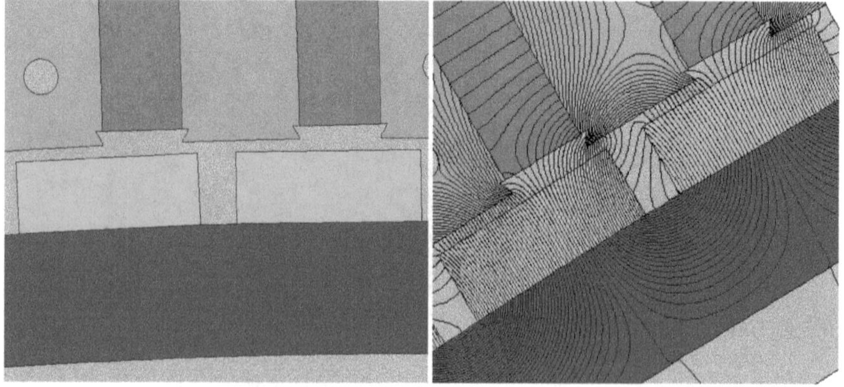

*Figure 3.46 Right: rotor with laminated iron yoke and surface-mounted magnets. Left: rotor field distribution at load $I_s = I_q = 3200$ A, $T_{mag} = 80$ °C.*

*Table 3.10 Surface-mounted magnet properties*

| | | |
|---|---|---|
| Single magnet dimension | $b \times l \times h$ | 25 × 45 × 33 mm |
| Number of magnets in the rotor | – | 7680 pcs. |
| Total magnet mass | $m_{\text{mag}}$ | 2157 kg |
| Machine torque | $T_{el}$ | 1919 kNm |
| Machine air gap power | $P_d$ | 2652 kW |
| Power factor | $\cos(\varphi)$ | −0.69 |
| Generator terminal voltage | $U_{LL}$ | 676 V |
| Short circuit peak torque | $Tsc$ | 2665 kNm |
| Short circuit peak torque ratio | – | 1.21 p.u. |
| Short circuit peak current | Isc | 11743 A |
| Short circuit peak current ratio | – | 3.67 p.u. |

into account compared to the spoke magnet design within Section 3.5.2. The magnets are simply glued on top of the laminated iron yoke and may be further secured with a glass fiber reinforced bandage. Here, the huge resulting air gap – the Permeability of magnet material is typically around $\mu_0$ – helps to attenuate the parasitic sub-harmonics generated by the considered stator winding to a minimum acceptable level. All other machine dimensions are kept constant with respect to the previous chapter. The rotor yoke must be laminated, in order to not drive the rotor losses beyond a certain level and to limit the rotor temperature to any appropriate value. In particular a laminated rotor is needed if a single layer concentrated coil topology on the stator side is chosen which has an inherent amount of sub-harmonic field waves which can penetrate deep into structural machine parts.

But the question remains how to ensure that the glued magnets keep glued on the laminated rotor yoke over the entire machine lifetime? Serious concerns of magnet flaking during operation within harsh environments such as tropical conditions have been noticed in similar rotor structures. Therefore it is not advisable at all to follow up with this approach. From the eddy current content point of view within the magnet material, which is higher compared to other promising designs, the magnets need to be divided into many considerable smaller parts (segmentation) to get rid of those eddy currents. Furthermore, there exists only a limited adaption possibility. Since the pole surface is fixed, the magnet material can be adjusted in its height only to change the flux linkage within the stator winding. This measure is not economically in terms of magnet cost. Also modifying or playing with the pole coverage does not offer huge advantages. The corresponding model properties are listed in Table 3.10.

## 3.5.4 Simple interior magnets

To overcome the issue with the glued magnets which has been mentioned within Section 3.5.3, an approach wherein the magnets are covered with a thin layer of the laminated rotor iron is investigated here as second promising design.

### 3.5.4.1 Rotor geometry and properties

Figure 3.47 depicts the geometry details which have been considered for this approach. Again the magnet dimensions and its internal segmentation remain unchanged compared to the models discussed previously. A thin laminated iron layer above the magnets is protecting the magnets against flaking and corrosion. This approach may ensure a better corrosion protection of the magnet material, if the remaining air within the magnet slot is filled with some suitable casting material or potting compound after the magnets have been inserted.

Because the laminated rotor iron is entirely covering the magnets, any subharmonics generated within the considered stator winding are amplified because of the presence of rotor iron within the air gap region in a combination with a continuous rotor yoke. This leads to a distortion within the intended magnetic circuit and affects the resulting machine characteristic to a considerable extent.

The main drawback is that, because of this magnet coverage with electrical steel, the rotor stray field becomes more and therefore the utilization of the magnet material is low. A huge amount of magnet weight is required, if the same flux linkage and machine torque compared to the spoke magnet design has to be generated in the stator winding.

However, this approach yields to the lowest radial force exposure in terms of changes within the air gap with, which is even lower compared to the previously discussed surface-mounted magnet design. As within the previously discussed topology, here the adaption possibilities are also limited because of the same reasons. The corresponding model properties are listed within Table 3.11.

In order to limit the considerable amount of leakage flux and to push the magnet utilization toward higher economical values the buried magnets inside the laminated rotor iron need to be rearranged to some other shape. This approach will be discussed in the next section.

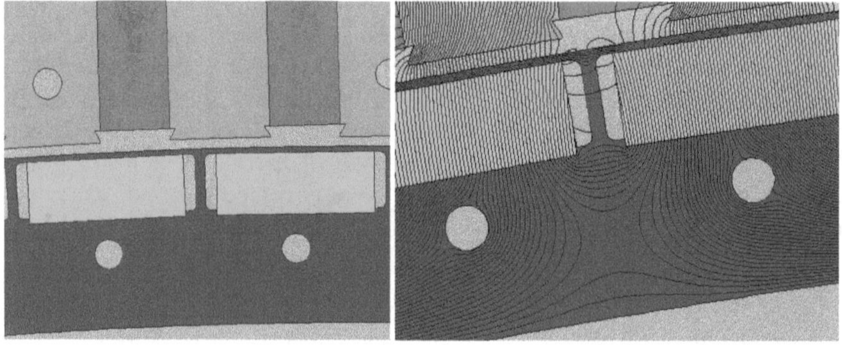

*Figure 3.47 Right: rotor with laminated iron yoke and interior magnets. Left: rotor field distribution at load $I_s = I_q = 3200$ A, $T_{mag} = 80$ °C.*

*Table 3.11 Interior magnet properties*

| | | |
|---|---|---|
| Single magnet dimension | $b \times l \times h$ | 25 × 45 × 33 mm |
| Number of magnets in the rotor | – | 7680 pcs. |
| Total magnet mass | $m_{mag}$ | 2157 kg |
| Machine torque | $T_{el}$ | 1706 kNm |
| Machine air gap power | $P_d$ | 2358 kW |
| Power factor | $\cos(\varphi)$ | −0.56 |
| Generator terminal voltage | $U_{LL}$ | 732 V |
| Short circuit peak torque | $Tsc$ | 2500 kNm |
| Short circuit peak torque ratio | – | 1.14 p.u. |
| Short circuit peak current | Isc | 6672 A |
| Short circuit peak current ratio | – | 2.09 p.u. |

## *3.5.5 V-shape interior magnets*

This section deals with a V-shape magnet arrangement within the rotor structure. A V-shape magnet arrangement is basically a derivative of the simple interior magnet design which has been discussed in the chapter before. Since this approach offers and combines various advantages compared to all previously introduced models, it is considered and discussed here finally.

### 3.5.5.1 Rotor geometry in general

A V-shape magnet arrangement according to Figure 3.48 inside the rotor structure combines mainly the benefits of a spoke magnet and the interior magnet design

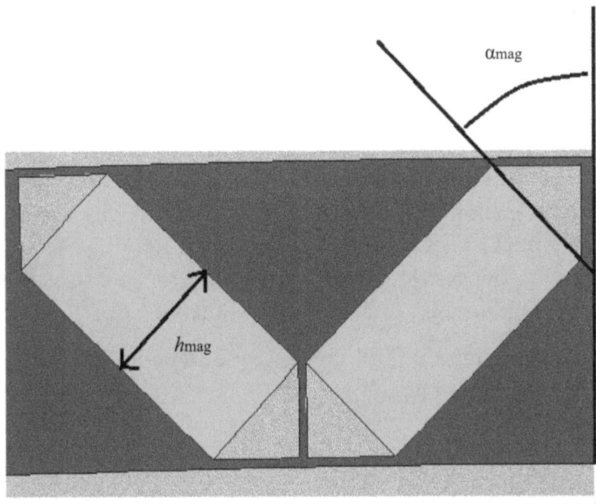


*Figure 3.48 Parameters to be considered to get a suitable design with interior V-shape magnets*

which have been discussed in Sections 3.5.2 and 3.5.4, respectively. The main drawbacks of glued surface magnets such as magnet flaking within Section 3.5.3 do not exist within this design in addition.

By means of adjusting the magnet tilting angle $\alpha_{mag}$ as well as the magnet thickness $h_{mag}$ the flux concentration level can be adapted over a wide range. With small tilting angles for almost no flux concentration up to highest possible flux concentration if $\alpha_{mag}$ is equal to 90°. If the resulting stray field in the yoke is neglected here, every desired level can be freely adjusted. The flux concentration level is directly linked to the resulting radial magnetic force exposure between stator and rotor. Therefore the radial magnetic force can be adapted to a given structural stiffness to some extent, if required. Further to that the V-shape approach leads to a minimum leakage flux inside the rotor compared to other promising designs, if the two parameters $\alpha_{mag}$ and $h_{mag}$ are adapted properly to the given pole area and the stray bridges in the rotor iron have been designed accordingly. The magnet thickness is mainly adapted to a value where the design is withstanding the material-specific demagnetization level. The magnet material can be best utilized, if the magnet thickness is as small as possible at larger magnet tilting angles. However, the thinner the magnets the more critical the irreversible demagnetization in case of a sudden short circuit within the stator winding. This has to be avoided under all circumstances since a running machine with partially demagnetized magnets will output less torque and power and is prone to noise emissions.

## *3.5.6 Modified air gap dimension for V-shape magnets*

Within all previously discussed rotor models the mechanical air gap was considered to be 4 mm. However, for this design the machine air gap has been increased by 1 mm in addition to 5 mm in total, since it seems to be the most promising design to the given structural stiffness. After seeing all the efforts and obstacles we had to take into account within the workshop during machine assembly and machine testing with spoke type magnet rotor, this measure of increasing the air gap further by 1 mm in addition is mandatory.

### 3.5.6.1 Two promising models with V-shape interior magnets

As before all stator geometry parameters besides the air gap geometry are kept constant and only the magnet arrangement inside the rotor has been modified. For the comparison with the previously discussed rotor models two different configurations are taken into account in the following analysis:

- Lowest flux concentration for a given torque value in order to keep the radial magnetic load as low as possible
- Lowest magnet weight for a given torque value in order to determine the lowest possible magnet weight

The values of magnet tilting angle $\alpha_{mag}$ and magnet thickness $h_{mag}$ that fulfill these boundary conditions have been determined by an optimization method which is introduced within Section 3.6.1. The considered torque for this optimized magnet

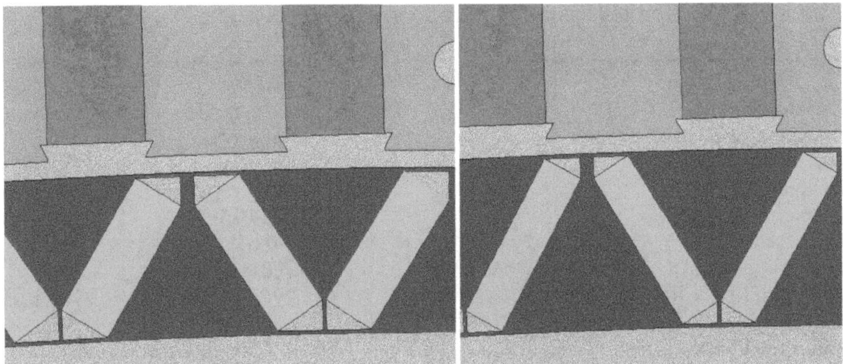

*Figure 3.49* *V-shape rotor magnet arrangement. Boundary conditions:* $T_{el}$ = *2000 kNm, n = 13.2 rpm,* $I_s = I_{s,q}$ = *3200 A,* $I_{s,d}$ = *0 A,* $T_{mag}$ = *80 °C. Right: lowest possible flux concentration. Left: lowest magnet weight with acceptable demagnetization risk.*

shape is 2000 kNm at 3200 Amperes phase current. The results for magnet tilting angle and resulting magnet thickness are given below:

- The lowest flux concentration at: $\alpha_{\text{mag}} = 57.75°$ and $h_{\text{mag}} = 16$ mm
- The lowest magnet weight at: $\alpha_{\text{mag}} = 59.88°$ and $h_{\text{mag}} = 12.6$ mm

Figure 3.49 depicts these two optimized rotor models to the boundary conditions which have been mentioned above. The boundary condition leads already to a considerable magnet tilting angle with a corresponding higher radial attraction force exposure. As within the simple interior magnet design, the remaining air pockets inside the magnet slots which is necessary to reduce the leakage flux can be filled with some casting material after the magnet has been inserted.

However, the resulting delicate latticed continuous rotor lamination structure has the drawback that this has to be handled with care during production since it tends to get bend easily.

The corresponding properties of these two models are listed and compared in Table 3.12.

## 3.5.7 *Comparison of the results*

Within this section, the results of the different rotor models which have been investigated within the previous sections will be compared.

### 3.5.7.1 Radial force as a function of air gap width

The resulting radial force – especially the steepness of the increase in terms of air gap width – is determining the structural design and its stiffness of the generator housing significantly. Further to that it answers the question whether a generator may be equipped with two bearing brackets or whether a solution with one bearing

*Table 3.12 V-shape interior magnet properties*

| | | | |
|---|---|---|---|
| Magnet tilting angle | $\alpha_{mag}$ | 57.75° | 59.88° |
| Magnet thickness | $h_{mag}$ | 16 mm | 12.6 mm |
| Number of magnets in the rotor | – | 7680 pcs. | 7680 pcs. |
| Total magnet mass | $m_{mag}$ | 1836 kg | 1698 kg |
| Machine torque | $T_{el}$ | 2000 kNm | 2000 kNm |
| Machine air gap power | $P_d$ | 2764 kW | 2764 kW |
| Power factor | $\cos(\varphi)$ | −0.67 | −0.65 |
| Generator terminal voltage | $U_{LL}$ | 708 V | 726 V |
| Short circuit peak torque | $T_{sc}$ | 2975 kNm | 2745 kNm |
| Short circuit peak torque ratio | – | 1.49 p.u. | 1.37 p.u. |
| Short circuit peak current | Isc | 10650 A | 9840 A |
| Short circuit peak current ratio | – | 3.33 p.u. | 3.08 p.u. |

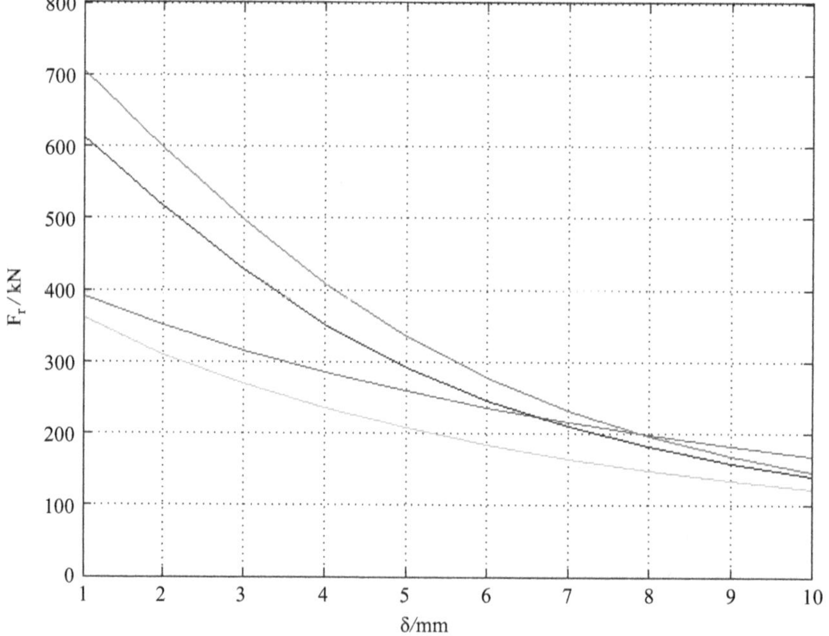


*Figure 3.50 Radial force in terms of air gap width for $I_s = 0$ A, $T_{mag} = 80°$ C. Spoke magnet arrangement (red), V-shape magnet arrangement 57.75°/16 mm (black), surface magnets (blue), simple interior magnet arrangement (green).*

bracket is sufficient. In the following, the radial loads of all investigated permanent magnet rotor types are compared in terms of air gap width in order to find out advantages and disadvantages of each discussed design.

The results of this comparison are illustrated in Figure 3.50. As expected the rotor with spoke magnet arrangement has the highest radial force and the steepest increase in terms of air gap width for the given design. This is mainly because of the non-negligible amount of leakage flux in the rotor yoke which contributes to the total flux linkage in case of small air gaps.

The V-shape magnet arrangement has an inherently lower amount of leakage flux and therefore the steepness is decreased to some extent compared to the spoke magnet design.

The surface-mounted magnet and the simple interior magnet design leading to the lowest increase in radial force in terms of air gap with. This is due to the fact that the total air gap within the magnetic circuit is larger.

### 3.5.7.2 Comparison of machine properties

Table 3.13 compares all discussed rotor models which have been introduced within Sections 3.5.2–3.5.5. As mentioned before the spoke magnet design is the reference

*Table 3.13 Comparison of all previously discussed promising rotor types. Boundary condition: $I_s = I_q = const. = 3200\ A$, $T_{mag} = 80\ °C$.*

| | | **Spoke magnets** | **Surface magnets** | **Interior magnets** | **V-magnets $\alpha_{mag}$ 57.75° $h_{mag}$ 16 mm** | **V-magnets $\alpha_{mag}$ 59.88° $h_{mag}$ 12.6 mm** |
|---|---|---|---|---|---|---|
| Machine air gap | $\delta$/mm | 4 | 4 | 4 | 5 | 5 |
| Magnet weight | $m_{mag}$/kg | 2110 | 2110 | 2110 | 1836 | 1698 |
| Machine speed | $n$/rpm | 13.2 | 13.2 | 13.2 | 13.2 | 13.2 |
| **No load** | | | | | | |
| Terminal voltage | $U_{LL}$/V | 612 | 497 | 465 | 533 | 542 |
| Cogging torque[c] | $T_{cog,18}$/kNm | 13 | 0.8 | 13 | 0.9 | 17 |
| Short circuit peak current[a] | $I_{sk}$/A | 8760 | 8880 | 7260 | 8448 | 7992 |
| | p.u. | 2.74 | 2.78 | 2.27 | 2.64 | 2.50 |
| Short circuit peak torque[a] | $T_{sk}$/kNm | 2438 | 1914 | 1446 | 2128 | 2038 |
| | p.u. | 1.11 | 0.99 | 0.85 | 1.06 | 1.02 |
| **Rated load** | | | | | | |
| Terminal voltage | $U_{LL}$/V | 747 | 676 | 732 | 708 | 726 |
| Phase current | $I_s$/A | 3200 | 3200 | 3200 | 3200 | 3200 |
| Load torque[c] | $T_{el,0}$/kNm | 2204 | 1919 | 1706 | 2000 | 2000 |
| | $T_{el,6}$/kNm | 8.7 | 21 | 23 | 9.9 | 2.5 |
| Power factor | $\cos(\varphi)$ | −0.72 | −0.69 | −0.56 | −0.67 | −0.65 |
| Load angle | $\vartheta$/° | 44 | 45 | 49 | 44 | 45 |
| Magnet losses | $P_m$/kW | 7 | 22 | 14 | 15 | 19 |
| Short circuit peak current[b] | $I_{sk}$/A | 10286 | 11743 | 6672 | 10650 | 9840 |
| | p.u. | 3.21 | 3.67 | 2.09 | 3.33 | 3.08 |
| Short circuit peak torque[b] | $T_{sk}$/kNm | 2984 | 2665 | 2500 | 2975 | 2745 |
| | p.u. | 1.35 | 1.39 | 1.47 | 1.49 | 1.37 |

[a] 3-phase short circuit current from no-load, occurring at voltage zero-crossing.
[b] 3-phase short circuit current from rated load, occurring at voltage zero-crossing.
[c] Subscript number in formula is the harmonic order.

where a prototype machine has been devised and the other promising topologies will be compared to this in the following:

The surface magnet approach creates approximately 10% less torque as the reference design with spoke magnets. The main benefit is the considerable small steepness within the radial force, which is depicted in Section 3.5.7.1. However, the magnet losses are higher and if the short circuit behavior is compared, the peak short circuit current is slightly higher. But the resulting short circuit peak torque is lower. This is because of the much smaller machine inductance due to the large air gap width. A drawback in surface-mounted magnets of this size is always the difficulty to get the magnets reliable for the entire machine life of 20–30 years fixed onto the surface and to prevent corrosion of magnets. Various magnet gluing techniques have been devised in industry. But whether gluing of magnets is a reliable solution to ensure the design machine life time is always the question.

The simple interior magnet approach has the worst magnet utilization in terms of torque for the given phase current of 3200 A and magnet volume used. The reactive power component is the highest of all competing topologies. This is because of the higher machine inductance. That is the reason why the terminal voltage has more or less the same value at the rated load compared to the reference, even if the back EMF is the smallest of all competing designs. Due to the smallest flux linkage the short circuit peak current as well as the peak torque has the smallest values. This design has a larger load angle $\vartheta$ and therefore the risk of machine instability is higher. The only benefit at the end is the lowest radial force exposure and that the magnets are safely installed in pockets. Corrosion and magnet pop-up cannot happen here and the magnets are safe against demagnetization.

The V-shape interior magnet rotor was designed to create also approximately 10% less power. This has to be balanced by a higher phase current in order to meet the values of the spoke magnet design. **Since V-shape interior magnet is the most promising design, the air gap has been directly increased to 5 mm which has become a design requirement for the series generator system.** Compared to the reference design, the magnet material is reduced and the air gap has been increased, but nevertheless this approach is capable to deliver 2000 kNm at 3200 A phase current! From the terminal voltage and the reactive power point of view, the V-shape approach can very well compete with the reference design. Even the short circuit behavior such as peak torque and peak current is within the same range compared to the reference. However, the magnet losses are quite high. But suitable remedies such as further subdividing the magnet material can be applied, if required. The radial force dependency in terms of air gap with is almost comparable to the reference design with spoke magnets so no additional reinforcement of the corresponding carrying structure is needed here.

## 3.6 Optimization strategies on V-shape magnets

In a later stage, the V-shape magnets were investigated in deep on their further potential. Interesting results were found in addition that's why we would like to have a dedicated chapter on this as well. In the following the optimization potentials and strategies will be presented.

### 3.6.1 Deriving an appropriate solution with interior V-shape magnets

The following pages describe and depict how to obtain a desired and best suitable rotor design out of a variety of possible combinations in terms of magnet thickness $h_{mag}$ and magnet tilting angle $\alpha_{mag}$ as shown in Figure 3.48. At the beginning a set of generator models with various V-shape magnet structures is computed with the finite element approach, whose magnet tilting angle and magnet thickness might be suitable for the intended design. The calculated model properties out of the finite element analysis – hereinafter called raw data – are subsequently processed to find an optimum design in regards to various criteria.

Figures 3.51–3.58 depict the post processed raw data which has been determined from finite element calculations. In order to refine the mesh of the $\alpha_{mag}$–$h_{mag}$ plane, a 2D cubic spline interpolation method is applied. The stator geometry as well as the slot and pole count is taken from the reference machine that is described in Section 3.5.5. If for example the goal would be to find the machine torque of 2000 kNm at the smallest possible magnet tilting angle, all possible

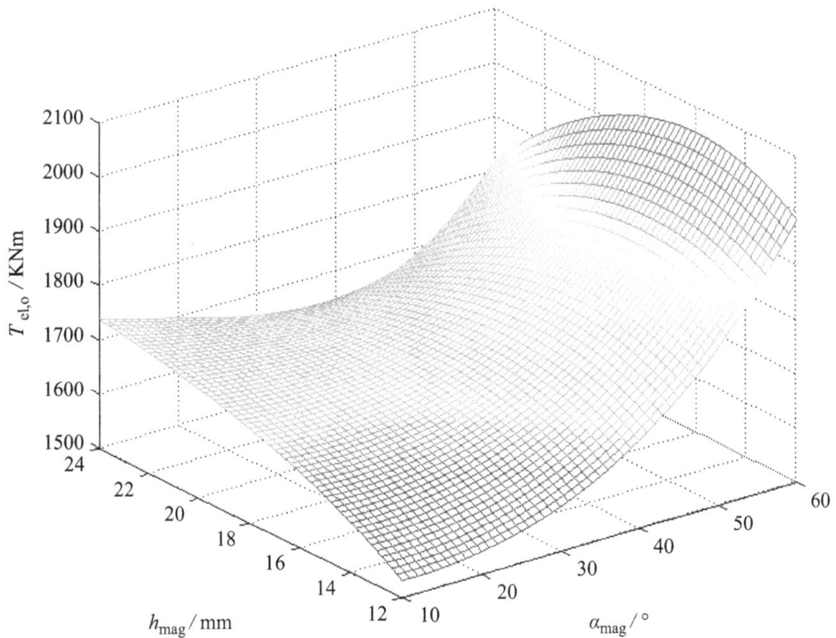


*Figure 3.51 Torque (DC component or 0. harmonic order) in terms of magnet tilting angle $\alpha_{mag}$ and magnet thickness $h_{mag}$, $n = 13.2$ rpm, $I_s = I_q = 3200$ A, $I_d = 0$ A, $T_{mag} = 80$ °C*

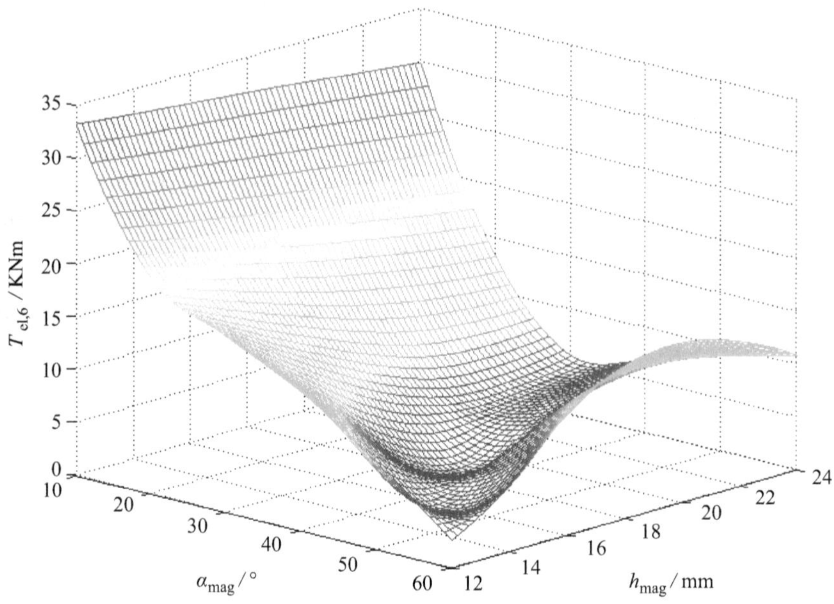


*Figure 3.52 Torque ripple (6th harmonic order) in terms of magnet tilting angle $\alpha_{mag}$ and magnet thickness $h_{mag}$, $n = 13.2$ rpm, $I_s = I_q = 3200$ A, $I_d = 0$ A, $T_{mag} = 80$ °C*

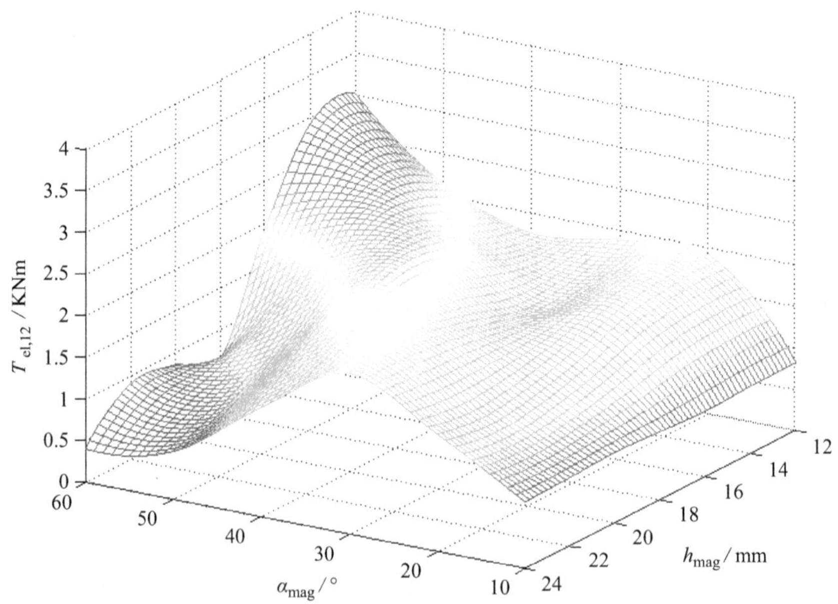


*Figure 3.53 Torque ripple (12th harmonic order) in terms of magnet tilting angle $\alpha_{mag}$ and magnet thickness $h_{mag}$, $n = 13.2$ rpm, $I_s = I_q = 3200$ A, $I_d = 0$ A, $T_{mag} = 80$ °C*

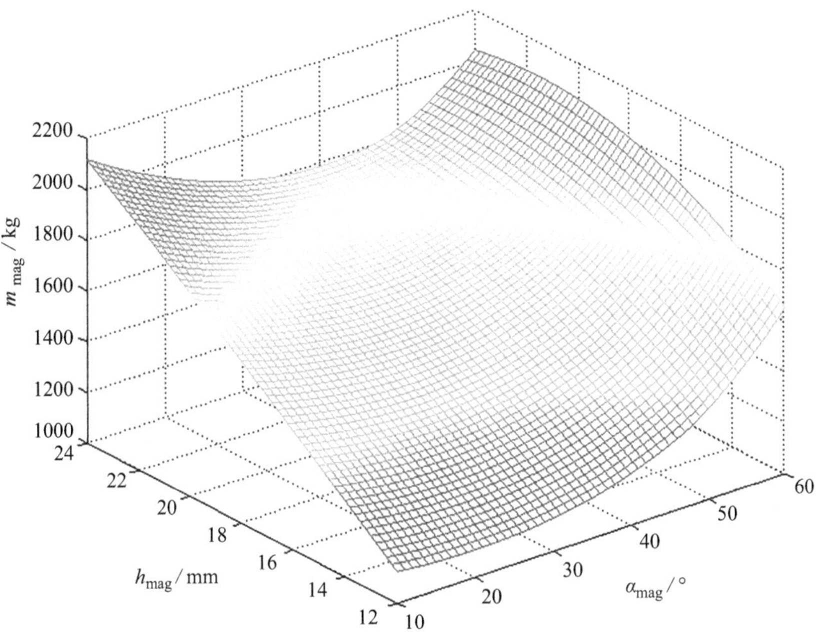


*Figure 3.54 Total magnet weight in terms of magnet tilting angle $\alpha_{mag}$ and magnet thickness $h_{mag}$*

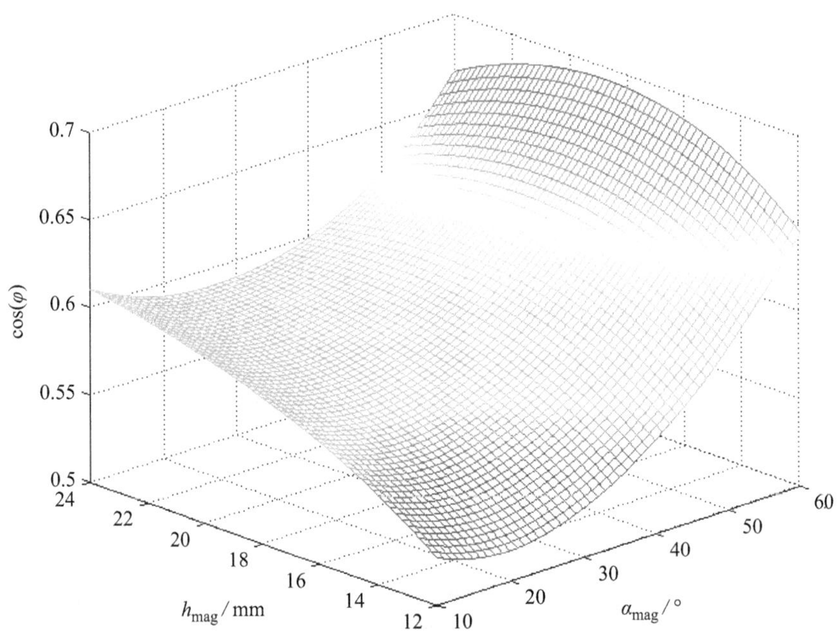


*Figure 3.55 Power factor in terms of magnet tilting angle $\alpha_{mag}$ and magnet thickness $h_{mag}$, $n = 13.2$ rpm, $I_s = I_q = 3200$ A, $I_d = 0$ A, $T_{mag} = 80\,°C$*

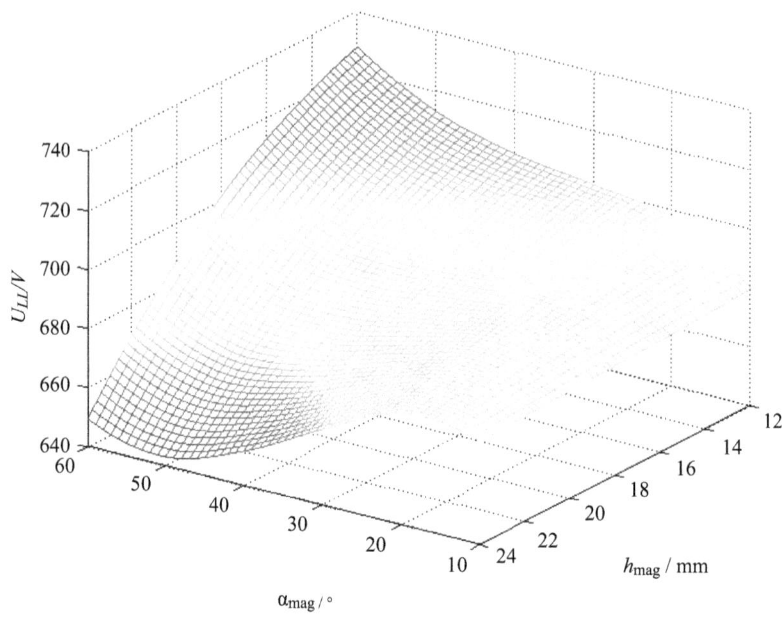


*Figure 3.56 Terminal voltage in terms of magnet tilting angle $\alpha_{mag}$ and magnet thickness $h_{mag}$, $n = 13.2$ rpm, $I_s = I_q = 3200$ A, $I_d = 0$ A, $T_{mag} = 80$ °C*

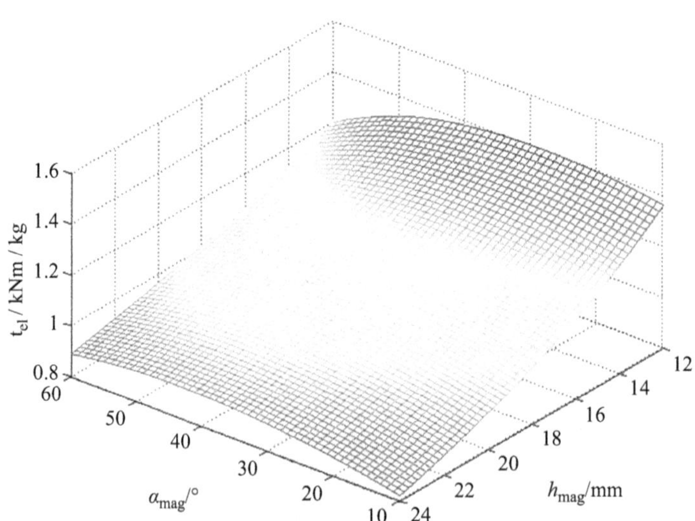


*Figure 3.57 Specific torque (torque which refers to magnet weight ) in terms of magnet tilting angle $\alpha_{mag}$ and magnet thickness $h_{mag}$, $n = 13.2$ rpm, $I_s = I_q = 3200$ A, $I_d = 0$ A, $T_{mag} = 80$ °C*

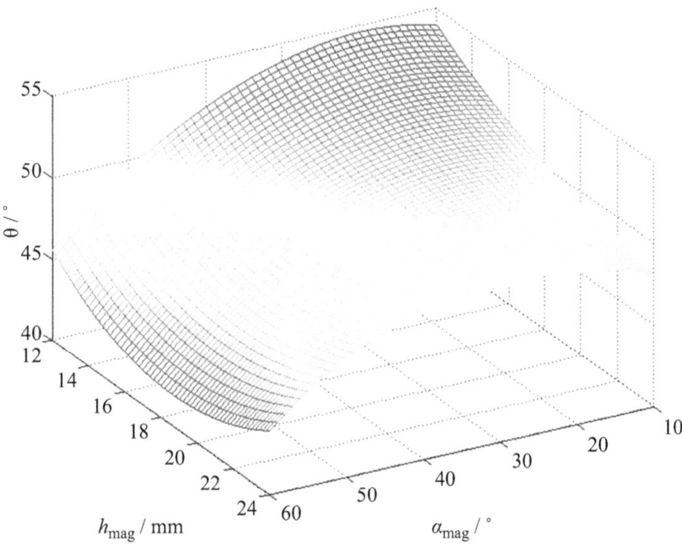


*Figure 3.58 Load angle in terms of magnet tilting angle $\alpha_{mag}$ and magnet thickness $h_{mag}$, $n = 13.2$ rpm, $I_s = I_q = 3200$ A, $I_d = 0$ A, $T_{mag} = 80$ °C*

combinations leading to the desired torque are taken into account out of Figure 3.51. The smallest tilting angle of the set of combinations is calculated subsequently by means of various interpolations within the existing plane. Once the corresponding values for magnet tilting angle $\alpha_{\text{mag}}$ and magnet thickness $h_{\text{mag}}$ have been determined, all the other corresponding machine properties can be determined out of the planes depicted within Sections 3.5.2–3.5.8 by means of simply finding the corresponding value to the given determined magnet tilting angle and magnet height.

Figures 3.52 and 3.53 depict a subset of typical major harmonic torque components within the generated load torque. These harmonic torque components are often a root cause for the generation of vibrations and noise emissions. Therefore a look on these torque components might be useful as well during machine optimization. The chosen design should have, if possible, inherent low harmonic torque content. In particular within Figure 3.52 it can be seen that with the right magnet tilting angle and magnet thickness the sixth order torque ripple can be drastically reduced, if a parameter set which lies in the valley is chosen. It can be seen that especially for small magnet tilting angles which leads to a rotor design close to simple interior magnets, which has been described in Section 3.5.4 the 6th order torque ripple is large. However the 12th order torque component is for this stator and rotor design of minor influence since the values are not changing much.

With regard to magnet utilization: Figures 3.54 and 3.57 depict the total machine magnet weight dependency as well as the produced torque within the machine referred to its magnet weight used. The later expression is given in the unit machine torque in kNm per used magnet weight in kg. It can be seen that the best magnet utilization appears for moderate magnet tilting angles in the range of 30 degrees for a resulting magnet thickness of approximately 24 mm. This is the most economical way of using rate earth magnet material in permanent magnet synchronous generators. Since the magnet material cost is the dominating cost driver in large permanent magnet synchronous machines this variant would lead to the most cost-effective generators from the magnet utilization point of view. However, from the produced machine torque in Figure 3.51 it can be seen that the produced machine torque is not the highest.

### *3.6.2 Combining a continuous rotor iron yoke with concentrated stator windings*

As described previously, a continuous laminated rotor iron yoke – typically applied for an interior magnet configuration as described in Section 3.5.4 – sometimes has to have a certain yoke height. Depending on the pole count and the magnet arrangement within the rotor structure it is sometimes required to guide the working field harmonic properly through a rather big laminated rotor iron yoke. However, this measure degrades the machine characteristics significantly, if a single layer stator winding with concentrated coils is considered. A continuous laminated rotor iron of a certain height amplifies any present sub-harmonics which are a non-negligible property for concentrated single layer stator windings. These sub-harmonics are disturbing the intended magnetic circuit inside the active parts to any undesired extend.

Based on the interior V-shape magnet configuration it is explained in the following, how to find a suitable continuously laminated rotor iron shape which has the capability to attenuate these sub-harmonics to a minimum acceptable level. However, this approach can be applied to any interior magnet topology, if a rotor yoke of a certain height is required.

#### 3.6.2.1 Rotor yoke shape adaptation for a V-shape interior magnets

During the finite element simulations, some curious field distribution within a continuously laminated rotor iron yoke has been observed within some cases at load wherein the rotor iron yoke has to have a certain height in order to guide the rotor field properly. This is in particular the case for small magnet tilting angles $\alpha_{mag}$, if a V-shape magnet arrangement is considered.

Figure 3.59 compares the field lines at no load and load respectively. It can be seen clearly that the field lines within the laminated rotor iron yoke look quite disturbed at rated load and it is difficult to draw conclusions on the pole count from the rotor field distribution. Figure 3.60 depicts the initial parametric design for small and large magnet tilting angles which has been presented in Section 3.5.5. The yoke height underneath the magnets is adapted in terms of magnet tilting angle and magnet thickness and is continuously decreasing for increasing tilting angles $\alpha_{mag}$. For tilting angles above a certain value a constant thin layer as magnet slot

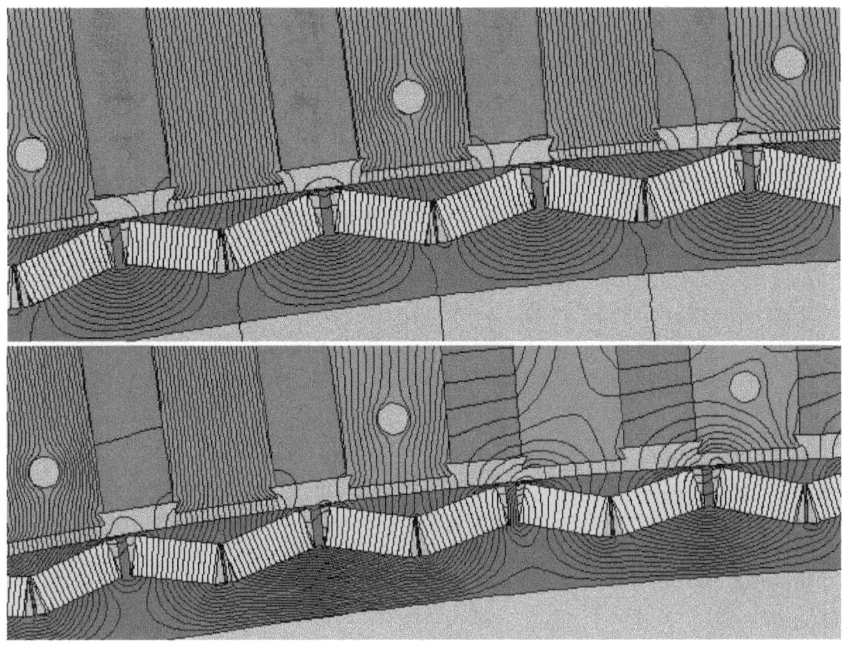

*Figure 3.59* Model with $\alpha_{mag}$ = 15° and $h_{mag}$ = 16 mm. Upper picture: field lines at no load ($I_s$ = 0 A). Lower picture: field lines at current loading, $I_s = I_q$ = 3200 A.

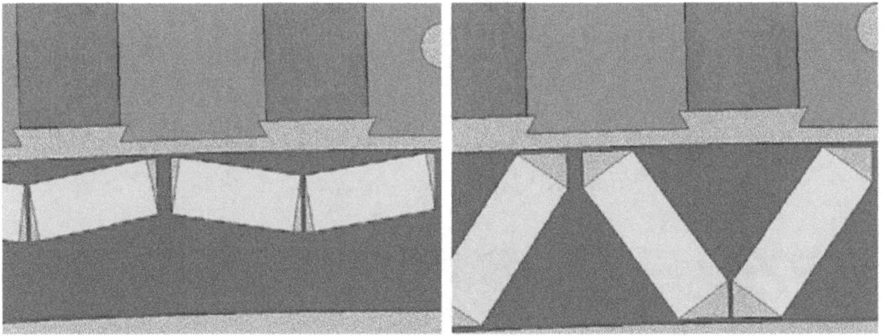

*Figure 3.60* Left: rotor with considerable yoke height for small magnet tilting angles $\alpha_{mag}$. Right: rotor with a thin layer as yoke for magnet tilting angles above a certain value.

enclosure of 1 mm thickness remains within the rotor lamination only for structural reasons. But for small magnet tiling angles there is a considerable yoke thickness required to guide the magnet flux between the magnets.

To overcome this problem in case small magnet tilting angles are desired, the following remedies have been introduced:

- If this additional rotor yoke height is required within the magnetic circuit – here in case of small magnet tilting angles $\alpha_{\text{mag}}$ – it is only added partially with an appropriate thickness in order to guide only the working harmonics and to suppress the field amplification of any parasitic sub-harmonics as good as possible.
- An elliptical shape of the rotor yoke for small magnet tilting angles as depicted in Figure 3.61 is the best choice. For larger magnet tilting angles this elliptical shape slowly merges into its initial shape.
- The lamination area within the yoke-side pole gap should be made of a thin latticed area whose function is basically to obtain a small leakage flux within the yoke area. In addition this structure acts as magnet slot enclosure to protect the magnet material and to get a continuous lamination which simplifies the production.

As an example Figure 3.62 depicts the rotor field at load. Compared to Figure 3.59 the flux lines of the working harmonic can be noticed without any distortion and the present sub-harmonics are successful attenuated to a minimum acceptable level.

Figures 3.63 and 3.64 compare the machine characteristics with initial and optimized laminated rotor iron. It can be noticed that the amplification of

*Figure 3.61 Left: rotor with improved yoke geometry for small tilting angles $\alpha_{mag}$. Middle: decreased elliptical shape for increasing tilting angles. Right: no elliptical shape for sufficient high tilting angles required.*

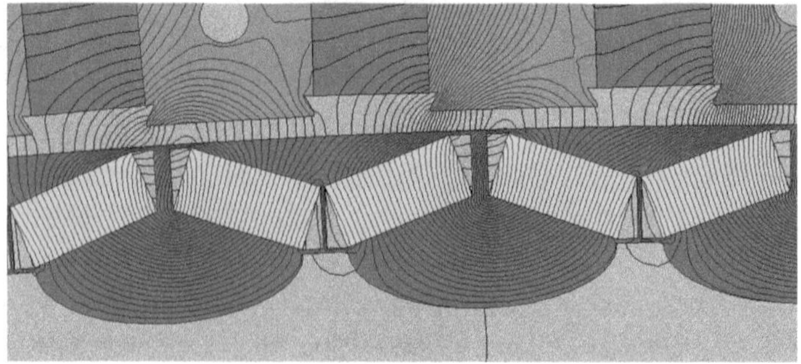

*Figure 3.62 Rotor field at load within the improved rotor iron lamination structure, $I_s = I_q = 3200$ A, $\alpha_{mag}=20°$, $h_{mag} = 16$ mm*

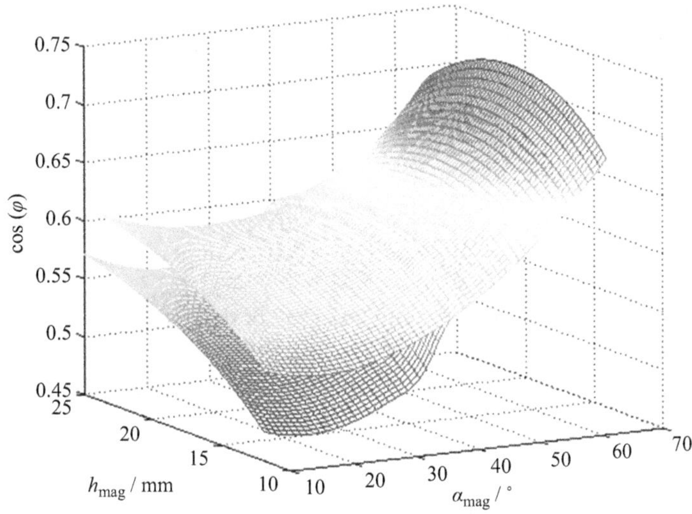


*Figure 3.63 Comparison of the power factor cos(φ) in terms of magnet tilting angle and magnet thickness between initial and optimized laminated rotor iron yoke. $T_{mag}$ = 80 °C, $I_s = I_q$ = 3200 A*

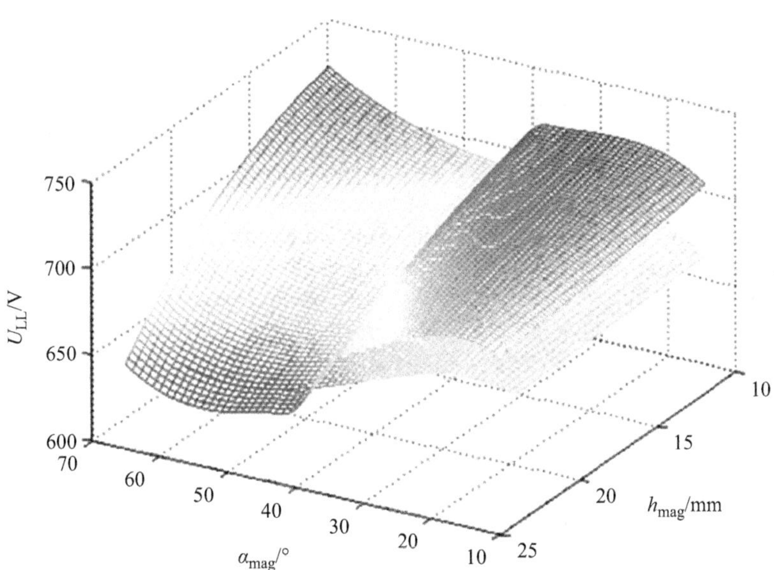


*Figure 3.64 Comparison of the terminal voltage $U_{LL}$ in terms of magnet tilting angle and magnet thickness between initial and optimized laminated rotor iron yoke $T_{mag}$ = 80 °C, $I_s = I_q$ = 3200 A*

any sub-harmonic due to an inapplicable rotor iron shape has a considerable impact on the machine properties such as power factor $\cos(\varphi)$ and the terminal voltage $U_{LL}$. The machine reactance is increasing because of the additional harmonic reactance component and thus the power factor is lower and the terminal voltage is increasing for magnet tilting angles smaller than 40°, if an improper rotor yoke shape as initially introduced is chosen.

## 3.7 Summary

Starting from the geometrical scaling rules of designing direct-drive generator systems an overview of suitable stator winding systems for large direct drive generators is given. From this a most promising stator winding configuration for a modular permanent magnet synchronous machines is derived. Subsequently, an overview of different rotor configurations is given in which a most promising rotor magnet configuration is devised. Finally, an optimization approach for V-shape magnets is depicted; furthermore, a way to optimize the rotor iron is given.

## 3.8 Outlook

The direct-drive generator examples presented in this section of the book have been devised and commissioned in the years from 2010 to 2015 at *wind-direct GmbH*. In the meantime, many boundary conditions have changed.

### *3.8.1 Evolution of gearboxes and medium-speed concepts*

Gearboxes for geared drive drains have shown remarkable technical evolutions over the past 20 years, especially regarding the application of journal bearings instead of roller bearings for supporting the individual planets. This led not only to the omission of failure-prone parts, but could also be used to implement more planets into the first stage, up to eight planets are currently used. Increasing the number of planets per gear stage to more than eight would be technically possible, but would come with serious penalties as the achievable step-up ratio would go down. [EizedeVries: NGC]. However, these developments resulted in an increase of the achievable torque density of wind power gearboxes from levels in the 125–150 Nm/kg range to values of 220 Nm/kg today – being a technology leap of 50%. Also the application of innovative lubrication concepts instead of simple splash lubrication yielded advantages regarding partial load efficiency. However, it is unlikely that another "game changer" such as journal bearings may appear, so that future leaps in torque density – without compromising quality – may be rather unlikely. The focus may be rather on further cost reductions through "economies of scale."

The combination of gearboxes and medium-speed permanent-magnet generators – sometimes called "Hybrid drive" or "semi-direct drive" is seen as the most favorable compromise for wind turbine drivetrains by some of the wind turbine

OEMs. *Vestas* can be regarded as the pioneering company on this route. Some of the OEM developing direct-drive switched technology recently, one of them being the leading Chinese company *Goldwind*. Main driver behind this move may be the challenging logistics of single-piece direct drives: above 4–5 MW the masses of economic designs reach values of 80 t and diameters of more than 5 m.

In the meantime, there is a serious competition between direct-drive and geared generator systems on the market. Direct drive generator systems have to be well designed and properly integrated into the overall nacelle structure to achieve cost savings as much as possible in order to compete with geared drive train concepts. However, at the time of writing this book, current studies show that the direct-drive generator concept may still be able to compete with geared drive generator concepts, especially when proper segmentation comes into picture.

### *3.8.2 Raw materials*

Aluminum instead of copper? Every time when the copper prices are increasing there is always the call on alternative stator winding materials. Typically aluminum is then taken into consideration as an alternative then. Many calculations have been made with aluminum stator windings and concepts are available. But if there is not a serious need for aluminum from the copper cost point of view the copper winding will stay in place. Due to the fact that the electrical conductivity of aluminum is around 2/3 of the conductivity of copper the machine losses will increase for the same conductor cross section which is then resulting into issues in cooling. If on the other hand the cross section of the aluminum conductors is increased according to the difference in electrical conductivity this is resulting in a more bulky and overall heavier electrical machine due to the increase of carrying structure weight and additional electrical steel demand and permanent magnets. In the end, several designs with different winding materials would be calculated and compared with respect to their LCOE: however, at current price levels for copper (medium), electrical steel (high), and magnets (medium), aluminum design cannot compete with copper.

A first and very important technical boundary condition is the evolution of rare earth permanent magnets. 10–15 years back, OEMs had been limited to the magnet grade N38SH or N38H for the magnets in the rotor. The number in the magnet grades refers to the max. magnet energy product. Nowadays the energy density of rare earth permanent magnets has been increased up to the grades N52H and higher and it still keeps increasing. Having magnet material with higher energy density available enables to increase the power and torque density of large scale direct drive generators even further or results in more compact machines.

In addition to this many attempts have been made to reduce the heavy rare earth materials like Dysprosium (Dy), Terbium (Tb), and Holmium (Ho) witch are much more expensive compared to classical Neodymium (Nd) and Praseodym (Pd) to make the magnets more affordable and cost attractive.

Most recent developments aim to substitute even Nd and Pd by some amount of Cerium (Ce). Typically 30% of the total magnet mass besides Iron is a Nd/Pd mixture. Ce is a side product which is occurring if the above-mentioned elements are extracted from the ore.

However, the question of rare earth availability is a major one, since one country (China) still supplies 90% of the current market: recent political initiatives "de-risking strategy" would favor to look for alternative supplies. Most likely, this would lead to increasing magnet prices, so that any generator design must aim to minimize the amount of magnet material used.

### *3.8.3 Alternatives to permanent-magnet excitation – conventional direct-current excitation?*

As stated at the beginning of this chapter, the history of direct-drive generator for wind power started with current-excitation developed mainly by the German pioneering company *ENERCON* 30 years ago. Would this technology return in case of soaring rare earth prices? Competitive designs for direct-drive PMG need approximately 400–500 kg of magnets per MW of rated power. Experience has shown that even at magnet price levels of more than 100 €/kg (tripled from the time of writing this chapter), permanent-magnet would be at par with conventional current-excited designs.

### *3.8.4 Alternatives to permanent-magnet excitation – high-temperature super-conductivity (HTS)?*

The outlook of Chapter 5 "Drivetrain concepts and developments" in Volume 1 already touches upon the subject "High-temperature super-conductivity (HTS)": could this become the future of the direct drive? As has been already mentioned there, the successful field testing of the Envision 3.6 MW wind turbine prototype (sometimes also called the "EcoSwing demonstrator") as depicted in Figures 3.65 and 3.66 certainly took this technology to the next level [11].

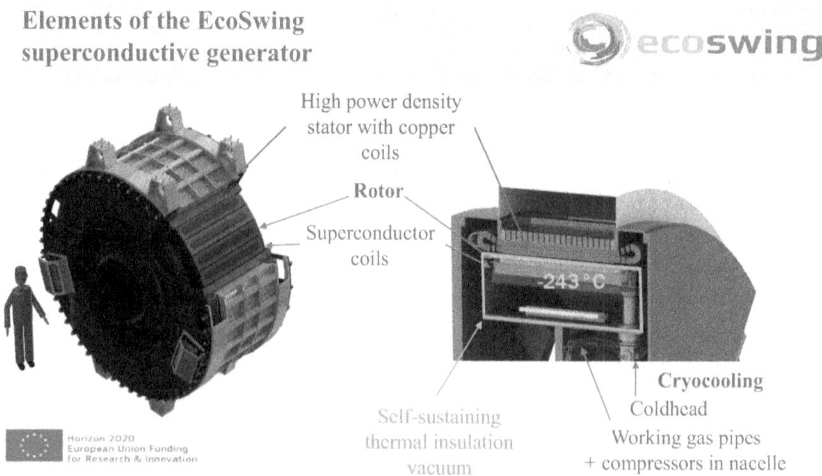


*Figure 3.65 The EcoSwing HTS generator [11,12]*

Horizon 2020
European Union Funding
for Research & Innovation
"EcoSwing has received funding from the European Union's Horizon 2020 research and innovation programme under grant agreement No 656024." "Herein we reflect only the author's view. The Commission is not responsible for any use that may be made of the information it contains."


*Figure 3.66 A side-by-side view of the PM generator (left) and the HTS generator (right) [12]*

However, there remain huge challenges if the goal is to compete with current wind turbine drive trains regarding LCOE. One major point already raised in Volume 1 is the (almost) constant power requirement for cooling the rotor safely to the required cryogenic temperatures, in the EcoSwing case −243 °C. Due to this, the partial load efficiency – which had been one major advantage of direct-drives versus geared drive trains – is heavily reduced (Figures 3.67 and 3.68). One reason for this very high cooling power is that the rotor torque in a direct-drive application is quite high – especially in a short-circuit case! This yields massive structures between cold and hot components which conduct the heat and lead to low cooling efficiency. This disadvantage could be heavily improved in the case of a geared drive train combined with a HTS generator.

The authors believe that HTS may come in long-term future (>20 years from now), however, the near and mid-term technology for direct-drive wind generators would still be with rare earth permanent magnets. HTS will certainly be developed further and will start to play a certain role in electrical drives, but most likely NOT for direct-drive (but maybe medium-speed?) wind power – at least not as preferred first application.

Finally, the main reasons for this statement and major obstacles seen as of today shall be summarized:

- Very high generator cost (today approximately double cost compared to current DD-PMG, still 120%–150% if HTS components reduce to 1/3 of current cost level)

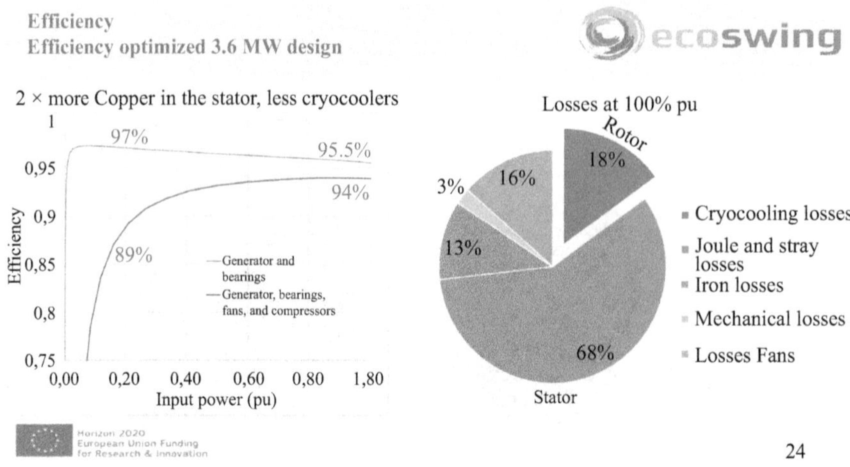


*Figure 3.67 EcoSwing generator efficiency and losses for already "optimized design" [12]*

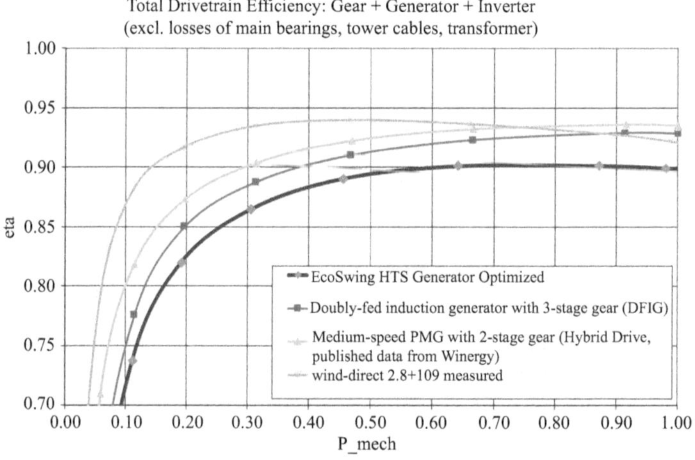


*Figure 3.68 Efficiency comparison of different DT concepts (by the author)*

- Poor overall efficiency, especially at low power (mainly due to cryogenic cooling): approximately 3%–5% reduced AEP compared to current DD-PMG, which is very much [12]. Even 1%–2% less AEP than current DD-ESG.
- Long downtimes for cooling down the rotor (several days), further AEP losses as consequence
- Very high short-circuit torque due to small inductance: torque peak of double-rated torque may cause severe blade damages

- General questions regarding reliability of this technology – especially when being compared to the extremely simple and robust permanent-magnet concept

## References

[1] Jöckel S. *Calculation of Different Generator Systems for Wind Turbines with Particular Reference to Low-Speed Permanent-Magnet Machines (Berichte aus der Energietechnik) Paperback*, Shaker Verlag, 2003.

[2] Klinger F. and Müller L. State of the Art and New Technologies of Direct Drive Wind Turbines. *International Renewable Energy Conference and Exhibition (IRENEC)*, 2012.

[3] Richter R. *Elektrische Maschinen II – Synchronmaschinen und Einankerumformer*. Birkhäuser-Verlag, Basel/Stuttgart, 1953.

[4] Decker T., Cardaun M., Mülder C., *et al.* Experimental Validation of a Structure-Borne Sound Model of a Direct Drive Wind Turbine Generator. *Forsch Ingenieurwes* 87:3–12, 2023.

[5] Cardaun M., Mülder C., Decker T., *et al.* Multi-Physical Simulation Toolchain for the Prediction of Acoustic Emissions of Direct Drive Wind Turbines. *Journal of Physics: Conference Series 2265*, 2022.

[6] Jöckel S. Design and Prototype Testing of the wd2.8+109 PMG DD, *5th International Conference on Drivetrain Concepts for Wind Turbines*, 2014.

[7] Vries E. de *Stator Segments: the Golden Triad of Modularity, Scalability and Simplicity*. Windpower Monthly. Feature 12-07-2023. Available from https://www.windpowermonthly.com/article/1829671

[8] Crabtree C. *Survey of Commercially Available Condition Monitoring Systems for Wind Turbines*, 2010.

[9] Hidding E. and Jöckel A. *Giants of the Sea: Where Do We Go in 10 Years from Now?* Winergy Webcast Spring 2021, Available from https://www.winergy-group.com

[10] Muik T. Vergleichender Entwurf von Ständerwicklungen für einen permanentmagneterregten Synchrongenerator in einem Untersee-Gezeitenkraftwerk, Loher GmbH/TU Darmstadt, 2019.

[11] Song X., Bührer C., Molgaard A., *et al.* Commissioning of the World's First Full-Scale MW-Class Superconducting Generator on a Direct Drive Wind Turbine, *IEEE Transactions on Energy Conversion* 35(3), 2020.

[12] Bauer M., Tiemo W., Carsten B., *et al.* EcoSwing Generator – World's first superconducting wind power generator supplying power to the grid, *Presentation in Thyboron* 2019.

# Chapter 4
# Main converter design

*Holger Groke*[1] *and Norbert Hennchen*[2]

The main converter (MC) of a wind turbine (WT) is a key component of modern, efficient multi-megawatt turbines. On the one hand, it must withstand the fluctuating stresses and loads on the rotor-side of the WT, and on the other hand, it must meet the needs of the grid on the other side of the drive train, at the same time.

After a brief historical review from the point of view of power electronics, this chapter provides an overview of state-of-the-art electrical drive trains (EDT) of modern WT. This overview covers basic design rules for converter subsystems such as inductors, filters and transformers, as well as advantages and disadvantages of design criteria for different grounding measures and other topics.

After an introduction to the EDT the focus will be on the state-of-the-art assembly of main converters for WT. General issues and boundary conditions such as the installation location of the converter, different approaches to pulse generation, two- or multi-voltage-level approaches as well as the influence on service-life of power electronics will be addressed.

The chapter ends with a reference design of a six-megawatt main converter for a WT introduced by FREQCON GmbH. The reference provides a detailed design and layout of the previously introduced EDT from a manufacturer's point of view. Finally, a brief outlook is given at the end of this chapter.

## 4.1 Brief historical background

In the second half of the 1970s to the early 1980s, when the main power switches were thyristor-based power semiconductor devices, the application of power converters for WTs began as a novel approach. Before that time, the Danish concept with a squirrel-cage asynchronous generator directly connected to the grid and passively stall-driven rotor blades had been state-of-the-art since the 1950s. In the mid-80s of the last century, a novel conceptual approach was presented with a synchronous machine with electrical excitation, a power converter system and pitch-controlled rotor blades. For example, the Krogmann-15/50 WT combined all

[1]Institute for Electrical Drives, Power Electronics and Devices, University of Bremen, Germany
[2]Freqcon GmbH, Germany

these features. It was developed as part of a research project in 1987 and funded by the German government. At that time the rated electrical power output of 50 kW was in the normal range for WT. One author of this chapter was part of this project and developed the MC for this system. The basic concept is still the most commonly used EDT type in modern multi-megawatt turbines today.

To diminish faults, the electrical excitation of the synchronous generator is replaced by permanent magnet excitation in modern drive trains in order to reduce the number of electronic circuits and cables and thus the costs. The main advantage of installing an electrical power converter is that in principle it basically operates like an electrical gear box. On the one hand, it decouples the varying frequency of the stator currents originating to the varying rotational speed of the generator, and the fixed frequency of the electrical grid or network on the other hand. The pitch-controlled blades were needed to restrict power conversion by limiting the rotational speed of the rotor in case of high wind speeds.

Due to further development in the field of semiconductor power devices, the thyristor-based switch was more and more detached by modern Insulated Gate Bipolar Transistor (IGBT) devices in the mid-90s of the last century. Nowadays thyristor-based devices are installed in HVDC-stations when it comes to energy technology. The IGBT is still the most common device in all modern electrical power converter systems. The main difference between a thyristor-based and an IGBT-based converter system is the pulse generation unit. The IGBT must be actively switched on and off in contrast to a thyristor, which must be ignited to on-state. The thyristor is self-extinguishing when the grid current is crossing zero point. In the 1990s, the average rated power of WTs increased rapidly to a few hundred kilowatts.

At the turn of the millennium change many turbine manufacturers reached the one-megawatt class. Up-scaling the WT to increase efficiency, performance and power output was the main goal for most manufactures. During the first decade of the year 2000, the term multi-megawatt-class became established. With constant grid voltage the increasing rated power of the WT yields to increasing electrical currents which must be handled by the MC. Since the rated current of IGBT switches in the 1700 V class is limited to round about 1000 A, most manufacturers have started to parallel entire converter systems or, alternatively, to parallel the power semiconductor components and their peripherals. An issue at that time was to synchronize all gate signals for the parallel IGBT devices in the microsecond range.

In the early 2000s, the drive train concept with asynchronous induction machines was also extended by installing doubly-fed induction machines instead of squirrel caged types. The stator-windings of this machine type are directly connected to the grid in on-state, whereas the rotor-windings are connected to slip rings and fed by the main converter. This concept seemed to have at least one major advantage over that of a synchronous machine. The MC may only be designed for the slip-power range in the over-synchronous operation mode of the machine. This means that the rated power of the MC is almost a third of the rated power of the WT and the MC only feeds the rotor windings. A major problem with this concept is to cross the synchronous-point of the machine without injecting currents with a very low frequency or even DC-currents into the rotor windings via the MC. The thermal load of the IGBT switches must be carefully observed for this operating-point.

Switching high currents with the MC has a negative impact on the power quality of the grid. Since the early 2000s, this matter has become increasingly important and is still a design issue of the EDT of WTs today. The harmonic distortion level of grid voltages and grid currents are hardly regulated worldwide by a grid connection policy. Another topic that has gained more and more importance since the early 2000s is grid codes (GC). Since 2008, WTs in Germany have not been allowed to disconnect from the grid immediately in the event of a grid failure, such as a short circuit. For the MC of the EDT with a doubly-fed induction machine installed, this regulatory requirement is extremely problematic from a design point of view. The problem is to keep the inducted currents in the rotor windings in a feasible range in case of dynamic changes of the stator currents and to protect the IGBT switches from over-currents and over-voltage. Whereas the MC of the EDT with a synchronous machine is designed for the full rated power of the WT, delivering grid services is quite easier from the design point of view. Since then, national and international grid codes have been regularly updated and further legal requirements such as high-voltage-ride-through have been added. The technical constraints to meet all these legal regulations are the main reason for the disappearance of the EDT with doubly-fed induction machines in our days.

Today, the synchronization of decentralized renewable energy stations, not only wind-parks, is a big topic again. For example, with time dynamic grouping of real plants to virtual power plants (VPP). In the future, not only wind farms, but almost all renewable energy plants will have to offer and provide even more grid services in order to replace conventional power plants. In VPPs, all of them can be combined at the same time. Therefore, the legal and technical implementation of services like fast frequency response (FFR) will be an issue in the very near future.

## 4.2 Electrical drive train

First, a look at state-of-the-art electrical drive trains, the most common variants and their pros and cons are considered to get an overview of the challenges in designing main converters for WTs. So, this section, therefore, summarizes the basic setup of the power section for almost all common EDTs relevant to WTs. In addition to indicating the main functional description of each component of the EDT, the main design principles are explained. For a detailed look on generator design refer to the specific chapter of this book. In this chapter, the focus is on the terminal behavior of the generator, not on its internals.

Following up, grounding concepts are presented in relation to the operation of the MC. The grounding concept is not only relevant for the MC, but for all systems and subsystems of the WT with any electrical connection. The section ends with a summary of cooling concepts, protection systems, and unit certification.

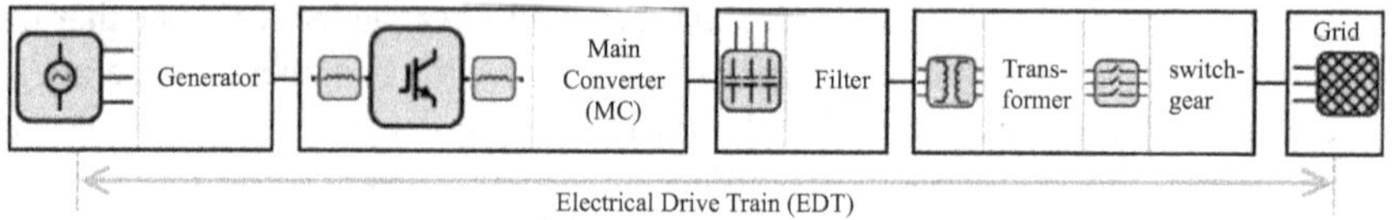


*Figure 4.1 General assembly of the electrical drive train, © 2023 Freqcon GmbH*

### 4.2.1 General assembly

Regarding the design of the power section of the EDT from the functional point of view, the electrical drive train commonly appears like the one sketched in Figure 4.1, ignoring components or subsystems for controlling, communication purposes and even power cables.

Over the past years, as the rated power of WTs reached the ten-megawatt margin, almost all manufactures pursued the objective of integrating the MC together with the transformer and the switchgear inside the nacelle. The disadvantage of a larger nacelle and heavy components on top of the tower is balanced by a much better system performance in many different aspects. Mentionable advantages are seen in terms of electromagnetic compatibility (EMC), lower costs for cables and even more available space inside the nacelle when considering the ratio of available space to rated power for different placements of the components. Another advantage of this approach is that there is only one medium voltage cable installed inside the tower of the WT and that the nacelle can completely be assembled in a factory. Therefore, all components of the EDT can be tested and tuned individually in conjunction with all main subsystems and components before the nacelle leaves the factory.

Prior to this trend the transformer and switchgear are either placed inside an outdoor housing right next to the tower of the WT or the transformer is placed inside the foundation of the tower and the switchgear is placed inside the entrance area of the tower. The MC is either placed somewhere in the lower third of the tower or it is placed inside the nacelle. In all these cases, the number of all low voltage cables, the diameter of these cables and the effort to install them is much higher compared to the installation inside the nacelle.

### 4.2.2 Generator

There are two different types of induction machines commonly found in electrical drive trains of WTs, doubly-fed induction generators (DFIG) and synchronous generators respectively permanent magnet excited synchronous machines (PMSM), as shown in Figure 4.2.

Today there is only one globally operating manufacturer left who designs WTs in the multi-megawatt class with a DFIG-concept. The most significant differences when comparing both concepts from the MC point of view are:

- The rated power of the main converter for a DFIG can be limited to round about one-third of the rated power of the generator, whereas the main converter

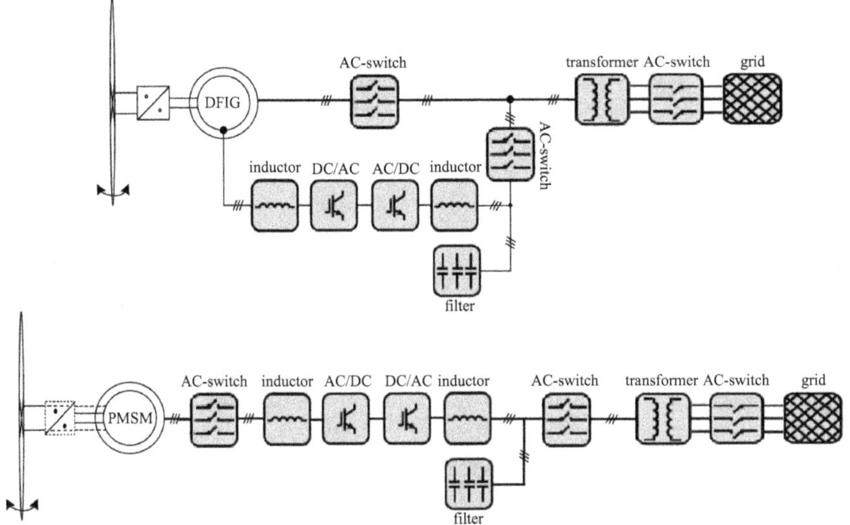


*Figure 4.2 Typical EDT of a DFIG-concept (top) and a permanent magnet excited synchronous generator (bottom), © 2023 Freqcon GmbH*

for an EDT concept with an SG must be designed for the full rated power of the machine at minimum.

- The stator-windings of a DFIG can be designed to be directly connected to medium voltage level, whereas the rotor-winding can still be fed on low voltage level via slip-rings connected to the MC. A negative impact is that the machine requires reactive power due to the magnetization.
- The rated power of the MC has a direct impact on the harmonic distortion level of grid currents and voltages and, therefore, the filter design. The distortion level is higher for EDT-design with SG than for DFIG concepts, respectively. The situation gets even more complicated in case of connecting several turbines with SG to the park network or the grid. Due to same or even identical switching patterns and frequencies the total harmonic distortion level for same frequencies increases with the number of turbines installed.
- Controlling the DFIG according to grid codes is a big challenge as well as controlling the machine near the synchronous point. At synchronous point the MC feeds the rotor-windings with DC-currents. The cooling concept of the MC is usually not designed for the high unbalanced, unsymmetrical thermal stress regarding the IGBT switch-states at that operating point. To fulfill grid codes with a converter designed for the full rated power of the generator and beyond decoupling generator and grid or wind park in principle is much easier under many constraints.

### 4.2.3 Main converter

The MC usually consists of two IGBT-based so-called 4-quadrant-chopper circuits (4-QC) for three phase-voltages. Both circuits are connected via a DC-link voltage, which is typically controlled by the 4-QC with grid connection. Therefore, it operates like an AC/DC-converter with power factor correction (PFC). A PFC-active-frontend is a must-have for modern converter systems with grid connection. The inductor in each line of the grid is essential for the PFC and is typically installed inside the cabinet of the MC. The key feature of a PFC-active-frontend is the ability to set active- and reactive power independently via the control-unit, but of course in the range of the rated power of the converter. For further reading on PFC-active-frontend and its mode of operation, see [1–4]. In Figure 4.3 the AC/DC-conversion is shown for each line of the grid. Also shown are the switchgear, the transformer, the circuit breaker (CB), and the filter. Each AC/DC-converter consists of a so-called half-bridge circuit built by two IGBT-switches connected in series. The three-phase AC/DC-converter therefore counts six IGBT-switches which can be triggered independently of each other via the appropriate gate-signals of the IGBT-devices.

On the market there are many different standard packages and different types of housing for IGBT-switches are available across all power-scales. The product-tree of most manufacturers ranges from packages with a single IGBT-switch, a half-bridge circuit with two switches, up to packages with six IGBTs for a full three-phase 4-QC with corresponding diodes. Figure 4.4 shows an IGBT-based power-stack with six half-bridge circuits including integrated gate-units for a water-cooled MC.

In most cases, the free-wheeling diode of every switch is also integrated in each of the packages. Typically, most manufacturers also offer specialized IGBT-packages for the wind-sector, the photovoltaic-sector, etc. For further reading on these topics have a look at [5–7].

The second 4-QC circuit is connected to the generator of the EDT. The parting of the MC into two 4-QC electrically decouples the electro-magnetic energy

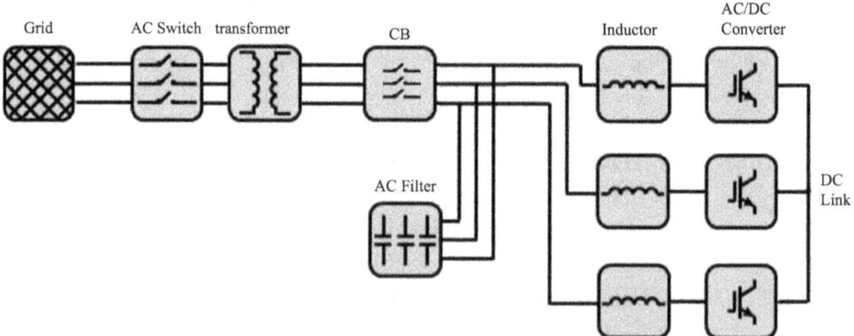


*Figure 4.3 Principle grid-side connection of the main converter with PFC-inductor, filter, transformer and switchgear, © 2018 Freqcon GmbH*

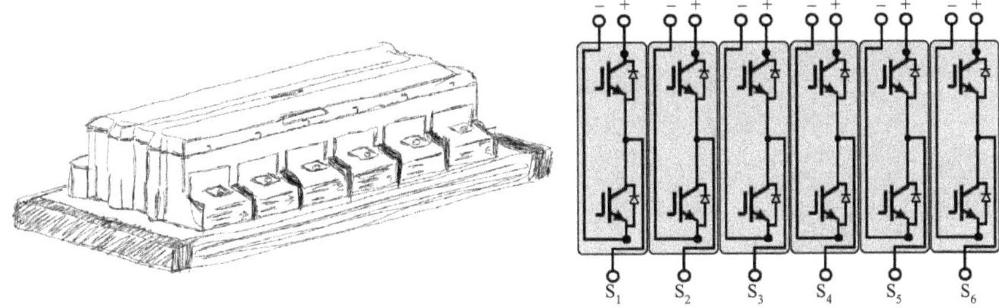


*Figure 4.4 Stack with six IGBT half bridge-circuits, integrated gate units and cooling plate from the manufacturer Semikron GmbH for AC-DC/DC-AC and chopper applications*

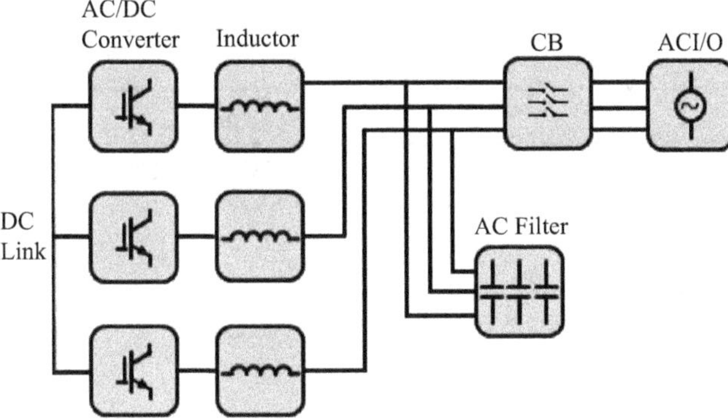


*Figure 4.5 Principle of load-side part of the main converter to feed a three-phase generator, © 2018 Freqcon GmbH*

conversion part of the EDT and the energy supply of the grid. This procedure is essential because of the completely opposite load conditions on the grid-side and on the generator-side. The DC-link voltage is stabilized via a high capacitance which also has a positive impact on the decoupling of both 4-QC and therefore supports the independent operability of both parts of the MC. Figure 4.6 shows the 4-QC for the generator-side of the MC for a common three-phase generator. From a structural perspective the generator-side converter is built the same way as the grid-side AC/DC-converter. The design parameters of both 4-QC might differ, but in general both sides are typically identical. This is also true for filter-, inductor-, and switch-parameters which are selected for their specific power consumption and, therefore, for the rated current of each line with respect to the given voltage-level of the grid and the nominal DC-link voltage of the MC.

Most manufacturers of semiconductor switches offer IGBT devices for various nominal currents with the same footprint and voltage-level of the devices. Therefore, most manufacturers of converter systems also offer a variety of up-scaled converter-systems. Most manufacturers of filters, DC-links, inductors,

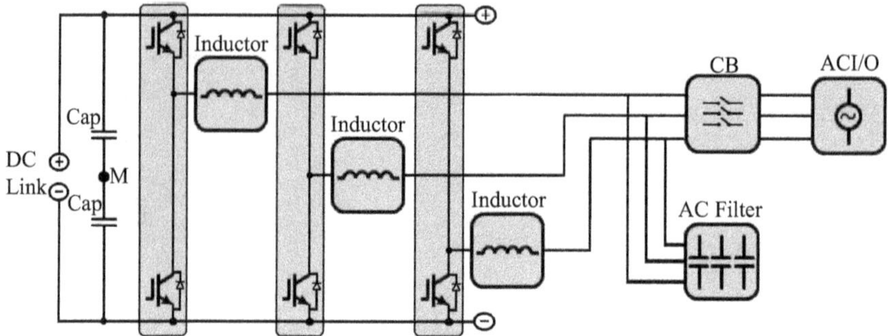


*Figure 4.6 Load-side of the main converter designed with three half-bridge circuits, © 2023 Freqcon GmbH*

*Table 4.1 Rated active power $P_r$ @ cos $\varphi$=1 for the MSC series with 1700 V IGBT*

| | AC-Output | | | | |
|---|---|---|---|---|---|
| $P_r$ [1] | 250 kW | 500 kW | 750 kW | 1000 kW | 1500 kW |
| $S_r$ [2] | 268 kVA | 537 kVA | 805 kVA | 1074 kVA | 1611 kVA |
| $I_r$ [3] | 250 A | 500 A | 750 A | 1000 A | 1500 A |
| $I_{max}$ [4] (0,5 s) | 300 A | 600 A | 900 A | 1200 A | 1800 A |

[1]Rated AC active power $P_r$ @ cos $\varphi$ = 1
[2]Rated AC apparent power $S_r$
[3]AC rated current $I_r$
[4]Maximal AC output current $I_{max}$ (0,5 s)

circuit breakers, etc. also offer up-scaled components with the same physical dimensions of the components. This design strategy has many positive side effects for manufacturers of MC, e.g., cost-saving for redesign and re-certification of converters. Table 4.1 shows a converter-family offered by FREQCON GmbH, covering a wide range of up-scaled converter units.

Figure 4.6 shows a more common representation of the circuits shown in Figure 4.5. It gives an impression of the physical connections of the three half-bridge circuits. To drive high currents of many hundred Amperes at a typical DC-link voltage-level of 1200 V for low-voltage MC two and more half-bridge circuits are connected in parallel.

The set-point of the DC-link voltage of MCs for WTs is typically in the range of 950 to 1300 V for low-voltage systems, depending on the low-voltage grid connection, the dielectric strength of the DC-link capacitors, the chosen semiconductor-switches, the needs and design, respectively. The line-to-line voltage of the primary-side of the transformer which connects the plant to the medium- or high-voltage grid, typically varies between 600 and 660 V. Almost all onshore-WTs have a low-voltage connection for cost and service reasons. This is also true for most converters for offshore-WTs, but nowadays MCs with medium DC-link

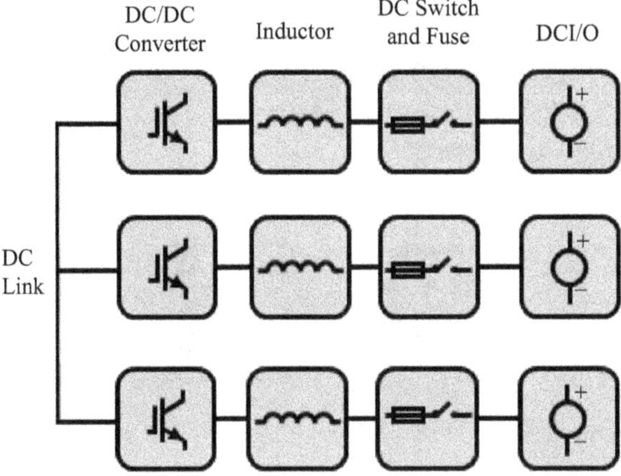


*Figure 4.7 Generator-side connection of the main converter for three DC-inputs or three DC-outputs, © 2013 Freqcon GmbH*

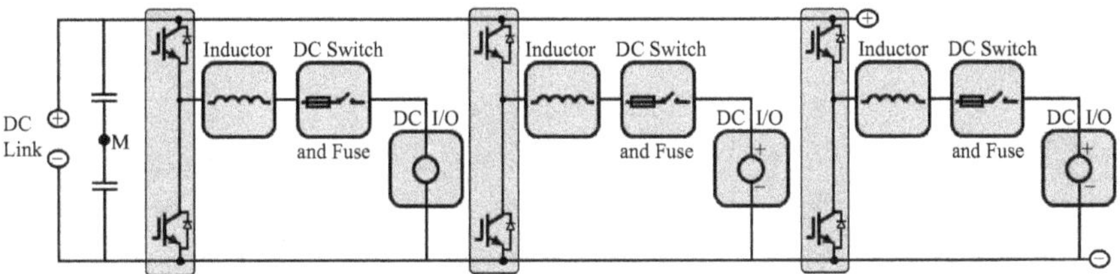


*Figure 4.8 Generator-side of the main converter for connection of DC-loads*

voltage and a direct medium voltage grid connection become more and more popular for WTs in the multi-megawatt-class [8,9]. ABB has already delivered round about 250 main converters with medium voltage grid connection for the north- and east-sea offshore wind parks as well as for offshore-WTs in China in the year 2022 [10].

Instead of connecting a three-phase generator to the 4-QC at the load side, each half-bridge circuit can be connected to one DC-input or to one DC-output (DC-I/O), as can be seen in Figures 4.7 and 4.8.

The DC/DC-converter can also be extended by more half-bridge circuits to support more than three DC-I/Os. The DC/DC-converter on the load-side also gives the possibility to connect, e.g., two three-strand windings of a synchronous machine via a rectifier, as shown in Figure 4.9, or battery-systems.

#### 4.2.3.1 Control system of the converter

Up to this point only the energy conversion part of the MC is mentioned. However, an equally important part of the MC is the control part and its subsystems like analogue- and digital sensor-systems, their placing and the appropriate signal

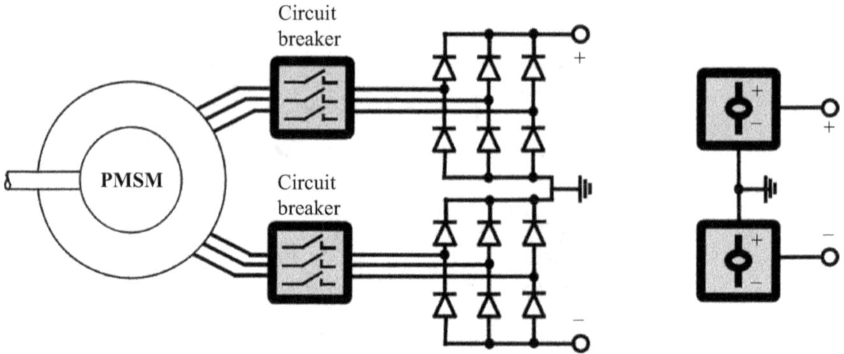


*Figure 4.9 Rectification of voltages of a six-phase permanent-magnet excited synchronous machine (left) and simplified circuit diagram (right)*

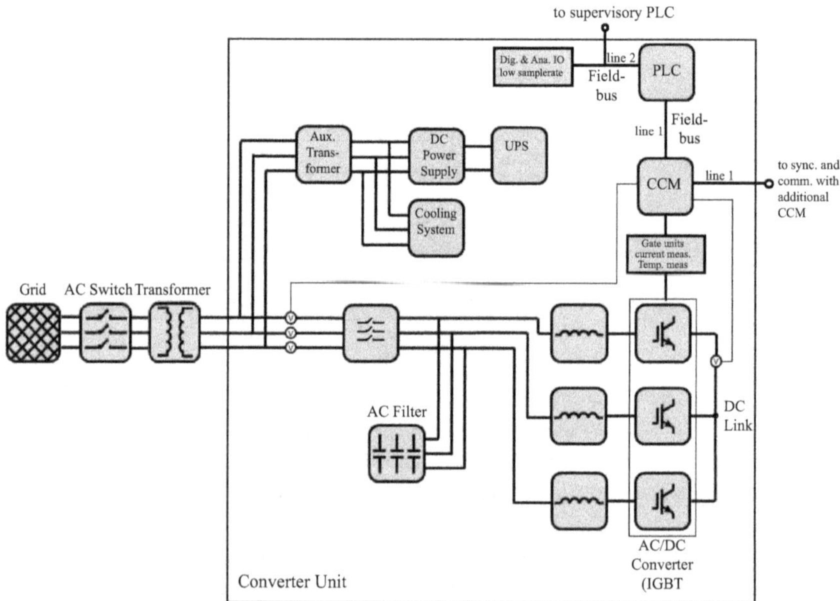


*Figure 4.10 Grid-side converter unit with control and power conversion chain, © 2013 Freqcon GmbH*

conditioning of each sensor-channel. Therefore, most of the applied interfaces in the MC are standard fieldbus systems with programmable logic controller (PLC) and industry-PCs (IPCs) as the main controller for hard real-time constraints. In Figure 4.10 only a rough sketch of the control part of the MC is shown. For further reading on applying fieldbus communications see [11,12].

Communication with the controller of the main converter is typically possible via internet and standard protocols like OPC, PROFINET, ECAT, etc. In general, the control-system as well as the safety-system of the MC is connected to an uninterruptable power supply (UPS). In case of a grid voltage loss, the UPS can keep the system up for at least 30 s, which is enough time to shut down the control- and power chain of the MC. In case of a malfunction of the UPS the MC shuts down abruptly and the switch is opened automatically for safety reasons. The UPS is also an important subsystem to keep the cooling system alive during a fault-ride-through (FRT) event.

Depending on the computing power of the control unit, the number of I/O-channels for analog and digital signals, the signal processing technology of the applied unit, e.g., FPGA-, PLCs, IPC-, or DSP-based technology and others, the number of control units varies and depends on the application. Most manufacturers subdivide between hard real-time requirements with high sample-rates on the one hand and hard real-time requirements with low sample-rates on the other hand. This leads nearly automatically to the approach of applying two separated fieldbus-lines.

The points of voltage measurements are also shown in Figure 4.10 and must belong to the fieldbus-line which is synchronized with the gate-units of the IGBTs. The measurement has a typical accuracy of min. $\pm$ 5 % based on the total measuring range.

A current measurement is typically placed directly at the AC-output of the IGBT-devices. For example, in case of SEMIKRON-devices the current sensors are already integrated in most devices, like the IPM-devices (see Figure 4.4), and they can easily be adopted to an analog or a digital input of the converter control module (CCM). The measured currents are typically used for control algorithms of the converter unit (CU) and for monitoring purposes. The PLC and the CCM, as shown in Figure 4.10, communicate via the PROFINET-fieldbus in hard real-time. All top-level control functions like AC/DC- and DC/AC-conversion applications are typically program-tasks of this PLC.

The set-points are transferred via one fieldbus-line to the CCM which computes the "fire-signals." These signals are start- and stop-timestamps with a relation to the timestamp of the fieldbus-clock and are an output of a PWM-, a tolerance-band- (TBM), or any other modulation algorithm. The fire-signals are output signals of a current-controller which represents the fastest control-loop in general. This applies to both the grid-side and the generator-side. The current-control is typically clocked synchronously with the fieldbus-clock, e.g., 250 microseconds. Placing the current-controller task on the CCM avoids excessive utilization of the fieldbus-interface and saves computing-power on PLC-side.

By serializing more CCMs in one and the same fieldbus-line, more fire-signals can be calculated by the other CCMs. In this way, all CCMs connected to one and the same fieldbus-line can be synchronized precisely in the range of at least a microsecond.

When the time-stamps for all fire-signals for all the IGBTs of the MC are calculated on the top-level PLC and the computed time-stamps are transferred to the gate-units directly over the fieldbus, the voltage- and, respectively, the instantaneous current-values must be transferred via synchronized bus-connected measurement-devices. This approach could also be realized, for example, by applying ETHER-CAT fieldbus-devices.

An advantage of the ETHER-CAT system is given by the fact that one top-level PLC can be synchronized under hard-real-time-constraints with another or even more PLCs. This is a very helpful feature for large-scale systems with many fieldbus-devices, hard-real-time constraints and high or very high task-frequencies. In other words, in a PROFINET-driven fieldbus system, there can only be one fieldbus-line with precisely synchronized CCMs, but no more. For example, in an ETHER-CAT-driven fieldbus system the top-level PLCs can be precisely synchronized and thus also two or even several fieldbus-lines and thus, also fieldbus-devices of different lines can be synchronized.

The control and protection concepts are also designed in a modular way. The program-code of the control-functions of a converter unit is typically divided into individual function-blocks, each with its own version number. All function-blocks are independent of each other, so the adaptions of the control of the CU from the user-side are independent from protection and control software-modules which are certified software components and must not be changed under any circumstances. So, a change of parameters which have a direct influence on certified modules cannot be done from the user-side due to this encapsulation.

In general, two or even several CUs are connected in parallel to deliver the desired rated power of the whole system as can be seen in Figure 4.11. The figure provides a view inside of a 9 MW WT-converter with two times three synchronized CUs connected in parallel. Each CU has a rated output power of 1.5 MW. In this case, the synchronization of the CCMs is a key-feature which is based on the synchronicity of the used fieldbus-system. Optionally the main converter can be connected to a wind-park controller which controls some individual top-level set-points, e.g., reactive or active power of the main converter.

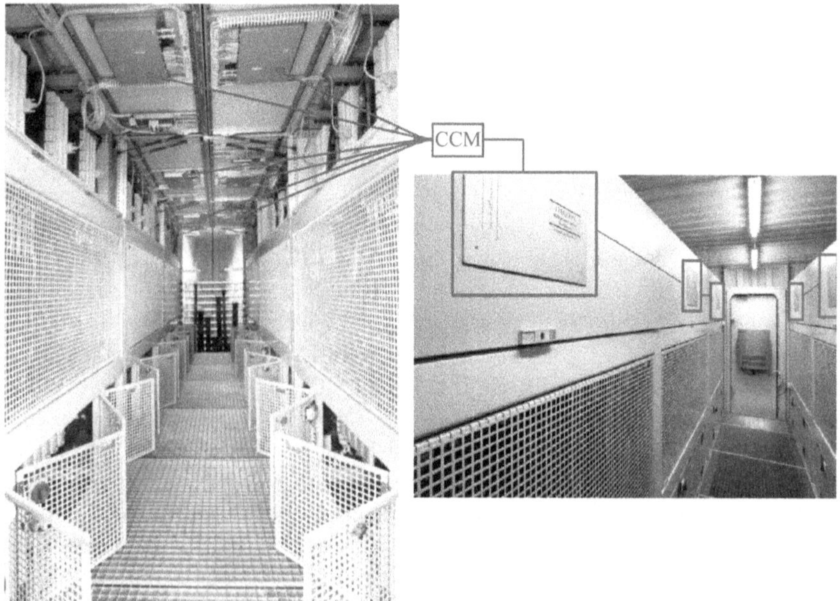


*Figure 4.11 View inside of a 9 MW WT-converter with installed PLCs, © 2013 Freqcon GmbH*

### 4.2.3.2 Power stack

The power stack of the converter unit consists of two 4-QC and the DC-link between them as main components. The DC-link is typically composed of two or more capacitors connected in series for voltage-resistance reasons. A minimum number of capacitors must be connected in parallel to carry the ripple-current without thermal overload. In addition to the condition for the minimum number of capacitors, respectively the minimum DC-link capacitance, the controllability of the DC-link voltage must also be considered. The larger the DC-link capacitance is selected, the better the DC-link voltage can be controlled. Nowadays, DC-links are constructed of copper or aluminum material. Flat, rigid plates are pressed together with a thin isolating plastic material in between. This approach has proven itself in practice and minimizes the leakage inductance of a rather large DC-link.

The maximum possible number of capacitors of the DC-link is limited by the space available in the cabinet of the power-stack. Film-capacitors with a rated voltage of $U_r = 750V$ and a nominal capacitance of $C_N = 1500\,\mu F$ have a permissible ripple current in the range of typically 100 A to 120 A. For example, each power-stack of the prototype MC shown in Figure 4.12 is built with ten capacitors in parallel.

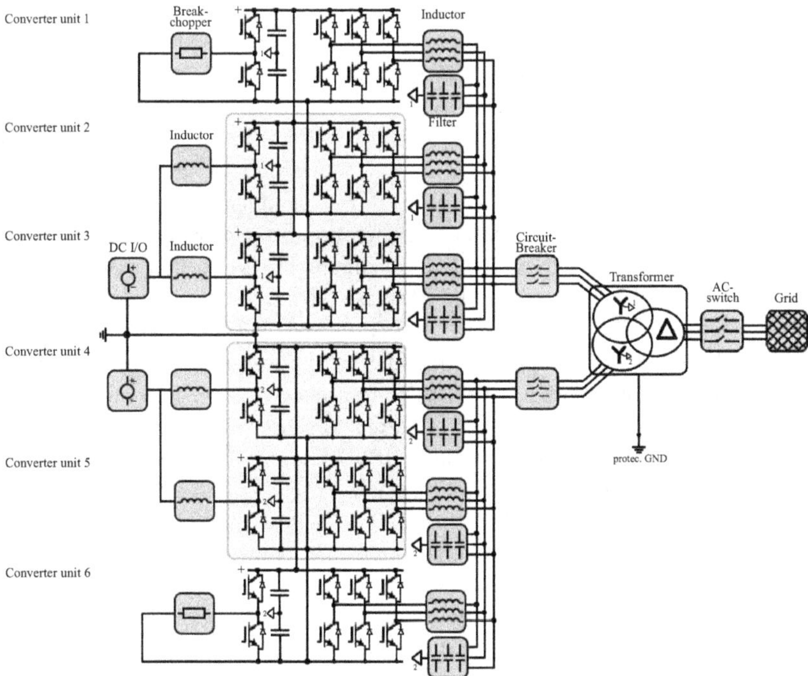


*Figure 4.12 Principle equivalent circuit diagram of a prototype main converter with a capacity of 9 MW installed in a 20 ft. container and consisting of six CUs, designed to correspond to a 6 MW offshore WT*

The output signals of the control unit are pulse-width-modulated (PWM) signals or alternatively tolerance-band modulated (TBM) signals in most cases. There are several different techniques to generate PWM-signals. For further reading on PWM-signal generation see [13,3]. Typical voltage-levels of the PWM-signals are 3.3 V or 5 V, depending on the applied technology of the signal processing controller and the appropriate fieldbus system applied. The PWM-signals are the input signals for the gate units. The gate units shift the input voltage level up to a symmetric output voltage of $\pm 15$ V or in other cases up to +15 V and $-10$ V or $-5$ V, depending on the specific device chosen by the circuit designer or developer. The gate units are designed to provide high gate-currents and must take into account other constraints of the IGBT devices in conjunction with the system behavior. Some gate units provide a temperature measurement for IGBT-devices with integrated temperature sensors, like resistors with a negative temperature coefficient (NTC) or other technologies. Common gate units detect and trigger on a high voltage-rise as function of the device-current with a so-called $V_{CE}$-detecting-circuit to protect the device of a damage.

Snubber-circuits are also a common method to protect the power stack and its connected load of unwanted resonant effects and over-voltage-peaks. The snubber-circuit basically damps resonant frequencies via resistors and capacitors connected in series. For further reading on gate-signal generation and snubber-circuits see [14–17].

Depending on the manufacturer of WTs and the desired vertical production depth of the main converter the whole spectrum of single IGBT devices from all manufacturers like INFINEON, SEMIKRON, MITSUBISHI, etc., up to complete power stacks are applied. For example, SEMIKRON offers complete ready-to-use (plug-and-play) power stacks with different semiconductor technologies. Other manufacturers like INFINEON offer all the devices and hand out application notes with design examples for gate-units, snubber-circuits, etc. From the application designer's point of view, the SEMIKRON approach is advantageous for rapid prototyping and small series, when the focus is on solving the application problem. WT-manufacturers like ENERCON, VESTAS, and NORDEX, GE-Energy and SIEMENS have separated departments for circuit designs and layouts or they outsource the development to appropriate companies which are specialized on these developments or on the development of complete power stacks.

With the intelligent power module (IPM) devices from SEMIKRON it is possible, to build a main converter with a rated power of up to 1.5 MW for an AC-grid voltage of 600 V with only one SKiiP-IPM power stack. If a higher-rated output power is needed, full converter units are connected in parallel, which requires some further considerations and features which will be discussed later on.

Figure 4.12 shows a built prototype MC which is realized to correspond to a 6 MW offshore WT. The currents of two three-strand-windings of a permanent-magnet-excited synchronous generator are rectified, as can be seen in Figure 4.9. For cost and feasibility reasons, a circuit breaker is chosen for the AC-side of the EDT between generator terminal box and the rectifier-circuit. After rectification the phase-voltages are boosted via the DC/DC-converter to the DC-voltage levels of two PFC-active-frontend converters, which feed the grid via a three-windings transformer with a [Dyn] vector-group. The transformer has a voltage-ratio of (2 × 600 V)/30 kV. The MC is designed with a capacity of 9 MW to ensure redundancy of the converter system. Thus, it is possible to switch-off single converter sections remotely should a failure occur. It consists of various components and modules that have been assembled together to create one unit. The converter system is split-up into the following individual components as can be seen in Figure 4.12:

- The rectifier converts the 6-phase AC currents of the generator into DC-current.
- The boost converter increases the DC voltage level for the grid-side converter.
- The brake chopper burns energy which cannot be fed into the grid, e.g., during grid failures.
- The grid-side PFC-active-frontend converts the DC currents into three-phase AC-currents corresponding to the grid connection.

An excerpt from the specifications of the 9 MVA prototype converter system is given in Table 4.2.

The main components of the MC are installed inside a 20 ft. container which is to be installed on a base frame besides the WT-tower. Six converter units (CUs) are installed inside the container. Each has a rated AC-power output of 1.5 MW and each one is equipped with its own converter control module. The MC is independently controlled from a central main PLC and no manual operation by personnel is required during operation.

As mentioned before, the prototype MC has a capacity of 9 MW for redundancy purposes, the generator itself has a rated AC-output power of 6 MW. In case of a

*Table 4.2 Excerpt from the general technical data of the 9 MW prototype MC for an offshore-WT*

| Operating parameters | |
|---|---|
| DC voltage range | 2 × 1100 – 1200 V |
| Rated DC current | 3 × 2750 A |
| Rated output power | 3 × 3000 kW |
| Rated AC voltage | 2 × 600 V |
| Rated AC current | 6 × 1400 A |
| Rated AC frequency | 50 Hz |
| Frequency range | 47.5 – 51.5 Hz |
| Efficiency | ≥98.25% Body text2 |
| Rated power factor | 1 |
| Power factor range according to grid code demands | 0.925 cap. – 0.925 ind. |
| THD @ rated power | ≤4% |
| Variation in grid voltage | ≤5% |
| Cooling principle | Water/air cooling |
| Maximum cooling water inlet temperature | 45 °C |

failure, it is possible to switch off individual power-lines on the grid-side converter units as well as on the generator-side; in total one CU is redundant on each side of the EDT. In addition, this feature has a positive impact on the life-time of the converter as well. It should be noted that the grounding concept of the converter unit, or the main converter must to be considered separately. The return-conductor-line is used for the star-point-connection of filters and for single-phase-loads. The protective ground must be connected to all touchable metal-objects, e.g., the converter cabinets, the container itself, etc., in case of rigidly- or low-impedance grounded grid connection. Grounding concepts are discussed in more detail in Section 4.2.3.5.

The IGBT-devices used for this prototype are IPM devices with a system AC-current of 500 $A_{RMS}$ per AC-terminal (sinusoidal current) and a total nominal-current rating of 3600 A. The devices are based on 1700 V IGBTs and their corresponding and integrated free-wheeling diodes. For further reading see [18]. Each device consists of six half-bridge-circuits as shown in Figure 4.4. The manufacturer has already installed the parallelization of all six half-bridge circuits by internal shortening of the terminals. In this way, one IPM device can feed up to 3600 Ampere of DC-current into the DC-link via the DC/DC-boost converter on the generator-side of the EDT. Therefore, one IPM device can drive the DC-current of one string of the generator or, in other words, half of the generator's rated power.

As can be seen in Figure 4.13 (left), the converter cabinet interior essentially consists of four IPM devices, placed in the upper area of the cabinet. All six

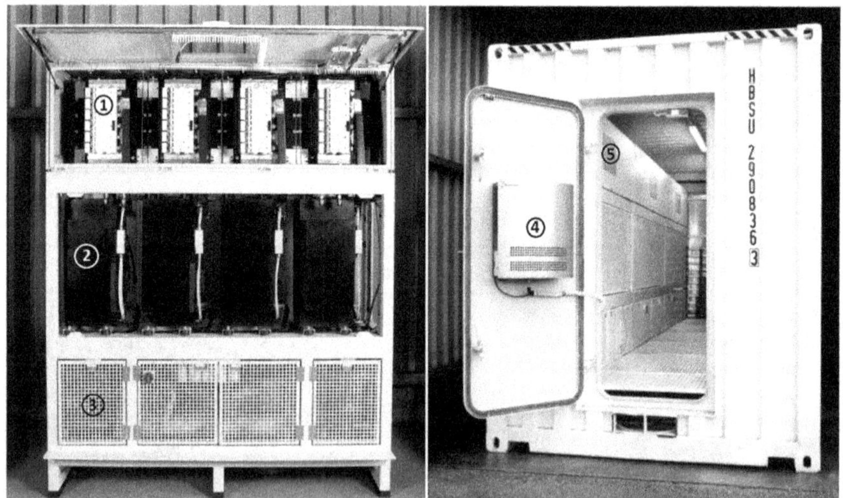


*Figure 4.13 A 9 MW main converter consisting of six synchronized 1.5 MW CUs in a 20 ft. container housing, © 2013 Freqcon GmbH, converter cabinet interior, front view (left) with (1): Water-cooled IPM-devices, (2): water-cooled inductors, (3): Water-pumps, switches, etc., and converter container side view (right) with (4): supervisory PLC as well as (5): one of six CCMs placed in the upper area of the cabinet*

converter units have essentially the same structure, whereby two of them are connected in parallel, but not in back-to-back (B2B) configuration, because of the installation inside the container. In the middle area of the cabinet, the water-cooled inductors are placed. The DC-link voltage is set to 1100–1200 V and it is controlled by the grid-side AC/DC-converter. For each three-strand winding of the generator, a brake-chopper-circuit is installed. It converts generated energy to heat, e.g. during grid failures.

In each converter unit, one CCM is installed to control the cabinet and therefore one unit. All six CCMs can be controlled via a PLC. This PLC is placed at the front door of the 20 ft. container, as can be seen in Figure 4.13 (right). Figure 4.13 (left) shows the interior of one converter unit. All the main components of the power conversion chain of the MC are water-cooled. Also shown in Figure 4.13 (right) is the MC of the 6 MW offshore WT installed in its 20 ft. container which is air-conditioned.

#### 4.2.3.3 Protection system

The protection system must be implemented according to technical requirements for the connection and operation of customer installations to the high-voltage grid network. This is proven or verified by numerous certified institutions in or for the country where the plant is set up. The most relevant law-, ordinance-, normative-, and technical guidelines are cited and briefly presented in Section 4.2.6.

*Grid-side protection of the converter*

The grid protection is typically an integrated component within the main converter system. For example, for those already mentioned MSC-series converter units (see Table 4.1), the voltage and frequency measurements consist of an analogue HVIO board for signal matching and a subsequent PROFINET interface for signal processing to the main control unit, including the grid protection parameters, see Figures 4.10 and 4.14. The MCs of type MSC 250-X, MSC 352-X and MSC 500-X are disconnected from the grid by a SIEMENS-3VA circuit breaker. The MCs of type MSC 750-X, MSC 1000-X, and MSC 1500-X are disconnected by the SIEMENS-3WL circuit breaker. The delay-time of the SIEMENS-3WL circuit breaker is 81 milliseconds. For the 3VA-type of circuit breaker the disconnecting time without delay of the circuit breaker is in the range of 20–40 ms. The grid protection functions are implemented within the inverter control unit. The parameters of the grid protection can typically be read and set by using the specified user interfaces of the control unit. The monitored voltages are the phase-to-neutral terminal voltages of the MC. The voltage measurements for triggering off-the-grid protection are linked by logical OR.

A description of the parameters responsible for the grid protection function are shown in Table 4.3.

The manufacturer must declare that the specific grid protection tasks operate independently from the main control. An uninterrupted power supply unit ensures the continuous operation of the control unit (thus the protection functions) for at least 30 s after losing grid supply. A failure of the UPS is followed by an immediate

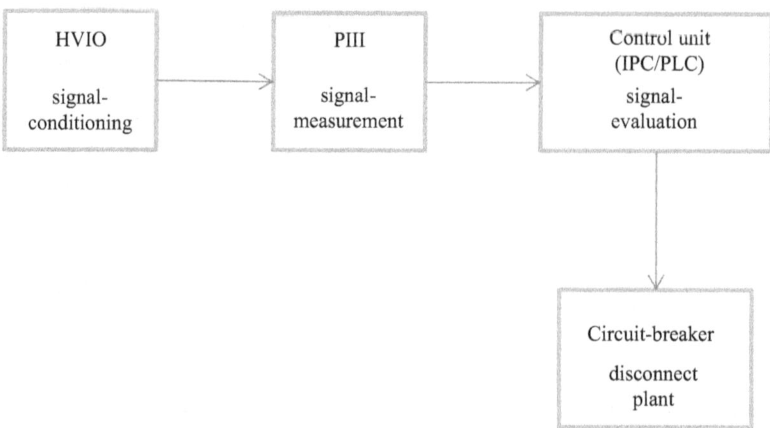


*Figure 4.14 Signal path of the voltage and frequency measurement as well as the release of the circuit breaker, © 2018 Freqcon GmbH*

*Table 4.3 Parameters responsible for the grid protection of the MSC-Series*

| **Description** | **Unit** | **Min.** | **Default** | **Max.** |
|---|---|---|---|---|
| Over-voltage protection U> | – | 1.00 | 1.20 | 1.50 |
| Delay time of U> | s | 0 | 0.16 | 1.00 |
| Over-voltage protection U>> | – | 1.00 | 1.10 | 1.50 |
| Delay time of U>> | s | 0 | 1.00 | 1.00 |
| Under-voltage protection U< | – | 0 | 0.87 | 1.00 |
| Delay time of U< | S | 0 | 2.00 | 5.00 |
| Under-voltage protection U<< | – | 0 | 0.50 | 1.00 |
| Delay time of U<< | S | 0 | 1.50 | 5.00 |
| Under-voltage protection U<<< | – | 0 | 0.10 | 1.00 |
| Delay time of U<<< | S | 0 | 0.40 | 5.00 |
| Over-frequency protection f> | Hz | 40.00 | 51.50 | 70.00 |
| Delay time of f> | S | 0 | 0.3 | 5.00 |
| Under-frequency protection f< | Hz | 40.00 | 47.50 | 70.00 |
| Delay time of f< | S | 0 | 0.30 | 5.00 |

switch-off of the MC. The protection parameters are readable without additional equipment. A test terminal is installed for the determination of grid protection settings without disconnecting any wires. The accuracy of the under-voltage and over-voltage protection should be less than 1%. The accuracy of the under-frequency and over-frequency protection should be less than 0.1 Hz. The fallback ratios of 0.98% and 1.02% are maintained. The values presented in Table 4.3 correspond to regulations standards in Germany that are enumerated in Section 4.2.6.

*Self-protection of the converter*

A self-protection of the main converter is also implemented via trigger values in the processing of the main control. The exceedance of set-values causes an immediate shut down of the main converter. It is not intended to change these settings. The self-protection functions must not interfere with the requirements concerning static voltage maintenance and dynamic grid support, as presented in Section 4.2.5.

#### 4.2.3.4 Cooling concept

Fully air-cooled converter-systems are reasonably applicable for AC-power outputs of a few hundred kilowatts in general. For multi-megawatt WTs, the overall-efficiency with reference to less installation-space, feasibility and costs are the main reasons for the application of water-cooled systems. Therefore, the focus of this chapter is only on water-cooled converter-systems.

A positive side-effect of applying water-cooled systems is that not only the power-stack but also the subsystems, e.g., the inductor-circuit or other peripherals, can be tempered on a set-value. Another advantage of a water-cooled systems is furthermore that the surrounding control cabinets and other switch-cabinets do not necessarily have to have other cooling-devices, e.g., an air-conditioning.

A practical design approach starts by summing-up the overall heat loss performance of the IGBT-devices, choke-devices, diodes, break-chopper, etc., and all other components and subsystems with relevant heat losses. Besides these waste-heat-producing parts, the operating temperature is dependent on numerous environmental conditions, like weather-conditions, humidity, etc., as well. Depending on the climate-zone, there are different guidelines, standards like VDI 2078 and other restrictions to estimate a mean average ambient temperature.

For a rough estimate of the mean resulting waste-heat output of the MC for a multi-megawatt WT that has to be dissipated, a reasonable value is in the vicinity of three to four percent of the total electrical power-output of the WT. The design of a water-cooling system starts with the sum of the waste-heat power consumption. This total waste-heat output can be split into two main parts of the cooling principle that takes place. Firstly, the power-stack and all other subsystems of the MC indicate heat via the principle of convection, and secondly, heat-dissipation via a coolant. The coolant-liquid of the cooling-system is typically a mixture of water and a glycol. A typical coolant to be used for IPM devices is a mixture of 50% water and 50% glycol. The maximum temperature-rise of the coolant $\Delta T_c$ is 6 °C for the IPM devices used for the prototype MC of the 6 MW offshore-WT, as shown in Figure 4.12. The maximum coolant inlet temperature for that MC is 45°C. A rough estimation of the total waste-heat dissipation $P_{\text{heat}_{\text{M}}\text{C}}$ is possible with the efficiency value for the MC, in case the value is not specified in more detail. The efficiency value should be a given value by the general technical data of the MC.

Experience shows that a share of approx. 80% of the total waste-heat power can be discharged via the coolant and ca. 20% via air-convection to the surrounding environment. Therefore, round about 80% of the total waste-heat of the MC and its subsystems must be transported via the fluid to the heat-exchanger which typically gives off this heat to the ambient air (outside the plant) by an efficiency of 85% to 99% [19].

In general, the design-parameters are:

- maximum temperature $T_{\text{inlet}}$ of the fluid medium (water-coolant-mixture) at the input of the main converter
- minimum flow-rate $\dot{d}_c$ for the cooler of the power-stack (first device in cooling-chain)
- heat output of the main converter $P_{\text{heat}_\text{M}\text{C}}$ (e.g., 80% of total waste-heat dissipation)
- ambient temperature $T_\text{a}$

With the given parameter values, the maximum coolant temperature after the MC can be calculated. Based on the datasheet of the coolant-fluid the specific gravity $\rho_c$ of the fluid-mixture and its specific heat-capacity $c_c$ it can be determined. By the given coolant-flow-rate $\dot{d}$ of the power-stack and all other subsystems the mass-flow $\dot{m}$ of the cooling-system can be determined. The coolant-flow-rate of the power-stack for the prototype MC of the 6 MW WT is 16 liters per minute for instance. Put it all together and you get:

$$\dot{m}_c = \dot{d}_c \cdot \rho_c$$

The maximum outlet temperature of the coolant can then be calculated in the following way:

$$P_{\text{heat}_\text{M}\text{C}} = \dot{m}_c \cdot c_c \cdot \Delta T_c$$

$$\rightarrow \Delta T_c = \frac{P_{\text{heat}_\text{M}\text{C}}}{\dot{m}_c} \cdot c_c$$

$$T_{\text{outlet}} = T_{\text{inlet}} + \Delta T_c$$

Under the specification of the maximum temperature at the output, the water-to-air heat-exchanger and the knowledge of the ambient temperature, the calculation of the cooling-system can be finalized. Therefore, the formerly determined value of $T_{\text{outlet}}$ goes over to the value $T_{\text{inlet}}$ and the new value for $T_{\text{outlet}}$ goes over to the value of the ambient temperature $T_\text{a}$. The datasheet of the heat-exchanger-system delivers the parameter for cooling capacity (typically in kW) versus ambient temperature ($T_\text{a}$). Figure 4.15 shows the layout of the cooling system of the 6 MW prototype main converter (shown in Figure 4.13) for an offshore WT.

#### 4.2.3.5 Grounding concept

The grounding concept is important in failure-situations, e.g., short-circuits (line-to-line) or a conductance between phase-lines and ground (line-to-ground). In these cases, a protection against over-voltage, high contact-voltage, neutral-point-shift, etc., is essential. Different neutral-point-treatments do not have any negative impacts on the energy-transfer in medium-grid networks. For cable-based medium-grid networks an isolated star-point of the grid-transformer and therefore, a

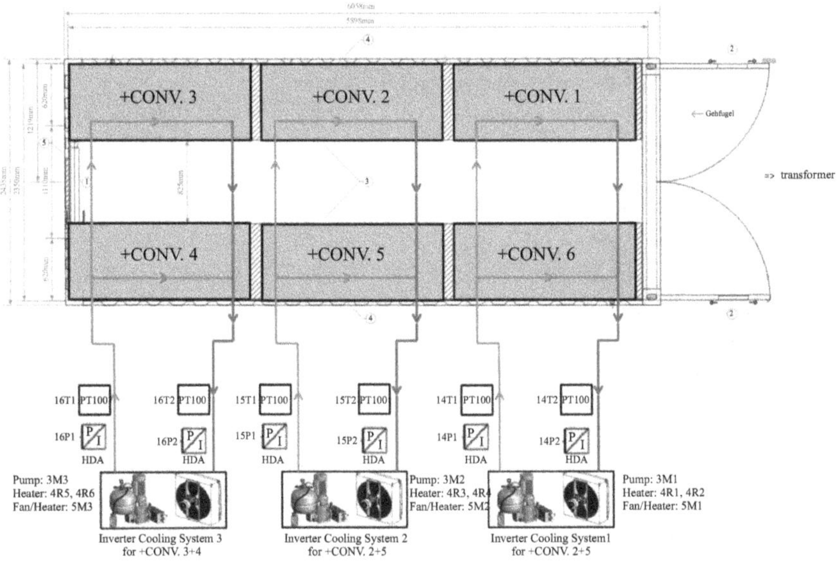


*Figure 4.15 Cooling concept of the 9 MW prototype MC for a 6 MW offshore WT © 2013 Freqcon GmbH*

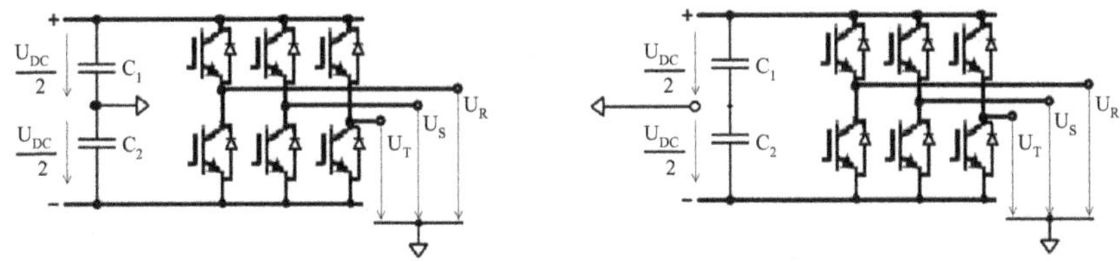


*Figure 4.16 Center-point treatment of DC-link, with fixed center-point reference-potential (left) and a drifting reference-potential (right)*

continuation as an isolated grid network is reasonable. Wind park-networks have requirements which can be easily combined with an isolated grid network, like, a star-network structure, limited extent, hardly varying magnitude of the grid impedance, no requirements to self-extinguish and enough protection signal level [20].

Focusing on the requirements of the power-stack there are two possibilities to treat the center-point of the DC-link, as can be seen in Figure 4.16. Both have advantages but also disadvantages. The set output-voltages are typically referenced to $\frac{1}{2}$ $U_{DC}$. Most manufacturers of converter-systems do not ground the center-point of the DC-link as shown in (Figure 4.16, right). The advantage is that no parasitic equalizing currents can find their way back via the DC-link. This may then result in charging or discharging of the DC-link, as can be seen looking in Figure 4.16 (left). Parasitic currents may arise in case of unsymmetrical load-feeding conditions and in dynamic-load feeding situations, depending on the state of the 4-QC switch-pattern. Dynamically arising circular currents must then be treated. In normal, balanced load situations the mean voltage of the center-point of the DC-link is of course in the range of the reference-potential.

By grounding the center-point of the DC-link (Figure 4.16, left) the reference-potential cannot drift anymore. However, this approach has consequences for the modulator-circuit. The modulator-circuit, no matter if it is realized in hard- or software, receives its input value from the PLC which calculates the instantaneous values for the control-output. This value is converted to a pulse-width-signal by the modulator-circuit and output to the gate-unit to fire the according IGBT of each half-bridge-circuit. In case of a three-phase system each calculated instantaneous output-value is phase-shifted by 120°. Therefore, the maximum achievable peak-level of the output-voltage cannot be reached for the phase-to-phase voltages with a pure sine-delta modulation. By adding the third-harmonic-value to each control-value or by implementing a vector modulation instead of a sine-delta-modulation, it is possible to increase the line-to-line-voltage by ca. 15%. Due to a reference-potential shift of up to $\pm 0.5 \cdot U_{DC}$, neither the modulation principle of adding the third-harmonic nor the vector-modulation for DC-links with fixed center-point grounding are applicable in these cases.

From the design point of view, the lower set-able peak-value results in a higher current rating of the IGBTs for the converter unit when comparing systems with the same power output. The main advantage when grounding the center-point of the DC-link is given by a significant advantage of suppression of common-mode currents and voltages. By keeping the symmetry of the hardware-structure regarding the reference-potential and, if symmetrical output-signals are generated, no common-mode components can propagate neither in the systems nor in the subsystems. Another advantage is that the real line-to-neutral or line-to-earth voltage can directly be measured which is not possible in case of a floating DC-link center-point.

### 4.2.4 Filter components

As described in Section 4.2.7, the inductance of the transformer is part of the filter design and builds a part of the grid-side filter-inductance. A so-called LCL-filter, a filter of third order, is a standard filter component for almost all relevant EDTs for WTs. The filter consists of a capacitive part, the line filter capacitors, and an inductive part, the line filter inductors. For damping purposes, the capacitor is followed by a low-ohmic series resistor for damping purposes. The line-inductance is divided into two partial inductances with the capacitance part of the filter in the mid-position between both inductors. The filter has two resonance-frequencies in general, whereby the first, main resonance frequency is set between the modulated fundamental frequency and the (lowest) switching frequency of the IGBTs.

Due to the hard switching IGBT devices with a fixed or variable switching-frequency, the sinusoidal current has a fundamental frequency of 50 or 60 Hz. Unfortunately, numerous harmonics of the fundamental current frequency, the switching frequency and mixed frequencies of both fundamentals are superimposed.

Based on measurements, experience and proven practicability the output current of the MC can be described by the sum of the rated current $I_r$ and an additional current $I_h$ (distortion-current) on top of $I_r$, which represents the harmonic part of the current. Alternatively, the rated current $I_r$ can be multiplied by a factor $n$ greater than one.

$$I_r(@50\text{ Hz } or \text{ } 60\text{ Hz}) = n_f \cdot I_r \text{ typical} : n_f > 1.0 \text{ and } n_f < 1.2$$

$$I_r(@f_{\text{switch}}) = n_s \cdot I_r \text{ typical} : n_S > 0.05 \text{ and } n_S < 0.4$$

By setting the maximum tolerable ratio of ripple-current to grid-current (e.g., a ratio of $\frac{I_{\text{ripple}}}{I_r} = 0.1$ has proven itself) on the one hand, the maximum distortion-current for each phase of the line-inductor can be calculated as follows:

$$\Delta I_{L,\text{max.}} = \frac{I_{\text{ripple}}}{I_r} \cdot \sqrt{\frac{2}{3}} \cdot \frac{P_r}{U_r}$$

whereas $P_r$ is the total active power output of the converter for each line (phase) and $U_r$ is the line-to-neutral voltage. If the calculated distortion-current $\Delta I_{L,\text{max.}}$ is inserted into the formula for calculating the current-distortion caused by the hard switching of the IGBT-devices of the MC, on the other hand, the total inductance of the line filter can be calculated for each phase as follows [21]:

$$\Delta I_{L,\text{max.}} = \frac{U_{DC} \cdot T_{\text{switch}}}{L_{\text{line}}}$$

$$\rightarrow L_{\text{Filter}} = \frac{U_{DC}}{k \cdot f_{switch} \cdot \Delta I_{L,\text{max.}}}$$

whereas the distortion-current $\Delta I_{L,\text{max.}}$ caused by the MC is depending on the converter-topology and the type of modulation. For instance, for a conventional, classical PWM-modulation without added third harmonic and a 4-QC-circuit the factor $k$ is 16. With added third harmonic or a vector-modulation, $k$ is equal to eight [21].

The calculated inductance of the line-inductor of the LCL-filter is typically distributed equally on both sides of the capacitance, but the inductance of the transformer should be considered as part of the grid-side inductance of the filter. In some cases, the inductance of the transformer is already enough for the grid-side part of the LCL-filter, so only an LC-filter is installed inside the MC.

The dimensioning of the line filter capacity $C_{\text{Filter}}$ is based on the resonant circuit of the line filter capacitors and the primary inductance of the grid-side transformer $L_T$. The required capacitance of the line filter capacitors can be calculated using Thomson's equation:

$$f_0 = \frac{1}{\omega \cdot \sqrt{L_T \cdot C_{\text{Filter}}}}$$

The resonant circuit of the transformer inductance and the line filter capacitors should have a resonant frequency $f_0$ of approx. 1075 Hz. This value has been proven in practice. It is below the switching frequency of the IGBT devices, but above the frequency of the typical harmonics in the range of 150–650 Hz.

A positive side effect is given by the choice of a [Dyn]-vector-group of the transformer, because no frequencies with a multiple of three are transformed from the star-side to the delta-side of the transformer. The star-side is typically the low-voltage side with connection to the MC and the delta-side is the connection-side of the grid.

For the prototype MC with a rated output power of 9 MW which corresponds to a 6 MW offshore-WT, as specified in Table 4.2, the line-filter inductance can be calculated as follows (see Figure 4.12 and Table 4.2):

$$\Delta I_{L,\max.} = \frac{I_{\text{ripple}}}{I_r} \cdot \sqrt{\frac{2}{3}} \cdot \frac{P_r}{U_r} = 0.1 \cdot \sqrt{\frac{2}{3}} \cdot 1400\,A \cdot 0.925 = 105\,A$$

The factor of 0.925 must be considered, according to grid code demands. The power-factor decreases to $\pm 0.925$ according to maximum capacitive or inductive demand by the grid. For a 1.5-MW converter unit the inductance has a value of:

$$L_{\text{Filter}} = \frac{U_{DC}}{k \cdot f_{\text{switch}} \cdot \Delta I_{L,\max.}} \approx \frac{1200\text{ V}}{16 \cdot 4000\text{ Hz} \cdot 105\,A} \approx 179\,\mu\text{H}$$

For a 3-MW converter unit (two times 1.5 MW B2B, see Figure 4.12) the inductance has a value of 89 μH due to a factor of two for $P_r$ respectively $\Delta I_{L,\max.}$. Therefore, the selected inductivity for the line-filter inductor has a value of $L_{\text{Filter}} = 90\,\mu\text{H}$ for a 3-MW converter unit.

For the calculation of the capacitance of the LCL-filter only the inductance of the transformer is taken into account. It is the only significant inductance on the grid-side of the filter, and the line-inductivity of the filter is achieved exclusively by the converter-side inductor.

$$C_{\text{Filter}} = \left. \frac{1}{(2 \cdot \pi \cdot 1075\text{ Hz})^2 \cdot L_T} \right|_{L_T = 24\,\mu\text{H}} \approx 913\,\mu\text{F}$$

Therefore, the selected capacity for the line-filter capacitor has a value of $C_{\text{Filter}} = 900\,\mu\text{F}$ for each 3-MW converter unit in B2B setup (see Figure 4.12). The resonance-frequency is shifted to 1079 Hz. For the calculation of value $L_T$, refer to Section 4.2.7.

### *4.2.5 Active- and reactive-power feeding*

Wind Turbines that are connected to the public grid must comply with the required laws, standards, technical guidelines and normative regulations, as roughly examined in Section 4.2.6. The main converter typically has a set of operating modes regarding grid support options which are stored in the main controller of the MC. The measurement and monitoring of voltage- and frequency-values of the grid-lines is part of the control system of the MC. Therefore, the observance the regulations of these rules are also implemented in the controller of the MC. The main control of the MC is typically connected to a supervised PLC, e.g., park-controller or other remote stations. The set-values of the control for power output of the MC can be changed by supervisory PLC to coordinate the active and reactive power output of the WT.

#### 4.2.5.1 Active-power feed-in

Depending on whether the turbine is connected to the medium- or the high-voltage grid, the setting for the mode of operation of the MC must be adopted. The operating modes for Germany are descripted below. In a quasi-stationary mode of operation, the converter typically must have the capability to operate continuously at the grid in a frequency range of 45–55 Hz. The limits of the reconnection can be set in the control software of the converter. The minimum requirements for connecting to German medium- and high-voltage grid networks are summarized below.

**Requirements according to the BDEW-MSR**

The reconnection takes only place in the frequency range 47.5 Hz $< f <$ 50.05 Hz and at a voltage $U_{N,grid} > 0.95U_r$. In case of increasing grid frequency, the active-power output of the WT can be decreased by parametrization depending on the slope of the change in frequency. The reduction of active-power output of the MC is reset when the grid frequency falls below a parameterized value to return into control for maximum power output. The decrease in active-power output for this operating mode is shown in Figure 4.17.

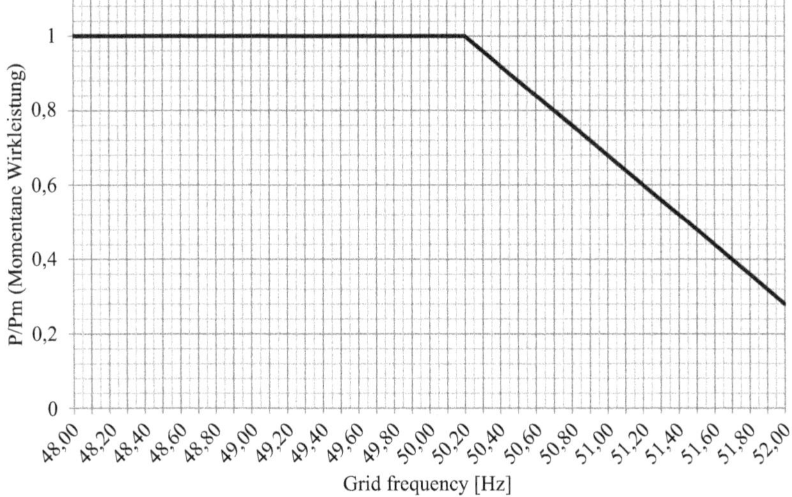


*Figure 4.17 Reduction of active-power output conform to BDEW-MSR, © 2013 Freqcon GmbH*

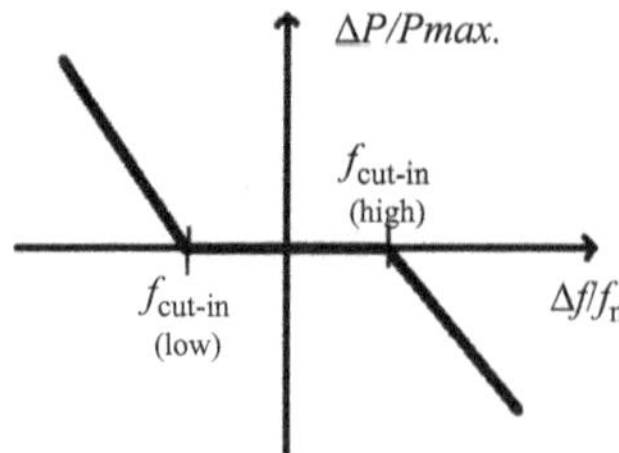


*Figure 4.18 Reduction of active-power output conform to VDE-AR-N 4120, © 2018 Freqcon GmbH*

**Requirements according to the TC2007**

The reconnection takes only place in the frequency range 48.5 Hz $< f <$ 50.05 Hz. The requirements according to the TC2007 can be realized by changing parameters in the according main control functions of the MC.

**Requirements according to the VDE-AR-N 4120**

The same is true for a defined frequency range of 48.5 Hz $< f <$ 50.05 Hz conform to the VDE-AR-N-4120 for connections to high-voltage grids in Germany. The manufacturer can also declare the requirements according to the VDE-AR-N 4120 that can also be realized by simply changing the parameter settings in the according main control of the MC. Furthermore, the reconnection to the grid is only possible with a simultaneous clearance signal. This signal is only evaluated if the converter's operating mode is set to VDE-AR-N 4120.

The main difference between this operation mode compared to TC2007 or BDEW-MSR is that the maximum active-power output of the MC is permanently coupled to the instantaneous frequency of the grid network. If the grid frequency is above a parameterized value, the active-power output is limited by a parameterized gradient, e.g., $5\% \frac{P_N}{Min.}$. The slope can also be set in the main control functions of the MC. In the VDE-AR-N-4120 a maximum gradient of $10\% \frac{P_N}{Min.}$ is allowed for positive and the negative slope as can be seen in Figure 4.18.

The reduction of active-power output is reset if the grid frequency falls below a parameterized value in the control software. In that the case, the active-power output can be increased again. This functionality is also parameterized and stored in the main control of the MC.

### 4.2.5.2 Reactive-power feed-in

As with the active-power control of the MC, different modes of operation are also supported for the reactive-power control. The basic control mode of operation is realized by a static-value for the set-value of reactive-power controller. Depending on the (maximum) rated reactive-power output of the MC, two more values are settable via the main control of the MC, to automatically adopt the reactive-power output depending on the instantaneous voltage-level of the grid (Q(U)-control).

The first parameter sets the dynamic of the reactive-power control. In other words, this parameter defines the settling-time of the reactive-power output in case of a set-value change. The second parameter sets the desired percentage of the maximum reactive-power as set-value in the controller function.

The active-power reduction is depending on the reactive-power output:

$$Q_{\text{desired}} = U \cdot I_q, \text{with} : |I_q| \leq I_r$$

$$P_{\text{desired}} = U \cdot I_p \text{with} : |I_p| \leq \left\{ \frac{I_r}{\sqrt{I_r^2 - I_q^2}} \right.$$

$$I_r = 1.13 \cdot \frac{S}{U_r} \text{with} : S = \sqrt{P^2 + Q^2}$$

The Q(U)-control must be implemented as a subordinate control in relation to a static and therefore fixed reactive-power output setting. In other words, the set-value for the Q(U)-control must not have any influence on the power-output of the MC in case of a static reactive-power output set. The following formula describes the calculation of the voltage deviation for an MSC-family converter as presented in Section 4.2.3 as: $\Delta U = \frac{U - U_r}{U_r}$. Figure 4.19 shows the according Q(U)-control for the MSC-family converter.

Typically, the power output performance of the MC is documented and declared by the manufacturer, as given in Figure 4.20.

#### 4.2.5.3 Dynamic grid support

The grid connection rules demand voltage support by a generating unit like a WT during fault-ride-through (FRT) events. In the control unit of the MC, the FRT behavior for the compliance of the guidelines can be parameterized. In general, the reactive current is a function of the voltage (phase to neutral-point) before and during the voltage dip. The detection of an FRT event occurs by falling below a

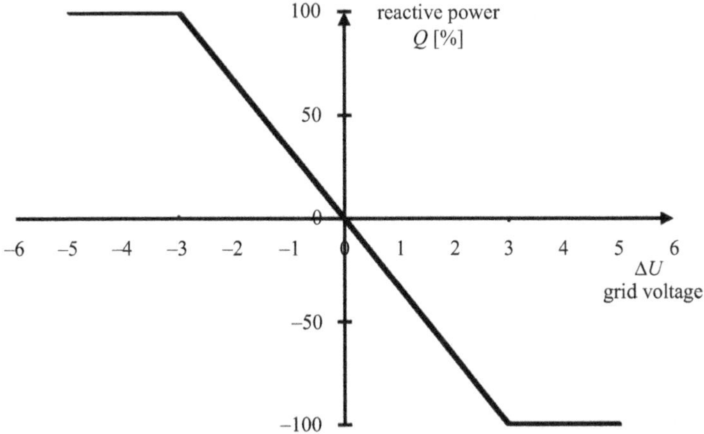


*Figure 4.19 Characteristic curve Q(U) of MSC-Series, © 2018 Freqcon GmbH Negative values = capacitive behavior, referenced to Q, positive values = inductive behavior, referenced to Q*

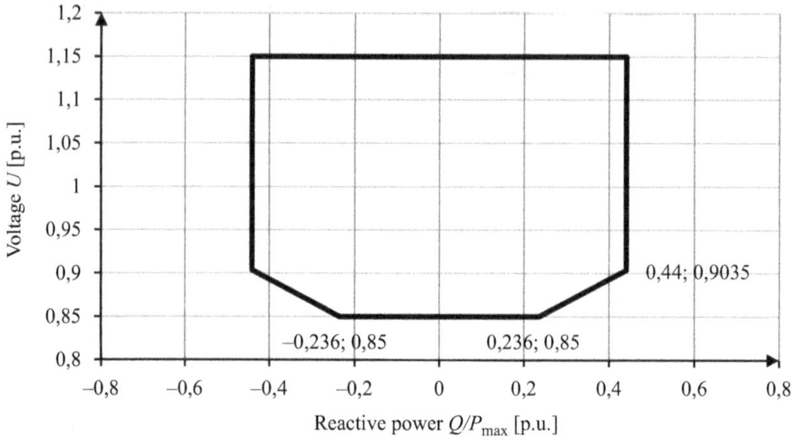


*Figure 4.20 Characteristic curve Q(U) of MSC-Series, © 2018 Freqcon GmbH*

set-voltage value for a low-voltage-ride-through (LVRT) or, by exceeding a set-voltage value for a high-voltage-ride-through (HVRT). Fallback thresholds are defined via other parameters.

In addition, the VDE-AR-N 4120 requires the detection of an FRT event with a sudden voltage change greater than or equal to 5 %. To meet the requirements of the BDEW-MSR and TC2007, the detection of sudden voltage change must be deactivated. There are no requirements for the behavior during OVRT and unbalanced voltage dips within the BDEW-MSR and TC2007, yet.

### 4.2.6 Relevant standards

In Germany WTs with a connection to the public grid network in the medium- or high-voltage range have to fulfill the standards described below. The standards are the basis for an accredited company by the Deutsche Akkreditierungsstelle (DAkkS) in Germany to evaluate the conformity:

- "Generating Plants Connected to the Medium-Voltage Networks"
- Guideline for generating plants' connection to and parallel operation with the medium-voltage network, BDEW (BDEW-MSR).
- Technical Guidelines for Generator Units of the FGW, Part 3: "Determining the Electrical Properties of Generator Units at Medium-, High- and Extra-High-Voltage," issued by: Fördergesellschaft Windenergie und andere Erneuerbaren Energien, FGW e.V., Germany.
- Technical Guidelines for Generator Units of the FGW, Part 4: "Requirements to Modelling and Validation of Simulation Tools of the Electrical Behaviour of generation units and power plants," issued by: Fördergesellschaft Windenergie und andere Erneuerbaren Energien, FGW e.V., Germany.

- Technical Guidelines for Generator Units of the FGW, Part 8: "Certification of Generator Units and Generation Plants at Medium-, High- and Extra-High-Voltage Grids," issued by: Fördergesellschaft Windenergie und andere Erneuerbaren Energien, FGW e.V., Germany.
- IEC 61400-21, Ed. 2, "Wind turbine generator systems – Part 21: Measurement and assessment of power quality characteristics of grid connected wind turbines."
- Transmission Code 2007, "Network and System Rules of the German Transmission System Operators."
- VDE-AR-N 4120, "Technical Requirements for the connection and operation of customer installations to the high voltage network" (TAB high voltage), January 2015, issued by: VDE.

### *4.2.7 Transformer and switchgear*

The most frequently used transformer to feed the medium- and high-voltage grid network with the power generated by the WT is a transformer with star-delta-changeover. The star-side is used on the low-voltage side and the delta-winding configuration is placed on the medium- or high-voltage level. An example of a three-winding transformer is shown in Figure 4.12. The line-to-line voltage of the low-voltage side of the transformer typically varies between 600 and 660 V.

The according switchgear must be able to connect and disconnect the system safely on the low-voltage side as well as on the medium- or the high-voltage side. The switchgear is either designed as a gas-isolated gear or as a classic air-isolated one. The gas-isolated switchgear is more cos-intensive, but the installation space is only half as large compared to an air-isolated switchgear. In general, the circuit breaker on the medium-voltage grid-side is designed for the rated current of the transformer:

$$I_{r,\text{transf.}} = \frac{S_{\text{transf.}}}{\sqrt{3} \cdot U_N} \text{with } : U_N \text{ line-to-line voltage}$$

The minimum relative short-circuit-voltage ($u_k$) of the transformer with connection to the medium-voltage grid is 4% according to regulations. A low value of $u_k$ results in less power-losses and provokes a lower voltage-drop on the MC-side in case of rated load. On the other hand, a low $u_k$ results in a high short-circuit-current on the WT-side, due to a lower voltage-drop over the inductance of the transformer.

The design of all converter-systems discussed in this chapter are based on a mean value for $u_k$ of 6%, which turns out to be a good compromise in everyday practice. The advantage of this approach is that the converter-system under design is independent of any constraints which could be given at the connection point to the grid. All components and subsystems of the MC are independent of the exact knowledge of the medium- or high-voltage grid and operate in the same way. The setting of $u_k$ also sets the value of the grid-side inductor of the LCL-filter component, or at least a part of it. The rated current of a transformer can be calculated with type-plate data as follows:

$$U_l = \frac{U_N}{\sqrt{3}}$$

$$U_k = u_k \cdot U_l$$

$$I_r = \frac{S}{\sqrt{3} \cdot U_N}$$

whereas $U_l$ represents the converter-side line-to-neutral voltage, $U_N$ represents the phase-to-phase-voltage on the converter-side of the transformer, and $I_r$ represents the rated phase-current (RMS-value).

In case of neglecting ohmic losses of the transformer, the impedance of a transformer can be estimated as follows. This approximation is even more permissible, the higher the transformer's rated apparent power gets.

$Z_{transf.} \approx \omega \cdot L \approx 2 \cdot \pi \cdot f \cdot L$ with f: frequency of grid-voltage

$$Z_{transf.} \approx \frac{U_k}{I_r}$$

$$\rightarrow L \approx \frac{U_k}{I_r} \cdot \frac{1}{2 \cdot \pi \cdot f}$$

For the prototype MC with a rated output power of 9 MW which corresponds to a 6-MW offshore-WT, as specified in Table 4.2, the inductance $L_T$ is related to one strand of the transformer (see Figure 4.12) and can be calculated with $U_k \approx 21V$ as follows:

$$Z_{transf.} \approx \omega \cdot L \approx 2 \cdot \pi \cdot 50\ \text{Hz} \cdot L$$

$$Z_{transf.} \approx \frac{U_k}{I_r} = \frac{0.06 \cdot \frac{600\ V}{\sqrt{3}}}{2.800\ A} \approx 7,42\ \text{m}\Omega$$

$$\rightarrow L_T \approx 7,42\ \text{m}\Omega \cdot \frac{1}{2 \cdot \pi \cdot 50} \approx 24\ \mu\text{H}$$

## 4.3 Design example

For a 6-MW offshore WT for a Chinese manufacturer of converter-systems, FREQCON GmbH designed a prototype of a full power grid-feed converter for offshore WT applications. The MC can convert the AC-input-power generated by an offshore WT into grid compliant AC-power and can feed in the transformed energy to a connected public or non-public grid. For reasons of redundancy and to avoid down-times, the MC is split up into two identical 3-MW converter units, in the following referred to as converter units 1 and 2. In other words, each converter unit (units 1 and 2) includes a 3-MW generator-side and a 3-MW grid-side unit.

Each converter unit is equipped with its own main controller. Therefore, the two converter units can operate completely independent of each other. Each unit consists of nine cabinets which include the grid or machine connections, filters and the power-stacks. All components are installed inside the power converter cabinets.

The MC is designed as direct-drive full-power converter and can adapt to the connected permanent magnet excited synchronous generator (PMSG). The main power circuit combines the machine-side and grid-side converter through DC-links in a B2B topology as shown in Figure 4.21.

The different cabinets are housing the main components as enumerated in Table 4.4. An overview of the internal components of the cabinets is shown in Figures 4.21–4.23. The internal components of CU 1-1 and CU 1-2 are listed

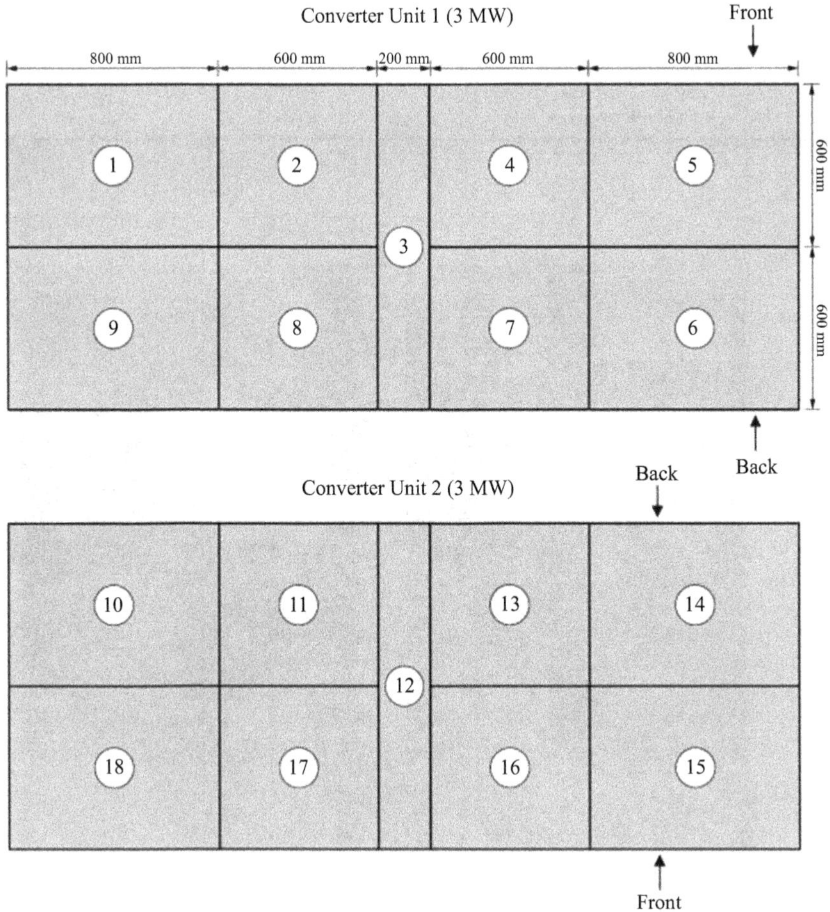


*Figure 4.21 Cabinet overview of a 6-MW prototype MC in B2B-topology, © 2018 Freqcon GmbH*

*Table 4.4 Cabinet overview of 6 MW (2 × 3 MW) main converter*

| Position | Cabinet name | Abbreviation | Converter unit |
|---|---|---|---|
| 1 | Converter cabinet 1.1 | CON 1.1 | 1 |
| 2 | Inductor cabinet 1.1 | IND 1.1 | 1 |
| 3 | Distribution panel 1 | DP 1 | 1 |
| 4 | Generator circuit breaker cabinet 1 | GCB 1 | 1 |
| 5 | Control cabinet 1 | CTRL 1 | 1 |
| 6 | Brake chopper cabinet 1 | CHOP 1 | 1 |
| 7 | Main circuit breaker cabinet 1 | MCB 1 | 1 |
| 8 | Inductor cabinet 1.2 | IND 1.2 | 1 |
| 9 | Converter cabinet 1.2 | CON 1.2 | 1 |
| 10 | Converter cabinet 2.1 | CON 2.1 | 2 |
| 11 | Inductor cabinet 2.1 | IND 2.1 | 2 |
| 12 | Distribution panel 2 | DP 2 | 2 |
| 13 | Main circuit breaker cabinet 2 | MCB 2 | 2 |
| 14 | Brake chopper cabinet 2 | CHOP 2 | 2 |
| 15 | Control cabinet 2 | CTRL 2 | 2 |
| 16 | Generator circuit breaker cabinet 2 | GCB 2 | 2 |
| 17 | Inductor cabinet 2.2 | IND 2.2 | 2 |
| 18 | Converter cabinet 2.2 | CON 2.2 | 2 |

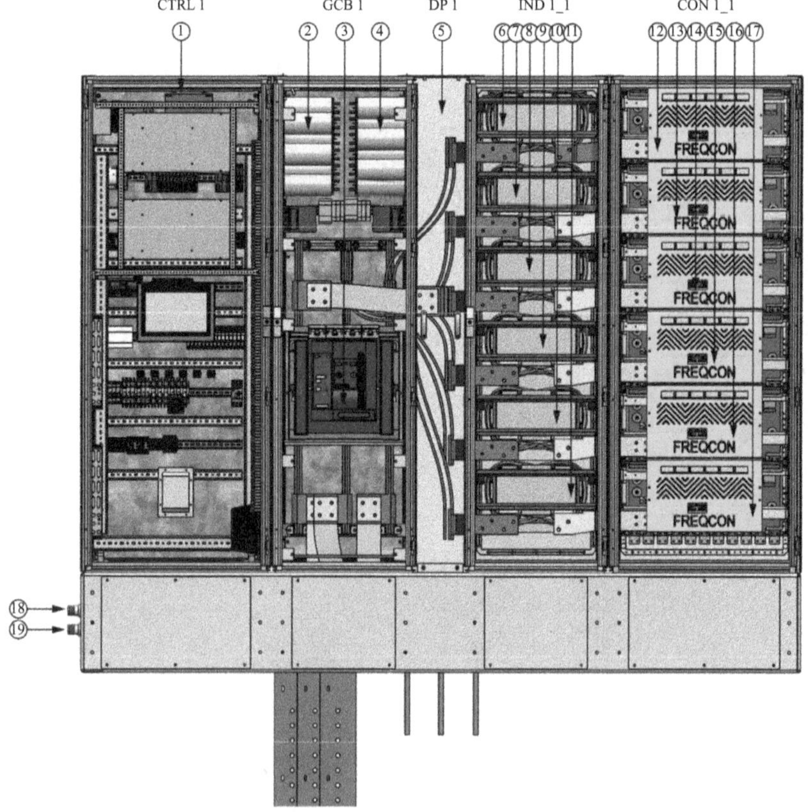


*Figure 4.22 Front view of on internal components of converter unit 1, © 2018 Freqcon GmbH*

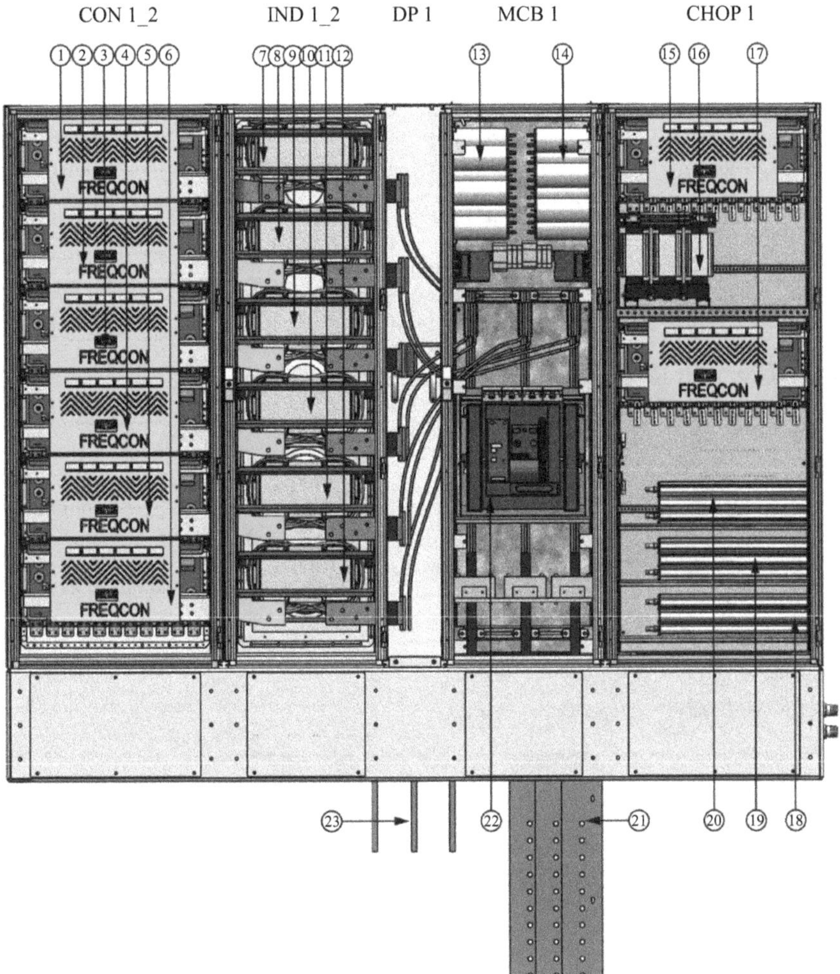


*Figure 4.23 Back view of on internal components of converter unit 1, © 2018 Freqcon GmbH*

in Tables 4.5 and 4.6. The enumeration corresponds to Figures 4.21 and 4.22. An overview of the internal components of the control cabinet is given in Table 4.7.

The power generated by the generator is rectified by the machine-side converter and then fed into grid by the grid-side converter. An LCL respectively LC-filter is arranged between grid and grid-side converter to suppress voltage distortions and current harmonics.

The MC is connected to the grid via a circuit breaker. This circuit breaker is also equipped with an over-current protection. The WT auxiliary transformer is used to supply the MC. The converter is able to measure voltages, currents and power of both grid- and generator-lines. The system operates at low-voltage level and is intended for operation on public and non-public grid networks.

*Table 4.5 Internal components of converter unit 1-1*

| Position | Component |
|---|---|
| 1 | Control unit |
| 2 | AC-filter 1500 A, 3 × 900 μF |
| 3 | AC-filter 1500 A, 3 × 900 μF |
| 4 | Circuit breaker, $I_N$=4.000 A, up to 690 V (generator-side) |
| 5 | Distribution panel |
| 6 | Inductor 1500 A, L1-1 |
| 7 | Inductor 1500 A, L2-1 |
| 8 | Inductor 1500 A, L3-1 |
| 9 | Inductor 1500 A, L1-2 |
| 10 | Inductor 1500 A, L2-2 |
| 11 | Inductor 1500 A, L3-2 |
| 12 | Power stack AC/DC, 1500 A, L1-1 |
| 13 | Power stack AC/DC, 1500 A, L2-1 |
| 14 | Power stack AC/DC, 1500 A, L3-1 |
| 15 | Power stack AC/DC, 1500 A, L1-2 |
| 16 | Power stack AC/DC, 1500 A, L2-2 |
| 17 | Power stack AC/DC, 1500 A, L3-2 |
| 18 | Cooling water outlet, hose connection 1½″ |
| 19 | Cooling water inlet, hose connection 1½″ |

*Table 4.6 Internal components of converter unit 1-2*

| Position | Component |
|---|---|
| 1 | Power stack AC/DC, 1500 A, L1-1 |
| 2 | Power stack AC/DC, 1500 A, L2-1 |
| 3 | Power stack AC/DC, 1500 A, L3-1 |
| 4 | Power stack AC/DC, 1500 A, L1-2 |
| 5 | Power stack AC/DC, 1500 A, L2-2 |
| 6 | Power stack AC/DC, 1500 A, L3-2 |
| 7 | Inductor 1500 A, L1-1 |
| 8 | Inductor 1500 A, L2-1 |
| 9 | Inductor 1500 A, L3-2 |
| 10 | Inductor 1500 A, L1-2 |
| 11 | Inductor 1500 A, L2-2 |
| 12 | Inductor 1500 A, L3-2 |
| 13 | AC-filter 1500 A, 3 × 900 μF |
| 14 | AC-filter 1500 A, 3 × 900 μF |
| 15 | Power stack AC/DC, 1500 A, break chopper 1 |
| 16 | Transformer (for control cabinet supply) |
| 17 | Power stack AC/DC, 1500 A, break chopper 2 |
| 18 | Break chopper resistors |
| 19 | Break chopper resistors |
| 20 | Break chopper resistors |
| 21 | WT generator AC connection terminals bus |
| 22 | Circuit breaker $I_N$=4000 A, up to 690 V (AC grid-side) |
| 23 | AC transformer connection terminals bus |

*Table 4.7 Internal components of the control cabinet*

| Position | Component |
|---|---|
| 1 | Enclosure thermostat (low) |
| 2 | Enclosure thermostat (high) |
| 3 | SCALANCE X005 IE Entry Level Switch |
| 4 | Power supply unit for UPS |
| 5 | Uninterruptible power supply (UPS) |
| 6 | Power storage device for UPS |
| 7 | Resistors |
| 8 | High voltage measuring module 2.0 |
| 9 | Resistors |
| 10 | High voltage measuring module 2.0 |
| 11 | Cabinet heater with fan (800 W) |
| 12 | PLC – Simotion P320-4 |
| 13 | 2 × Converter Control Module (CCM) |

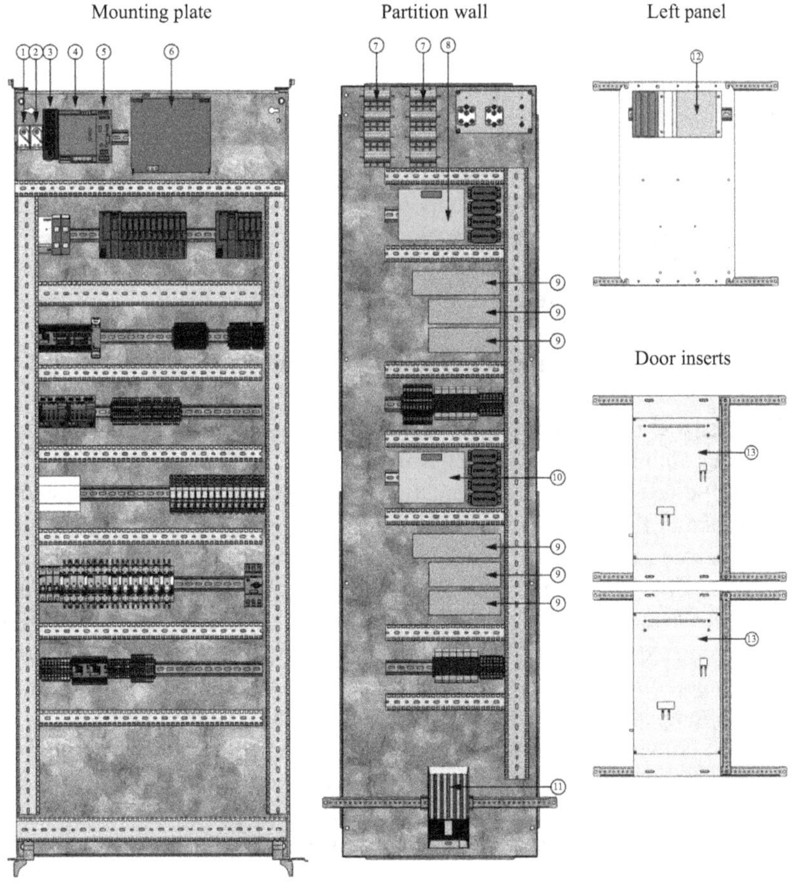


*Figure 4.24 Internal components of control cabinet, © 2018 Freqcon GmbH*

Cooling of the power modules, brake chopper resistors and inductors is realized by a water-to-air cooling circuit. Each converter unit is equipped with an individual cooling circuit. An external cooling system must be connected to the cooling circuits of both converter units.

The following functionality is given by the presented MC:

- Active power control
- Reactive power control
- DC bus voltage control
- Fault detection and monitoring
- Active anti-Islanding
- System fault clearing time according to IEEE 1547
- Low Voltage Ride-Through (LVRT)
- High Voltage Ride-Through (HVRT)
- Q-SET control
- P(f) control
- Derating capability
- Grid monitoring
- Start conditions definition

In grid connected operations, the 4-QC main converter is capable to operate in all four quadrants of apparent power. After commissioning and parameterization, the main converter operates automatically. The system design provides remote monitoring and access to the converter functions and set point adjustment via web. The following flow-chart (see Figure 4.25) provides information regarding the different MC machine-states. The individual converter main states are depicted.

Table 4.8 gives an overview of the general technical data of the 6 MW main converter.

To dissipate the heat produced by the power-stacks, inductors and brake-choppers, a cooling circuit is installed inside the individual converter units. The water-glycol-circle (WGC) is a circulation system for closed cooling systems. A water-glycol mix flows through the cooling modules of the converter and dissipates the heat produced by the IGBT devices, brake-chopper resistors and the inductors. The IGBT-devices need a minimum temperature to operate properly. To heat up the IGBT devices from a cold start, the coolant provided by an external cooling system must be pre-heated. The pre-heating of the coolant must be performed until a coolant temperature of 25 °C has been reached. After the IGBT modules have reached the temperature of about 25 °C, the temperature is maintained for 30 minutes and the converter is permitted to begin its normal operation. The pre-heater used in the external cooling system must be switched off automatically by a control unit at a temperature of >28 °C. In general, the coolant recommended by the manufacturer is a mix of 60% water and 40% antifreeze (example for "Antifrogen® N" coolant concentrate).

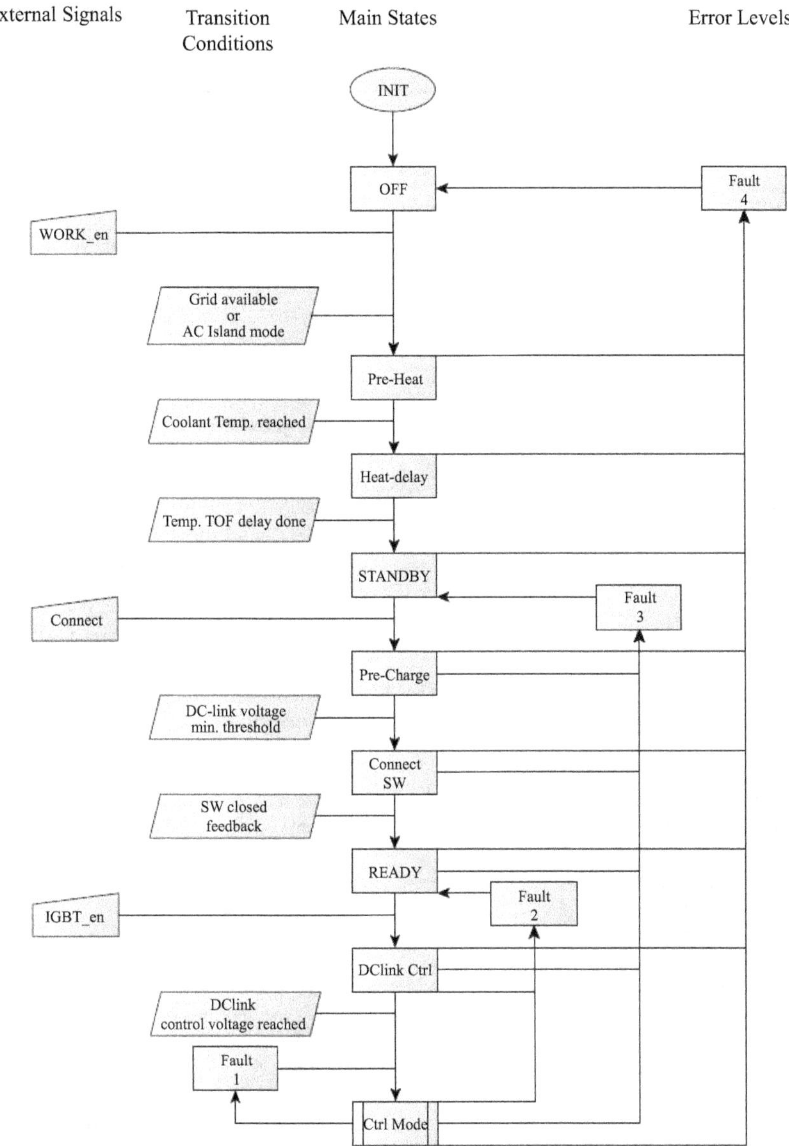


*Figure 4.25 Main converter machine-states, © 2018 Freqcon GmbH*

The requirements for the external cooling system which must be connected to the converter units are listed in Table 4.9.

The MC is designed for indoor installation on a flat and ground surface inside the nacelle of a wind turbine. Figure 4.26 gives a rough impression of the installation site of the MC inside the nacelle.

*Table 4.8 General technical data of a 6-MW main converter for offshore-WT applications*

| General Data | Description |
|---|---|
| Number of cabinets | 18 (/9 cabinets per unit) |
| Protection class | IP 54 |
| Service life (rated condition) | 20 years |
| Noise level | <70 db(A) measured @ 1m |
| Overvoltage category | Type I and II SPD per IEC61643-11 |
| Temperature range (in operation) | −30 °C to + 50 °C |
| Temperature range (storage) | −40 °C to + 70 °C |
| Weight CU 1 | Approx. 5060 kg |
| Weight CU 2 | Approx. 5060 kg |
| Dimensions CU 1 (WxHxD) | 3000 mm × 2400 mm × 1200 mm |
| Dimensions CU 2 (WxHxD) | 3000 mm × 2400 mm × 1200 mm |
| **Converter** | **Description** |
| Principle | IGBT Inverter (28 IGBT-devices) |
| Converter type | Direct-Drive Full Power Converter |
| Application | PMSG wind turbine generator |
| Controlling | Four quadrant vector control |
| Electrical levels | Two levels |
| Topology | Two PWM in back-to-back |
| IGBT-type | SKiiP 3614 GB17E4-6DUW |
| IGBT switching frequency | 5 kHz |
| Over temperature protection | All relevant temp. meas. by PLC |
| **AC-input (WT-generator)** | **Description** |
| Rated input voltage | 690 V AC |
| Rated input current | 6 phases, 2 × 3200 A |
| Rated input power | 6000 kW |
| **AC-output (grid)** | **Description** |
| Rated output voltage | 690 V AC |
| Rated power output | 6000 kW |
| Rated output current | 2 × 3200 A |
| Rated output frequency | 50 Hz |
| Output frequency range | 47.5 to 52.5 Hz |
| Rated DC-link voltage | 1200 V DC |
| Power factor range | 0.9 … 1 leading or lagging |
| THD @ rated power output | ≤3% |
| Dissymmetry in 3-phase voltage | ≤5% |
| Max. efficiency | ≥97% |
| AC current protection | 2 × 3200 A circuit breaker |
| Short-circuit rating | 42 kA for 1.0 s |
| AC disconnection method | Circuit breaker |
| **Controller** | **Description** |
| Main controller | Siemens Simotion P320-4 |
| Control Software | FREQCON framework |
| Internal Communication | Profinet |
| Ext. communication | Ethernet, TCP/IP, analogue 4…20mA |
| Response time | <20 ms |
| **Water cooling system** | **Description** |
| Cooling principle | Liquid to air cooled, active climate ctrl. |

*(Continues)*

*Table 4.8 (Continued)*

| **General Data** | **Description** |
|---|---|
| Cooling media flow rate | 2 × 150 l/min. |
| Max. cooling water inlet temp. | 65 °C (width derating 75 °C) |
| Cooling system's operating pressure | 2...3 bar |
| Coolant capacity | 75 l |

*Table 4.9 Requirements for external cooling-system*

| **Requirements** | **Description** |
|---|---|
| System pressure loss of entire plant | $\Delta p = 1.6$ bar |
| Heat quantity per 6 MW unit | 192.8 kW |
| Coolant inlet temperature | 65 °C (width derating 75 °C) |
| Coolant hose connection | 2 × 1½″ |

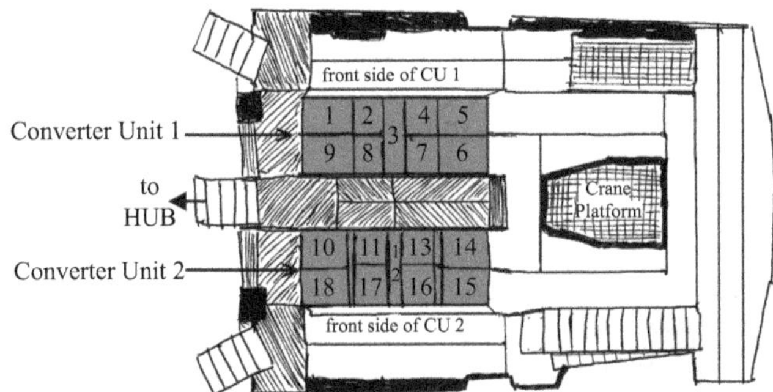


*Figure 4.26 Installation site inside the nacelle of the WT, © 2018 Freqcon GmbH*

## 4.4 Future trends and developments

As outlined in Section 4.1 the breakthrough in the development of main converters for WT was supported by the market launch of the IGBT devices. Today, the IGBT devices are increasingly replaced by semiconductor devices based on silicon carbide (SIC). The advantage of these switching devices compared to the silicon-based devices, is lower switching-losses at the same switching frequency. Therefore, SIC devices can be switched at a higher frequency while keeping the switching losses the same compared to silicon-devices. Ultimately, SIC-based switches have a higher current-rating due to lower losses compared to silicon devices. Almost all manufacturers of semiconductor devices announce SIC-based devices with a voltage range of up to 2.3 kV. The intension is to replace the standard 1.7 kV devices in the active-frontend converters and to apply a DC-link voltage of up to 1500 V which is the maximum permissible voltage for low-voltage systems. All this potentially leads to higher power ratings for main converters not only for wind turbines but for all power-generating units.

There is also a trend towards hybrid system designs that combine a WT and a battery-storage system. The combination of both systems helps to support grid network requirements. During times when the WT must not feed-in electrical power into the grid, the battery systems can be charged and at least a small amount of energy per WT can be stored. An advanced concept is the combination of wind turbines with electrolyzers, especially in offshore applications.

The current trend combines all three systems mentioned, the wind turbine, the battery-storage and the electrolyzer in one unit. These generating units are built without an electrical grid connection. The WT must be capable of black-start that is supported by the battery-storage system. By adding a fuel-cell on the offshore platform the unit can generate electrical power. The produced hydrogen is transported either via a pipeline or by ship to an onshore station.

Potential development trends can also be seen in the sector of network services for electrical grids. All large conventional power plants must be substituted by renewable energies in order to ultimately operate an electrical grid powered one hundred percent using renewable energies. Therefore, wind turbines must also deliver instantaneously all network services on demand. A novel approach to deliver instantaneous-reserve power, for example, is to control the main converter in such a way that its clamping behavior is identical to the clamping behavior of a synchronous machine. This control of the MC is based on a mathematical-physical model of a synchronous machine with less or even the same rated electrical power. Therefore, the concept is called the virtual synchronous machine concept (VSM-concept) [22,23]. Of course, the MC cannot replicate all the characteristics of a synchronous machine. For example, a short circuit current can easily be delivered by a synchronous machine, but not by a converter system based on semiconductor devices without enormous oversizing of the silicon devices. Fortunately, most of the characteristics can be emulated with the converter. For example, delivering instantaneous-reserve power on grid demand is one of them. The advantage of this approach is that all interventions to deliver network services to the grid take place automatically by the converter and all the existing network services for the electrical grid could persist.

## References


[1] C. I. Odeh, "A Three-Phase Boost DC-AC Converter," *Niger. J. Technol.*, vol. 30, no. 1.

[2] F. Zach, *Leistungselektronik: ein Handbuch. Band 2, 6.*, Überarbeitete und erweiterte Auflage. Wiesbaden [Heidelberg]: Springer Vieweg, 2022.

[3] D. Schröder and R. Marquardt (eds.), *Leistungselektronische Schaltungen: Funktion, Auslegung und Anwendung*. Berlin, Heidelberg: Springer Berlin Heidelberg, 2019. doi:10.1007/978-3-662-55325-1.

[4] H. Groke, *Regelung eines permanentmagneterregten Transversalflussgenerators für direkt angetriebene Windenergieanlagen*, Als Ms. gedruckt. in Fortschritt-Berichte VDI Reihe 21, Elektrotechnik, no. 405. Düsseldorf: VDI-Verl, 2013.

[5] I. T. AG, "IGBT Modules – Infineon Technologies." https://www.infineon.com/cms/en/product/power/igbt/igbt-modules/ (accessed Jun. 23, 2023).

[6] "IGBT Power Modules." https://www.danfoss.com/en/products/dsp/power-modules/igbt-power-modules/ (accessed Jun. 23, 2023).

[7] "MITSUBISHI ELECTRIC Semiconductors & Devices: Applications." https://www.mitsubishielectric.com/semiconductors/application/index.html (accessed Jun. 23, 2023).

[8] M. Eichler, "Umrichter für den Offshore-Einsatz," p. 6, Mar. 2008.

[9] "Umrichter für weltweit größten Offshore-Windpark," https://www.industr.com. https://www.industr.com/de/umrichter-fuer-weltweit-groessten-offshore-windpark-2608086 (accessed Jun. 23, 2023).

[10] "ABB to Deliver Power Converters for the World's Largest Offshore Wind Farm," Jun. 10, 2021. https://new.abb.com/news/detail/79198/abb-to-deliver-power-converters-for-the-worlds-largest-offshore-wind-farm (accessed Jun. 23, 2023).

[11] G. Schnell and B. Wiedemann, Eds., *Bussysteme in der Automatisierungs- und Prozesstechnik: Grundlagen, Systeme und Anwendungen der industriellen Kommunikation*. Wiesbaden: Springer Fachmedien Wiesbaden, 2019. doi: 10.1007/978-3-658-23688-5.

[12] B. Scherff, E. Haese, and H. R. Wenzek, *Feldbussysteme in der Praxis*. Berlin, Heidelberg: Springer Berlin Heidelberg, 1999. doi:10.1007/978-3-642-59854-8.

[13] D. Jiang, Z. Shen, Q. Li, J. Chen, and Z. Liu, *Advanced Pulse-Width-Modulation: With Freedom to Optimize Power Electronics Converters*. in CPSS Power Electronics Series. Singapore: Springer Singapore, 2021. doi:10.1007/978-981-33-4385-6.

[14] "Designing RC Snubbers," vol. 2023, 2023.

[15] "Detail | Semikron Danfoss." https://www.semikron-danfoss.com/service-support/downloads/detail/semikron-application-note-gate-driver-basics-en-2021-07-22-rev-01.html (accessed Jun. 22, 2023).

[16] "REH984e_07.pdf" (accessed Jun. 23, 2023). [Online]. Available: https://www.fujielectric.com/products/semiconductor/model/igbt/application/box/doc/pdf/REH984e/REH984e_07.pdf

[17] ROHM Co., Ltd., "Snubber Circuit Design Methods", no. 62, AN037E Rev.002, 2020.

[18] A. Wintrich, U. Nicolai, W. Tursky, and T. Reimann, *Applikationshandbuch Leistungshalbleiter*, 2., Überarbeitete Auflage. Ilmenau: ISLE Verlag, 2015.

[19] "Gegenstromwärmetauscher: Prinzip und Wirkungsgrad," *Kesselheld*. https://www.kesselheld.de/gegenstromwaermetauscher/ (accessed Jun. 23, 2023).

[20] M. Wurm, L. Fickert, and K. Groher, "Erdschlusschutz in Windparknetzen mit isoliertem Sternpunkt," *E Elektrotechnik Informationstechnik*, vol. 121, no. 10, pp. 343–350, Oct. 2004. doi: 10.1007/BF03055473.

[21] J. W. Kolar, H. Ertl, and F. C. Zach, "Influence of the modulation method on the conduction and switching losses of a PWM converter system," in *Conference Record of the 1990 IEEE Industry Applications Society Annual Meeting*, Oct. 1990, pp. 502–512, vol.1. doi: 10.1109/IAS.1990.152232.

[22] A. Ernst, W. Holzke, D. Koczy, N. Kaminski, and B. Orlik, "Model-based Converter Control for the Emulation of a Wind Turbine Drive Train," in *2022 24th European Conference on Power Electronics and Applications (EPE'22 ECCE Europe)*, Sep. 2022, pp. 1–10.

[23] A. Ernst, D. Matthies, W. Holzke, and B. Orlik, "Validation of a Generator-Side Boost Converter with Load by a Fictitious Synchronous Machine," in *PCIM Europe Digital Days 2021; International Exhibition and Conference for Power Electronics, Intelligent Motion, Renewable Energy and Energy Management*, May 2021, pp. 1–7.

# Chapter 5

# Grid compliance and electrical system characterization

*Torben Jersch*[1]

The chapter describes the development of grid integration tests for wind turbines and classifies them in terms of development methodology. The basic aspects of grid integration tests in the field are explained and the common practice as well as the challenges of field tests are described.

A classification of these tests in the development and validation process, which is based on the development method according to the V-model, is given. This argumentation forms the basis for the development of the nacelle test bench DyNaLab as well as its further development, the Hil-GridCoP test bench. The setup is explained as well as essential design points are shown. The results obtained on the test benches open up the prerequisite for future grid integration tests to continue to be carried out on test benches. Current challenges, both technological and economic, as well as new requirements in grid operation are described. Additional test benches and concepts will also be explained, which will contribute to solving the problem of grid integration when setting up a decentralized energy supply.

## 5.1 State-of-the-art grid compliance

To ensure the safe and efficient operation of the power grid, transmission system operators set requirements for power generation units. These requirements must be proven by means of grid compliance tests to ensure that all power generation units, such as wind turbines, meet these requirements. Parameters that contribute to the grid quality, such as voltage regulation, frequency maintenance, and emissive harmonics, are evaluated during the test. This is done to ensure that the installations do not cause interference or negatively affect the power quality.

Grid Code Compliance of a single wind turbine is not critical for grid operation and grid stability, however, with the increasing expansion of renewable energy feed-in, hundreds of identical turbines are installed, which have a decisive impact on grid operation due to their number and the associated total power. Conformity

[1]Fraunhofer-IWES, Fraunhofer-Institute for Wind Energy Systems, Germany

tests on wind turbines are performed as unit tests. Entire wind farms, also known as Power Park Modules (PPM), must demonstrate grid compliance for each farm.

Due to the high importance of grid compliance, extensive tests and test procedures are described in various standards to ensure a minimum standard for testing. These tests are usually performed on the first prototypes of wind turbines in the field. In some countries, including Germany, these tests even become part of a certification process, so that the tests essentially correspond to an extended and external quality assurance process.

In the following subsection, mainly the test procedures according to IEC 61400-21-1 are used. In some countries, specific guidelines exist for the description of measurement methods, such as FGW TR3 in Germany. In essence, however, they do not differ from the methods described in IEC.

The tests to be performed are presented below in accordance with IEC 61400-21-1 in order to provide an overview.

## 5.2 Scope of grid code compliance testing

### *5.2.1 Power quality*

Harmonic measurement, also called network analysis, is a method for investigating electrical power generating units. It involves analyzing and evaluating the various harmonic frequencies generated by non-linear loads. Since high harmonic currents lead to considerable heating of the grid equipment and these are not taken into account in the design of the components, they can be damaged. The aim of the measurement is to identify and eliminate possible faults and problems in order to ensure a stable and efficient power supply,

The aim of the measurement according to IEC 61400-21-1 is to display the power generation as a function of the wind speed during the period of the power quality measurement. The measurement requires seven wind bins each, which are grouped over a period of 10 min.

The power quality category also includes flicker measurements in the area of grid disturbances. Due to modern technology, the electrical power generators built today have very low flicker coefficients, which are up to a factor of 10 below the limit values, so these tests are practically no longer relevant for proving grid code compliance and are therefore not dealt with further here.

### *5.2.2 Steady state operation*

The measurements to check the continuous operation behavior are used to determine or prove the active power feed-in at changing wind speeds and thus represent the basic function of the wind turbine.

In the determination of the maximum power as well as the determination of the reactive power control behavior. Here, too, the power is plotted as a function of the wind speed. For the determination of the maximum power, the obtained data are determined to the highest powers over a period of 200 ms, 60 s, and 600 s. These values are needed for the planning and design of the grid resources as well as the grid connection of a wind farm. But also the grid operator requires the compliance with the maximum values.

The third point in this category is to check whether the plant behaves inductively or capacitively on the grid with different active power feeds. This behavior must be taken into account when designing the grids.

### *5.2.3 Capability*

The reactive power control is also tested under the category control quality. In the case of reactive power provision, the quality and dynamics of the control system are again tested. In both tests, the operating state of the wind turbine is changed and documented by specifying setpoints. The tests are carried out under different wind conditions and operating points of the wind turbine.

### *5.2.4 Active power control behavior*

In the event of grid restrictions, wind farms must reduce the active power in accordance with the grid operator's instructions and thus operate in power-limited mode despite sufficient wind supply. The tests to check the control quality of the wind turbine are to show that it is capable of dynamically reducing the power and achieving and maintaining the required power range with the required accuracy. In addition, the settling time is checked when the target power is changed.

The last point to be checked is whether the wind turbines comply with the required power gradient ramps; to this end, tests are carried out on a slow and a fast ramp.

These tests can demonstrate that the wind turbine can be used in all conceivable operating modes. From the perspective of today's modern converter systems, these tests do not appear to be particularly demanding; however, the first tests were standardized in the mid-2000s, at which time other system configurations were installed to be measured by these standards.

### *5.2.5 Power of the active power control with frequency increase*

In order to stabilize the grid frequency in the event of an overfrequency in the grid, the feed-in power must be reduced; this is a fundamental instrument in the stabilization of the grids. Therefore, wind turbines must also make a contribution.

The reaction to an underfrequency is called the provision of synthetic inertia. In response to an underfrequency, more active power must be fed in. As a rule, however, wind turbines are operated at the optimum operating point, so that the power cannot be increased when the corresponding wind supply is available.

For this purpose, tests are carried out at low partial load, high partial load and in regulated operation. Since in the past the grid frequency was not adjustable in the field, these tests can also be performed by changing the setpoints in the controller or by generic frequency inputs. The latest version of International Standard IEC 61400-21-1 also recommends the use of grid emulators.

### *5.2.6 Behavior in the event of grid faults*

Grid faults occur regularly in medium-voltage and high-voltage grids but are quickly cleared by protective devices. The power generation units must remain on the grid despite short voltage sags to prevent a cascading collapse of the power grids. In Germany, supporting reactive power must be fed in within 60 ms during a grid fault to stabilize the grid voltage and reduce the effects of the grid fault locally.

Voltage sags are often performed by FRT containers in the field, as described in detail later. The objective of the tests is to determine the dynamic behavior during overvoltage and undervoltage events, to simulate the ability to ride through grid faults and to determine the feed-in behavior of active and reactive power.

These evidence methods in the field are of utmost importance for the development of wind turbines (WT), as the main power components (generator, converter, transformer) must be correctly designed for overcurrents that occur. However, design is often at a limit without corresponding safety margins. Furthermore, all turbine and converter controllers must be correctly matched to each other, as the voltage dips usually affect both the electrical and the mechanical system.

### *5.2.7 Disconnection from grid*

These tests are used to check the turbine's protection systems during appropriate grid excitation, especially undervoltage events that are too long, overvoltage events or frequency deviations. In summary, the TSOs use grid compliance tests to comply with grid rules and thus create planning certainty. Since power generation plants that meet the technical requirements are seamlessly connected to the grid according to a fixed process.

Performing measurements of the operating behavior in the event of setpoint changes poses few problems in the field, as these can be performed by measurements during operation, and other setpoints are communicated to the WTs. Challenging from a development methodology point of view are the FRT tests, which cannot be carried out earlier in the development. The execution of the frequency variation tests in field tests is not feasible in reality, so that alternative methods have become established in the industry, other presetting of controller setpoints and limits.

The basis for carrying out these tests was laid almost 20 years ago, when renewable energies only contributed sporadically to grid operation. With the increasing integration of renewables, TSOs will require other compliance tests to ensure the seamless integration of power generation plants into the grid in the future.

## 5.3 Methods for compliance testing – field test

### *5.3.1 Field test – detailed view of the FRT tests*

In order to be able to connect wind turbines to the grid, approval from the transmission system operator is required. The requirements for this can differ from country to country. In many cases, an accredited test report attesting to compliance with the country-specific requirements according to IEC61400-21 is sufficient.

Sometimes a certification body is involved, which grants a certificate on the basis of the accredited test report in order to independently guarantee compliance with the grid requirements.

In a field test campaign, a prototype plant is set up, and fully commissioned and the auxiliary units and software are adjusted and tested together. The first tests are carried out step by step in partial load operation and the plant is monitored manually around the clock. This phase is referred to as "run-in" and is used for tuning and initial commissioning tests. After successful commissioning, the wind turbine is released to operate at full load and in unmanned mode. This is typically followed by pre-tests, which mainly consist of FRT tests. For this purpose, an FRT container is set up and connected between the medium-voltage cables from the WT to the substation. In the event of a fault, the inductive voltage divider can be used to generate voltage sags corresponding to the set ratio of the inductances. For this purpose, a switchgear switches between the predefined configurations. To change the fault depths, i.e., the residual voltages in the event of a fault, the inductors must be converted manually. This allows voltage dips to be changed, for which the entire container must be de-energized. These transitions represent a time bottleneck and are tried to be reduced as much as possible in the field.

The subsequent measurement campaign is accompanied by an external accredited measurement body and the measurement data is recorded with their measurement equipment. The evaluation and assessment of the tests is carried out in cooperation with a certification company according to objective criteria, in Germany, for example, according to FGW TR8. The test campaign can last up to 42 weeks, as it is completely dependent on weather conditions, especially wind conditions.

### *5.3.2 Classification of the field tests to the overall development of the WT*

This method of field testing corresponds to the overall system test according to the V-model development methodology presented in the first book. Testing is performed very late in the wind turbine development process, even after commissioning. However, late testing and validation has significant disadvantages, as errors are discovered late in the development process and subsequent developments can be associated with high costs and longer project times. In this process, development, construction, and validation proceed sequentially. In the past, the application of this approach resulted in the overall development times of new WT developments being extended by up to 50%, thus massively delaying the product launch of the new WT.

An important aspect of the development of large wind farm projects is the planning of the grid connection. The wind farm operator is obliged to ensure that all specifications of the grid connection guidelines are met for the entire wind farm.

This requires integrative planning of the wind turbines, the park cabling and the planning of grid devices such as filters or reactive power compensation units. These individual components all have long delivery times, and the next and larger generation of wind turbines is often already being planned and purchased for new wind farm projects.

Since the planning of the wind farm and the development of the wind turbine have to take place at the same time, the development of the grid connection is carried out on a model-based development with wind turbines that have not yet been fully developed. Therefore, a defined electrical behavior, which is described by means of the model, is part of the contract. Thus, there is a special focus on the compliance with the assured electrical characteristics. However, in terms of development methodology, this conflicts with validation through field testing.

These challenges mean that validation methods that accelerate the entire validation process or enable validation at a lower level of testing are extremely attractive in the industry. The reason is that significant cost savings potential can be expected and accelerated product introduction is made possible.

## 5.4 Test benches for grid compliance testing

Due to the verifiable methodological disadvantages of field tests, the DyNaLab nacelle test stand was set up in 2015. The aim of this project was to relocate grid integration tests of wind turbines from the field to the laboratory.

### *5.4.1 Emulate realistic conditions on test benches*

Until 2015, conducting tests on test rigs was not an established method to replace field tests. To replace testing, comparable test conditions should be created to allow direct comparability to field testing and thus promote acceptance. Tests often demonstrate not only the inverter-specific control behavior, but also that of the overall turbine behavior. This is most evident in tests of active power injection and control performance. It is obvious that for power control, the pitch system and the overall system dynamics are critical.

There is also a strong interaction between the grid and the turbine during voltage sag tests. Here it becomes clear that in the event of a grid fault with a residual voltage of 0 V, no more power can be delivered to the grid. The brake chopper is designed depending on the turbine type – either type 3 or type 4 – which means that the torque can no longer be generated when grid faults occur. This represents a disturbance excitation for the wind turbine and excites its natural oscillations. The FRT tests usually excite the first drive train oscillation and the tower oscillation. Both are actively damped by the turbine control system and are reflected in the output power of the wind turbine and thus in the current and voltage measurements.

The grid integration tests for active power control are used to determine the settling time or to prove that a tolerance band can be reached and the power output can be maintained. In the event of voltage dips, the active power must be restored to the pre-fault level within 0.5–2 s. To verify this, an exact replication of the aeroelastic properties of the wind turbines on test rigs is required.

In addition, it must be taken into account that the conformity and certification tests are only valid together with always a specific controller version. Therefore, it is important to perform the test bench tests with the original controller as well. The simulation of the wind conditions must be precise so that the turbine controller also functions without errors on the test rig with virtual simulation of the aeroelastic turbine behavior; an example of this is the monitoring of the blade root bending moments, the values of which must be within the expected limits.

### *5.4.2 Design basics of DyNaLab drive unit*

Grid integration tests therefore require testing with the original turbine controller, the correct emulation of the wind turbine behavior and their dynamic properties. This is realized in DyNaLab using the mHiL method, which is described in Chapter 6.

Replicating the dynamics of wind turbines during grid faults is a key design requirement for the motor, drive train and converter unit of a total system test rig. The rotors of the drive motors, the shaft of the test rig and the rotor of the generator form a rotating system, which is a multi-mass oscillator. In order to dynamically control required torques at the flange of the nacelle, a stiff mechanical structure of the drive train is mandatory. For this reason, the overall length of the drive train was designed to be as short as possible and the two motors of the DyNaLab are mounted on one shaft. The test specimen and the test rig are connected via a homokinetic coupling with high stiffness, which enables torque transmission without feedback.

Without active injection of control energy into the system test bench-DUT, the torque at the flange can only be controlled with the dynamics of the torsional natural frequency. The test specimen and the motor excite the system to oscillate with the first torsional natural frequency of the test stand. In order to appropriately control the torque at the flange during step-like torque changes and to damp the drive train, the power of the converters was dimensioned significantly above the rated power of the motors.

The nominal power of the test stand is 10 MW and is achieved with two motors of 5 MW each at a rotational speed of 11 rpm. The inverter systems have a total power of $2 \times 13$ MW. The currents are limited to a maximum current of 175% to allow highly dynamic control of the motors and the drive train. Figure 5.1 shows the overall structure of the DyNaLab consisting of the drive unit described here and the grid emulator described below.

### *5.4.3 Electrical design*

In order to ensure maximum flexibility in the use of the grid emulator and to be able to perform all grid code tests, the design of the grid emulator was based on a complete emulation of medium-voltage grids using power electronics. Design criteria included a highly dynamic voltage control of 1 kHz to emulate FRT events, as well as the ability to carry occurring short-circuit currents without physical feedback and without active current limiting. In Chapter 6 of this book, the topology of the grid emulator was presented. It consists of three sinusoidal voltage sources that are synchronized exclusively by software and thus also enable unbalanced grid conditions.

The power electronics of the power emulator is based on 4 ABB ACS6000 converter cabinet and uses a DC link voltage U_DC+- of 4800 V. These modules are designed as a 3-level NPC topology. Typically, the modulated output voltage (phase-neutral) of a half-bridge has a fundamental of 1800 V and a continuous

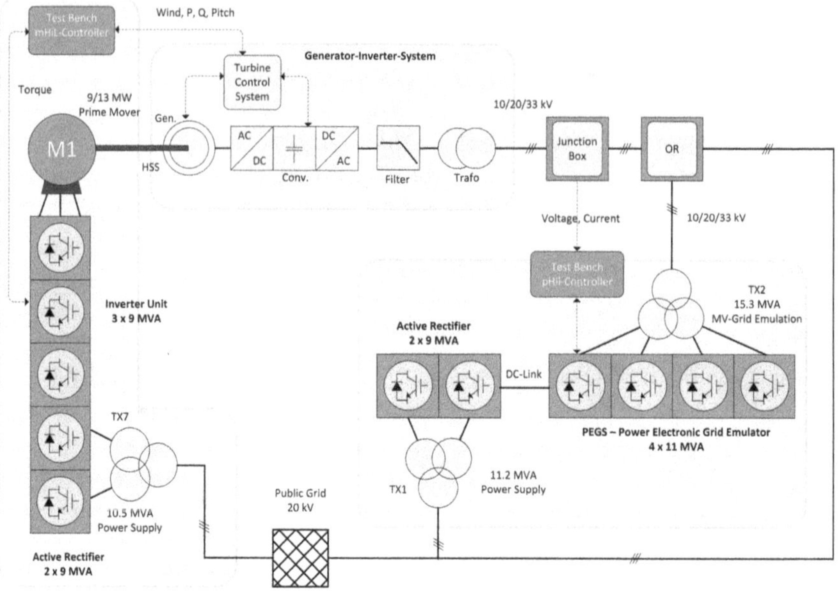


*Figure 5.1 DyNaLab test bench*

current capability of 1300 A, as well as a current of 2050 A for a short time of up to 2 s. The grid emulator can thus provide up to 11 MVA per converter cabinet and up to 44 MVA total power.

The structure of the DynaLab grid emulator is outlined below.

Figure 5.2 shows that two half-bridges are connected together to form an H-bridge to create a 5-level topology, with an undervoltage winding of the step-up transformer between the half-bridges. Each transformer core has two undervoltage windings, adding both magnetic fluxes doubles the voltages on the top voltage winding. The interleave switching of the half-bridges and the coupling of the two winding systems results in a 9-phase topology on the high-voltage side. Synchronization of all phases and 120° phase shift of the single-phase grid emulators results in a classic three-phase system. In the consideration phase to phase, the grid emulator represents a 17-level topology.

The topology increases the switching frequency of individual half-bridges from 350 Hz to 2.8 kHz of the grid emulator. Since the currents of the DUTs are low compared to the peak power of the grid emulator of 44 MVA, this also results in a low thermal load on the semiconductors. Thus, the switching frequency can be additionally increased for tests in continuous operation, as in harmonic measurements.

This topology combines two features that are crucial for performing FRT tests. The high clock frequency allows fast voltage changes with a dynamic range of one kHz and thus also dynamic control of the system. Thanks to the high installed inverter power, fault currents can be conducted without the current being limited actively by the current limitation of the grid emulator.

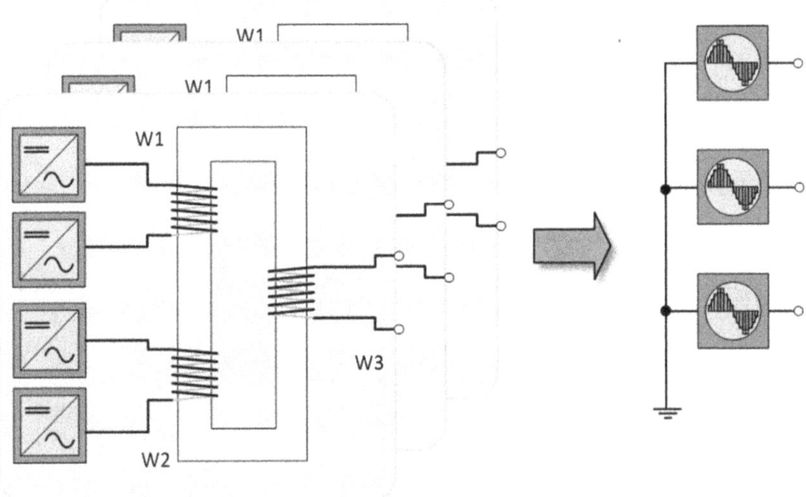


*Figure 5.2 Topology and transformer configuration of the 44 MVA grid emulator of the DyNaLab*

When emulating 2-phase faults, the amplitude of the voltages must be controlled separately and the phase position of two individual phases must be shifted up to 180°, for faults with residual voltage of 0 V. The topology of the separate grid emulators allows this by simple setpoint provision. Grid emulators with 3- or 5-leg transformers must always have a subordinate current control for voltage and phase control in order to compensate for the magnetic flux. For this reason, the DyNaLab grid emulator demonstrates significantly better control behavior for emulating FRT faults compared to other grid emulators.

### *5.4.4 Evaluation of the DyNaLab tests*

The hardware design, consisting of the drive unit and grid emulator, has proven to be suitable for simulating field tests with the control methods and HiL operating modes described in Chapter 6. In recent years, this has been successfully demonstrated on an onshore wind turbine from Enercon and an offshore turbine from SGRE, as well as seven other test campaigns.

Typically, one of the first five nacelles produced is taken to the test bench as a test specimen and tested in the laboratory in parallel with the field test. By conducting tests in parallel in the field and on the test rig, important goals for reducing development times and accelerating time to market can be met.

Full market penetration of this test method has not yet been achieved. Performing testing and validation according to the V-model at a lower level is not sufficient, as the construction of a complete nacelle cannot be completed significantly earlier compared to the erection of a wind turbine and considerable additional costs are incurred. Test campaigns at complete system test rigs such as DyNaLab or the nacelle test rig at LORC (Denmark) incur costs of €250–€350 k per month. The high logistic effort (weight of the test specimen are 300–400 t), the integration of the nacelle by means of test specimen specific adapters weighing several tons into the test rig and the setup of numerous auxiliary systems. A complete test campaign typically last 8–12 months.

## 5.5 Methods for compliance testing – HiL-GridCoP

The HiL-GridCoP project was launched to perform grid integration tests at an even lower test level to accelerate validations and reduce costs. Instead of testing the entire nacelle, only the components in the main power flow are tested. These include the generator converter system, the grid transformer, and the original turbine control cabinet and controller, which is also referred to as the minimum system in the following. Using the real-time model, the aeroelastic properties of the wind turbine are emulated and extended to include the mechanical properties of the nacelle drive train to reproduce the field behavior on the test rig and perform grid integration tests. Thus, grid integration tests are performed on the test bench reproducing the field behavior of the entire turbine.

In the HiL-GridCoP project, test bench test results were directly compared with field tests to demonstrate feasibility and improve testing methodology. The objective of this project was to describe these tests in the IEC-61400-21-4 (Wind energy generation systems – Part 21-4: Measurement and assessment of electrical characteristics – Wind turbine components and subsystems) standardization as an alternative to field and whole system testing, thereby increasing confidence that this test method can be adopted by industry with minimal risk in their product validation strategies.

### *5.5.1 Drivetrain design of Hil-GridCoP test bench*

The HiL-GridCoP test bench, see Figure 5.3, differs from the nacelle test bench by the drive unit, a high-speed motor with a continuous power of 9 MW and a peak power of up to 13 MW for half an hour. The speed range is 0–2200 rpm and operates in the nominal speed range from 1000 to 1800 rpm.

Wind turbines in the field exhibit changes in speed of up to 50 rpm in a few milliseconds in the event of grid faults on the fast-running generator. To enable these dynamic speed changes and at the same time transmit the required torques, high power and dynamics of the drive unit are required. In particular, the design of the drive unit must take into account the energy required to accelerate or brake the inertial mass of the drive motor.

Due to these requirements, higher demands were placed on the internal control system with regard to very low delay times in the inverter-internal signal processing. In addition, the drive unit was constructed from three converter systems connected in parallel to allow higher switching frequencies and therefore reduce delay times. The system thus has a total of 27 MVA of converter power.

Figure 5.4 shows the setup of a DUT in the HiL-GridCoP test bench. The transformer on the right side is connected to the grid emulator via the red 20 kV

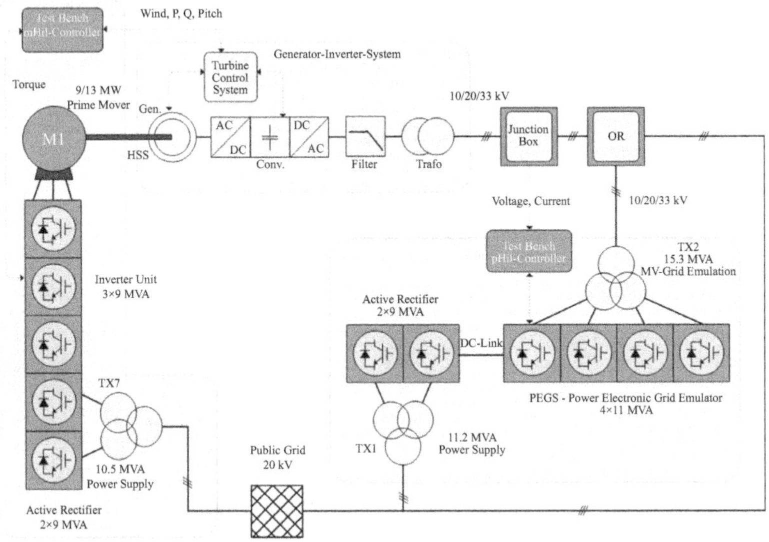


*Figure 5.3 Hil-GridCoP test bench single line diagram*

*Figure 5.4 Set-up of a 5.6 MW test item in the IWES HiL-GridCoP test bench*

cables of the air-insulated switchgear. The generator is connected to the drive unit by a multi-plate clutch. The test bench was developed as part of the Hil-GridCop research project. Further details can be found in [5].

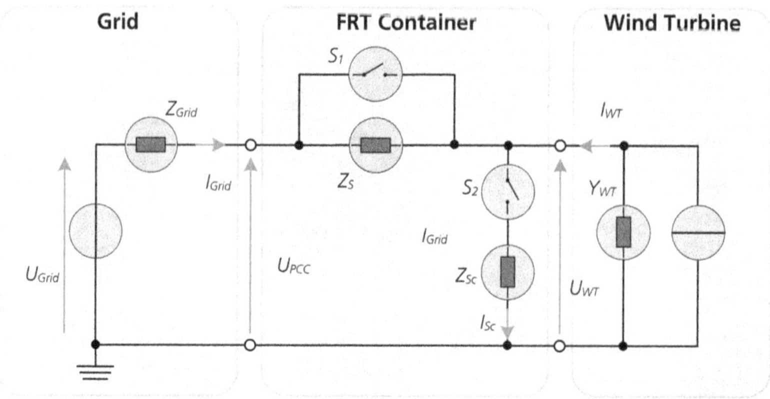


*Figure 5.5 Illustration of a UVRT container*

### 5.5.2 Electrical design and impedance emulation

The grid emulator of the nacelle test bench is also used for the Hil-GridCoP test bench. The power flow does not circulate via the DC link, but via the 20 kV public grid. In addition, an option was created to also operate the DUTs on the 20 kV public grid.

As further developments compared to the test execution at DyNaLab, the simulation of mains faults was revised. The norms and standards, such as IEC 61400-21-1, only require the specification of amplitudes for voltage dip tests, but the method is practically only carried out with a voltage dip container, i.e. inductive voltage dividers. In addition to the intended voltage dip, this method also inherently exhibits a phase jump in the 3-phase system and a change in impedance. Both must be taken into account for a reproduction of field tests on test benches.

The structure of an FRT container is shown in Figure 5.5. The grid is represented as voltage source U_Grid behind the grid impedance Z_Grid. The FRT container consists of an inductive voltage divider, consisting of the series impedance Zs, the short-circuit impedance Z_Sc, and the switches S_1 and S_2.

The impedance Z_Sc is used to short-circuit the grid. The series impedance Z_S is required to reduce the short-circuit current flowing from the grid. If a UVRT is to be tested, switch S1 is opened so that the wind turbine is connected to the grid via series impedance. To effect the actual grid fault, switch S2 is closed for typically 200 ms to 2 s and then opened again. The open-circuit voltage U_Wt at the wind turbine terminals adjusts according to the impedance divider ratio. In the post-fault case, the series impedance Zs remains effective as in the pre-fault case. After a total of approx. 10 s, the switch S1 is closed again so that the turbine is normally connected to the grid again.

### 5.5.3 Superposition by the current of the turbine

The grid emulator sets a specific voltage and emulates the desired impedance of the grid. To emulate the operation of the DUT on the normal grid, the grid voltage U_grid is set and the current-dependent voltage drop across the grid impedance is added to the voltage. After opening switch S1, the sum of the grid voltage and the current-dependent voltage drop via Z_grid and Zs is set as the voltage. In the pre-fault and post-fault case, the impedance of the entire system is Z_Total = Z_Grid + Z_S. The series impedance used to reduce the feedback effects of the grid fault is typically very large, so it results in very weak grid conditions for the wind turbine. In case of a fault, the grid emulator sets the open circuit voltage according to the inductive voltage divider and the current-dependent voltage drop from the parallel connection of $(Z_{Grid} + Z_S)||Z_{Sc}$. In particular, for FRTs with a low residual voltage, the impedance to which the wind turbine feeds becomes very small. The turbine must therefore withstand a voltage jump, a phase jump and a sudden change in impedance during an FRT.

In the following, exemplary test results of a 5.X MW wind turbine are shown. The tests were carried out almost simultaneously on the test bench and in the field and were part of the Hil-GridCoP research project.

The evaluations in the following graphs were made according to IEC 61400-21-1 for the calculation of the Positive, negative and zero sequence quantities from the measurement of the currents and voltages in the rotating field system.

Figure 5.6 shows the voltage curve in a comparison of test bench tests with field tests. It can be seen that the series impedance is switched on approximately 3 s before the grid fault. In both cases, this leads to a drop in the voltages with very similar behavior. In the pre-fault case, there is a quasi-static deviation of 1%–2%.

During the emulation of the actual fault case, both measurements show very good correlation. Slight deviations can be seen during voltage recovery. Here are two aspects that come into play: When switching off currents by means of circuit breakers, these always switch close to the current zero crossing. Thus, first one circuit breaker switches off and the fault changes from a symmetrical to an unsymmetrical fault. When the next current zero crossing is reached, the other two circuit breakers switch off. The discussion of the results showed that this switching behavior should be implemented in the real-time simulation. Furthermore, a

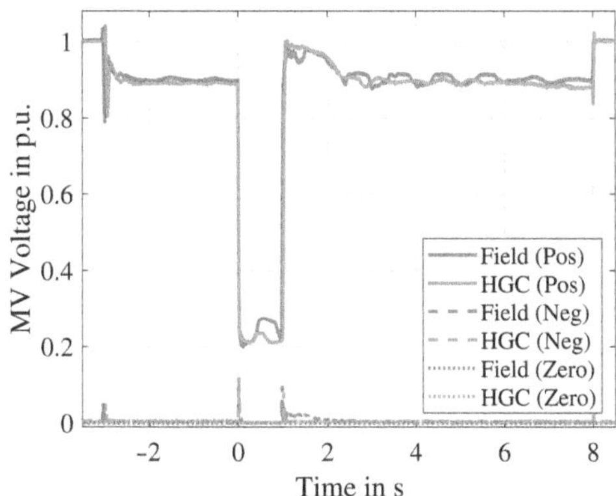


*Figure 5.6 3 phase FRT with 20% residual voltage – comparison of voltages in field tests and on the HiL-GridCoP test bench*

method must be found to minimize the internal magnetic flux control, since this was activated during some tests.

Figure 5.7 shows the active current behavior of fault in the field and the test bench test. Characteristic is the return of the current or power after the FRT. It can be seen that the two behaviors correlate to a high degree.

Figure 5.8 shows the comparison of the reactive current injection of both tests. At the time of the fault, the reactive current must be injected according to the fault depth due to grid code requirements. This can be seen well in the figure. The current levels in the pre-fault case and post-fault case agree very well, which shows that the determined FRT-container parameters can be reproduced very well.

Figure 5.9 shows the emulation of a two-phase fault. Again, the correlation of the voltage waveforms is very good. When simulating two-phase faults, it must be noted that the points between which the faults are generated and the other the

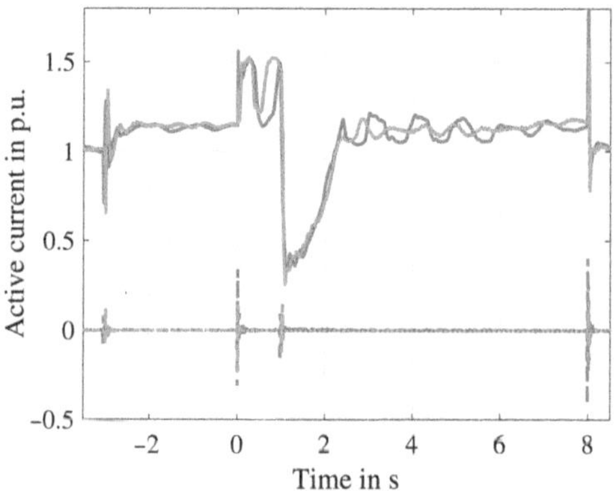


*Figure 5.7 Three-phase FRT with 20% residual voltage – comparison of active currents from field tests and HiL-GridCoP bench tests*

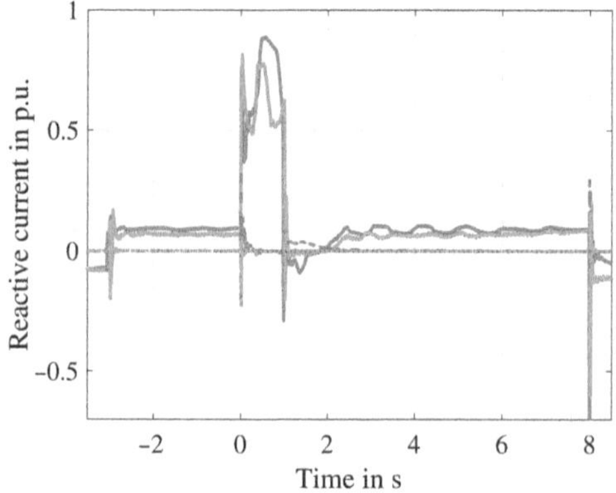


*Figure 5.8 Three-phase FRT with 20% residual voltage – comparison of active currents from field tests and HiL-GridCoP bench tests*

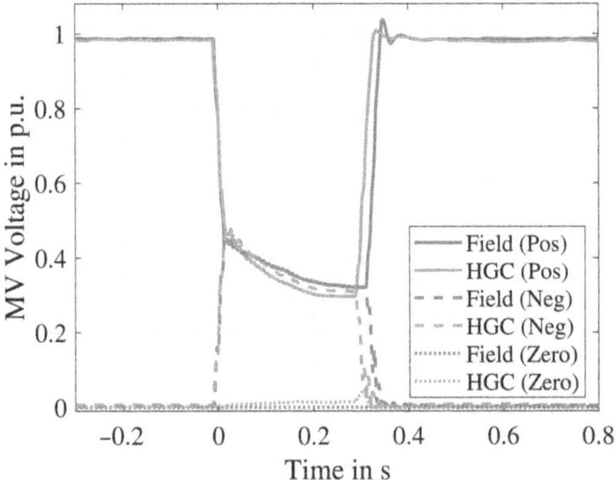


*Figure 5.9 Two-phase FRT – comparison of active currents from field tests and HiL-GridCoP bench tests*

impedances and voltage levels are not identical. By topology now only another setpoint must be given, but there is no magnetic influence.

Figure 5.9 shows the simulation of a two-phase fault with a residual voltage of 0 V. Here the agreement of the voltage curves is very high. When emulating two-phase faults, it must be noted that the phases between which the faults are generated and the fault-free phase differ in their impedances and voltage levels. With the DyNaLab mains emulator, it is possible to set the voltages and impedances individually, so that each phase has its own setpoint.

### *5.5.4 Conclusion*

In the Hil-GridCoP project, it was shown that equivalent results to those obtained in field tests can be achieved by means of test bench testing. This was also confirmed by an accompanying certification body.

In contrast to the nacelles weighing up to 400 tons for testing on a complete system test bench, the equipment required for a minimal system can be transported on a truck. The assembly of the minimal system takes only 8 to 10 days, in contrast to the 6 to 8 weeks integration of the nacelle into a total system test bench.

Thanks to the lower complexity of the test specimen, the bench tests on the minimal system were able to speed up the test execution considerably. In this way, it was possible to carry out a complete measurement campaign in only 3 months. This corresponds to an acceleration by a factor of 3. The costs are also lower by a factor of 5–6 compared to the tests on the DyNaLab.

In addition, validation of the minimum system can be carried out at an early stage of the development process. The construction and procurement of these components can be carried out separately and only brought together at the test bench. This allows validation tests to be carried out in full at a lower level and is thus in line with the methodology of the V-model. In addition, test benches offer the advantage of plannability and reproducibility, which is why they can already be used accompanying the development phase. Both enable companies to optimize internal processes.

### 5.5.5 Resume

By using a high-speed drive motor, it was possible to provide evidence for type 3 wind turbines. The control methods and validation procedures used can be transferred one-to-one to medium-speed topologies, which allows these procedures to be used for these topologies as well.

Important for the future development of modern validation and certification procedures for grid code conformity testing is the demonstration of the step from field testing to bench testing described as state of the art in IEC-61400-21-4.

Important for the future development of modern validation and certification procedures for grid code compliance was the demonstration of the equivalence of bench testing and field testing. The test procedure with the minimum system will be described as state of the art in IEC-61400-21-4.

This will allow test methods that be developed beyond traditional field testing. These include component-based unit certification, validation of new other grid faults, and validation of power systems.

## 5.6 Where is grid compliance testing heading?

The activities described so far are aimed at mapping the grid integration test on test stands in accordance with the current standardization. However, there are various developments that make further development of grid integration testing necessary.

To date, the test results of grid integration tests are used for model validation of RMS models. In the current situation, in which renewable energy generation units are increasingly integrated into the power grids and thus conventional power plants with synchronous generators are replaced by a decentralized energy supply, a secure grid operation can no longer be planned and ensured on the basis of the RMS models. RMS models of wind turbines are assembled into overall wind farm models by simple aggregation rules [4]. Aggregation rules are also no longer sufficient since possible interactions between wind turbines or the wind farm cannot be represented and are not taken into account.

Therefore, EMT-based simulation studies will become increasingly important for transmission system operators as a complement to grid codes.

Various challenges arise from this development. Although initial requirements for EMT models are available, validation regulations are currently still lacking. The transition to grids dominated by inverters also requires fundamentally different control concepts for wind turbines and other power generation units. The exact definition of the requirements for grid forming units is not yet finalized and their validation or even the proof that a wind turbine is operating in grid forming mode is not yet rudimentary. New validation procedures and measurement rules for validating the so-called grid forming control architecture need to be developed, but they need to be made practical for large-scale deployment. These developments are outlined in this chapter.

The growth of wind turbines is progressing continuously. But the validation possibilities, be it through test benches or field tests, are reaching their performance limits. The locations that offer good wind conditions and a grid with high short-circuit power are becoming increasingly rare or are no longer available at all. That's why new solutions are needed. Grid operators are resisting tests in their grid area because they have a significant impact on the overall electrical near-field environment due to grid feedback, especially that from FRT tests. Consequently, finding suitable locations for field tests is becoming increasingly difficult.

System test benches such as DyNaLab have been built in the 10 MW power class. However, onshore wind turbines in this power range are already under development or will be in a few years. Offshore, the power range of the test benches was already exceeded several years ago, so new test rigs are needed. It is advisable to additionally consider these new aspects during the development of the test benches with regard to their functionality and necessary future applicability. However, these test benches should be able to continue the development trend toward testing at lower V-model test levels. The industry hopes that component-based unit certification will further reduce costs and provide earlier evidence of grid code conformity.

Thirdly, it should be mentioned that the current technology used for the grid emulators, which were commissioned in the period 2013–2017, has technical limitations. Due to the slow-switching IGCT semiconductor modules, the grid emulators only have a limited control bandwidth in the range of a few hundred Hertz. However, this is far from sufficient to demonstrate high-frequency EMT model validation, harmonic, and impedance characteristics. Consequently, it is necessary to further develop the grid simulation technology as well.

## 5.7 Testing on lower V-levels to reduce costs

As already mentioned, there is a high-cost pressure in the wind industry and thus also the desire for cost savings for non-stationary tests. As already outlined in the chapter on the DyNaLab nacelle test rig, this approach of system-level testing will not be continued in the future due to the disproportionate test times, complexity, and therefore cost. As shown in the chapter on the Hil-GridCoP test bench, costs and time can be reduced by testing on smaller test benches without compromising the quality of the tests. It is of great importance for industry to continue this development trend and perform validation at a lower test level. This is accompanied by the increasing virtualization of tests to emulate realistic wind turbine behavior. So-called X-HIL methods (refer to chapter 6) must be used, as demonstrated on the Hil-GridCoP test bench. It is imperative that these X-HIL methods are used consistently across all test levels.

Another reason for continuing the chosen path is the high investment costs for new, more powerful full system test benches in the power range from 25 megawatts to 30 megawatts. Torques of up to 30 meganewton meters can be expected for tests on entire nacelles and with the required drive power of modern offshore wind turbines at low rotational speeds. The dynamic mode of operation, as shown above, of these test rigs excludes the use of gearboxes. This results in a lower natural frequency of the mechanical drive train, so that the dynamic requirements can hardly be realized or would not be practical. Direct drives with correspondingly large diameters and weight are the better choice.

Drive units in this power class usually have large dimensions and high forces and torques have to be supported. This requires correspondingly large and cost-intensive foundations for the test benches. Cost estimates from 2018 showed that an upgrade of the motors of the DyNaLab test bench to a drive power of 20 MW could only be realized with a total investment sum of €35 million. A further power increase to up to 30 MW, including new construction of the foundation, hall, and other auxiliary systems, would cost €120–€150 million today.

The operation of such test rigs becomes correspondingly more complex and expensive. The costs for tests on the 25 MW LORC test rig, which is designed exclusively for mechanical load tests of gear units, are about twice as high as on a conventional total system test rig in the power range of 10–15 MW [6]. Therefore, a realistic use for a research facility is hardly feasible and an economic operation for the industry with corresponding amortization costs is hardly presentable.

The development of specialized component tests that can be performed at a lower level is in line with current developments in the industry.

## 5.8 Harmonic measurement methods

In addition to the ability to ride through dynamic faults in the energy system, in order to obtain the unit certificate for EZE, requirements are also imposed by the grid codes on the system perturbations during continuous operation (also referred to as power quality). Grid perturbations are understood to include not only harmonics but also switching operations, flicker and unbalances during continuous operation of the plant.

For example, the TAR High Voltage [7] specifies limits for permissible harmonic currents at the PCC in the 110 kV grid for each ordinal number. In order to be able to comply with these harmonic limits, the aim of today's measurement methods, as described in IEC 61400-21-1 [1], is to determine the harmonic values of the energy supply unit in continuous operation. For this purpose, the measurement is performed according to IEC 61000-4-7:2008 in measurement periods of 10 min for each active power bin. It is crucial to ensure that the measurement of the interconnected system is carried out with minimum distortion, without being able to make or adjust targeted changes.

According to IEC 61400-4-7, the harmonic currents are measured at the PCC, and the measurement result is a superposition of different harmonic currents, as shown in Figure 5.10.

The converter of a wind turbine is current controlled and injects the fundamental current but also harmonic currents into the grid. The grid voltage is not free of harmonics and thus represents a voltage source with the harmonic behind a grid impedance. This harmonic bias of the grid voltage leads to harmonic currents at the

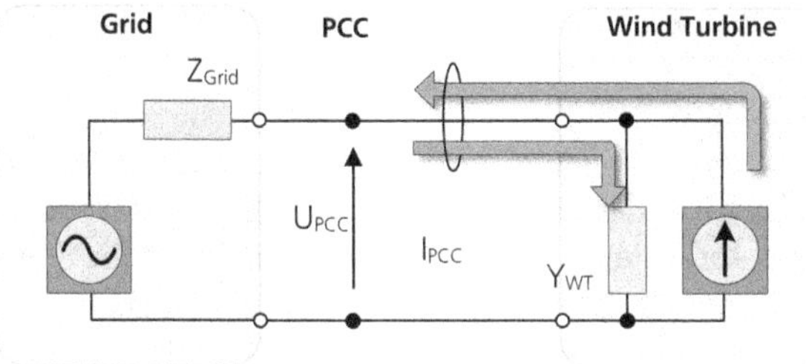


*Figure 5.10 Illustration of the harmonic current in wind turbines*

PCC in accordance with the frequency-dependent grid impedance and the frequency-dependent admittance of the wind turbine as well as the distortion itself.

The harmonic preload on the grid side cannot be determined, since with known impedances or admittances the voltage harmonics can technically only be measured at the terminals of the wind turbine. However, these disturbances cannot be clearly assigned to the grid due to the harmonic currents of the wind turbine. Therefore, the share of the disturbances caused by the grid cannot be determined precisely.

In addition, harmonics can occur, which are caused by interactions between the grid and one or more wind turbines due to unfavorable impedance conditions. These disturbances are called resonance disturbances and lead to an increase of the harmonic currents.

During the measurement at the PCC, the summation of all these harmonic currents is recorded. With the current measurement methods, which perform a point-by-point measurement at the PCC of the WT, it is not possible to distinguish between the different harmonic currents. Therefore, the results of a site-specific measurement are not generally valid and cannot be transferred to other sites. Thus, a methodological problem exists. Time of day-dependent changes of harmonic and frequency-dependent impedances can be measured in mains operation.

## 5.9 Grid emulators – technology limits for high frequencies

The technology of the inverters is based on slow switching IGCT semiconductor modules with a clock frequency of 350 Hz, which can however be increased to 2.8 kHz as described in Section 5.10. For harmonic measurements, the clock frequencies can be increased up to 4.4 kHz.

The harmonic behavior has a special characteristic of the inverter operation, which is dominated by the clock frequency of the grid emulator. The other harmonic components change in the high-frequency range the characteristic only with change of the modulation degree but not undeterministically in the course of time.

The MV grid of the grid emulator has special filters for reducing high-frequency harmonics. In normal operation (2.8 kHz switching frequency), they allow THD-U_50 reduction to as low as 1% to 3%. When operating at higher switching frequencies (4.4 kHz), a voltage quality better than 1% THD_U_50 can be achieved.

The very good reproducibility of tests on test benches allows precise repetition of the harmonic measurement at the same operating points or under the same wind conditions. By changing the clock frequency of the power emulator, the characteristic frequency spectrum can be manipulated, and thus the dominant switching frequencies and harmonics can be filtered out of the measurement by comparing two measurements.

By filtering and testing two measurements, a nearly distortion-free measurement is achieved, which meets the requirements of a harmonic measurement according to IEC 61400-4-7.

As previously explained, the harmonic currents of the wind turbine also depend on the impedance of the grid or, in the case of bench testing, on that of the grid emulator.

Although the impedance of the grid emulator can be controlled, the control bandwidth for impedance emulation can only be controlled in a range of a few hundred Hertz due to internal delay times of 255 μs of the grid emulator.

It is also possible to characterize, by means of testing and experience, the impedance of grid emulators in the high-frequency range more precisely. However, this does not extend beyond a good estimate and thus does not correspond to any validation. Therefore, harmonic measurements on grid emulators can give better results than in the field, but they do not solve the methodological problem of harmonic measurement.

## 5.10 Harmonic measurement on test benches – PQ4Wind test bench

For the measurement of harmonic currents, validation of harmonic models and determination of the frequency-dependent impedance of wind turbines by measurements on test benches, test facilities must have a substantially different design. The test facility will be able to eliminate or selectively adjust the interfering "background" harmonics and the frequency-dependent impedances by specific emulation of grid characteristics in high quality. The aim is to decouple the measurements by giving the grid side a known behavior and thus to test or measure the specimen accordingly.

As explained in the previous chapter, decoupling between the grid and the object under test, or in this case between the grid emulator, is necessary to ensure correct measurement and validation of the harmonic models.

In contrast to the current grid emulator, which focuses on testing FRT capability, the design of the PQ4Wind converter test bench focused on precise measurement of the high-frequency behavior on test benches. The test bench is designed for testing converter systems with a power rating of up to 8 MW. Thus, the entire power range of common onshore wind energy systems can be covered and it also offers the possibility to test existing offshore wind energy systems.

The test bench is designed for low-voltage systems, the actual common voltage class of inverters in wind turbines. In view of the trend toward designing wind converters with higher voltages, current models have voltages of 750–850 V. With this copper losses and cable cross-sections can be reduced. In order to be able to test low-voltage converters in the future, the test setup is designed for voltages U_PP of up to 1000 V.

The test bench consists of five identical modules, each with a capacity of 4 MW, which are operated on a common DC link. Two converter modules act as generator emulators, while two others serve as grid emulators. The fifth module serves as a rectifier unit to the grid and thus covers the power requirements of the entire test stand, the power dissipation of the test equipment and the DUT.

For decoupling, the test bench will be designed to allow the highest possible control dynamics and precise imprinting of harmonics up to 10 kHz. To achieve this goal, each 4MW module will be operated with a cumulative clock frequency of 128 kHz. This is achieved by connecting 8 half-bridges per phase in parallel, which in turn are clocked at 16 kHz. The topology is shown in Figure 5.11.

The half bridges consist of a three-level NPC circuit consisting of two 1200 V IGBT – 2-level modules are built. Figure 5.11 shows the structure of one phase of a module. This allows to raise the DC link voltage up to 1700 V and thus to generate the required output voltage of up to 1000 V.

To further reduce harmonics, the fast-switching IGBT modules are controlled to operate with minimal bridge delay times. This allows precise control of the voltages near zero crossing of voltages. A LC filter with a high cut-off frequency is used to filter the switching frequencies of 128 kHz. This results in a THD of less than 0.5% at a nominal voltage of 690 V. Thus, a near-perfect voltage is achieved, with test-stand-specific distortion only above the range under consideration.

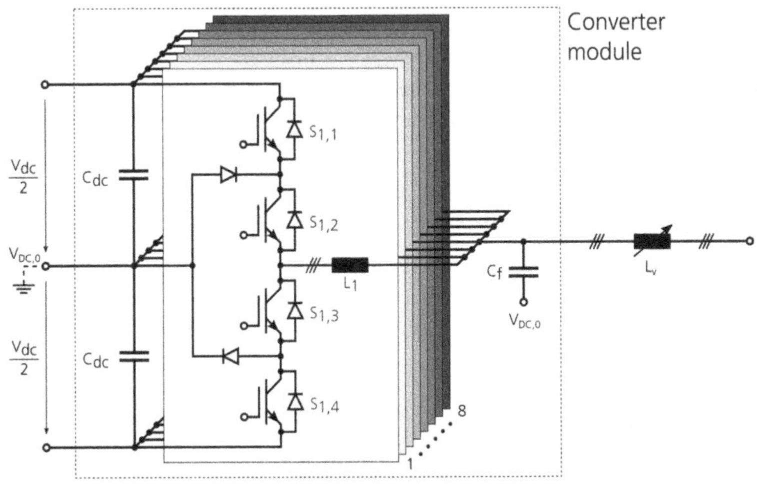


*Figure 5.11 PQ4Wind power electronic topology*

To ensure dynamic controllability in pHil operation, it is necessary to minimize the delay times in the control loop. External current and voltage setpoint provisions are sent to the inverter via the real-time interface with a sampling time of one microsecond. The signal processing of these, to control all half-bridges, is done on an FPGA controller board within the inverter unit. This is also operated with a sampling rate of 1 μs.

The signal propagation times between the setpoint input and the IGBT control are thus optimized so that minimal and deterministic delay times occur in the control loop.

The delay times resulting from the switching operation of the converters are significantly minimized by the cumulative switching frequencies. Thanks to the high bandwidth and minimal delay times, an extremely high control bandwidth can be achieved and the emulation of impedances using pHil control is made possible very precisely.

### *5.10.1 Control the impedances of the test bench*

The two design fundamentals are to generate ideal sinusoidal oscillations and to precisely adjust the frequency-dependent impedance of an emulated grid over a wider frequency range.

To generate a location-independent harmonic model of the wind turbine, the frequency-dependent impedances of the test specimens must be determined.

The highly dynamic controllability of the voltages enables the imposition of harmonic voltages on the fundamental sine wave up to 10 kHz. When specimen are selectively superimposed with mono-frequency oscillations, they respond with a harmonic current. From their measurements of current and voltage, the impedance can be calculated. Using an automated execution, a frequency sweep is performed from a few hertz to 10 kHz to determine the frequency-dependent impedance of the specimen. Since the internal impedance of wind turbines depends on the operating point, these measurements must be repeated for different operating points. By accurately determining the harmonic currents as well as the impedance, the location-independent harmonic model of wind turbines can be determined. These capabilities for validating high-frequency properties also allow new requirements to be placed on EMT models and impedance models.

## 5.11 New requirements for wind farm models

In addition to the requirements placed on grid integration processes by OEMs and the technological requirements, the transformation of energy grids toward a decentralized energy supply is also increasing the requirements placed on grid integration processes by transmission system operators.

According to the current process, a PPM is approved for connection to the grid if it is proven that the Grid Code requirements are met. In the future, each grid connection will be individually examined to determine whether there are any interactions between the grid and the new wind farm. Grid connection approval will be granted if the result of the investigation is positive.

Currently, only RMS models are validated with the measurement results of the grid integration tests on wind turbines. The models have a validity range up to a frequency of 15 Hz. To create wind farm models, individual strings of wind farms are aggregated by simply adding the power of the turbines. There are also very simple aggregation rules for cable and transformer models, which are essentially based on averaging the input impedances. Actual validation requirements for wind farm models are limited to simple power and harmonic measurements at the grid interconnection point.

These modeling methods neither allow the minimization nor the assessment of risks that may arise from interactions within the park itself. Also, interactions with the upstream grid cannot be determined simulatively in this way. They are also not suitable for further more comprehensive planning of other grid resources. Other modeling methods must therefore be used to compensate for these weaknesses.

In contrast, the use of EMT simulations has been established as state-of-the-art in the development of power electronics and electrical systems for many years. These can also correctly reproduce high-frequency phenomena.

In the future, it will be important to apply EMT simulations to the calculation of the behavior of large electrical systems or entire grid regions in order to identify interactions.

The models for such interaction studies must be of high accuracy and valid over a wide frequency range to cover the expected interactions. Therefore, new requirements for the EMT model of wind turbines and wind farms are currently being developed by the transmission system operators and adopted in the grid connection guideline.

## 5.12 EMT model requirements

The requirements for EMT models are currently being specified. For example, there is a position paper of the four German transmission system operators [3] in which it is explicitly pointed out that the previous standard modeling is not sufficient for so-called high-frequency interactions and in which it is also described that EMT model requirements are set up to 2.5 kHz. Clearly going beyond this, it is described by Expert Group Interaction Studies and Simulation Models in [2] that EMT model requirements exist not only for single systems, but also for PPM, i.e., for entire wind farms. In addition, numerous sub-requirements are mentioned, such as the validity of the modeling for positive and negative sequences, validity for different operating points, and behavior in case of symmetric and asymmetric faults in the grid. The model must include the controller and its functionalities. Modeling of cables, modeling of transformers including saturation effects, modeling of filters, damping resistors, switching in case of voltage protection, and modeling of all relevant operating points. The model shall be suitable to reproduce the frequency-dependent impedance of the wind farm.

From the position paper quoted above, the application of these models is additionally very clear. Stability problems, in particular controller interactions, are to be identified which differ significantly from the classical stability phenomena in transmission grids. It is planned to create a data set or database of appropriate models. From the data set, the individual models can be automatically built up to the respective grid model and the interaction studies for the wind farm can be carried out. Through these simulation studies and by means of the EMT simulation models to be supplied, risks due to interactions between current directions are to be identified at an early stage. Changes are to be planned at an early stage. This process shall also be extended to the operation of PPM. Therefore, a quasi-continuous update of these models will be carried out, e.g., in case of controller adjustments or in case of decisive changes in the wind turbine components. It is thus an iterative process that extends over the entire life cycle from planning to construction to operation of wind farms in order to be able to ensure safe grid operation.

As a consequence of such an approach, it can be expected that the importance of compliance or certification tests will further increase. On the one hand, the requirements for grid integration functionalities must be fulfilled, which result from the corresponding grid study for the wind farm. This may require customization of the development, which happens late in the development process. Simple and manageable validation methods should be available here to enable cost-effective conformance testing. There will be a high demand for compliance verification due to the continuous updating of models in TSO databases. In view of the expected increase in the number of wind turbines over the next few decades, a method should also be used here that allows the amount of work to be managed.

## 5.13 Component-based units certification

The consequent combination of development trends and challenges to be addressed consists in the component-based unit certification. This provides for the grid integration tests to be performed on the basis of validated EMT models and the engineering tests to be decoupled from them. For this reason, the V-model consisting of three legs was developed, see Figure 5.12.

As described above, the measurement of the electrical characteristics of wind turbines is increasingly coming up against technical or economic limits on system test benches and in the field according to the current state of the art.

The number of available test sites is decreasing and the existing test rigs are no longer powerful enough to perform the tests for offshore wind turbines (15 MW and more) in the laboratory on the known system test benches. Industry is demanding the ability to test and validate even earlier in the development process. In addition, new requirements are constantly being defined for the grid. Reference should be made here to the introduction of voltage-forming control methods, the implementation of which will play a decisive role in decentralized power supply grids.

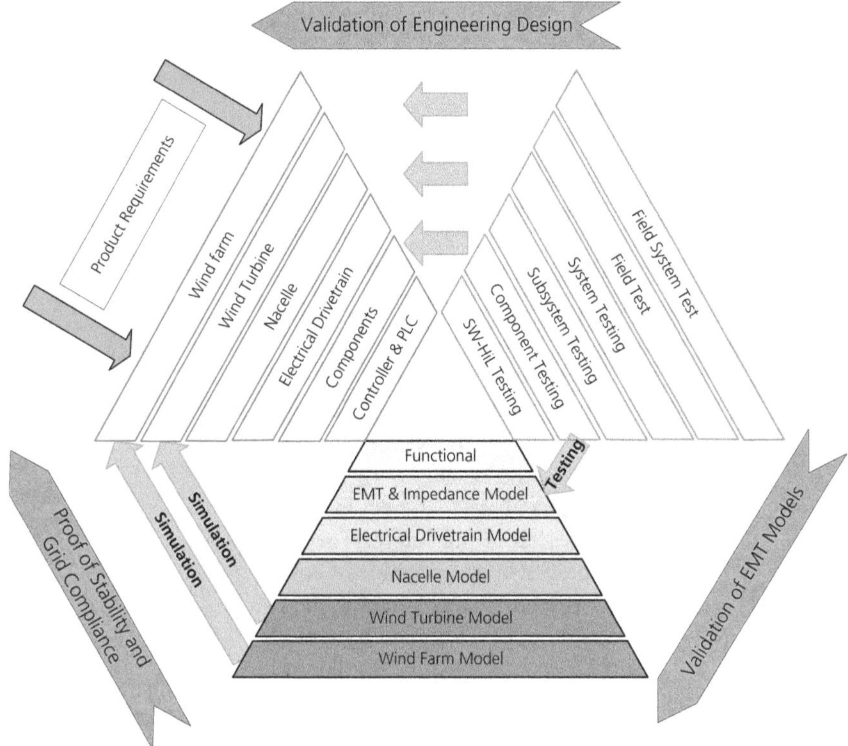


*Figure 5.12 Extended V-model for the topic of grid integration of wind energy. Top view of the classic V-model with an additional leg of virtualized testing.*

However, the requirements definition for these methods has only just begun and will probably require several revisions and also show significant differences in the countries.

While for wind turbine manufacturers the determination of the electrical properties of a single wind turbine is still crucial for a type certification, for wind farm developers and operators the properties of an entire wind farm and potential interactions are increasingly relevant. The increasing importance after the validation of the electrical characteristics and the model validation of ever larger PPMs therefore require an adapted validation process, the component certification.

Figure 5.12 shows the development methodology according to the extended V-model, taking component certification into account. The direct validation of large WTs as well as wind farms is not possible in the classical way (PPMs on field test, WT on system test bench). Therefore, the electrical properties and the EMT model validation are to be determined component by component using component test benches. The validation of the overall systems is then only carried out at the model level as a virtual test bench, but no longer directly.

EMT models describe the electrical characteristic and functions of the components. The models of the WT components are combined into an overall model and test scenarios are simulated at WT or wind farm level. Based on the simulation results obtained, the electrical properties are evaluated. For this validation process, the validity and accuracy of the individual EMT models must be ensured by specific test bench tests.

The models validated in this way can be seamlessly integrated into the interaction studies of the transmission system operators. This enables a continuous validation process across the entire value chain of grid integration.

## 5.14 Outlook

The transformation of grid to a decentralized energy supply will bring with it some new requirements for grid operation. However, new requirements and functionalities will only be introduced on the basis of verified findings. However, since there are currently a lot of developments in this area, many changes in the grid requirement guidelines can be expected in the future. The implementation of validation methods should be as comprehensive as possible to ensure the security of the entire power system. Although this leads to a cost factor and can thus be perceived as an obstacle to the energy transition, these validation methods are indispensable. Future processes must therefore be able to dynamically adapt to the state of the art. It is important that validation methods are further developed in parallel with the development of new grid operational methods. Test bench testing and the corresponding infrastructure are currently important to ensure flexibility in achieving these goals.

As a final point, it should be noted that the present system development is strongly focused on compliance with grid requirements and test procedures. However, this also means that systems are planned and optimized according to grid codes and applicable test procedures. However, this does not lead to the most grid-serving behavior in all cases. A technical overfulfilment of the requirements would be possible in some operating points without causing damage. These limit ranges are not exhausted from the design, but the main focus is on obtaining the corresponding certificate.

This requires an extended quality assurance process that is carried out independently and that takes into account the real grid behavior in the field.

## References


[1] *IEC 61400-21-1; Wind Energy Generation Systems – Part 21-1: Measurement and Assessment of Electrical Characteristics – Wind Turbines*

[2] *Expert Group Interaction Studies and Simulation Models (EG ISSM)*, 2021, https://www.entsoe.eu/documents/nc/GC%20ESC/ISSM/EG_ISSM_Final_Report_211001.pdf

[3] *EMT-Modellanforderungen und Bedarf an standardisierten Schnittstellen für Interaktionsstudien*, Positionspapier der deutschen Übertragungsnetzbetreiber, 2023, https://www.netztransparenz.de/Weitere-Veroeffentlichungen/Interaktionsstudien-und-Modellanforderungen
[4] *IEC FDIS 61400-27-1; Wind Energy Generation Systems – Part 27-1: Electrical Simulation Models – Generic Models*
[5] G. Curioni, and T. Jersch, "*HiL-Grid-CoP – Hardware-in-the-Loop Prüfung der elektrischen Netzverträglichkeit von Muti-Megawatt Windenergieanlagen mit schnelllaufenden Generatorsystemen: Schlussbericht*," BMWK Project Final Report, Fraunhofer Institute for Wind Energy Systems, Bremerhaven, 2022, BMWK Project Funding Code 0324170A-B + D
[6] Lindø Offshore Renewables Center, Governance, https://www.lorc.dk/governance
[7] VDE-AR-N 4120 Anwendungsregel:2018-11, "Technical requirements for the connection and operation of customer installations to the high voltage network (TAR high voltage)," VDE VERLAG GmbH

*Chapter 6*

# X-Hardware-in-the-loop test methods for validation

*Adam Zuga[1], Mohsen Neshati[1,2], Florian Hans[1], Nils Johannsen[3] and Oliver Feindt[1]*

This chapter discusses the topic of using in-the-loop testing methodologies originating from model-based design in the wind turbine development process to achieve accelerated time to market by performing verification and validation tests on suitable test benches in an early stage. The first section introduces the topic with a description of relevant definitions and essential methodologies for testing wind energy systems. Section 6.2 describes the method of mechanical hardware-in-the-loop as well as the resulting test scenarios and shows results of tests of a wind turbine nacelle on a system test rig. Section 6.3 introduces further test possibilities of electrical components with the power hardware-in-the-loop method and deals with the essential structure, stability, and practical aspects of these tests. Section 6.4 provides essential test concepts and the resulting various options for implementing controller hardware-in-the-loop setups. Section 6.5 rounds off the topic with an outlook on future developments and challenges of hardware-in-the-loop methods in the wind energy sector.

## 6.1 Introduction

Performing field tests with products in fully functional prototype status is an important and promising method for verifying correct and safe overall functionality and obtaining valid test results. Nevertheless, this test method is usually associated with high costs and can only take place at the end of the product development process, which has certain disadvantages. On the one hand, several test scenarios cannot be verified due to their infrequent occurrence in the field or the risk of critical operating conditions that could lead to damage or failure of the prototype. On the other hand, covering a necessary operating or measurement range can be time-consuming and therefore be associated with high costs. To illustrate this point, consider an example from the wind industry where wind conditions, because of their difficult-to-predict, highly stochastic behavior, have a

[1]Fraunhofer IWES, Fraunhofer Institute for Wind Energy Systems, Bremerhaven, Germany
[2]Siemens Gamesa Renewable Energy A/S, Brande, Denmark
[3]Beckhoff Automation GmbH & Co. KG, Lübeck, Germany

direct impact on the duration of wind turbine tests in the field to perform different tests for a range of wind speeds. It is certainly easy to see why the "wait for wind" statement familiar from sailing has also found its way into these tests. Another disadvantage of testing at the prototype stage is the timing of the development process itself. Design problems, malfunctions or critical operating states that could not be identified in the simulation due to missing representation can only be detected very late in this way. At this point, a complete remedy is often very cost-intensive or even not possible without a complete redesign. All these points work against cost reductions and the shortening of time-to-market, which in today's times, e.g., with regard to the energy transition, are important and indispensable prerequisites for its early and successful achievement. In product development processes of the automotive, aerospace, and aeronautics sectors, development lifecycle models known from software development, such as the V-model [1] (cf. Chapter 9.4, p. 421 in [2]), have been used for a long time. In addition to the development phases specified there, tests are defined not only at prototype level but also at the system and component level, which leads to earlier assurance of the desired functionality and thus to an increase in product quality. The rapid increase in the computing power of high-performance computers enables the modeling as well as calculation of increasingly complex mathematical simulation models of the systems to be developed. This can be used in the context of model-based design (MBD) to support the development cycle. Therefore, simulation models can be built in this way with parts or the entire system and simulated in a virtual environment without any risks. As a result, important simulation investigations are already carried out in the design phase (left-hand side of the V-model in Figure 6.1), which primarily answer important questions regarding layout, design, or functionality and secondarily contribute to cost and time savings. In the automotive sector, for example, the source code for electronic control units (ECUs) is tested

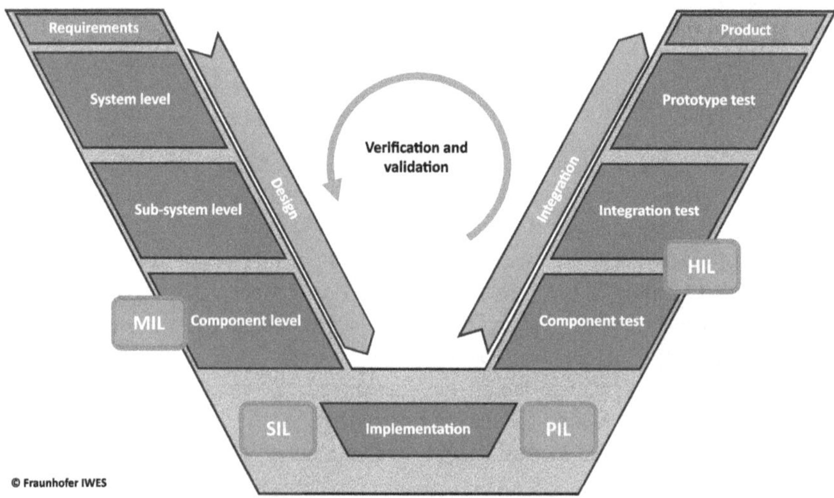


*Figure 6.1 V-model according to [1] with assignment of MBD methods to the respective level with reference to [4]*

before the actual hardware is available, as noted in [3,4]. As soon as the control hardware is accessible, a component is delivered or built, or a subsystem is constructed, it seems obvious to couple it with existing simulation models and test at an early stage. For example, control algorithms can be tested on actual hardware or components in interaction with a realistic environment (virtual simulation) before a prototype is completed. The models are then executed in real time and connected to the respective hardware in a closed-loop manner, which is called hardware-in-the-loop or HIL for short. This approach allows components to be tested under realistic conditions before the full system test in the field. In addition, the measurements on the real component can be used to validate the virtual model, thus increasing the accuracy of the overall model at an early stage.

For the design and certification process of wind turbines, aero-elastic load simulations with partly already complex models and integrated control algorithms have been used for more than 20 years. Likewise, turbine controllers have also been tested for many years using HIL and real-time models, initially with simplified models, but recently increasingly with models at load calculation level. For a long time, component tests were only carried out by manufacturers, mainly without HIL and not realistic load assumptions. It is only since more and more component and system test benches have been built (within the last 10 years) that there has been an increasing trend towards HiL testing at component and system level. But, before the relevant types of HIL tests in the wind energy sector are explained by the following subsections in more detail, let us first go into the most essential definitions and terminology specific to the related application in this area.

As mentioned in the above, the two main reasons for such a development and testing philosophy are an early risk mitigation and a quick and accelerated market rollout. The sooner this is done in the product development cycle, the more effective. For the models used for simulation in the design phase, whose complexity increases in the direction of the component design (from upper left to bottom in the V-model in Figure 6.1), two of the most essential methods of the MBD [5] are used (cf. Figure 6.2), model-in-the-loop (MIL) and software-in-the-loop (SIL).

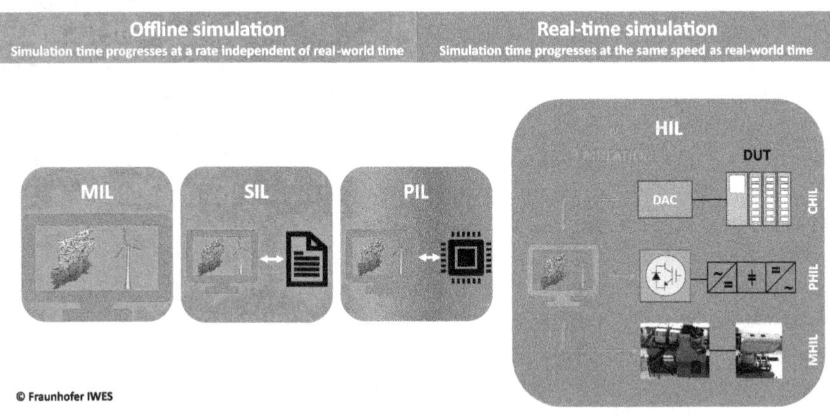


Figure 6.2 Common MBD "in-the-loop" test methods (inspired by [6])

As already mentioned, these methods are predominantly used, e.g., in the automotive sector for the integration and testing of control algorithms. However, in the wind domain, it makes sense to use them also for component models (e.g., mechatronic pitch actuator model or electrical generator-inverter model). In MIL, the plant model and the controller model or special component models are mapped and simulated with physical and mathematical models mainly in one simulation environment (e.g., MATLAB® Simulink®). The calculation can be performed with available processor speed with the shortest possible calculation time for the respective step size, this is generally referred to as offline simulation, and with all suitable solution algorithms. SIL differs from MIL only in that the model is now available in the source code (e.g., C code) that will later be used on the target hardware. This approach is useful for testing control algorithms. However, the principle can also be used for coupling with, e.g., a load calculation simulation program of WTs, where the model, with either controller algorithms or a component representation, is also converted into source code and usually integrated into the simulation as an external component by means of a compiled object code (e.g., a dynamically linked library). The purpose for this is the protection of intellectual property sensitive information such as advanced control or model algorithms in the so-called object code or as a black box model. The execution of SIL simulations is also performed offline, but with a fixed step size due to the connected discrete source code. As a preliminary stage to HIL, processor-in the-loop (PIL) should be additionally mentioned at this point, although it is used less frequently and is therefore not widespread. In PIL, the source code used in SIL is now executed on the target processor or similar hardware and linked to the model via a suitable interface. Here, it is not yet a matter of execution under real-time conditions, but rather of analyzing the code on the target platform regarding runtime behavior and the equivalence of the results compared to the model. This method can also be used not only for controllers but also for real-time models of the wind turbine for example, in order to be able to examine them on the real-time hardware platform before they are connected to the real component. In terms of classification, PIL is already on the right side of the V-model (cf. Figure 6.1) since the target hardware is already partially tested here. The next stage is finally the HIL method, which is used in different stages and methods depending on the application. Classically, the test of control algorithms on the target hardware must be mentioned here, which is connected to the plant model simulated in real time. This type of HIL method is called controller-HIL (CHIL) or signal-based HIL. Here, the data exchange between the virtual model and the tested hardware only needs to be replicated at the signal level, since this is also the case in the real system. Depending on the goal of the test, the original interfaces used in the future (terminals for the respective bus system or analog and digital I/O's), or other suitable interfaces can be used to connect to the real-time model. The former is certainly recommended, since the technology used later in the field is also tested in terms of signal runtimes, resolution, and susceptibility to interference. In addition, no changes to the signal mapping should be made, so the structure of the controller remains unchanged. This is particularly important and desirable with regard to the relevant national and international standards and guidelines (e.g., DIBt, IEC 61400-22/IECRE OD-501, DNVGL) in the context of type certification. As the control system of a wind turbine no longer consist of one single control unit, but has a nested and complex structure, it is very important to the manufacturer to test it consequently in the same setup in which it will later be operated in the field. Advanced replica systems with all controllers, interfaces, bus systems, safety systems, and emulations of sensors and actuators are set up to test as faithfully and comprehensively as possible. Such systems are also used for testing of component models, minimal systems, or subsystems on test benches. As soon as a mechanical or electrical component is tested in a real context, i.e., with coupling to the remaining part of the full system model, the type of interface changes from signal-based to physical, via which either mechanical or electrical signals must be applied and sensed, depending on the intersection of the system. To do this, additional actuators and sensors must be added to the test scenario. These components are part of the test bench to which the device under test (DUT), either a real hardware device, component or system, is connected mechanically, electrically or on both sides. These test bench components have their own dynamic behavior and in combination with the DUT this results in a changed system behavior compared to the real system. To be able to emulate the original behavior, additional control algorithms are required that can apply the desired dynamics and damp the disturbing dynamics of the test bench. This is generally referred to as HIL control. The virtual real-time models to which the DUT is connected by means of mechanical-HIL (MHIL) (cf. [2]) or power-HIL (PHIL) methods also require suitable interfaces to output signals to the test bench as setpoints on the intersection of the model and to be able to return measured values. This is the only way to ensure that the virtual model and the real DUT are tightly coupled. The HIL controller must ensure that the setpoints can be applied to the DUT as accurate as possible and within a short delay, so that this physically closed loop remains stable and emulates the coupling of the real system. By additional control components (e.g., the turbine controller of the WT), which are part of the DUT on the test bench but generate setpoints to the real-time model and receive calculated values as measurements, a second, signal-based loop between model and controller must then be closed (cf. Figure 6.8 from Section 6.2).

To carry out HIL tests of wind turbines, a real-time capable system model as well as a real-time simulator with corresponding interfaces and certain properties are required in addition to the test specimen and the test bench. Let us give a broad overview of these two points before we move on to the following subsections.

Reliable and valid models are required that are calculated in real-time (the simulation time progresses at exactly the same speed as real-world time), to be able to perform realistic HIL tests in the laboratory that correspond to field tests. However, there are two issues to consider here. First, real-time capability (strict maintenance of a fixed cycle time with moderate central processing unit (CPU) load to avoid overflows) and model fidelity (very high accuracy of the results compared to the reference model) are contradictory development goals. Second, for prototypes that are tested in the laboratory before being tested in the field, there are no or very limited measurement data available for model validation. The following two approaches can resolve this dilemma. On the one hand, simple models can be used where the results can be verified using analytical methods. However, when using such models, the test possibilities are limited to a few operating points or frequencies, which requires the creation of different models for the respective test purpose and can, therefore, be very time-consuming. Or additional adjustments have to be made, for example, in the form of simulation results generated offline (e.g., look-up tables), which in turn limit the informative value of these models. On the other hand, it is possible to use more complex models that cover a wider range of applicability. In the wind sector, for example, complex load calculation models are used to map the aero-elastic behavior of wind turbines, which are used for load assessment in the design process. The models of commercial, free, or manufacturer-specific calculation tools (e.g., alaska/Wind, Bladed, HAWC2, MoWiT,[1] OpenFAST, Flex5, and SIMPACK) have now been verified using many simulation benchmarks and validated using a large amount of measurement data from numerous prototype measurements. These models are therefore very well suited for obtaining reliable simulation results over a wide operating range. However, due to their code structure, some of these tools cannot be compiled on real-time platforms and, therefore do not fulfill hard real-time requirements. Some preliminary steps are necessary before these models can be transferred to the real-time simulator hardware. These steps are shown in the upper part of Figure 6.3 for creating a real-time model for MHIL tests on the Fraunhofer IWES nacelle test bench. First, the parameters of the original model (e.g., reliable manufacturer model in the format of a load calculation tool used by the manufacturer) are transferred to the real-time simulation tool (in the case of Fraunhofer IWES MoWiT is used). It has proven useful to carry this out automatically using calculation scripts to rule out manual transfer mistakes as much as possible. To be able to maintain a fixed cycle time with a suitable CPU load when calculating the model on the real-time simulator, the model complexity must be reduced in most cases. For the aero-elastic models mentioned here, there are mainly two ways of achieving this. First, the number of mostly iterative calculation algorithms of the aerodynamics per blade can be reduced, which influences the CPU load linearly in the valid range of calculation. Second, modal reduction [7] is used to limit the structural degrees of freedom with a focus on the important and most contributing modes for the test purpose. Before the model is compiled on the real-time platform, it must be verified whether the model delivers consistent and reliable results compared to the manufacturer's trusted load calculation tool. For this purpose, it is advisable to simulate and compare selected load cases in the time domain

[1]http://www.mowit.info/

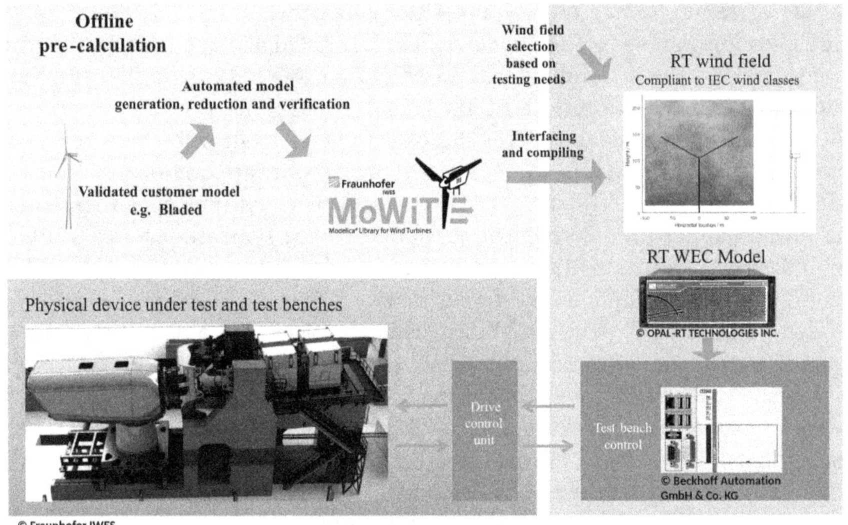


*Figure 6.3 HIL testing workflow and communication path at DyNaLab, Fraunhofer IWES*

(proven methods in this regard can be found in [8]). Also, the comparison of coupled modes in the frequency domain by means of Campbell diagrams and fast-Fourier-transform (FFT) or power spectral density (PSD) plots of the calculated as well as special load cases (e.g., fault ride-through (FRT) events) is very important. This allows verification that the dynamic behaviour is mapped as accurately as possible. As soon as a sufficiently good match is achieved (the rule of thumb is below 10% deviation, whereby other limit values can also be selected depending on the reference value and application), the model can be ported to the real-time simulator system.

Real-time simulators are computer-aided systems that can run virtual plant models on CPUs and/or field programmable gate arrays (FPGAs) under hard real-time requirements. For this purpose, these systems are equipped with real-time capable hardware for low latency and precise time measurement and control, a real-time operating system (RTOS) with real-time planning and deterministic scheduling as well as special real-time compilers including code optimization techniques. All these components guarantee deterministic execution of the model code by strictly complying with real-time conditions, minimization of execution times, and prioritization of tasks, including efficient and dynamic resource and memory management. Depending on the requirements, different software and hardware systems are offered by various manufacturers (e.g., dSPACE, OPAL RT, RTDS, Speedgoat, and Typhoon HIL). The systems include corresponding software packages or libraries for existing modeling or programming tools (e.g., MATLAB® and Simulink®). Automatic code generation is also supported as part of rapid control prototyping (RCP) to generate suitable source code (e.g., C code) from the model, which can be compiled and executed on the real-time simulator. The majority of today's programmable logic controller (PLC) or industrial PC manufacturers (e.g., Bachmann, Beckhoff,[2] B&R, Siemens, and Rockwell) also offer solutions for the execution of models for hard real-time conditions on their devices by means of special software add-ons or libraries (e.g., Beckhoff TwinCAT 3 Target for Simulink). The real-time operating system and functionalities for deterministic execution are the standard features of PLCs and industrial PCs with a PLC runtime. Figure 6.4 shows systems by some of the mentioned manufacturers of real-time simulators.

Real-time capable, hardware-based interfaces also play a significant role in establishing a suitable data connection between the model and the external systems to be connected, such as control hardware or test specimen throughout a test bench. Depending on the specific requirements of the application and availability, either analog/digital IOs or bus systems are utilized. To ensure stable coupling of the simulated, virtual systems with the real hardware, the interfaces used should have short cycle times, low latencies, and low jitter. As the coupled systems interact with and influence each other, any dead times should be kept to a minimum through appropriate synchronization to implement consistent and realistic couplings between the systems. For systems that are coupled with each other through different automation levels (cf. Figure 6.3 in the lower part – communication from OPAL RTS via Beckhoff-embedded PC to drive control unit), this can be crucial for stable control and operation.

In the following sections, the previously mentioned HIL methods are discussed in more detail, including the description of different possibilities of testing scenarios and setups on test benches with the necessary interfaces, actuators, control techniques as well as practical aspects. The chapter is concluded by an outlook on future requirements and challenges for test benches, models, and HIL systems.

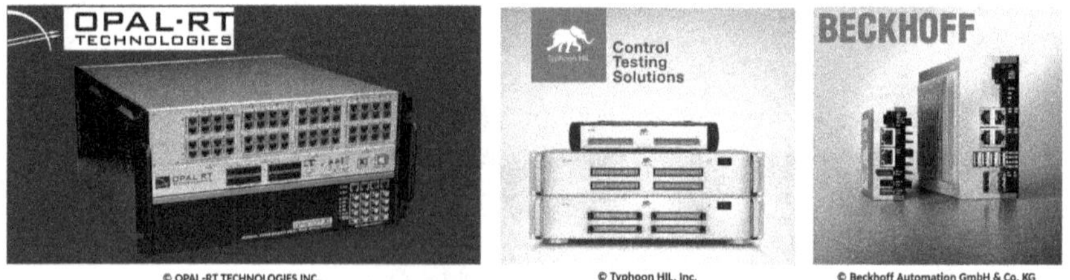


*Figure 6.4 Real-time simulator hardware - f.l.t.r. OPAL RT, Typhoon HIL, Beckhoff*

[2]Please refer to Chapter 7 for more information on Beckhoff hardware.

## 6.2 Mechanical hardware-in-the-loop (MHIL)

Hardware-in-the-loop systems enable mutual interaction between real hardware and a real-time simulation in a multi-physics testing environment. In the case of MHIL systems, the simulation represents a mechanical system with corresponding environmental effects, and the exchanged measures through the interface with the hardware are mechanical quantities. The objective of MHIL is to emulate additional mechanical effects and hardware to execute tests that closely resemble real conditions and enhance the significance of measurement results. As a result, MHIL systems have become widely used for testing purposes across various sectors, from automotive to wind energy.

The automotive sector utilizes MHIL to conduct tests on the complete vehicle or just the suspension system. The interaction between the wheels and the road surface, including friction and multidimensional forces, is emulated to obtain accurate results. Similarly, in the wind energy sector, MHIL is used to conduct experiments on the complete wind turbine nacelle or the generator-converter system. The MHIL framework in this case replicates the environmental conditions such as the wind, blades, aerodynamics, and the tower to emulate real-world scenarios.

To further clarify, consider the block diagram of the original system shown in Figure 6.5(a), which introduces a plant with its associated controller. In a MHIL setup for test and validation purposes, the plant is split into two components: the DUT represented by P12, and the residual part emulated using a real-time simulation represented by P11. Furthermore, additional actuators are utilized in a test bench to apply the effect of P11 on the DUT, which introduces additional hardware and control units. As a result, the original system is modified by adding additional cascaded loops, as illustrated in Figure 6.5(b), where the actuator system and the connection between hardware and simulation form the inner loop. Overall, it is the original plant controller that governs the overall system performance through the outer loop. In general, during validation, the system operation on the test bench is compared to that of the original system operating under normal conditions in the field. This in fact can be challenging since the presence of additional actuators in the

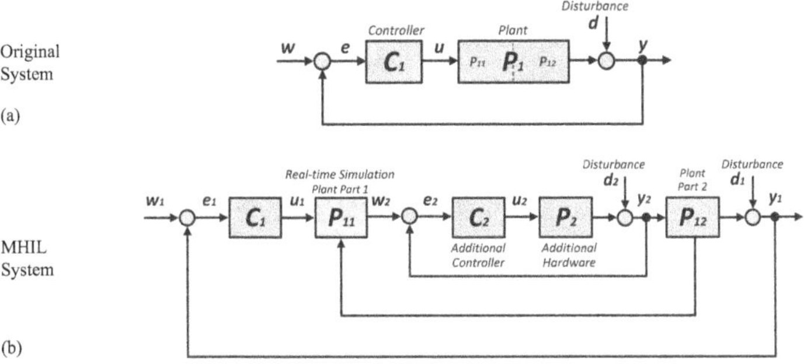


*Figure 6.5 Generic block diagram comparison of a plant and a HIL system*

MHIL setup introduces additional inherent dynamics and parasitic effects that may deteriorate the overall system performance [9].

A further illustration of MHIL systems in this section is provided based on an example from test and validation facilities in the wind energy sector, as illustrated in Figure 6.6. The rising penetration of renewable energy sources, such as wind power, has resulted in a demand for more stringent grid compliance regulations and a rapid surge in the power output of modern wind turbines. Consequently, there is a need for novel testing facilities to ensure efficient and reliable integration into the electrical grid. The measurement and assessment of the electrical properties of wind turbines are conventionally conducted on complete prototypes in the field. Nonetheless, ground test benches offer an alternative to field testing for validating and certifying WT subsystems, such as the nacelle or power train, under adjustable, reproducible, and deterministic test conditions.

To replicate the real-world conditions in the laboratory, high-performance actuators and additional HIL test methods are required. These methods ensure a realistic emulation of the turbulent inflow and missing components, such as rotor blades, pitch system, and the tower. Moreover, the accurate adjustment of electrical properties, such as grid strength, during grid events like voltage sags or frequency changes is crucial.

From this perspective, a DUT is the intersection of mechanical and electrical inputs, and a mutual interaction between rotor-side dynamics and grid-side events is necessary to obtain accurate measurements. To this end, the implementation regarding emulation of inflow and missing mechanical components is referred to as

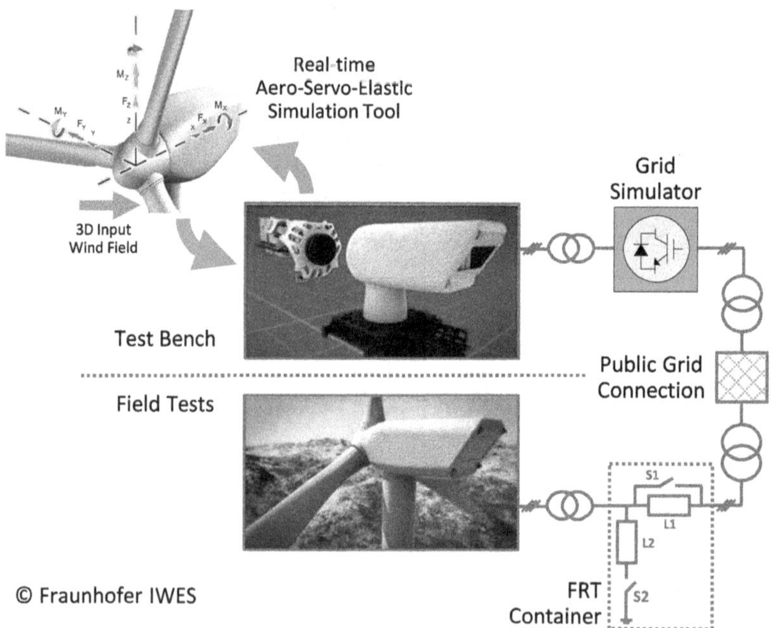


*Figure 6.6 Wind turbine system test benches as an alternative to field tests using UVRT container, for assessment of electrical properties*

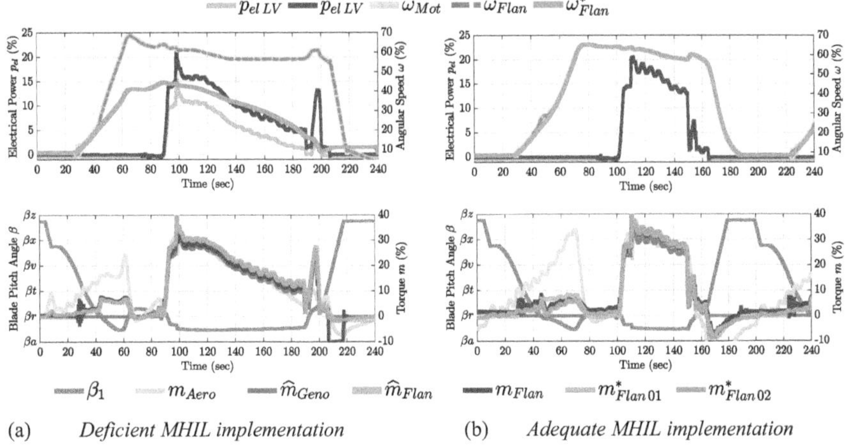


(a) *Deficient MHIL implementation* (b) *Adequate MHIL implementation*

*Figure 6.7 Measurement during system startup for a constant wind input of 5 m/s [11]*

the mechanical-hardware-in-the-loop framework. This includes incorporating a real-time aero-servo-elastic simulation tool coupled with test bench actuators [10]. In addition, a designated control system is necessary to minimize the influence of the test bench's actuators on system response.

To elaborate on the impact of MHIL frameworks, example measurements for a turbine start-up procedure are illustrated above from a nacelle test bench application. Figure 6.7(a) illustrates an unsuccessful start-up, while Figure 6.7(b) demonstrates the corresponding correct operation [11]. In both cases, the DUT initiates the pitch-in at $t = 0$ s by decreasing the pitch angle $\beta 1$ and ramps up the HIL setup to the required speed for power generation. However, for the inefficient MHIL implementation in Figure 6.7(a), the speed of the DUT hardware $\omega_{Mot}$ deviates from the desired reference $\omega^*_{Flan}$ in simulation from $t = 40$ s. Consequently, system operation is terminated at $t = 200$ s by the DUT controller, which monitors feasible operation and commands to pitch-out and shut down the turbine in case of out-ranging deviations. This deviation from the desired behavior is caused due to malfunction of the MHIL implementation, being unable to synchronize operation on the test bench with that from the simulation tool.

## Mechanical hardware-in-the-loop systems

Hardware-in-the-loop systems allow for the interaction between real hardware and a real-time simulation in a multi-physics testing environment. If the simulation represents a mechanical system with corresponding environmental effects, and the interface with the hardware governs mechanical quantities, the setup is referred to as a MHIL system. Among mechanical measures are, angular speed or torsional torque, strain, force, or acceleration in rotational or non-rotational directions. The purpose of MHIL is to emulate additional mechanical effects and hardware, to execute tests under conditions that are as close to real-world conditions as possible, and to enhance the significance of measurement results. These systems are widely used for testing purposes in various sectors, including the automotive and wind energy industries.

In contrast, Figure 6.7(b) presents an efficient performance, where the DUT starts generating from $t$ = 95 s and continues to operate until it is intentionally stopped by the user at $t$ = 150 s. The start-up procedure is then restarted at $t$ = 200 s. This example highlights the critical requirement of an adequate coupling of the simulation tool with the hardware, as the DUT controller incorporates extensive monitoring functionalities that can easily detect any deviation from field conditions.

The implemented MHIL framework for the previously demonstrated measurements is presented in the schematic in Figure 6.8, for Dynamic Nacelle Testing Laboratory (DyNaLab) [12], a ground test bench for full-scale wind turbine nacelles. This framework is designed in consistency with WT control and operation and provides high-fidelity and adequate modularity in the support of type certification of utility-scale WTs according to the standard. This framework consists of three main components, namely, virtual rotor and tower emulation, real-time automation system, and HIL control system.

For the emulation of the missing WT components in the laboratory, a real-time implementation of a fully coupled aero-servo-elastic simulation tool (MoWiT) is utilized [10,13]. This tool is developed by Fraunhofer IWES and provides the corresponding rotor response as set-points for test bench operation, based on the input wind field, pitch angle, and actual measurement feedback from the hardware. Furthermore, it enables normal operation by providing the required rotor side measures received by the DUT controller, such as acceleration or strain measures, as well as anemometer-measured wind speed, among others.

The test bench automation system handles data flow and provides subsystems with the required sensor, actuator, and status signals. It incorporates real-time

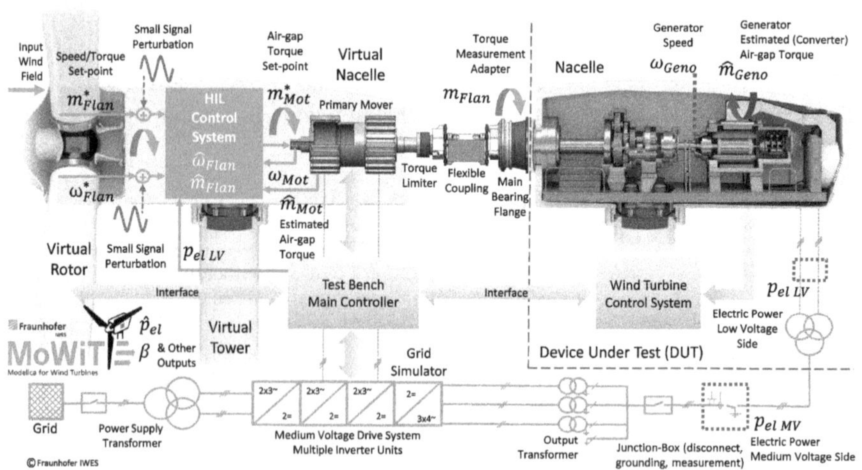


*Figure 6.8 Schematic of the MHIL framework at DyNaLab [13]*

capable field-bus interfaces and therefore enhances system controllability, by realizing short cycle times and low communication jitter.

In this context, a sensor-less test bench control system is implemented, referred to as HIL controller, having the duty to apply and maintain the desired operating point at the DUT main bearing (flange), during static as well as dynamic conditions [9,13]. This includes rotor side dynamics as well as those induced due to grid transient events, both influencing system response simultaneously. This is designed and implemented for the complete drive train present on the test bench, providing a high dynamic performance enabling a realistic mutual interaction of a virtual rotor with the nacelle. Figure 6.8 illustrates the HIL framework and allocation of measurement points and stimulative measures, as well as the notation of variables.

DyNaLab has been developed by Fraunhofer IWES in Bremerhaven, Germany, for testing WTs up to 9 MW, with the main system components illustrated in Figure 6.9. Since the inauguration of DyNaLab in 2015, ten DUTs from on-shore as well as off-shore prototypes in the range of 2.5–9 MW have been subject to extensive experiments at Fraunhofer IWES, as shown in Figure 6.10.

The electrical grid connection at DyNaLab is realized by utilizing a medium voltage grid emulator (MVGE), capable of symmetric and asymmetric faults, as well as overvoltage and frequency changes. Furthermore, different grid conditions are realized by an emulation of grid strength. In this way, the DUT is supplied by the MVGE and decoupled from the public grid, therefore, grid compliance testing can be accomplished independently. The MVGE is based on the medium voltage converter technology, having a short-circuit capacity of 44 MVA.

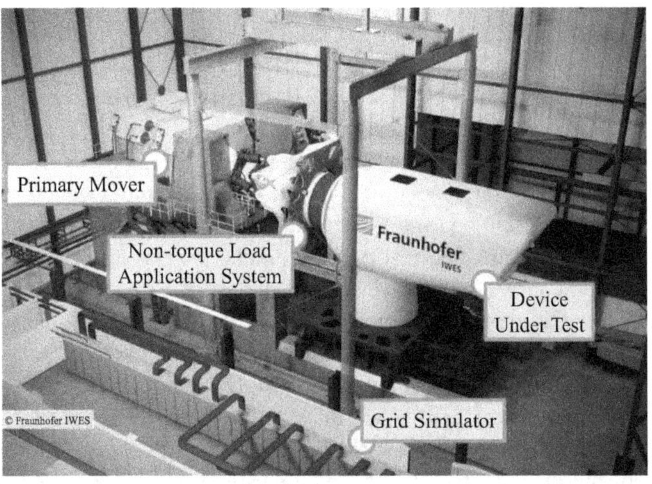


*Figure 6.9 Dynamic Nacelle Testing Laboratory (DyNaLab)*

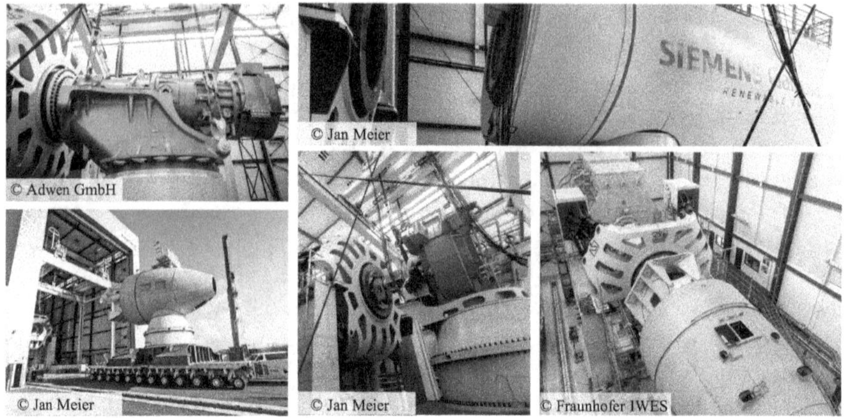


*Figure 6.10 Various wind turbine nacelle prototypes tested at DyNaLab*

The test bench drive train (TBDT) together with the hydraulic load application system (LAS) enables the reproduction of the rotor torque, as well as the corresponding forces and non-torque loads in six degrees of freedom.

The LAS is capable of applying forces up to 2 MN and bending moments of up to 20 MNm, as depicted in Figure 6.11. Furthermore, primary movers of the TBDT are two direct-drive synchronous and externally excited electrical machines in a tandem configuration, which deliver a total nominal power of 10 MW. The unique design of the drive train, as illustrated in Figure 6.11, allows for the application of high-dynamic torque with a nominal value of 8600 kNm. The drive system has an overload capacity of up to 150% applicable over the complete operation range with a nominal speed of 11 rpm. In this context, the overriding drive train controller for HIL test purposes is implemented on ABB's Power Electronic Controller (PEC), providing a real-time platform with fast communication to the drive for set-point commanding and measurement feedbacks.

The direct drive machines together with the dedicatedly designed TBDT support system operation with high dynamics. However, due to the small inertia of the drive train in comparison with that of a WT's rotor, and the inherent dynamic of the underlying drive, the system operation is sensitive to rapid load changes during transient events. This occurs during grid events such as frequency changes or voltage dips. Such test procedures often result in a dramatic change of the active power on the DUT side. Therefore, a beyond state-of-the-art test bench control system is required to utilize the maximum performance of actuators and simultaneously compensate for the inertia and suppress any undesired dynamics from the drive train.

The MHIL framework described here employs a model-based driving torque control. The closed-loop system block diagram is shown in Figure 6.12, with the variables corresponding to the measures presented in Figure 6.8. Reference values are indicated by the superscript "*" while estimated, unmeasured values are designated with the superscript "^". To take advantage of controller design in the

*Figure 6.11 DyNaLab primary movers, drive train, and load application system [9]*

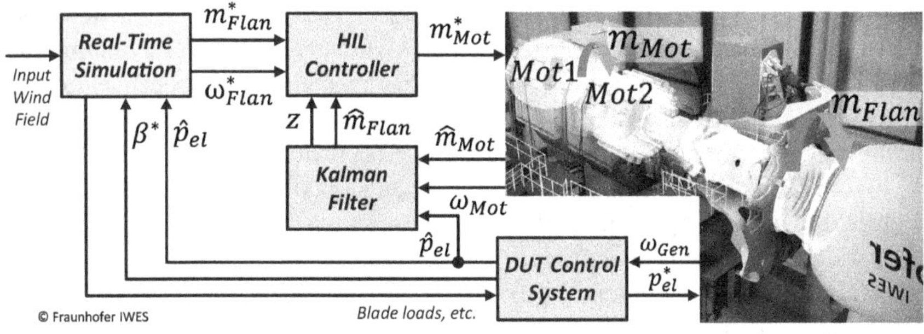


*Figure 6.12 Block diagram for test bench drive train control [13]*

frequency domain, an H-∞ robust controller incorporating the mixed sensitivity approach is implemented [14]. The control problem is reformulated by augmenting the plant with additional sensitivity transfer functions. These weighting functions introduce costs on attributes of the closed-loop system response in the frequency domain. The control synthesis is completed by an optimization process that shapes the sensitivity and complementary sensitivity functions, as well as the actuating variable response. Furthermore, to estimate the unmeasured output and state variables effectively, a discrete time-varying Kalman filter is implemented. This is an optimal estimator that provides robustness in the presence of noise by considering the stochastic properties of noisy measurements and plant model perturbations during the design. The torque measurement in the multi MNm range is limited, and therefore, the Kalman filter enables control with minimal sensors. The MHIL control is designed in discrete time using MATLAB and implemented in the Simulink motor control (SMC) environment of the PEC, and executed with a sample time of 250 μs [9,13].

The utilized control system here is well consistent with the requirements in practice, since the sensor-less approach enables control with minimal measurements, including only the speed and input electrical torque signals. Furthermore, this approach is compatible with standard DUT operation, since it is still the DUT controller that governs system operation by controlling the speed, and the MHIL controller follows and maintains the commanded rotor torque through the TBDT, delivered at the flange. In fact, the flange torque is a control variable and what finally matters is the delivered power to the grid at the point of common coupling (PCC), which must be preserved in accordance with the WT power curve. The presented torque control approach enables a high dynamic performance since no additional control loops incompatible with the normal DUT operation are present.

More details on the control system design and the integration into MHIL simulation framework is presented in the literature for electrical certification purposes [13]. In addition, for performance analysis independent of any specimen, experiments in a back-to-back configuration are executed using the two primary movers at DyNaLab. This is to distinguish the plausible control bandwidth in the presence of dead-times, underlying drive dynamics, and the additional MHIL control loop under realistic load conditions. This is achieved by using small signal perturbations as test functions, applied under dynamic and static load conditions as reported in [9]. Only in this way, the system performance could be evaluated experimentally and independent of any DUT. These experiments evaluate the maximum possible dynamic capability of the system, demonstrating an effective control with a bandwidth of up to 30 Hz [9].

Figure 6.13 shows the results of measurements obtained from a turbulent input wind field, with a comparison of the electrical power measured in the field to the power measured during operation of the system on the test bench. These measurements are intended to illustrate the overall system operation and to show that it is identical to field conditions. The speed ($\omega$) and torque ($m$) signals displayed in the figure are available only from the test bench system and show actual values that accurately follow the setpoints. Typically, such measurements are obtained over a period of 600 s to assess the power quality properties of the system, requiring various measurements at different power bins across the entire operating range. However, a direct comparison between the test bench and field measurements in the time domain is not meaningful in this case. This is because of the limited accuracy of wind measurements, which is the most significant input to the system. Even the validation of complete WT simulation tools faces similar challenges due to the high tolerance of available measurements from field tests and the lack of an experimentally calibrated torque measurement in the multi MNm range.

To address the challenges previously mentioned, a comprehensive verification method is employed, utilizing the transparent MHIL system to provide simulated measures throughout the entire system chain, from input wind to output electrical

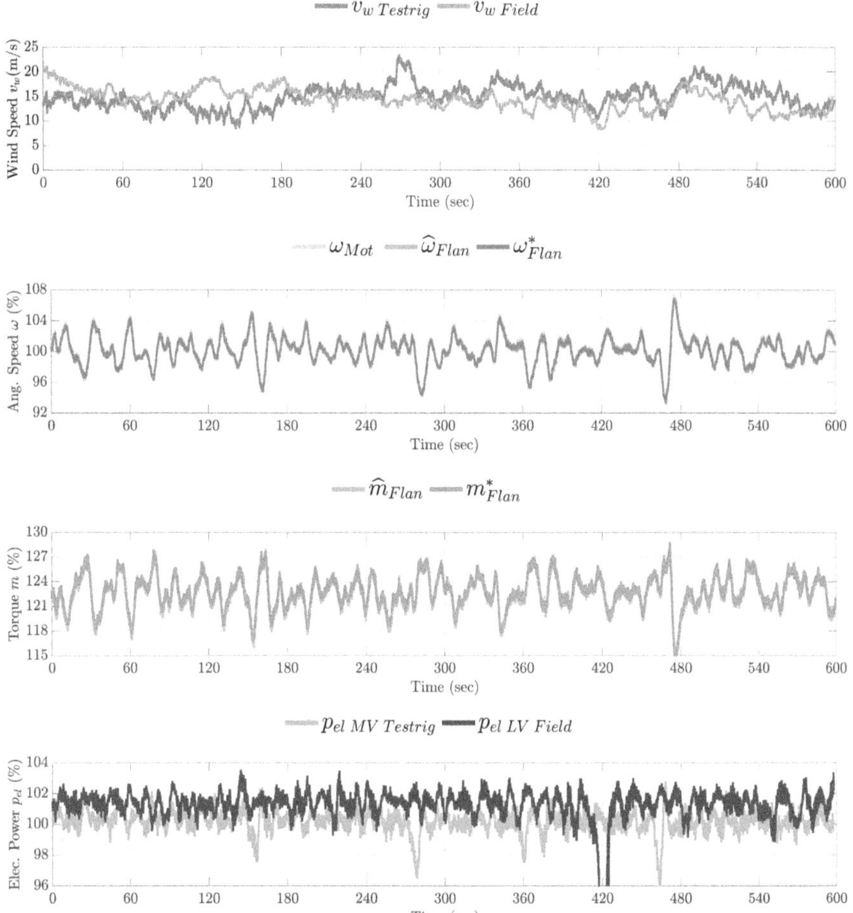


*Figure 6.13 Example measurement under turbulent inflow for the assessment of power quality properties*

power. The evaluation process consists of two levels: first, a feasibility analysis of the measurements taken on the test bench, and second, a comparison with simulation results provided by the manufacturer. The feasibility analysis provides additional value to the evaluation process and is facilitated by the MHIL setup and the installed torque measurement adapter. Figure 6.8 illustrates multiple sources of measurements and real-time simulation for a single quantity, such as speed, torque, or electric power, throughout the complete system chain. This supports the evaluation process, as the feasibility of the measured and simulated quantities can be verified in the first step, and any possible systematic errors in the MHIL setup can be quantified. The dedicated torque measurement adapter at DyNaLab is simulatively calibrated, providing the missing link for a complete evaluation, and therefore plays a significant role in the process. The allocation of measurement points and stimulative measures, as well as the notation of variables used for the evaluation process, are shown in Figure 6.8.

As the first step, the feasibility analysis is accomplished during deterministic load conditions. This is done based on results provided in Figures 6.14 and 6.15, for static and dynamic conditions, respectively. Figure 6.14(a) shows the measured rotation speed $\omega_{Mot}$ and power $p_{elLV}$ for a stepwise input wind, at various constant operating points. The measured values are well consistent with the corresponding reference ("*") and estimated values ("^"), indicating that the outputs of MoWiT are accurately aligned with the measurements, and that the desired operating points are well maintained on the test bench. In Figure 6.14(b), the torque curves allow for

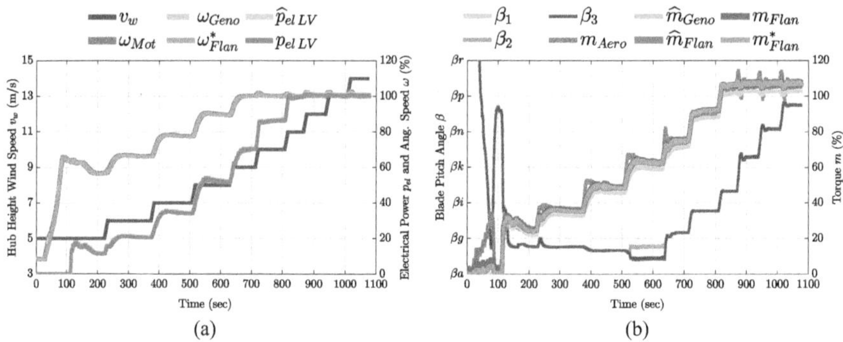


*Figure 6.14 Measurement results with a stepwise constant input wind from 5 to 14 m/s [11]; (a) average wind speed, electrical power, and angular speed and (b) torque and blade pitch angle*

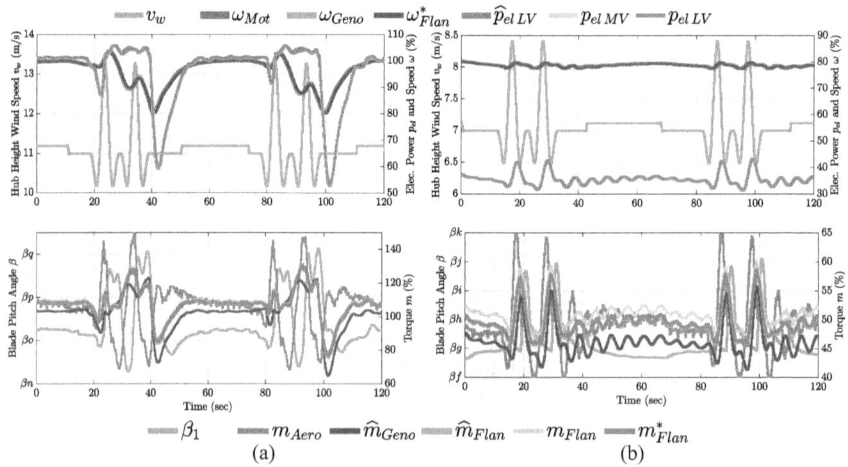


*Figure 6.15 Measurement results for an extreme operating gust (EOG) wind condition [11]; (a) full load conditions and (b) partial load conditions*

a comprehensive comparison of all the values through the chain from aerodynamic to generator torque. It is the aerodynamic torque $m_{\mathrm{Aero}}$ that drives the system as soon as the DUT decides to pitch-in. As shown, the steady state of all torque values corresponding to the flange ($\widehat{m}_{\mathrm{Flan}}$, $m_{\mathrm{Flan}}$) is well consistent with the aerodynamic torque, with the generator torque $\widehat{m}_{\mathrm{Geno}}$ naturally maintaining a slightly lower value due to losses. During this experiment, at steady state, the simulated (set points) and observed (actual values) values are in perfect agreement. The measured torque $m_{\mathrm{Flan}}$ is also consistent with the torque curves, with a maximum deviation of up to 2% observed during partial load. This deviation is due to the sensitivity of the torque measurement, causing a minimal variation at low torque levels [15]. All in all, a consistent operation of the MHIL setup with all underlying components and measurements is hereby demonstrated.

For an evaluation of system performance under dynamic conditions from the rotor side, measurements during extreme operating gust (EOG) wind conditions are obtained, as illustrated in Figure 6.15. The EOG profiles follow the IEC-61400-1 standard and have a turbulence intensity of 0.093 for full load operation at 11 m/s and 0.074 for operation at partial load with 7 m/s average hub-height wind speed. Each EOG wind profile is 10.5 s long and is consecutively applied four times during the experiment. Figure 6.15(a) illustrates system operation at full load, showing an accurate match between estimated ($\widehat{p}_{el\ LV}$)and measured electrical power ($p_{el\ LV}, p_{el\ MV}$) on the low and medium voltage side. Additionally, a comparison between the measured flange torque $m_{\mathrm{Flan}}$ with the simulated and observed values $\widehat{m}_{\mathrm{Flan}}$ demonstrates an accurate match, consistent with the aerodynamic torque $m_{\mathrm{Aero}}$. During this experiment at full load, the DUT exhibits sudden load changes with a gradient of fifty percent in three seconds. Overall, these tests with an input EOG provide a worst-case scenario of rotor-side-induced dynamics and demonstrate stable system operation under extreme conditions.

As the next step for a complete evaluation, the turbine specification (power curve and dynamic properties) and simulation results provided by the manufacturer are taken as the reference, and test bench measurements are finally compared off-line, with simulation results. This comparison is presented in Figure 6.16, where a laminar wind with a ramp from 3 to 17 m/s is used as input. The results demonstrate a high coherence between the simulation and experimental results, with the maximum deviation in speed, power, or torque being less than 2%. The difference observed only occurs partly during partial load, and no deviation is observed during full load. The feasibility study conducted earlier supports this analysis, as it reveals that the difference is not due to abnormal MHIL system operation. The discrepancy is attributed to the previously mentioned accuracy issues, which apply to both the measurement and simulation results. In conclusion, the evaluation of the MHIL setup with regards to mechanical measurements is complete at this stage.

Complete electrical tests according to national and international technical guidelines prescribe execution of various tests, namely, power quality evaluation (e.g., flicker and harmonics), over- and under-voltage-ride-through, reactive power capability, active power control, frequency change, and synthetic inertia. In the following, measurement results from the last three mentioned test cases are selected and described for further illustration. This is due to the observed high dynamic

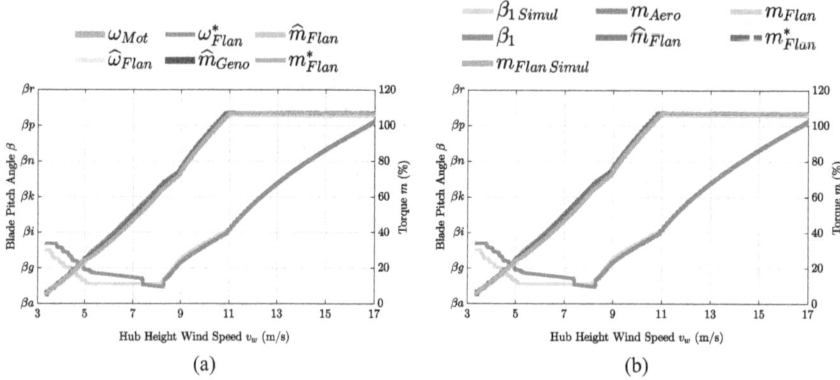

Figure 6.16 Measurements in comparison with simulation results for a ramped laminar wind [13]; (a) speed and power in accordance with wind and (b) torque and pitch angle in accordance with wind

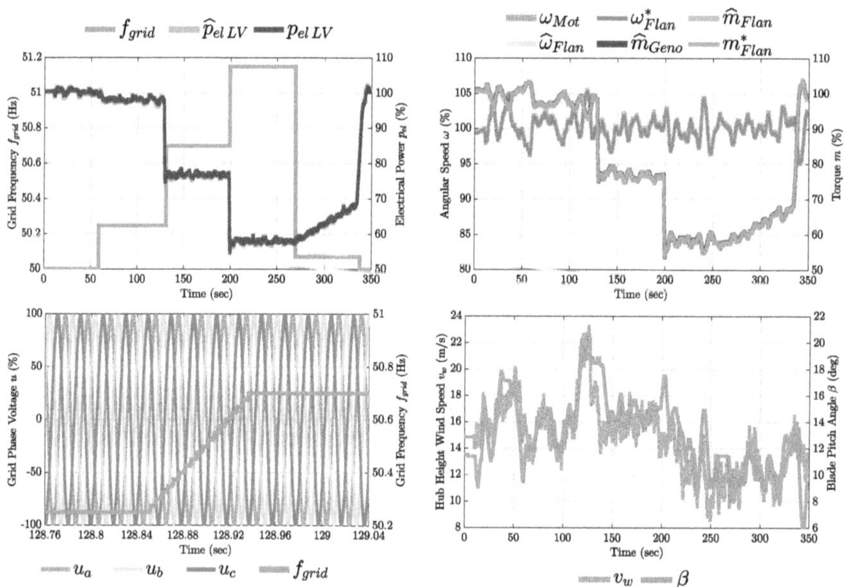

Figure 6.17 Measurements for output power limitation during grid frequency increase [13]

requirement that these test cases introduce for test bench MHIL operation. These tests aim to illustrate the system's response to grid side events acting as a disturbance for system operation and are complementary to the previously presented measurements resulting from rotor side effects.

The measurement results during a grid frequency increase are presented in Figure 6.17. This test evaluates the ability of WTs to provide grid support in response to a change in grid frequency. According to the standard, WTs must respond to frequency changes by adjusting the active power in a proportional manner and within the specified response time. The measurements are obtained using a turbulent wind field with an average wind speed of 14 m/s. The DUT is observed to respond to the frequency $f_{\text{grid}}$ increase by immediately reducing the active power $p_{el\ LV}$ through the electrical drive train. Despite the rapid reduction in active power, the MHIL setup follows and maintains the demanded torque $m^*_{\text{Flan}}$, ensuring that no undesired changes in speed are observed. This test places the highest demands on the operation of the test bench due to the high rate of change in active power.

Similar to the previously described experiment, the test Synthetic Inertia is also executed in order to examine the DUT performance in case of a grid frequency drop. Figure 6.18 illustrates the corresponding measurements during partial as well as nominal load conditions for a constant wind input, having a value of 8 m/s (Figure 6.18(a)) and 15 m/s (Figure 6.18(b)). The course of grid frequency $f_{\text{grid}}$ realized by the MVGE is implied to be identical to the variation observed in the field. During this experiment, the implemented inertia emulation functionality of the DUT temporarily increases the active power $p_{el\ LV}$. Here, the grid frequency is continuously estimated by the controller and the DUT reacts to a frequency drop according to a pre-defined characteristic.

The measurement results presented here for variable grid frequency events demonstrate successful test execution, which was made possible by the high dynamic response of the MHIL setup on the test bench. Once again, this experiment illustrates the efficient performance of the MHIL setup in the laboratory, with its ability to handle the high gradients required for power and torque steps. Furthermore, this test case highlights an additional advantage, since influencing the grid frequency is not feasible in the field. In addition, grid operators evaluate the aggregated response of all WTs on a farm level and not individual reactions. Therefore, the benefit of testing under constant wind conditions on the test bench is significant. It provides an indication of the cumulative response triggered by inertia response over a wide geographical area where local wind turbulence effects are reduced due to the aggregation [13].

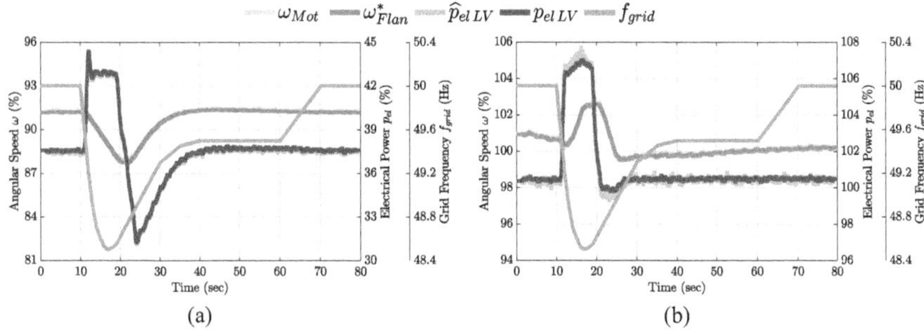


*Figure 6.18 Measurement results for the test Synthetic Inertia [13]; (a) partial load condition and (b) nominal load condition*

Figures 6.19 and 6.20 depict the measurement results obtained during the execution of under-voltage-ride-through (UVRT) and multiple-UVRT (M-UVRT) experiments, which represent a worst-case scenario for grid side events that impose the highest dynamic requirements on both the DUT and the test bench operation. As shown, the sudden drop in the simulated grid voltage causes an immediate

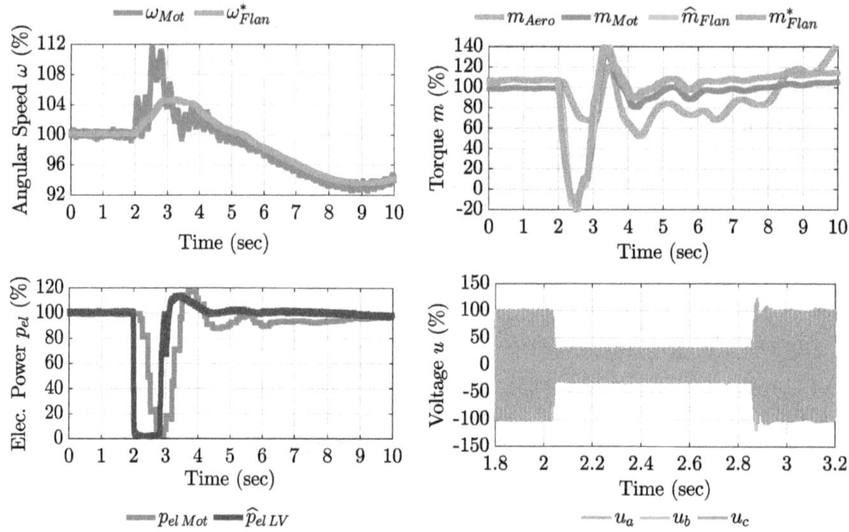


*Figure 6.19 Under-voltage-ride-through (UVRT) test results*

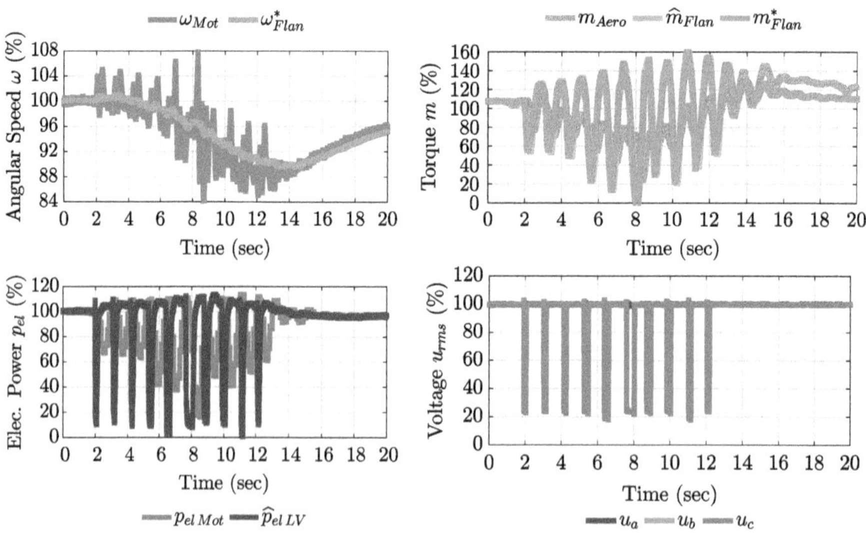


*Figure 6.20 Multiple under-voltage-ride-through (M-UVRT) test results*

reduction in the DUT side torque and power, from 100% to 0% within a few milliseconds, either in a single or repetitive manner. Such dynamic events impose requirements not only during the fault but also for the recovery phase that follows. In addition to withstanding such a fault, any extra vibrations introduced by the test bench must be damped, and the reference torsional behavior commanded by the simulation must be followed accurately.

## 6.3 Power hardware-in-the-loop (PHIL)

This chapter aims to provide a basic overview of the current state of the art in PHIL test facilities for validation, which represent an advanced application of conventional field and laboratory testing based on real-time simulation technologies. As PHIL is a rapidly evolving field of research in the wind sector, the focus is on the basic components of a PHIL setup and fundamental aspects of design. Essentially, the test benches are similar to the PHIL setups of smaller power classes, such as those that have been used since the early 2000s for testing battery systems in automobiles or for designing and testing small-scale photovoltaic systems. Prominent examples of PHIL systems for electrical testing of multi-megawatt wind turbines are the Dynamic Nacelle Testing Laboratory from Fraunhofer IWES (Germany, 10 MW), the wind turbine testing facilities at NREL (USA, 5MW), Clemson University (USA, 15 MW), ORE Catapult (United Kingdom, 15 MW), or the test bench at the Center for Wind Power Drives at RWTH Aachen University (Germany, 4 MW), to name just a few of the most powerful and largest laboratories constructed between 2013 and 2022. Compared to testing units in the free field, some advantages of PHIL laboratories are obvious, such as the good reproducibility of tests and the ability to test independently of external factors such as weather or wind conditions. For this reason, PHIL facilities are gaining increasing acceptance and more and more manufacturers are relying on this type of model validation [16].

### *6.3.1 Basic structure of a PHIL simulation*

In general, a PHIL integrates physical power equipment into a real-time simulation of other electrical components. In the context of wind turbine testing, the simulation can, for instance, be used to simulate a power grid or a virtual generator with the aero-elastic components of a turbine. Figure 6.21 illustrates an example of the structure of a DUT that is directly connected to a grid and the equivalent PHIL setup. To simulate the hypothetically optimal, naturally coupled system (NCS), the DUT is connected to a real-time platform via an interfacing algorithm (IA) including a power interface (PI) as well as a measurement and data acquisition system. The PI typically acts like a current or voltage source and aims to apply the reference obtained in the simulation to the terminals of the hardware. To close the loop, the injected current by the DUT (and/or applied voltage) is fed back to the real-time platform, where a model of the rest of the electrical system is simulated using the measured quantities.

In the following, the individual PHIL components and their key properties are described in more detail. In PHIL applications, the term HUT for hardware under

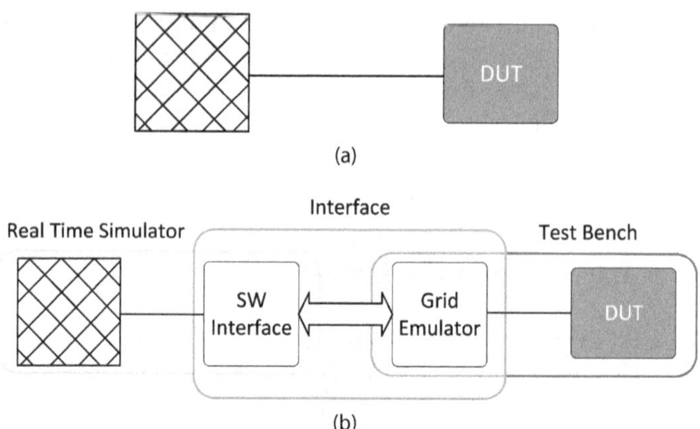


*Figure 6.21 Schematic structure of (a) a naturally coupled system and (b) its PHIL equivalent*

test is usually used instead of DUT, as is common in the wind industry. For reasons of consistency, DUT is used here as an abbreviation.

#### 6.3.1.1 Power interface

Depending on the nominal power and operation principle of the DUT, different PIs can be used to apply the currents or voltages from the real-time model to the terminals of the hardware. Mainly, three types of interfaces can be distinguished [17]:

1. PIs with linear power amplifiers
2. PIs with generator-type power amplifiers
3. PIs with switched-mode power amplifiers

From a system-theoretic point of view, PIs with linear power amplifiers are the most favorable units. Those units operate close at the linear regions of semiconductors, e.g., linear MOSFETs, which is also the origin of the name. Due to their high bandwidth (up to 20 kHz), fast response times, and wide linear operating ranges, linear power amplifiers introduce little distortion from their reference to the output, resulting in the fewest stability issues in closed-loop operation among the three solutions. On the other hand, they are comparatively expensive and often have high energy losses (up to one-third). However, the probably bigger disadvantage in the field of wind applications is that linear amplifiers are strongly limited in their output power rating until now. Thus, they are typically used only in PHIL systems with a power range of up to a few hundred kilowatts and are not used in applications in the megawatt range.

In contrast, PIs with generator-type power amplifiers are more scalable in their output power. These units essentially consist of regulated, three-phase synchronous generators that are driven by an ac or dc motor with or without a separate excitation system. Due to their structure, however, PIs with generator-type power amplifiers are limited in their application and a relatively large number of components plus a lot of space are required for this setup.

Much more scalable and flexible, on the other hand, are PIs with switched-mode power amplifiers. They can be used both in the low power range as well as in the multi-megawatt range and are therefore predestined for the construction of test facilities for wind turbines. Typical switched-mode amplifier topologies consist of front-end rectifier units (12-pulse active rectifiers), a dc link, and controllable back-end power converters. To achieve high efficiency (typically above 90%) and good quality of the PI output voltage, modern back-end power converters use cascaded or interleaved multi-level converter topologies that are interconnected via transformers or other passive output filters such as RC, LC, or LCL filters. Figure 6.22 shows the relevant components on the example of the DyNaLab MVGE at Fraunhofer IWES. Such a topology usually results in a total harmonic distortion (THD) of less than 5% for a 50 Hz fundamental reference signal and bandwidths in the range of 2–10 kHz. Instead of simple three-phase converter

*Figure 6.22 Medium voltage grid emulator (MVGE) of the DyNaLab including converter units, transformer, and ohmic-capacitive filter bank*

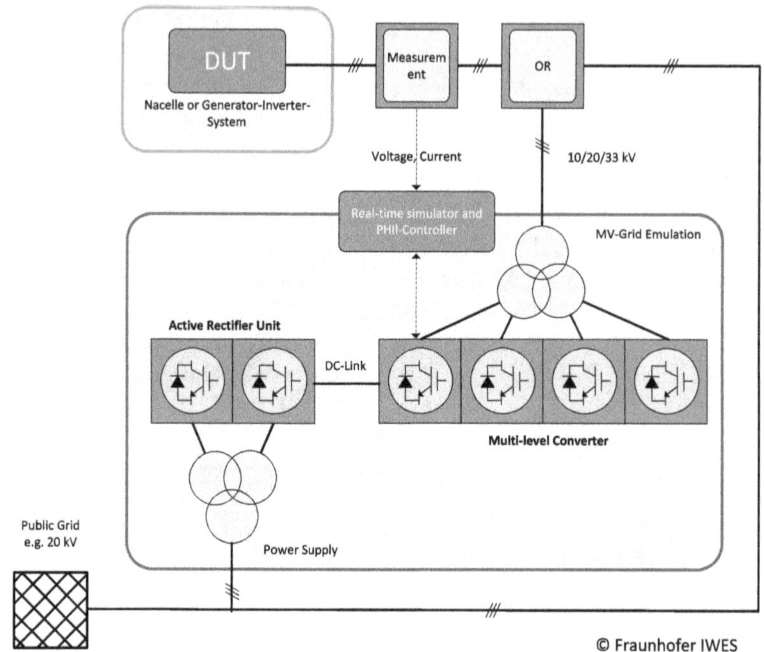


*Figure 6.23 Structure of a typical power interface for testing multi-megawatt wind turbines*

transformers, it is advantageous to install three single-phase step-up transformers in high-power and high-voltage applications. This design is particularly characterized by a low physical output impedance and allows easy switching between different nominal voltage levels, such as 10, 20, or 33 kV, and independent control of the three output phases. Figure 6.23 illustrates the resulting structure of a typical PI setup with switched-mode power amplifiers for testing large wind turbines. However, due to the saturation effects of the utilized transformers, one disadvantage of the PI is its highly non-linear behavior in different operating points. This must be considered in the design and operation. Moreover, conventional converter systems are in many cases limited in their control options. For example, it is often only possible to supply RMS phasor quantities and not to command instantaneous values. Advantages, on the other hand, are their large operating ranges and their capabilities to perform voltage amplitude changes or phase jumps with high gradients [18].

As it will be discussed in Section 6.3.2, the dynamical characteristics of the used PI predominately influence the closed-loop stability and achievable accuracy of the PHIL simulation. Besides the limited bandwidth, additional time delays or hardware filters at the output of the power interface lead to a non-ideal transfer behavior, introducing set-point deviations and an unavoidable phase lag. For the PHIL design, it is therefore important to have an a priori estimate of the PI's behavior in different operating points.

#### 6.3.1.2 Real-time simulators

One of the main advantages of PHIL testing is the freedom in modeling the grid side of the experiment. It is typically implemented purely on a real-time simulator (RTS) platform and therefore only limited by the simulation environment available on the simulator and its computing power. Compared to offline simulations, such as those performed with MATLAB, PSCAD, or PowerFactory from DIgSILENT, RTS systems have to be real-time capable, i.e., finish every simulation step within a predefined, mostly fixed time. Development of RTS is therefore pushing for smaller time steps to enable transient studies while still maintaining the possibility to model complex networks. Currently, step sizes in the range of hundreds of nanoseconds to several microseconds are common. This is realized by a platform consisting of one or more FPGAs for replicating linear circuit elements and CPUs with multiple parallel cores for performing slower, nonlinear operations. Network size and complexity of simulated devices can be increased by upgrading the respective hardware or combining multiple real-time simulators via high-speed digital interfaces. In addition, modern RTS systems also have several analog and digital interfaces that allow coupling with other periphery or measurement equipment and the power interface [19].

#### 6.3.1.3 PHIL interfacing algorithms

Depending on the type of the power amplifier, which either acts like a current or voltage source, the interface algorithm used to connect the hardware to the real-time simulator can be divided into current and voltage-type interfaces. Whereas DUTs with a grid-following control act as a current source and can thus be connected to a RTS via a voltage-type interface, DUTs that implement a grid-forming control can be connected via a current-type interface. Over the years, many different approaches were developed how to implement the respective software interface, differing in achievable accuracy and stability margins of the closed-loop system [20]. In research as well as in industrial practice, the ideal transformer model or method (ITM), the partial circuit duplication (PCD), and the damping impedance method (DIM) are the most frequently used interfacing algorithms.

*Ideal transformer model*

Probably the most popular method of coupling the real hardware with the model on the real-time simulator is to use a software interface that is based on the functioning of an ideal transformer. Considering a specific transformation ratio, an ideal transformer simply consists of an ideal current-controlled current source on the one side and an ideal voltage-controlled voltage source on the other side. Depending on the measured variable, i.e., current or voltage, this model represents exactly the desired function of an interface for PHIL applications. Figure 6.24 shows the basic single-phase circuit diagram of the voltage-type ITM.

When it comes to the implementation of the ITM on the real-time computer, it is important to note that it is usually not possible to simulate an inductance in series with an ideal current source or capacitance in parallel to an ideal voltage source. In the case of a voltage-type interface, for example, this is because the current source aims to force a current and the current through an inductor in turn defines its state,

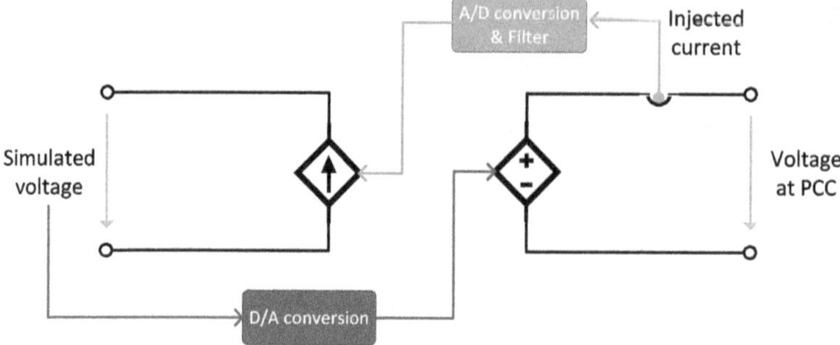


*Figure 6.24 Ideal transformer model interfacing algorithm*

which leads to a contradiction. To overcome this topological conflict and remove the resulting state degeneration, manufacturers of real-time simulators often recommended to implement snubber circuits, e.g., in the form of pure resistors or ohmic-inductive or ohmic-capacitive circuits in parallel to the simulated current source or in series to the simulated voltage source.

Two advantages of the ITM interface algorithm that make it so popular in academia and practice are its simplicity and accuracy. However, if practically unavoidable delays are taken into account, such as those caused by the acquisition of measurements, the analog-to-digital conversion or the effective delay of the PI, the ITM often suffers from stability problems. For this reason, it is most the time not possible to directly implement the ITM as shown in Figure 6.24 in practice. A straightforward approach for improving the stability properties of the method is to use additional (low-pass) filters, suppressing high-frequency components in the feedback signal. The additional filters can be realized either in the form of additional hardware or purely by software. However, the achievable accuracy then depends strongly on the selected cut-off frequency of the filters, which must be compared with the required stability margins. This extension of the ITM is sometimes also referred to as the Filtered Ideal Transformer Method (FTIM) or, if a voltage-type interface is used, the Feedback Current Filtering (FCF) method. Up to now, many test facilities for emulating frequency or phase angle changes and FRT events indirectly use the latter method as PHIL interface. In those tests that are performed to verify grid compliance, the focus is mainly on the evaluation of the wind turbine behavior at the fundamental frequency. For this reason, resonant filters with a low-pass characteristic in the high-frequency range are often used to detect the amplitude and phase angle of the measured currents and voltages at 50 Hz or 60 Hz before they are forwarded to the connected real-time system. The required PI reference voltage or current can then be calculated in a simplified way using complex phasors and applying Kirchhoff's circuit laws [18]. Although the resonant filters used are primarily for a different purpose, the interfacing algorithm exactly shows the structure of the FTIM or FCF method.

### *Partial circuit duplication*

If the hardware side contains a linking impedance or admittance, which is known in advance, the ITM can be modified by mirroring the component on the software side

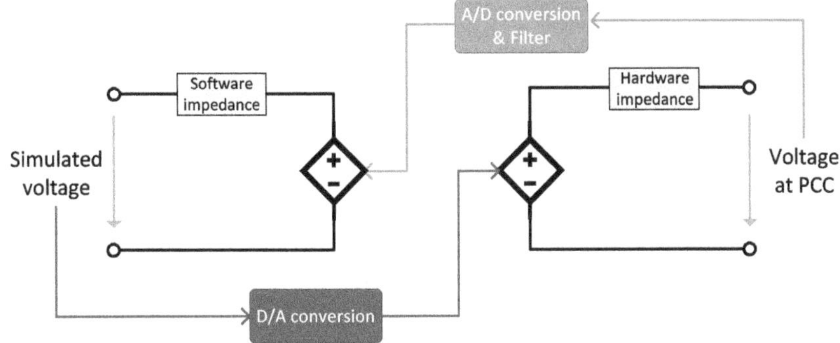


*Figure 6.25 Partial circuit duplication interfacing method*

and changing the virtual sources as depicted in Figure 6.25. Due to the virtually duplicated structure, this adaption is called Partial Circuit Duplication (PCD) method. The larger the coupling impedance, the better the stability limits of the setup, but the lower the simulation accuracy.

Considering the discussed structure of the switched-mode PI, most PHIL test benches designed for high power and output voltages have one or more transformers at the output of the converter units. With in-depth knowledge of the characteristics of these transformers, a virtual model can be provided on the real-time simulator when using the PCD method. On the one hand, this has the advantage that, as already explained, the stability limits of the system are improved, but also non-linear saturation effects can partly be compensated, and the accuracy of the PHIL simulation is increased.

*Advanced IA methods*

Combining the ITM and PCD method and additionally introducing a damping impedance, leads to the Damping Impedance Model (DIM), see e.g., [16,20]. This setup results in an absolutely stable system, if the virtual damping impedance is equal to the real hardware impedance of the DUT. In practical applications, however, the exact value or structure of the effective hardware impedance is usually not known during validation tests of wind turbines or cannot be determined beforehand. For this reason, complete decoupling is impossible most of the time, but estimating the impedance can nevertheless help to improve the stability limits and accuracy of the PHIL system.

In addition, more extensions and improvements of the ITM model or DIM can be found in the literature. For example, the Transmission Line Model (TLM), the Time-variant First-order Approximation (TFA), or the so-called Taganrog Algorithm. For further details, reference is made here to specific publications on the subject, see e.g., [20].

## 6.3.2 Stability and accuracy of PHIL-systems

Besides the usually not exactly known hardware impedance, one of the main difficulties in designing a PHIL system with one of the described interfaces is the non-ideal behavior of the PI. As described, linear amplifiers can be used for HIL setups with rather small powers, whereas switching power converters are very scalable in their rated power and output voltage and they are therefore predominantly used in test facilities for multi-megawatt wind turbines. The converter transformer units used only have linear behavior in certain working ranges and are usually controlled by means of pulse width modulation, space vector modulation methods or the use of advanced schemes, such as optimal pulse patterns. In combination with the time required for data acquisition and processing, an unavoidable time delay is introduced, causing a phase lag which is mainly responsible for stability problems that occur. To prevent and counteract possible problems during the design phase, it is advisable to analyze the PHIL system in advance. This is a challenging task, as typical PHIL systems are non-linear, hybrid systems consisting both of non-linear physical system parts, such as converter or amplifier units and hardware filters, as well as linear and non-linear system parts on the digital real-time computer. Even if non-linear effects are neglected, such systems cannot simply be described as linear, time-invariant systems from a control engineering point of view. Instead, they must be modeled by means of parametric transfer functions. But a detailed description of PHIL systems with this theory would go too far here and is usually not necessary for the initial design. Due to additional uncertainties, like the lack of knowledge about the exact DUT characteristics and because of the variety of possible PHIL topologies and interface algorithms, a simplified linear minimal setup is examined in more detail in the following as an example. In this case, well-known tools from linear control theory, such as frequency analyses and design methods or the root locus method, can be applied directly for PHIL design. Figure 6.26 shows the block diagram of the selected PHIL setup in the Laplace domain.

In the block diagram, $G_{PI}(s)$and $G_{flt}(s)$ represent the transfer functions of the PI and optional current feedback filter, respectively, whereas $Z_S(s)$ is the effective impedance of the Norton equivalent circuit simulated on the real-time system, and $Y_H(s)$ is the admittance of the DUT. With regard to the discussed power interfaces

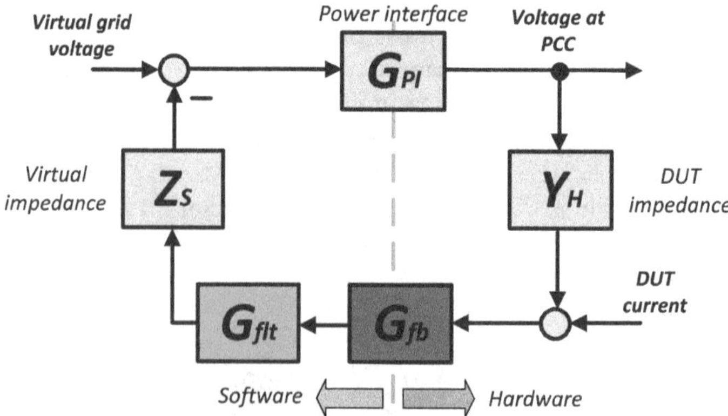


*Figure 6.26 Closed-loop block diagram of a voltage-type PHIL setup*

for multi-megawatt PHIL setups, $G_{PI}(s)$ is typically decomposed of a pure time delay of the form $e^{-sT_d}$ and some kind of low-pass filter. As described in the classic literature on modeling and control of power converters, the latter part of the PI might be approximated by a zero-order hold element or some advanced pulse-width modulation (PWM) model when ideal switches are assumed [21]. In addition, $G_{fb}(s)$ summarizes all other delays in the feedback path, e.g., introduced by the analog-to-digital conversion or communication. Then, the open-loop transfer function from the block diagram of Figure 6.26 is given by

$$G_{ol}(s) = G_{PI}(s)G_{flt}(s)G_{fb}(s)Z_S(s)Y_H(s). \tag{6.1}$$

Assuming that $Z_S(s) = L_S s + R_S$ and that the DUT effectively has a passive ohmic-inductive impedance in the high-frequency range, i.e., $Z_H(s) = 1/Y_H(s) = L_H s + R_H$, the term $Z_S(s)Y_H(s)$ basically represents a lead-lag filter. This filter provides static amplification of $R_H/R_S$ and introduces a phase lead if $L_S/R_S > L_H/R_H$ and a phase lag if $L_S/R_S < L_H/R_H$. A simplification often made at this point is to neglect all low-pass filtering elements and to summarize the time delays. This allows for a particularly simple analysis, which at the same time reveals some essential characteristics of common PHIL setups. For instance, using Nyquist plots for evaluation, three important conclusions can be derived:

1. For $R_S \leq R_H$ and $L_S \leq L_H$, the Nyquist plot starts at the point $(R_S/R_H + j0)$ and ends on a circle with radius $L_S/L_H$. According to the simplified Nyquist stability criterion, this implies that stability is always guaranteed.
2. For $L_S > L_H$, the Nyquist plot always ends on a circle which encloses the critical point $(-1 + j0)$, always resulting in an unstable closed-loop system.
3. For $R_S > R_H$ and $L_S < L_H$, the Nyquist plot starts outside the unit circle and ends on a circle with radius $L_S/L_H < 1$. In this case, stability can be ensured as long as the first intersection of the Nyquist plot with the negative real axis is to the right of the critical point $(-1 + j0)$. On the other hand, if the fist intersection is outside the unit circle, the closed-loop system is always unstable.

It becomes clear that especially in the second and third cases, suitable filters or modifications of the interface algorithm must be applied to set up the PHIL system. However, this always leads to a deterioration in simulation accuracy and a compromise must be found between the desired bandwidth and stability margins. The exact design depends strongly on the time delays of the system and the respective software and hardware circuit. In addition, the non-ideal behavior of the power interface influences the loop and, apart from attenuation, usually also causes a further phase decrease. As this generally also leads to a reduction in simulation accuracy, special filters should be used to counteract the discussed effects, such as those employed in the control of modern power converters.

In the design phase, the absolute or relative error of the closed-loop transfer functions of the NCS and the non-ideal PHIL setup is usually used to evaluate the accuracy of a system. This measure gives a clear indication of the accuracy (in steady-state operation) as a function of frequency. But such measures have the major disadvantage that the exact setup of the test facility and the scenarios to be simulated as well as the dynamic properties of the DUT must be known in advance, e.g., by prior tests. During operation, on the other hand, the errors in the currents and voltages on the hard- and software side of the implemented interface can be accessed. This does not require any prior knowledge, and a wide variety of accuracy measures, both for transient and steady-state operation can be derived on this basis [22,23]. For example, if the absolute error of a quantity before and after the interface is calculated for a certain frequency of interest and the result is related to the corresponding value on the software side, a simple measure of how well the oscillation is represented by the IA is obtained.

Apart from PHIL closed-loop stability and accuracy, the numerical stability of the integration algorithm or the discretization method on the RTS platform is also decisive. It is a well-known fact that stable systems may exhibit unstable behavior when simulated on a digital computer, if an inadequate simulation step size or integration algorithm is chosen. As far as numerical stability is concerned, there is again a compromise to be found, namely between simulation accuracy and computational burden. Frequently used fixed-step numerical integration algorithms are of the class of linear multistep (LMS) methods or of the class of Runge-Kutta (RK) methods and can further be divided into explicit and implicit methods. Explicit methods calculate the future state of a system only from the previous states, whereas implicit methods find a solution by solving an equation involving both the previous states as well as the future state itself. Explicit methods include, e.g., the forward Euler (ode1) method and the standard Runge–Kutta schemes such as RK4 (ode4) or the fifth-order method from Dormand–Prince pair (ode45 or Dopri). Contrary, methods from the backward differentiation (BDF) type, the Tustin method, or RK methods like the TR-BDRF2 (ode23tb or Radau) are common implicit integration algorithms. In general, numerical issues can mostly be avoided if low-order implicit solvers are used for simulating stiff systems. On the other hand, explicit methods are usually less computationally intensive and have advantages when hard real-time requirements must be met or when other physical domains or control schemes are to be integrated into the simulation.

Although the analysis of modern nonlinear solvers is not that straightforward and requires, e.g., the use of contractivity theory, the numerical stability properties can be estimated in a simplified way by evaluating the closed-loop poles in the $\lambda{\cdot}h$-plane, where $h$ is the step-size of the chosen integration algorithm. The simulation on a digital computer is absolutely numerically stable if the associated poles lie in the stability regions of the solver. More information on this topic can be found, for example, in [24].

### 6.3.3 *Practical aspects of PHIL testing*

The stability of the PHIL setup is a basic pre-requisite for the safe operation of any test facility for the validation of wind turbines. As described, the properties can often be analyzed and evaluated in a simplified way, and the impact of unmodeled time delays or varying software and hardware impedances can be assessed at the initial stage.

While communication delays or measurement inaccuracies of the sensors play a minor role for test facilities in the megawatt range, in practical operation there usually arise certain other hardware limitations that should be taken into account [18]. This particularly includes the controllability and accuracy of the power interfaces with switched-mode power amplifiers. Although it would theoretically be possible to operate the PI's converter units in a closed-control loop, which means that the output voltage or current is actively regulated in a feedback-loop, this structure is rarely used in practice. This is on the one hand due to the limited possibilities of the converter interfaces used, to a possible reduction of bandwidth, but also because of the integral states in the control units, making the PIs sensitive to oscillations that are more difficult to control. For this reason, many current PIs are only actuated as voltage sources without feedback when it comes to dynamic simulation of network events such as sudden voltage dips or phase angle jumps. But, since there is no comparison of reference and actual values at the terminals of the DUT, this approach reacts sensitively to output filters or transformers for increasing the applied voltage. The current flowing through these components causes a voltage drop, which can influence the simulation result. It is therefore advisable to consider the input impedance or admittance of the PI in the interface algorithm and compensate for any non-ideal behavior as far as possible. This also includes, for example, non-linear saturation effects of transformers, which mainly result from abrupt changes in the commanded voltage. Since the magnetic flux cannot change instantaneously, an additional anti-saturation control must usually be implemented for stable operation of the converter-transformer units. However, as the flux control also influences the wind turbine's power transformer, the measurements may differ in transient behavior compared to measuring devices in the free field. Therefore, the anti-saturation control must be properly adapted and transparently reported during test execution, so that the effects can also be presented in the best possible way in the model validation process. In [18], exemplary results of a measurement campaign with a DFIG wind turbine at Fraunhofer IWES are shown, in which FRT tests in the free field were compared with those of a test rig for minimal system tests. As can be verified, quasi-equivalent results can be achieved if appropriate controller settings are chosen and the above-mentioned aspects are taken into account when testing on PHIL test benches.

## 6.4 Controller hardware-in-the-loop (CHIL)

The control system of a wind turbine takes care of the overall system and turbine operation. It implements the control, monitoring and communication to all sensors, actuators, devices, and subsystems in the turbine. Thereby it takes care of overall decisions, detects failures, ensures a safe operation, and dynamically controls the turbine to enable a performant and optimized energy production.

While the control systems today are computer-based intelligent systems, their implementation exists in a wide variety of technologies and topologies: All sensors and actors can be connected to a single central controller by means of a fieldbus or by several distributed controllers that communicating to each other. More complex subsystems are often equipped with their own control system, for example, the pitch control system, and converter control system, which are actuated by the higher-level main control system. Each control system consists of a hardware platform with computer(s), input/output (I/O) modules as well as fieldbus communication and a software solution consisting of firmware, real-time environment, and the application software. Nowadays, industrial personal computers (industrial PCs) are often used, on which an operating system (OS) with a runtime environment and hardware interfaces is prepared for the implementation and execution of the application software. Further details on the control system in hardware and software are described in Chapter 7.

As the wind turbines get larger and more complex, so must the control system. It must interface more and more different kinds of sensors, enable smart operating strategies, integrate, and interact with complex subsystems. Even the logging and analysis of high-frequent signals to perform condition monitoring and predictive maintenance tasks, as well as the processing of artificial neural networks for anomaly detection or signal predictions are functionalities of a modern control system. While the turbines are often in volatile and remote environments, like the offshore turbines, it is essential that the control system can handle unexpected and undesirable situations. To ensure its reliable operation and increase its quality, it is therefore reasonable to subject the control system to intensive tests and assessments.

It is often sufficient to use models and simulations in pure virtual software environments for the testing of basic control algorithms for instance. But due to the real-time environment used in the field, delays and dead times occur during communication with the subsystems, as well as through subordinate cyclic processing and internal communication within the subsystems for instance. Such effects are usually not considered in the models and have a major impact on dynamic control algorithms. Among other things, therefore the utilization of a hardware-based test environment can become necessary.

The Guideline for the Certification of Wind Turbines [25] even requires in Chapter 4.5 the verification of load-relevant control and safety system functions by two alternatives. The first alternative requires to build up a model of the controller and test it against a simulation of the turbine to verify the load assumptions, while the second alternative describes the verification by functional testing. It describes testing of the control software with a high coverage of the system by real hardware components in a hardware-in-the-loop (HIL) environment.

### 6.4.1 Test concepts

There are many opportunities to validate, prove, and test the implementation of the control system. Due to the high proportion of effort in the application software, the use of software tests is common. During the implementation of the software application, modern integrated development environments (IDE) offer the implementation of software-based tests. In doing so, the methods from test-driven development are used for the application of unit tests, integration tests, and system tests. Individual components of the software application are validated by running them with specific inputs and parameters to check and validate expected outputs. Further, the combinations of several software units can be implemented and approved in so-called integration tests. These software tests have the advantage of fast execution and easy implementation during the development phase, without the requirement of a dedicated hardware environment. Although not yet common in industrial automation, concepts such as Continuous Integration (CI) with automated test procedures can be used in the development and maintenance of the application software to continuously increase and permanently guarantee the quality of software application.

A standard practice in the wind industry is to perform load simulations, where a model of the dynamic controller is evaluated against the structural loads of the mechanical components. In the so-called design load cases (DLC) (cf. Chapter 1.1 in [26]) the standard IEC 61400-1 [27] specifies such simulations, which are required for type certification and described in detail in the first chapter “Design loads of wind turbines” of Volume 1. This application of model-in-the-loop (MIL) is usually done on any computer and without the utilization of the hardware or platform as later used for the control system in the field. The VDI/VDE 3693 [28] describes MIL as the first step of three, to establish a complete virtual commissioning of a control system. As the second step software-in-the-loop (SIL) is outlined. The model of the controller will be converted into a productive state as it will be used on the control system. To test this controller together with the turbine model, an emulation of the control platform is used, but without the control system hardware or any real-time claim.

The third step is to test the control system in combination of the software with the hardware. This is where the implementations of hardware-in-the-loop (HIL) are applied. It aims to include the timing behavior and to test the overall system in a real-time environment. In that case the control system will be interfaced by a simulation of the turbine, whose will also be executed in real-time. A specialization of HIL would be processor-in-the-loop (PIL), where the execution of the software on the processor is tested to evaluate performance, timing, and numerical results.

Since there are no hard definitions and delimitations of HIL, as well as the wide variety of targets while testing of the different platforms and topologies, several kinds of setups are possible for its implementation. For this purpose, individual I/O modules, the entire I/O system with the fieldbus connection, the fieldbus interfaces to subsystems, up to tests of complete control systems and control cabinets can be realized. In the following sections, these different possibilities for implementing a controller hardware-in-the-loop (CHIL) setup are presented. With these various concepts of testing, it is always necessary to weigh up in advance what are the targets of the tests, which areas of the application, which parts of the hardware should be included in the testing and which setup is most sufficient for the purpose.

### *6.4.2 Software simulation*

Due to the increasing share of model-based developments for the control but also simulation of wind turbines, especially in the research and prototyping phase, the resulting models can be transferred and integrated into the control software as well [29]. By checking the controller against a model in the simulation environment beforehand, kind of software-based tests are also carried out. This is commonly used for dynamic and closed-loop control algorithms that will be transferred and reused in the real-time environment of the control system. But this procedure is useful as well, to transfer the physical models of components to the real-time environment of a simulation system for later testing purposes. The existing models can simply be reused and the controller that has already been evaluated in the design calculations can be integrated into the real application. Otherwise, the controllers and simulation models must be implemented once again on the specific platform with the potential of mistakes and thus the need for re-evaluation.

One approach is to use two computers and a communication interface between them. The first computer is of the same type as used for the control system in the turbine and executes the application software in real time. A second computer is used to execute a simulation of the turbine. As illustrated in Figure 6.27, this computer communicates with the control system software by means of for example an ethernet interface and protocol.

It is important to consider how the interface is implemented in the control system and on which level the data is exchanged with the simulation. Individual functions or areas of the application could be communicated with, tested, and validated as total or in isolation. The communication could take place at a level in the software, where the interface to inputs and outputs (I/O), to the hardware and fieldbus is implemented. Thus, checking the entire processing chain of the application is possible by emulating corresponding hardware signals and applying them to the software signals. On the other hand, the communication can also take place directly on the level of already scaled and evaluated physical signals. In that case, the signals to the hardware do not need to be emulated, but these parts of the

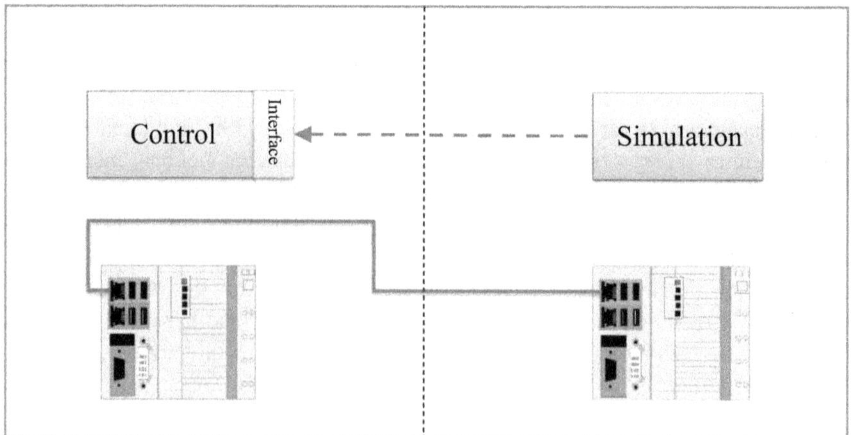


*Figure 6.27 Concept of a setup to connect a control system with a simulation system via an ethernet-based software interface*

processing are then also not tested. Additionally, a distinction could be made whether all I/Os are completely emulated and communicated or only a subset of them. Accordingly, different degrees of integration are achieved, also referred to as code coverage and usually expressed as a percentage value.

Beyond this, the implementation of communication and simulation must be considered. Are both real-time capable and thus correspond to the real conditions of the controller, or do some compromises have to be made here. Depending on the depth of the test required and the goal of the evaluation, different platforms and technologies need to be taken into account. Even if both technologies are implemented in real time, the synchronization between control and simulation still must be examined. For example, whether it is guaranteed that with each execution cycle of the control also the necessary execution of the simulation, as well as the communication via the interface, can be ensured.

An application of such a software interface is the combination of the TwinCAT 3 Runtime (XAR) with the Interface for Simulink (TE1410). TwinCAT 3 is a product from the company Beckhoff Automation and offers an automation software platform for PC-based Control with engineering tools, e.g., for the programing of PLC applications in IEC 61131-3, as well as a real-time capable runtime to execute the application on an Industrial-PC. The Interface for Simulink offers the exchange of data between Simulink and the TwinCAT runtime, while Simulink is distributed by the MathWorks company and offers a software environment for the model-based development and execution of physical simulations.

The Interface for Simulink as shown in Figure 6.28 includes a special functionality for the synchronization of the simulation with the runtime. The simulation waits for an event from the controller that it has completed the execution of a control cycle, then the simulation executes one cycle, exchanges the calculated values with the controller, and then waits for the next cycle. However, since

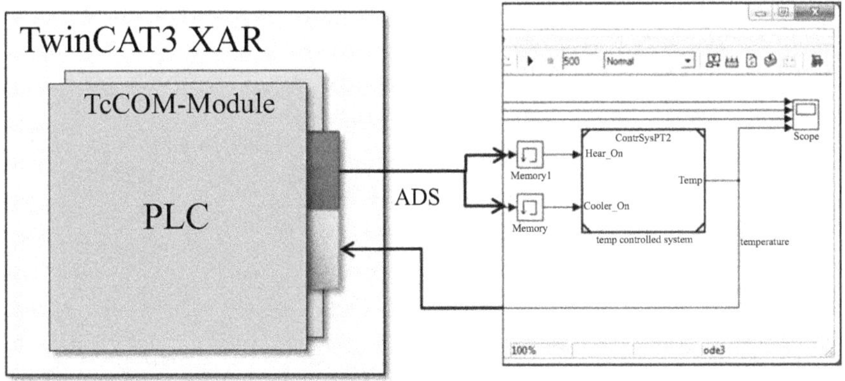


*Figure 6.28 Copyright © Beckhoff Automation GmbH & Co. KG. The interface for Simulink (right) allows a synchronized execution of the simulation and communication with the TwinCAT 3 Runtime (left).*

Simulink is not executed in real time, a powerful computer must be used to ensure that the simulation is executed sufficiently quickly so that no event or processing cycle of the controller is missed.

But with the TwinCAT 3 Target for Simulink (TE1400), another product of Beckhoff, it is possible to make use of the models, convert them into a TwinCAT module and execute them in real-time in the TwinCAT runtime. Thus, an Industrial-PC can be used as a simulation system to run the model and exchange data with the controller in real time.

Thus, it depends on the performance of the hardware and runtime of the simulation system, as well as the complexity of the model, whether continuous compliance with the cycle times is possible. If necessary, more performant hardware can be used or the degrees of freedom in the simulation models can be reduced to establish the real-time capability. The degrees of freedom used for the analysis of the structural loads are usually not necessary for the use of the models for testing purposes and thus can be reduced.

### 6.4.3 Hardwired simulation

If it is possible to run the simulation in real time, the communication via the ethernet interface can be implemented in real-time as well or replaced by an I/O interface. In this way, data can not only be exchanged with the software application but signals for the I/O modules of the control system can be generated and connected. A schema of such a setup is shown in Figure 6.29, with two controllers and their I/O modules, hardwired to each other. For each input to the controller, an output signal is generated from the simulation, and correspondingly for each output from the controller, this is used as an input for the simulation. This allows to test the whole processing chain of the control system, including the fieldbus, I/O modules

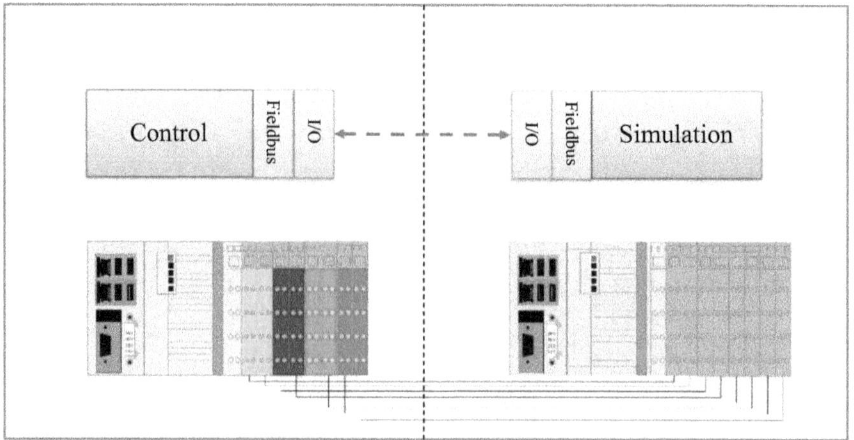


*Figure 6.29 Schema of a simulation system (right) whose uses I/O modules as hardwired interface to the I/O of the control system (left)*

as well as internal configurations and mappings, with the software and hardware used in the field. Depending on the test objectives, only some of the I/O signals, up to the complete control system with all its hardware interfaces can be built up.

Figure 6.30 shows such a full setup of a hardwired hardware-in-the loop installation. For every signal of the control system, a kind of mirrored signal exists in the simulation system and is wired to each other [30]. Installations like this, in the complete complexity of the control system, are leading to new challenges. For many signal types, from analog interfaces, pulse generators or speed signals to fieldbus interfaces, counterparts are offered in the form of I/O modules. But for some signals of sensors like resistance thermometers or strain gauge sensors usually no counterparts are available on the market.

Another often neglected but very elementary part of the control system for a wind turbine is the safety system. The safety system is required by guidelines for each machine with the need for redundancy and diversity to protect the machine as well as persons from any harm. If the safety system can be set up completely separately from the control system and uses only the common digital signals, a simple counterpart can be set up to check the situations and logics of the safety chain. For modern integrated safety systems, as shown in Figure 6.31, this may have to be considered when designing a counterpart and its simulation [31]. Due to the safety requirements, such safe I/O modules implement protective measures that

*Figure 6.30 The complete installation of a control system (left), as well as a complete counter station which emulates all signals of a wind turbine by means of a simulation (right). Copyright © Beckhoff Automation GmbH & Co. KG.*

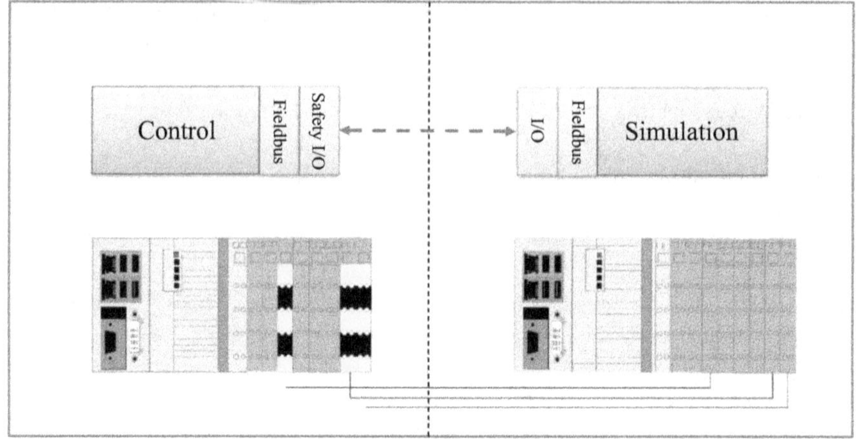


*Figure 6.31 A safety system implemented by safety I/O modules (yellow) as integrated part of the control system (left)*

require further arrangements for its signal connection. For example, the safety outputs might test the connected contactor with current pulses, which are not transmitted in the case of a representation and might therefore be detected as faulty.

In this setup, where every signal is hardwired and connected between the control system and the simulation system, all these signals need entirely to be simulated. Unfortunately, the usual simulations do not take care of the many tiny details that such a control system implements. The scaling of all analog signals on the I/O modules, the feedback from protection devices, contactors, and fuses in the control cabinet or the level of coolant in the cooling system are just a few of such details. The complete implementation of all these signals is a lot of effort and therefore it must be weighed up in advance what the goal of this test setup should be and whether the commonly purchased components really must be built up in hardware.

### 6.4.4 Fieldbus simulation

Any efforts and issues as described above for a simulation system with I/O modules built and wired in hardware can be avoided by emulating the fieldbus interface. Instead of connecting the real I/O to the fieldbus interface of the controller, a virtual fieldbus is used. Figure 6.32 demonstrates such a structure.

Depending on the fieldbus used, its physics and protocol, a fieldbus simulation can be implemented in special hardware or on an industrial computer. With an ethernet-based fieldbus such as EtherCAT or PROFINET, the network adapter of an Industrial-PC can be used. This adapter could be utilized by software which receives and evaluates the fieldbus telegrams. Corresponding this software generates the feedback and sends it back to the controller as specified by the fieldbus protocol. No real I/O exists, but the controller behaves like it is installed in a real environment and communicates with hardware modules. In this way, such a

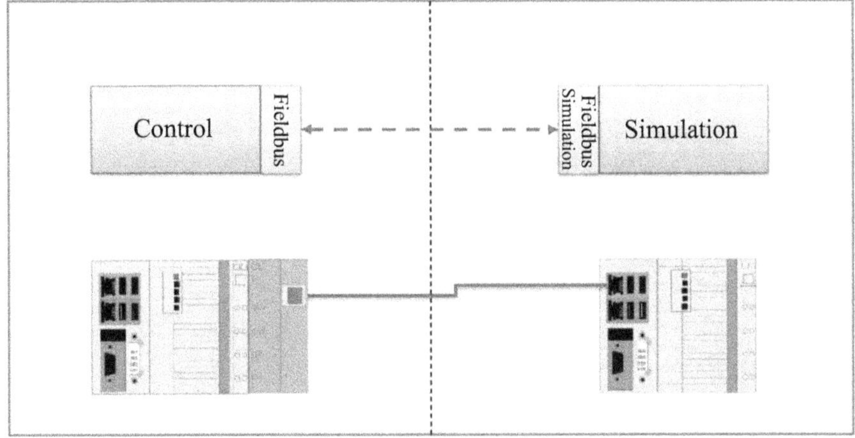


*Figure 6.32 Concept of a control system (left) whose implements a fieldbus interface and communicates with a simulated counterpart (right) of that interface*

fieldbus simulation allows to save the effort of procurement and installation of I/O modules as well as connecting all the wires. It also allows the easy implementation of the simulation of complex signals, up to safety signals and their logics, as well as any internal diagnostic information of the components.

A possible implementation is offered by the Real-Time EtherCAT Network Simulation of the company Acontis [32] or by the TwinCAT 3 EtherCAT Simulation from Beckhoff. It aims to implement a platform for virtual machine commissioning of any machine that is based on the EtherCAT fieldbus. The network adapter of an Industrial-PC can simulate the EtherCAT Slaves as I/O modules but also more complex interfaces to drives or subsystems. By executing the EtherCAT Simulation in the TwinCAT 3 runtime and with the combination of the already mentioned integration of simulation models into the real-time, an efficient hardware-in-the-loop environment can be set up.

## 6.4.5 Combined simulation

The depth of integration can be increased at will by combining these described technologies. This would allow to mix the use of real I/O modules with the emulation of other I/O modules via the fieldbus simulation. Further on, it would be possible as shown in Figure 6.33 to integrate real subsystems. That enables for example to integrate a real pitch system into the simulation environment of a wind turbine, combined with the control system in hardware. Even if such a subsystem is connected by an additional fieldbus interface or requires further hardwired I/O signals, this is feasible.

A major advantage of this structure is its consistency and proximity to real applications in the field. The controller uses the same configuration, interfaces, and

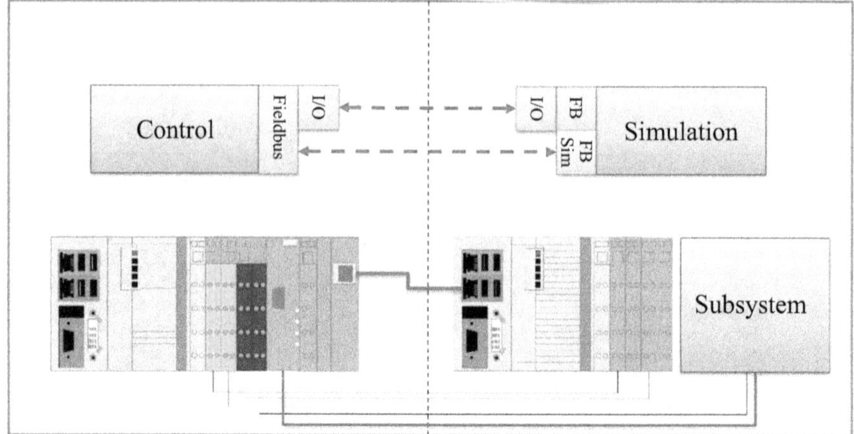


*Figure 6.33 The control system (left) interfaces real I/O modules to hardwired signals from a simulation (right), a fieldbus interface (FB) and further signals to a real subsystem, as well as virtual I/O modules by use of a fieldbus simulation (FB Sim)*

protocols as in the real application in a wind turbine. The goal should be that the controller could be disassembled from the simulation environment, installed into a turbine, started up and it will run without the need of any change. Then almost completely the same conditions are established in the test environment as in the field, apart from any environmental conditions such as the ambient air, temperature, or vibrations. However, this presupposes that all signals are simulated as they would occur in reality and that the simulation has been sufficiently validated, possibly compared with reality.

### 6.4.6 Farm simulation

Another use case of a hardware-in-the-loop setup for a controller is to evaluate and test a wind farm controller. The dynamic control of the power fed into the grid of a whole wind farm is highly dependent on influences like the delay and dead times in the communication between the wind farm controller and the wind turbine controllers. Therefore, the approvement of a stable closed-loop control is an essential part of the design of offshore wind farms and the wind farm controller. But also faults such as voltage dips in the power grid, as well as the circuit breaker positions for controlling current flows via cables and transformers, can be simulated and thus tested exactly when corresponding models are integrated.

Similar concepts as described for a wind turbine can be reused for a whole wind farm. A model of a wind turbine could be instantiated multiple times and processed in real-time to simulate a whole wind farm for instance. The emulation of the wind turbine controller communication interface to the wind farm controller is then implemented by means of utilizing this model. In addition, general external influences, and the influences of the turbines to each other must be modeled and their outcome must be exchanged in real-time between all the models. For an off-shore wind farm, these influences are the waves of the water, the wind itself, but also wake effects among the turbines. Depending on the detail level of the turbine model, wind and water models, an appropriate performant computing hardware must be chosen. It may be possible to execute several models on a single computer by utilizing performant processors, as well as their multiple cores for a parallel execution. That would simplify the implementation of communication among the models and reduce the need for interfaces and protocols to communicate across the boundaries of individual computers.

In Figure 6.34 such hardware setup for a wind farm simulation is shown. In total 10 computers are used, where eight of them simulate ten turbines each, so 80 turbines in total. One computer runs the wind and wake model and another one the electrical grid simulation. Together with the wind farm controller, the systems communicate with each other via multiple network interfaces to have the appropriate bandwidths available for data exchange in real time. In this case, the models were converted into TwinCAT modules, executed in the TwinCAT runtime, and using the EtherCAT Automation Protocol for communication [33].

The structure of the simulation environment, the models and their distribution on the computers is shown schematically in Figure 6.35. The single turbine was implemented using three physical models for mechanic, generator, and converter.

*Figure 6.34 Two racks equipped with industrial PCs and IT infrastructure for the simulation of a whole offshore wind farm. Copyright © Beckhoff Automation GmbH & Co. KG.*

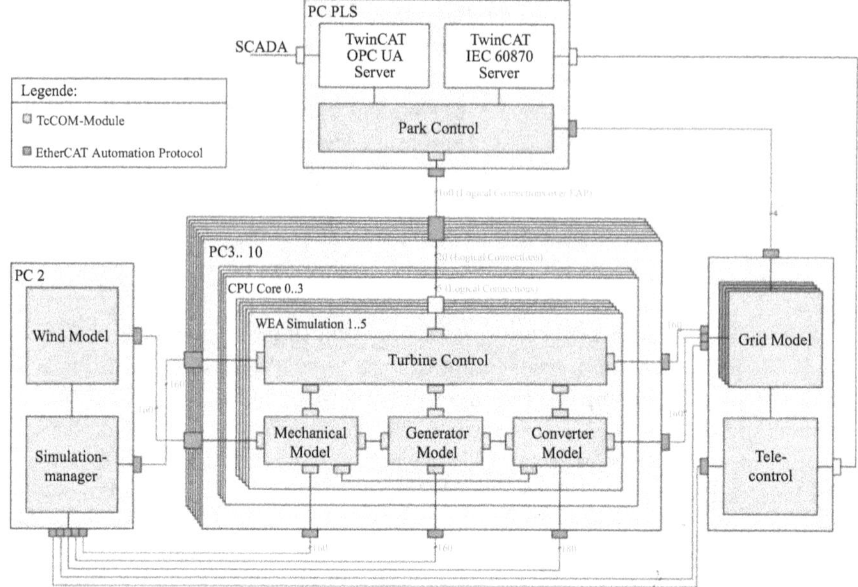


*Figure 6.35 The schematic of TwinCAT modules running on distributed PCs and communicating over the EtherCAT Automation Protocol to implement a real-time wind farm simulation. Copyright © Beckhoff Automation GmbH & Co. KG.*

A simplified representation of the control software was combined with these models. Thus, four modules were used for each turbine, five turbines per processor core and four cores on the processing unit of each computer. With this setup, it is possible to simulate up to 160 turbines. A total of 640 models would then be realized by TwinCAT modules, which would be executed and communicating in real-time above the 10 Industrial-PCs.

## 6.5 Outlook and future developments

This chapter aims to provide insights into the ongoing developments and future challenges in HIL testing for wind turbines, building upon the previous discussions on HIL testing concepts such as MHIL, PHIL, and CHIL. It begins with an overview of the potential opportunities for enhancing testing methods and subsequently delves into the testing of progressively larger wind turbines.

The field of HIL testing is constantly evolving, and [34] highlights several noteworthy developments. An important aspect of testing involves employing accurate real-time models. Currently, reduced-order models are utilized, varying in their level of detail based on the specific application. Research efforts are underway to automate the generation of reduced-order models from complex models, enabling high-frequency real-time simulations. As these methods advance, it becomes possible to integrate more intricate models into testing scenarios and to perform real-time simulations that address coupled multi-physics phenomena.

Another objective noted in [34] is to continually expand the bandwidth of HIL testing to capture and evaluate harmonics more effectively, particularly in PHIL testing. However, the existing constraints on controller stability hinder achieving higher bandwidth in this regard. Additionally, MHIL testing often faces limitations due to actuator constraints. Consequently, both actuators and control methods can be further enhanced to achieve better performance.

A promising method to enhance the accuracy and speed of simulation models as well as the ability to expand the bandwidth of HIL testing could be the application of Artificial Intelligence (AI) algorithms such as Artificial Neural Networks (ANN). While the utilization of ANNs in wind turbine testing is yet to be explored, it is worth examining previous and ongoing development in AI-supported HIL testing across other industry sectors. Some relevant studies shall be briefly introduced here. Turkson *et al.* [35] reviews the application of ANNs for the calibration of automotive engines and highlights the study [36]. This study, already conducted in 2005, utilized ANNs to semi-automatically reduce the model order of an engine, thereby accelerating simulations for HIL and SIL applications. Another noteworthy, more recent example is presented in [37], where an ANN is employed to model the nonlinearities of a permanent magnet synchronous motor (PMSM). This enables real-time execution of the motor model on a field-programmable gate array (FPGA) and is tested in a SIL setup, where a controller running on the same FPGA controls the PMSM.

The application of ANNs in wind turbine HIL testing holds similar potential and can be utilized for learning and approximating complex physical models. The learning process of ANNs aligns with the concept of generating reduced-order models discussed in [34] and can be integrated into ongoing research on automated model generation. Leveraging the computational power of graphics processing units (GPUs) or tensor processing units (TPUs), the training process of ANNs can be accelerated. Once trained, ANNs can be efficiently calculated on standard central processing units (CPUs). Additionally, PLCs can be employed for ANN calculations, enabling faster real-time simulations that can be synchronized with other simulation models or controllers within a HIL system.

The integration of AI in wind turbine testing offers the potential to increase the speed and accuracy of simulations, facilitating more comprehensive and realistic test scenarios. Moreover, the integration of AI-based methods opens up possibilities for optimizing test bench operation, controller performance and exploring advanced control strategies. Reviews of the application of AI in the field of wind energy are given in [38,39]. Both highlight the application of AI in wind turbine monitoring systems which could be transferred to support ongoing HIL tests. Furthermore, the automation of control systems as well as the optimization of controllers are underlined. Overall, there are numerous opportunities to harness the benefits of AI-based approaches in the context of wind turbine HIL testing. However, further research in this regard is necessary.

The following paragraphs of this chapter focus on another important challenge of wind turbine HIL testing. As wind energy converters continue to grow in size, the demand for updated nacelle test benches and advanced methodologies for HIL testing arises. Existing test benches face limitations in delivering the required high dynamic loads for testing larger wind turbines. Consequently, there is a need for larger motors, hydraulic activators, and stronger structures to fulfill testing requirements. However, the construction of larger nacelle test benches presents foreseeable challenges. Not only the cost for testing will increase with larger test benches, but also the execution of tests will become more difficult. The application of accurate dynamic loads and torques in the MNm range must be achieved while maintaining control stability for both the integrated mechanical system as well as the power control system [40]. Furthermore, the rapid development of larger wind turbines often renders current test benches inadequate within the short span of a few years. Therefore, it is of interest to explore alternatives to continuously constructing larger nacelle test benches.

There are already several potential solutions to these challenges. A notable trend is the move towards separate testing of smaller components. Initiatives like *HiL-Grid-CoP* [18,41] focus on the testing of the generator and power converter only, allowing for specific analysis, evaluation and validation of these components. Similarly, projects such as *PQ4Wind* [42] concentrate solely on testing the power converter, especially regarding its impact on grid stability. These smaller test benches enable faster development cycles and the opportunity to enable wind turbines to stabilize the grid by delivering power system inertia.

To address the challenges of testing and certifying fully integrated nacelles, the *VirtGondel* project [40,43,44] proposes improvements to existing infrastructure rather than the construction of new larger test benches. Drawing insights from the automotive, railroad, and aerospace sectors, which have dealt with similar testing challenges, can provide valuable solutions for nacelle testing. For instance, the automotive industry has emphasized the importance of laboratory tests for safety-critical issues while also advocating for the use of numerical and laboratory rig simulations [45]. These simulations allow for the optimization of test rig designs and better preparation of mandatory physical tests. Another consideration is the decreasing time available for development, as highlighted in [46], which suggests utilizing computer-aided engineering (CAE) tools to simulate some of the tests, thus reducing the number of physical tests.

In the context of wind turbine nacelle testing, the *VirtGondel* project integrates these concepts and tailors them to suit the specific needs of wind turbines. It introduces advanced methods and enables more efficient test campaigns through the virtual representation of a nacelle test bench. Essential aspects of HIL testing are the validation of the dynamic load application system and the accurate determination of the dynamic loads to be applied to the DUT. However, accurately measuring these large loads has been challenging. To address this issue, the project focuses on developing a validated virtual model of the DyNaLab test bench. This virtual model can improve the calculation of dynamic loads and, similar to the benefits highlighted in [45], support the planning of physical tests.

Moreover, there is the possibility of utilizing the simulation model to develop novel test methods. One of these methods introduced in [43], is called hybrid testing and combines physical and virtual testing to enhance the capabilities of an existing test bench. State-of-the-art simulation tools are used to enable the testing of wind turbine drivetrains with nominal power or load levels exceeding those of the test bench. The developed method suggests conducting the majority of tests below the maximum load of the test bench. Simulation methods are then employed to calculate the required results for test cases exceeding the test bench's limits. This combination of physical tests under partial load with simulation-based virtual tests for larger loads is called hybrid testing. The simulation model can be validated with partial load tests, although there are still limitations to be quantified, such as the modeling of non-linearities and complex measurement variables. Additionally, an important question considered in the project is to analyze how far the test bench capabilities can be extended with hybrid testing. Part of these questions, actual tests of the concept and potential solutions to improve the methodology are already addressed in [43].

Furthermore, efforts are being made to enhance test methodologies for complete wind turbines. One example is the project *Mobil-Grid-CoP* [47], which involves the utilization of containers equipped with PHIL setups. These containers can directly be delivered to a test turbine and enable the emulation of the grid on-site using a power converter system separating the public grid from the wind turbine. This way, different grid states can be emulated for the test turbine, e.g., the behavior in a low impedance high-voltage direct current (HVDC) grid can be tested. Overall, this approach significantly expands testing capabilities compared to standard fault ride through (FRT) containers. However, it does not benefit from the advantages of nacelle or component test benches, which offer reproducibility of test cases (i.e., wind speeds and load conditions).

Overall, the evolution of wind energy technology necessitates advances in test methods and test rigs. The increasing size of wind energy converters calls for enhanced facilities capable of delivering and handling high dynamic loads. As an alternative to building larger and larger test facilities, the move towards separate component testing and use of advanced testing methodologies taking advantage of physical and virtual testing have been highlighted. In addition, the development of mobile PHIL test rigs contributes to improved testing capabilities for wind turbines as a whole. Together, these advances play a crucial role in ensuring the reliability, efficiency, and performance of wind energy systems and offer the opportunity of faster development cycles.

## References


[1] Gräßler, I., Bruckmann, T., Dattner, M., *et al.* "VDI/VDE 2206: Development of Mechatronic and cyber-Physical Systems," VDI, 2021-11.

[2] Kyling, H., Wegner, A., Behnke, K., Rosemeier, M., and Antoniou, A., "Validation, Verification, and Full-Scale Testing," Chapter 9 in Wenske, J. (Ed.), *"Wind Turbine System Design, Vol. 1: Nacelles, Drivetrain, and*

*Verification,"* IET Energy Engineering Series 142A, The Institution of Engineering and Technology, London, United Kingdom, 2022.

[3] Joshi, A., "Automotive Applications of Hardware-in-the-Loop (HIL) simulation," SAE International, 2020.

[4] Reyes, R., "White Paper: Virtual Hardware 'in-the-Loop': Earlier Testing for Automotive Applications," *Synopsys, Inc.*, 2013.

[5] Tulpule, P., Rezaeian, A., Karumanchi, A., and Midlam-Mohler, S., "Model Based Design (MBD) and Hardware in the Loop (HIL) validation: Curriculum development," *2017 American Control Conference (ACC), Seattle, WA, USA*, 2017, pp. 5361–5366, https://doi.org/10.23919/ACC.2017.7963788

[6] Osório, C. R. D., Genic, A., and Costa, S., "Introduction to Typhoon HIL: Technology, Functionalities, and Applications," Chapter 1 in Tripathi, S. M., Gonzalez-Longatt, F. (eds.), *"Real-Time Simulation and Hardware-in-the-Loop Testing Using Typhoon HIL,"* Springer Nature Singapore Pte Ltd., 2023.

[7] Thomas, P., Gu, X., Samlaus, R., Hilmann, C., and Wihlfahrt, U., "The OneWind$^{TM}$ Modelica Library for Wind Turbine Simulation with Flexible Structure – Modal Reduction Method in Modelica," *The 10th International Modelica Conference*, Lund, Sweden, 2014, http://dx.doi.org/10.3384/ecp14096939.

[8] Huhn, M. L., and Popko, W., "Best Practice for Verification of Wind Turbine Numerical Models," *Journal of Physics: Conf. Ser.*, Vol. 1618, p. 052026, 2020, dx.doi.org/10.1088/1742-6596/1618/5/052026.

[9] Neshati, M., and Wenske, J., "Dynamometer Test Rig Drive Train Control with a High Dynamic Performance: Measurements and Experiences," *IFAC-PapersOnLine*, Vol. 53, Issue 2, pp. 12663–12668, 2020, https://doi.org/10.1016/j.ifacol.2020.12.1846.

[10] Feja, P. R., and Huhn, M. L., "Real Time Simulation of Wind Turbines for HiL Testing with MoWiT," *Wind Energy Science Conference 2019 (WESC 2019), Cork, Ireland, Zenodo*, https://doi.org/10.5281/zenodo.3518727.

[11] Neshati, M., Zhang, H., Thomas, P., Heller, M., Zuga, A., and Wenske, J., "Evaluation of a Hardware-in-the-loop Test Setup Using Mechanical Measurements with a DFIG Wind Turbine Nacelle," *Journal of Physics: Conference Series* 2265, 2022, http://dx.doi.org/10.1088/1742-6596/2265/2/022105.

[12] "Dynamic Nacelle Testing Laboratory (DyNaLab) – Date and Facts," Fraunhofer Institute for Wind Energy Systems IWES, Bremerhaven, August 2022, https://www.iwes.fraunhofer.de/content/dam/windenergie/en/documents/IWES_Datenblatt_DyNaLab_en-RZ_web.pdf

[13] Neshati, M., Feja, P. R., Zuga, A., Roettgers, H.T., Mendonça, Â., and Wenske, J., "Hardware-in-the-Loop Testing of Wind Turbine Nacelles for Electrical Certification on a Dynamometer Test Rig," *Journal of Physics: Conference Series* 1618, http://dx.doi.org/10.1088/1742-6596/1618/3/032042.

[14] Neshati, M., Curioni, G., Karimi, H.R., and Wenske, J, "H∞ Drive Train Control for Hardware-in-the-Loop Simulation with a Scaled Dynamometer Test Bench," *43rd Annual Conference of the IEEE Industrial Electronics Society, IECON,* 2017, https://doi.org/10.1109/IECON.2017.8217507.

[15] Zhang, H., Ortiz de Luna, R., Pilas, M., and Wenske, J., "A Study of Mechanical Torque Measurement on the Wind Turbine Drive Train—Ways and Feasibilities," *Wind Energy*, Vol. 21, Issue 12, pp. 1406–1422, 2018, https://doi.org/10.1002/we.2263.

[16] Jersch, T., Mehler, C., and Neshati, M., "Overview of Actual Development and Discussions of Electrical Certification of Wind Turbines on Test Benches," *16th International Wind Integration Workshop, Berlin, Germany*, October 2017.

[17] Lauss, G. F., Faruque, M. O., Schoder, K., Dufour, C., Viehweider, A., and Langston, J., "Characteristics and Design of Power Hardware-in-the-Loop Simulations for Electrical Power Systems," *IEEE Transactions on Industrial Electronics*, Vol. 63, No. 1, pp. 406–417, Jan. 2016, https://doi.org/10.1109/TIE.2015.2464308.

[18] Hans, F., Curioni, G., Jersch, T., *et al.*, "Towards Full Electrical Certification of Wind Turbines on Test Benches – Experiences Gained from the HiL-GridCoP Project," *21st Wind & Solar Integration Workshop*, October 2022, pp. 122–129, doi: 10.1049/icp.2022.2746.

[19] Omar Faruque, M. D., Strasser, T., Lauss, G., *et al.*, "Real-Time Simulation Technologies for Power Systems Design, Testing, and Analysis," *IEEE Power and Energy Technology Systems Journal* 2.2, pp. 63–73, 2015.

[20] Ren, W., Steurer, M., and Baldwin, T.L., "Improve the Stability and the Accuracy of Power Hardware-in-the-Loop Simulation by Selecting Appropriate Interface Algorithms," *IEEE Transactions on Industry Applications* 44.4, pp. 1286–1294, 2008.

[21] Hans, F., Oeltze, M., and Schumacher, W., "A Modified ZOH Model for Representing the Small-Signal PWM Behavior in Digital DC-AC Converter Systems," *IECON 2019-45th Annual Conference of the IEEE Industrial Electronics Society*, Lisbon, Portugal, October 2019, doi: 10.1109/IECON.2019.8927479.

[22] Hans, F., Haag, F., and Quistorf, G., "Vector-Based Accuracy Measures in Power-Hardware-in-the-Loop Simulations," *8th IEEE Workshop on the Electronic Grid*, Karlsruhe, Germany, October 2023.

[23] Lundstrom, B., and Salapaka, M. V., "Optimal Power Hardware-in-the-Loop Interfacing: Applying Modern Control for Design and Verification of High-Accuracy Interfaces," *IEEE Trans. Ind. Electron.*, Vol. 68, Issue 11, pp. 10388–10399, 2021, doi: 10.1109/TIE.2020.3032918.

[24] Cellier, F. E., and Kofman E., *Continuous System Simulation*. New York, NY: Springer, 2006, https://doi.org/10.1007/0-387-30260-3.

[25] GL 2010, "Guideline for the Certification of Wind Turbines," Edition 2010, Germanischer Lloyd Industrial Services GmbH, Hamburg, Germany.

[26] Thomas, P., Leimeister, M., Wegner, A., and Huhn, M. L., "Load calculation and load validation," in Wenske, J. (Ed.), *"Wind Turbine System Design, Vol. 1:*

*Nacelles, Drivetrain, and Verification,"* IET Energy Engineering Series 142A, The Institution of Engineering and Technology, London, United Kingdom, 2022.

[27] IEC 61400-1:2019 Edition 4, "Wind energy generation systems – Part 1: Design requirements," International Electrotechnical Commission, IEC, Genf, 2019-02.

[28] Barth, M., VDI/VDE 3693 "Virtual commissioning Part 1: Model types and glossary", VDI/VDE-Gesellschaft Mess- und Automatisierungstechnik (GMA), VDI Verlag, Düsseldorf, Germany, 2016, www.beuth.de.

[29] Bause, F., and Wallner, P., "Wie kommt die Intelligenz auf die Steuerung?," computer&automation Edition 2022-10, https://www.computer-automation.de/steuerungsebene/steuern-regeln/wie-kommt-die-intelligenz-auf-die-steuerung.199864/seite-2.html.

[30] "PC Control for an Offshore Wind Farm," PC Control 2008-02 https://www.pc-control.net/pdf/022008/solutions/pcc_0208_multibrid_e.pdf.

[31] Fischer, B., "Systematic Testing of Safety Relevant Controller Functions," E/E Conference 2013.

[32] "Real-Time EtherCAT Network Simulation," https://www.acontis.com/en/ethercat-simulation.html.

[33] "Full-Load Test Stand Ensures Smooth Commissioning of Offshore Wind Turbines," PC Control | Wind Special 2012, https://www.pc-control.net/pdf/special_wind_2012/solutions/pcc_special_wind_2012_areva_e.pdf.

[34] Millitzer, J., Mayer, D., Henke, C., *et al.*, "Recent Developments in Hardware-in-the-Loop Testing," *Model Validation and Uncertainty Quantification*, Vol. 3, pp. 65–73, Springer, Cham, 2018, https://doi.org/10.1007/978-3-319-74793-4_10.

[35] Turkson, R. F., Yan, F., Ali, M. K. A., and Hu, J., "Artificial Neural Network Applications in the Calibration of Spark-ignition Engines: An Overview," *Engineering Science and Technology*, Vol. 19, Issue 3, pp. 1346–1359, 2016, https://doi.org/10.1016/j.jestch.2016.03.003.

[36] Papadimitriou, I., Warner, M., Silvestri, J., Lennblad, J., and Tabar, S., "Neural Network Based Fast-Running Engine Models for Control-Oriented Applications," *SAE Technical Paper 2005-01-0072*, 2005, https://doi.org/10.4271/2005-01-0072.

[37] Bai, H., Liu, C., Breaz, E., and Gao, F., "Artificial Neural Network Aided Real-Time Simulation of Electric Traction System," *Energy and AI*, Vol. 1, 2020, https://doi.org/10.1016/j.egyai.2020.100010.

[38] García Márques, F. P., and Peinado Gonzalo, A., "A Comprehensive Review of Artificial Intelligence and Wind Energy," *Archives of Computational Methods in Engineering*, Vol. 29, pp. 2935–2958, 2021, http://dx.doi.org/10.1007/s11831-021-09678-4.

[39] Elyasichamazkoti, F., and Khajehpoor, A., "Application of Machine Learning for Wind Energy from Design to Energy – Water Nexus: A Survey," *Energy Nexus*, Vol. 2, 2021, https://doi.org/10.1016/j.nexus.2021.100011.

[40] Siddiqui, M. O., and Feja, P.R., "Nacelle Testing for Future Wind Turbines – What Lies Ahead?," Fraunhofer Institute for Wind Energy Systems, 2023, https://websites.fraunhofer.de/IWES-Blog/en/nacelle-testing-for-future-wind-turbines-what-lies-ahead/muhammad-omer-siddiqui.

[41] Wenske, J., Jersch, T., and Curioni, G., "HiL-Grid-CoP - Hardware-in-the-Loop Prüfung der elektrischen Netzverträglichkeit von Muti-Megawatt Windenergieanlagen mit schnelllaufenden Generatorsystemen: Schlussbericht," BMWK Project Final Report, Fraunhofer Institute for Wind Energy Systems, Bremerhaven, 2022, BMWK Project Funding Code 0324170A-B + D.

[42] Borowski, P., "Power Quality Test and Impedance model Validation for Main Converters of Wind Turbines – PQ4Wind," Fraunhofer Institute for Wind Energy Systems, BMWK Project Funding Code 03EE2009A, https://www.iwes.fraunhofer.de/en/research-projects/current-projects/pq4wind.html.

[43] Siddiqui, M. O., Nejad, A. R., and Wenske, J., "On a New Methodology for Testing Full Load Responses of Wind Turbine Drivetrains on a Test Bench," *Forschung im Ingenieurwesen* 87, pp. 173–184, 2023, https://doi.org/10.1007/s10010-023-00629-y.

[44] Feja, P. R., and Siddiqui, M. O., "The Step to the Digital Test Procedures: Hybrid Testing of Wind Turbines," Fraunhofer Institute for Wind Energy Systems, 2023, https://websites.fraunhofer.de/IWES-Blog/en/the-step-to-the-digital-test-procedures-hybrid-testing-of-wind-turbines/paul-robert-feja.

[45] Dressler, K., Speckert, M., and Bitsch, G., "Virtual Durability Test Rigs for Automotive Engineering," *Vehicle System Dynamics*, Vol. 47, Issue 4, pp. 387–401, 2009, https://doi.org/10.1080/00423110802056255.

[46] Huizinga, F. T. M. J. M., Van Ostaijen, R. A. A., and Van Oosten Slingeland, A., "A Practical Approach to Virtual Testing in Automotive Engineering," *Journal of Engineering Design*, Vol. 13, Issue 1, pp. 33–47, 2002, https://doi.org/10.1080/09544820110090304.

[47] Quistorf, G., "Mobil-Grid-CoP: Mobile Test Facility for Grid Compliance Tests," Fraunhofer Institute for Wind Energy Systems, BMWK Project Funding Code 03EE2014, https://www.iwes.fraunhofer.de/en/research-projects/current-projects/mobil-grid-cop.html.

*Chapter 7*

# Wind turbine control and automation

*Nils Johannsen*[1]

The design of the automation of wind turbines is the interdisciplinary task of realizing the interoperability of all components to ensure safe and reliable control for efficient energy production. This task is interdisciplinary because a variety of systems and aggregates are used in wind turbines:

- hydraulic systems with pumps and valves as for brakes
- cooling systems with fans and heaters as for the converter
- oil circuits with pumps and valves as for the gearbox
- lubrications systems with pumps and level sensors for bearings
- motors, drives, and positioning systems for pitch and yaw
- electrical systems and power electronics in converter and transformer

All components must interact with each other since the turbine can only be operated with all components. A fault in a single sensor can cause the turbine to stop if it is not designed redundantly. The temperature sensor of the coolant could be defective, so that the cooling of the converter and thus safe operation can no longer be guaranteed. A unique combination of requirements also makes the implementation of a control system an interdisciplinary task:

- variable environmental conditions with high winds and low temperatures
- aerodynamic control to regulate the power extraction from the wind
- nonlinear control to keep loads within a certain envelope
- power grid integration to manage the power feeding into the grid
- control system of turbines must comply with various safety standards
- outlying locations require remote monitoring to perform maintenance
- safety systems to protect people and the environment from harm
- overall high degree of complexity to ensure safe and reliable operation

Incorrect control can lead to damage followed by long downtimes. If the bearing lubrication is insufficient, this leads to higher friction with consequent corrosion. An unstable closed-loop control leads to an overspeed of the rotor or excitation of natural frequencies and thus to the destruction of the entire turbine.

[1]Beckhoff Automation GmbH & Co. KG, Germany

When designing the control system, conflicting objectives are aimed and must be balanced:

- maximize the production of electrical energy
- minimize loads and prevent structural damage
- optimize the quality of energy supplied to the grid
- avoid any downtime or shutdowns of the turbine
- fault-tolerant design to continue operation even after a malfunction

This overlap of the control system with almost every other field of the turbine leads to the requirement of an overall understanding to enable reliable control. The fine tuning of these systems, their characteristics, and mutual effects enable the optimization of the overall turbine operation.

## 7.1 Introduction

Automation systems and control systems consist of hardware and software. In the early days of the industry, the hardware was based on dedicated and specially for the purpose designed circuit boards with microcontrollers and circuits to evaluate the signals from sensors, as well as controlling actuators. Those analog signals were then digitalized and transferred to a microcontroller or another kind of processing unit, which is programmed by means of software to implement the logic and algorithms to control the machine. Nowadays, in most turbines these processing units are based on programmable logic controllers (PLCs) from the industrial automation sector. These controllers are implemented as industrial computers or embedded systems, whose application software is then programmed in programming languages like C/C++ and/or languages from the IEC 61131-3 standard [1]. Instead of using special circuit boards to read the so-called inputs from the sensors and interface the aggregates as outputs, the PLC suppliers also offer modular input/output (I/O) components. Usually, such I/O components are then prepared for an interface to the PLC, directly or by wires and couplers so that these inputs and outputs are usable in the software application simply by configuration. Nowadays most modern PLCs are based on industrial personal computers (PC), embedded systems or other kinds of single-board computers, with a processor, memory, drives, and an operating system. Thus, these industrial PCs have a similar architecture and components to ordinary PCs but are designed and specialized for industrial requirements and harsh environmental conditions. The operating system or a software part of it provides a real-time environment to execute the application software to control the machine, often also called PLC in software or Soft-PLC.

The term "control" of the wind turbine is used in industry and literature usually as a synonym for the closed-loop control of pitch and torque, also named dynamic control. More precisely, the dynamic control of the angle of attack on the rotor blade limits the extraction of energy from the wind, as well as the torque of the generator, which is applied by the converter to extract the energy and convert it into electricity. This closed-loop control is mainly responsible for the energy production as well as the structural fatigue load on large components such as the rotor, the drivetrain, and the tower. Since this dynamic control is so relevant and a special application of control engineering, it is a broad field of research and development, especially in the load calculation of wind turbines. Nevertheless, the concrete implementation of the control systems for the turbine via the sensors and actuators is essential for the operation, to avoid downtime and extreme loads. However, it is rather considered from the field of industrial automation and often not specifically researched for wind turbines [2].

Depending on the turbine design and philosophy of the conception, but also due to the production and company philosophy there might be multiple and optionally different kinds of control systems installed in the wind turbine or only a single superordinate control system. At the beginning of the development and the first designs of wind turbines, some manufacturers relied on existing components and obtained them from the industry. Such components like gearboxes or generators, which were already established in other fields and manufactured in a variety could easily be purchased without the need for development or manufacturing facilities. Due to special requirements from the wind industry, suppliers for this segment have additionally been established or founded for this purpose. Suppliers for pitch systems or converter systems were common and so the various companies formed a kind of own supplier industry. Accordingly, such complex systems were developed, manufactured, and delivered to the wind turbine manufacturer with a proprietary control system already in place. This resulted in the separation of various control systems in a single turbine, often named pitch control system, converter control system, yaw control system, and the superior main control system. In such designs, the main control system did not interface all the sensors and actuators itself but used an interface to the subordinate control systems of subsystems, usually by a fieldbus protocol.

Other companies, from the very beginning, implemented their own developments of such components, e.g., the generator or power electronics for special designs and applications in wind turbines. These companies also had their own development of control systems and own design of circuit boards with microcontrollers for different purposes in the turbine. But also due to the strong price pressure in the industry, there were many mergers of turbine manufacturers, as well as an increase in their in-house production depth. The sales and market for many sub-suppliers got lost, which led to a shakeout in the market. The turbine manufacturers established their own developments and manufacturing of such components instead. However, as it was often no longer economically viable to develop the electronics and basic software in-house, industrial control systems and other components were sourced from established suppliers in the market. While special large-scale components were still developed and manufactured by turbine manufacturers in-house. Since, in most cases, the resulting large companies have created internal enterprise structures, the control systems have not been merged again but have rather been separated by the thinking in terms of departments and responsibilities. This separation can lead to independently acting units, and a separation of concerns in hardware components. However, it requires individual interfaces between the systems, which might use different protocols, implementations, and even different automation systems from different manufactures. This significantly increases the complexity of the entire automation system, which can no longer be overseen by individual engineers, making it more difficult to maintain and service the equipment in the field. As previously described, this prevents a holistic view of the entire system. Consequently, individual, highly accurate decisions for maximum efficiency cannot easily be implemented in the control software anymore. The increased number of different components also increases the probability of outages or malfunctions with difficulties in the availability of the variety of spare parts and their storages. The increased number of intelligent systems is also complicating the implementation of the lately increasing requirements for cybersecurity as the intelligent systems all use their individual software components with their own configurations, interfaces, and software updates with the respective multiple concepts of maintenance. Therefore, a central system approach by an industrial PC and a decentral and modular I/O system is increasingly becoming the standard, as shown exemplary in Figure 7.1. Such modern control systems allow the separation of concerns and subsystems in a modular software structure with modern concepts

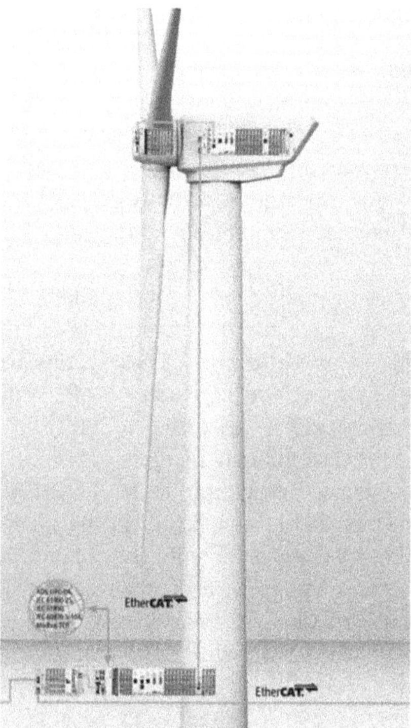


*Figure 7.1 The usual breakdown of control components and their positioning in the turbine: tower cabinet, nacelle cabinet, and in the hub. Copyright © Beckhoff Automation GmbH & Co. KG.*

for software development and the integration of automatically generated software modules from graphical programming, modeling, and simulation environments.

In the following, Section 7.2 describes the "Design Process" of the control system, followed by an overview of "Industrial Control Systems" in Section 7.3. To ensure safe operation of the turbine, a superior safety system must be implemented, which will be explained in Section 7.4 "Safety System." Section 7.5 "Control Cabinets" describes how the control system is installed in the turbine. Details about the implementation of the software application for the control system, the logic, and algorithms are described as the "Software Application" in Section 7.6. Details about the different kinds of sensors and actuators in the subsystems as well as how they are controlled are described in Section 7.7 "Subsystem Control." Finally, Section 7.8 introduces the "Monitoring and Maintenance" of turbines in the field, ending with "Conclusions and Outlook" in Section 7.9. Please be aware that these descriptions and information can only provide a generalization and overview of the topics. The actual implementation can vary and deviate, as the different and individual types of turbines as well as manufacturers each have their own singularities.

## 7.2 Design process

As all subsystems overlap with the automation, the whole development process of the wind turbine, the control system, and safety system should also be considered. From the initial design of the turbine and the chosen rotor and drivetrain, the objectives for the turbine and parameters like the rated speed or rated power already result. In addition, the dynamic control of pitch and torque must already be prepared for the load calculation to check the design load cases (DLC) [3]. The result of the load calculation is then the basic concept for operational management. That concept includes the operation states and modes, approved operating areas as well as stop, shutdown and parking conditions. This is how the intended nominal conditions are defined for starting and running the turbine, also the maximum allowed rotor speed and the maximum speed at which the turbine should be stopped, as well as which rotor speed triggers the overspeed detection resulting in an emergency stop and in case of a stop, which pitch speeds are used to move back the blades into the feather position. The active closed-loop control of pitch position and converter torque plays an essential role in maintaining the loads within a specific envelope to ensure a lifetime of 20 years and more. But also the behavior for stopping the turbine, as well as in the parked state, are increasingly actively controlled to eliminate loads or external influences in case of strong winds.

In addition, a safety concept must be drawn up, which defines the limits of the turbine. The exceeding of these limits will lead to a trip of the safety systems and the starting of an emergency stop. Such safety limits are usually the maximum speed of the rotor or generator, maximum current or power acquired from the generator, a maximum permitted jerk or shock from the machine carrier or tower top, as well as sometimes specific temperatures of components or other important values. With the selection of the large components and the main parts of the electrical system, further parameters are determined, such as the maximum possible cable twist in the tower top before damage to the cables could occur. This cable twist needs to be detected by an encoder with limit switches and avoided by tripping the safety system. The backlash, which may exist between the rotor to the generator by the different shafts, a gearbox and/or clutch requires a monitoring of the rotor speed in comparison to the generator speed to detect any slipping or runaway in the drivetrain. Furthermore, the commissioning, service, and maintenance of the turbine as well as the safety of the executing persons must be considered for the safety system: how, where, or by which components a person could come to harm, how often a person is on site and whether this circumstance could occur and consequently with which safety level it must be prevented.

This is followed by a selection of the subsystems and the associated sensors and actuators required to control them. The previously defined operating ranges and conditions must be matched and fully covered by the selected sensors. Any redundancies in sensors and actuators that are necessary for safety or availability purposes must also be taken into account here. The position of the devices as well as the mounting of, e.g., a sensor on the component must be defined. With the selection of the automation supplier and the control system or systems, the necessary I/O modules for the integration of the chosen sensors and actuators are then selected. Depending on the electrical components and their characteristics, their placement must be defined. Thus, all components of the control system can be installed in one single control cabinet into the nacelle and the sensors and actuators must be wired up to this cabinet. If the components are distributed and partly in the tower, a second control cabinet in the tower might be sufficient. In this case, a connection between the different control cabinets must be integrated. Depending on the chosen technology for the control system this might require further processing units or one processing unit in every cabinet, but modern systems allow to collect the digitalized information from the I/O modules by only a single communication cable between the cabinets.

Then the cable routing inside the nacelle must also be considered, as well as the possible influence of cables with high currents and their electromagnetic influence on the sensor cables. For this purpose, specially shielded cables are often required to dissipate such influences via the shielding. Many modern control systems also offer I/O solutions that do not require to be placed inside a control cabinet. Instead, they are protected against water and dust so that they can be mounted directly in the nacelle, near the sensors, actuators or maybe directly on the component. That reduces the necessary amount of wiring, as not all cables need to be wired from every sensor all over the nacelle to the cabinet, but just a short distance to the I/O module and a single I/O module will then communicate the information of all sensors and actuators by a single communication cable to the processing unit.

With this information about sensors, actuators, cables, and the dimensions as well as positions of the control cabinets, the internals of the cabinet can be designed. The electrical power supply fed in must be considered, as well as the distribution to all consumers and actuators. Within the control cabinet the processing unit with the I/O modules and the possible distribution of signals by internal wiring in the cabinet are planned. This results in the circuit diagram, which includes the single line diagram (SLD) to show the electrical power supply of the systems, as well as all cables, wires, and their connections between the systems, sensors, and actuators in the turbine as well as in the control cabinets.

The software development for the operational control of the turbine can be started after or already in parallel to the cabinet design. Reading and scaling of all inputs from the sensors, their evaluation and monitoring as well as the determination of the operating state and the control of actuators and subsystems must be implemented. The limits of the turbine, system, and subsystems as well as the operational management and safety concept must always be considered for this. Modern development tools then allow the integration of all or parts of the closed-loop control algorithms that were already developed for the load calculation.

Beyond that, if possible, first tests of the software can be carried out. Different approaches of software tests through unit tests or integration tests are used here, as well as internal or external models and simulations. These could represent single components, like the hydraulic system, a simplified abstraction of the rotor and drivetrain or the whole wind turbine including structural dynamics. By combination of simulation and control software so-called model-in-the-loop (MiL) or software-in-the-loop (SiL) tests are executed. The interface to more complex subsystems is usually tested for prototype development in an isolated environment in the factory or a laboratory. Next, the control systems will for example be connected to the processing unit of a pitch system or a complete pitch system and the communication as well as interaction are approved. Ideally, the subsystems that do not exist are simulated while one or more other subsystems are connected to implement a hardware-in-the-loop (HiL) setup. See Chapter 6 for more details on testing the controller.

With the control software as complete or advanced as possible, as well as the first set of control cabinets, individual parts of the turbine or the entire nacelle with all its components are then assembled in the factory hall. This allows the testing of individual systems as well as their interaction and interplay. Depending on the possibilities of the factory, the entire hub including the pitch systems with all drives and motors may be set up and tested. It is also possible to rotate the drivetrain by a motor or using the generator as a motor via the converter to carry out initial tests at speed for the lubrication, cooling, and speed sensors up to emergency stops due to overspeed or similar.

After assembling and erecting the complete prototype of the wind turbine, retesting of changed or reconnected wires, cables and interfaces, the aggregates and subsystems that are individually connected by the control system and put into operation. A complete pre-testing of the safety system must be carried out to ensure the safety of the people present during commissioning as well as to prevent any damage to the turbine in case of a malfunction. This allows first tests by manually operating the subsystems, followed by testing their automatic operation. After all auxiliary systems have been tested, a first spin can be allowed by unlocking the bolts holding the rotor in place. By manually changing the demanded pitch angle the first operating points can be approached at low rotation speed. An initial synchronization of the converter to the grid, switching on the generator and feeding power to the grid approves the entire system for the first time. After that increasingly higher speeds and powers can be demanded to test the stability of the dynamic control, drivetrain, and auxiliary aggregates under higher loads. Followed by the first automatic start of the systems and turbine, the various operation modes can be tested as well as the triggering of the various stop modes by inducing malfunctions. Some simple malfunctions can be tested for example by disconnecting sensors or by the manipulation of values in the control software. Convenient is also an implementation of prepared automated test procedures in the control software since they are easily reproducible and help commissioning of the hopefully following future turbines.

With the testing of all automatic modes and procedures, the operation at rated speed with nominal power and full load can be approached together with attempting a fully automatic and independent operation of the turbine. This is the beginning of continuous monitoring of the turbine during permanent operation. Many different environmental conditions which usually have not yet occurred during commissioning should be run through, especially the wind conditions across all operating ranges of partial and full load that only occur after several months of operation.

The next turbines, the first smaller series and turbines from series production, the wind turbines in the field are permanently monitored in the field. Continuous monitoring of operating conditions and evaluation of situations that may have led to a stop or unfortunate conditions allow an optimization of the turbine and its automation system. With changes and updates in the control software, permanent and continuous optimization of new wind turbines as well as the turbines in the field will be applied.

## 7.3 Industrial control systems

The invention of automation and control systems began mainly with the industrial revolution. Initial automation systems and their application started with temperature and pressure regulators for steam engines in the sixteenth century. In the wind industry, the first small turbines for experimental purposes in the 1960s and 1970s used only a few simple mechanical and electrical controls. With the increasing amount of wind turbines in the 1980s and 1990s, more advanced wind turbines with digital control systems began to appear. Circuits specially developed for the equipment of the turbine, utilizing microprocessors and electronics were used for the automation of major components such as pitch, converter, and yaw. The integration of sensors for the wind speed and direction as well as temperature and pressures were utilized to improve the overall performance and reliability of turbines.

The industrialization of wind energy in the early 2000s also created the need for automation systems ready for series production. Thus, the increasing use of industrial control systems began. Because of the remote location of turbines and for service as well as maintenance purposes the implementation of so-called "Supervisory Control and Data Acquisition" (SCADA) also became popular in the 2000s. There is no standard or unified definition of the term SCADA, but in common cases, this term is used for improved control software to establish remote communication interfaces and infrastructure. This includes data collection, remote monitoring, and interaction with the turbine, enabling operators to monitor and control multiple turbines from central control centers. As wind turbines became a relevant part of the energy supply in the 2010s, topics like grid integration and cyber-security came into focus of the development. The remote communication infrastructure of wind turbines and farms is usually based on the public internet and the cyber-security measures require firewalls, encryption, and authentication of the communication to prevent threats to the control systems.

Since the first series of wind turbines tended to be designed rather from mechanical engineering and with industrial equipment, the control systems tended to be used from industrial production machines. On the other hand, conventional power plants and the electrical grid are sourced from power engineering products. This separation of market segments started to change in the late 2010s and a blending of industrial technology into the energy sector seems to become apparent in the 2020s.

For the technology of industrial control systems, the International Electrotechnical Commission (IEC) established the standard IEC 1131 in 1993. With the advancement in the IEC 61331 for programmable controllers an increased standardization for control systems and especially with the part IEC 61131-3 the code standardization for the control software has come in place. The aim is to allow the implementation of hardware-independent control software with the five defined programming languages. These languages are either textual with "Instruction List" (IL) and "Structured Text" (ST) or graphical with "Ladder Diagram" (LD), "Function Block Diagram" (FB), and the "Sequential Function Chart" (SFC). With the 3rd Edition of the IEC 61131-3, the Instruction List was deprecated, but instead essential parts have been added to allow object-oriented programming in addition to the sequential programming. Thus, the control systems provide a development environment or engineering software as well as a runtime for programs developed in the languages.

With the evolvement of modern software technologies and the overlaps with technologies from the computer and information technology (IT) usually higher-level programming languages like C and C++ are also supported nowadays. Also supporting the integration of model-based developments (MBD) from tools like MATLAB®/Simulink® [4] or Modelica [5] based tools is common.

### *7.3.1 Industrial PCs*

The execution of the control software takes place on a processing unit. This unit is usually a programmable logic control (PLC) and ranges from small devices based on microcontrollers up to large devices with multiple processing units. Today, implementation is increasingly carried out by using computer-based systems. Those computer systems are commonly named by the term "PC" which stands for "personal computer" and therefore PC-based control became the usual terminus. For industrial usage such PCs were adapted to the requirements and used as so-called industrial PCs or IPCs – not to be confused with Individual Pitch Control which is also abbreviated as IPC. Figure 7.2 shows the industrial PC C6030, in which a single integrated computer board CB6263 is installed. The mainboard CB6263 integrates all components like CPU, RAM, and HDD of the computer in

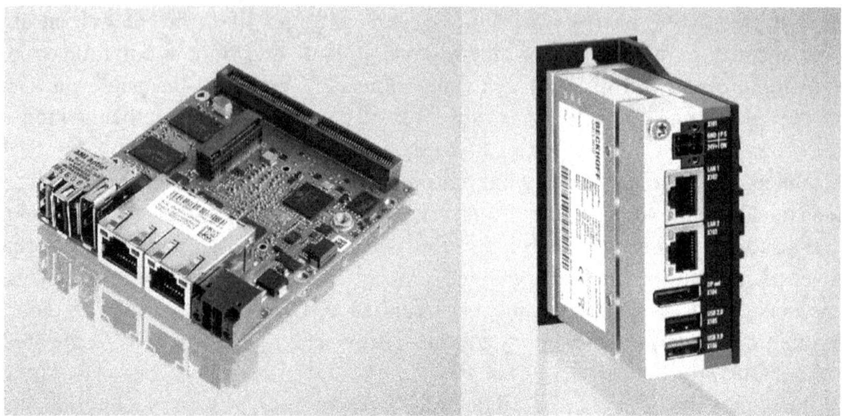

*Figure 7.2 The computer board CB6263 (left) is despite its compact format a fully functional mainboard as system on a chip (SoC). With the housing, it becomes the Industrial-PC C6015 (right) as a base for control systems. Copyright © Beckhoff Automation GmbH & Co. KG.*

one integrated circuit, which is also referred to as system-on-a-chip (SoC). Through the housing, the mainboard becomes the compact industrial PC C6030, which integrates all functionalities of a PC, is powered by 24 V, and provides the usual IT interfaces such as Ethernet, USB, and a Display Port.

While information technology (IT) evolves very quickly and by More's law "the number of transistors in integrated circuits doubles about every two years" [6], most of these parts are designed for the consumer market. Only some specific components are designed for embedded systems or the industrial market, which requires advanced properties concerning the temperature range, shock resistance, but also long-term availability for stable long-term production and accessibility of spare parts. By utilizing the extremely fast-growing computing power and performance of components from IT for a convenient price, the PC became an advanced technology for control systems. These systems provide a very flexible and scalable solution, high processing power and extensive memory capacity, with the latest advancements in communication and networking technologies. That allows to interface the PCs to various devices via standard communication protocols that are based on RS232/RS485 or Ethernet. Likewise, PCs offer a user-friendly and intuitive platform, commonly known in all areas of engineering. The implementation and integration of software, with extended services as well as human–machine interfaces (HMIs), through the integration of technologies that are already established in IT is enhancing the utilization for industrial control systems [7]. The constantly increasing computing power of PCs also contributes to this, making the integration and expansion of functionalities possible in the first place.

In the case of an industrial PC-based solution there are multiple software parts, often called layers of software: Firstly, there is the firmware of the IPC, which is called the "Basic Input/Output System" (BIOS) and used to boot the operating system (OS). The OS is the second layer, that provides the hardware abstraction and fundamental services. The core of an OS is the kernel, with device drivers to interface the hardware, handle signals and interrupts on the mainboard, execute programs on the "Central Processing Unit" (CPU), memory management to dynamically access the "Random-Access Memory" (RAM), and a file system to store data on the "Hard Disk Drive" (HDD) or the nowadays more common "Solid State Disk" (SSD). Depending on the operating system a few or a variety of services are applied in a higher-level of abstraction, called the "user mode" (also "user space" or "userland") in contrast to the "kernel mode." By the abstraction in the user mode, the OS prevents direct access to the hardware, requires using the kernel for such operations by its application programming interfaces (API), but also protects the hardware from inaccurate and malicious software in this way.

To implement and execute the specific control logic of the machine there is usually a kind of runtime service, offered by the automation supplier and installed into the OS, which is the third software layer in this definition. Such a runtime offers further interfaces to access the external I/O modules, implements fieldbus or other protocols and interfaces as well as the execution and scheduling of application software. This application software is implemented by the original equipment manufacturer (OEM) and is the fourth software layer, which adapts a generic and non-purpose control system into the specific wind turbine control system.

To guarantee safe control of the machine, hard real-time (RT) scheduling of the implemented logics is required, which is according to the DIN 44300 defined as follows: "Real-time operation is an operating mode of a computing system in which programs for the processing of data are continuously operational in such a way that the processing results are available within a specified period of time" [8]. So real-time presupposes an execution of software implementations in a predictable time and at defined points in time. This requirement can be met either by the exact scheduling of all processes in the entire operating system or by the runtime environment with real-time scheduling exclusively for the components necessary to control the machine. In the case of a real-time capable OS, even subordinate services with low priority are considered for the real-time scheduling. A runtime provides the real-time environment with the highest priority within the OS, but all other processes use a non-real-time scheduler asynchronous to the control logic.

### *7.3.2 I/O modules*

The control software utilizes the interfaces of the industrial PC to communicate with input and output (I/O) modules. I/O modules can be directly integrated as components of the PC and installed on the mainboard. Nowadays, more common are decentral I/O modules. These are separate devices that are interfaced from the PC by communication protocols. As common in IT, serial or Ethernet interfaces are used for this communication.

A large variety of sensors and actuators can be connected to the I/O modules. Several sensors in the industrial environments are simply using a digital signal that is low at 0 V, and high at 24 V and behaves as a digital contact interface. In case a defined situation has occurred the digital sensor switches from low to high signal – for example, when a contactor is reached, when a defined pressure, temperature, or other circumstances come in place the sensor switches on. Other types of sensors are integrated by simple analog signals, as either a current from 0 to 20 mA or 4 to 20 mA or a voltage from 0 to 10 V or −10 to +10 V. This analog value then corresponds to a range of the measured physical value. The 4 to 20 mA signal is most common because it is robust against electromagnetic influences and at the same time the minimum signal level of 4 mA allows an easy detection of a cable break or error in the sensor due to the resulting 0 mA signal. For example, a pressure sensor could output a signal of 4 mA at a pressure of 0 bar and 20 mA at a pressure of 10 bar, which is recalculated accordingly by interpolation during the evaluation of that signal. The same applies to the outputs: A relay could be used as a digital output with a low signal of 0 V and a high signal of 24 V to close the contactor. An example of an analog output is a valve, that fully closes at a current level of 0 mA, fully opens at 20 mA, and can be controlled freely from 0% to 100% in between. In the I/O modules these signals are then digitalized by an analog-to-digital (AD) or digital-to-analog (DA) converter. These digital values are then made available to the PLC. More complex sensors and actuators can have their own electronics and thus be interfaced by a communication protocol. Various I/O modules which are implemented as EtherCAT slaves are shown in Figure 7.3. These include digital I/O modules, the yellow modules are safe inputs and outputs to the safety system, the blue modules are intrinsically safe signals for connecting sensors in potentially explosive atmosphere, as well as gateways to subordinate fieldbus systems and measurement technology modules.

To allow many I/O channels on small space, the I/O modules usually include two, four, eight, or even more channels of the same type on a single module. Since most channels can only be used for specific signals, it is always a trade-off between the number of similar signals needed and the space and price required for the

*Figure 7.3 A modular I/O system as shown here using EtherCAT allows the decentralized integration of almost any signal, as well as the integration of fieldbus interfaces, the safety system (yellow) and intrinsically safe channels (blue). Copyright © Beckhoff Automation GmbH & Co. KG.*

modules. There are also more flexible modules that allow a wide variety of signals on a single connection through simple configuration, but for a correspondingly higher price. Next to that, there are several special signals, like for a speed or position signal by an encoder, for the detection of vibrations or high-precision measurement technology for strain gauge sensors. There are technologies where the I/O modules must be mounted on a special plate, the backplane or where it is flexibly possible to attach the modules to each other without additional components. Then the modules usually use some kind of internal connection via the backplane or a contact that is connected when plugged together. Couplers allow communication between segments of I/O modules over cables for the distribution of modules over higher distances. It also allows modular machine concepts where the stations are placed close to the process and thus to the sensors and actuators to be connected. Such decentralized positioning of I/O stations reduces the wiring effort and thus costs. To connect these resulting segments of I/O modules or distributed I/O stations with each other, as well as to connect the processing unit respectively the industrial PC, and to allow them to communicate in real-time a fieldbus interface and protocol is used.

### *7.3.3 Fieldbus*

Just like the execution of the software application, the communication by a fieldbus must exchange the digital signals with the PLC, respectively the PC, in real-time to ensure safe control. In case the serial interfaces or Ethernet interfaces of the PC are used, they also must be integrated into the real-time environment and implement the specific protocol to communicate with the I/O modules. This protocol must be real-time capable as well, just like the whole treatment inside the control system.

The serial interfaces were originally used to transmit the data via RS232 or RS485 based on protocols like CAN or PROFIBUS. Serial communication technology, although very common in those times, showed weaknesses when used in wind turbines. As shown in Figure 7.4, media converters are used to implement communication and pass the long distance between the tower base and the nacelle through the use of fiber optic cables. However, due to the susceptibility to electromagnetic interference, as well as the low performance and the difficulties in diagnosing faults in complex systems, this technology is no longer used as the main fieldbus in modern systems. More modern and common are I/O modules which are interfaced or integrated by Ethernet-based protocols like EtherCAT or PROFINET.

The capabilities of the industrial Ethernet protocol called "Process Field Network" or PROFINET depend on various factors, like the size of the network, the distance between devices, the desired fault tolerance, the scalability requirements, and cost considerations. The different standards, conformance classes and versions of PROFINET are maintained by the PROFIBUS & PROFINET International organization [10]. There is a major difference between PROFINET RT and PROFINET IRT, which allow different communication cycle times based on different kinds of components. The architecture of PROFINET is based on producers and consumers, so that multiple devices can provide data as well as

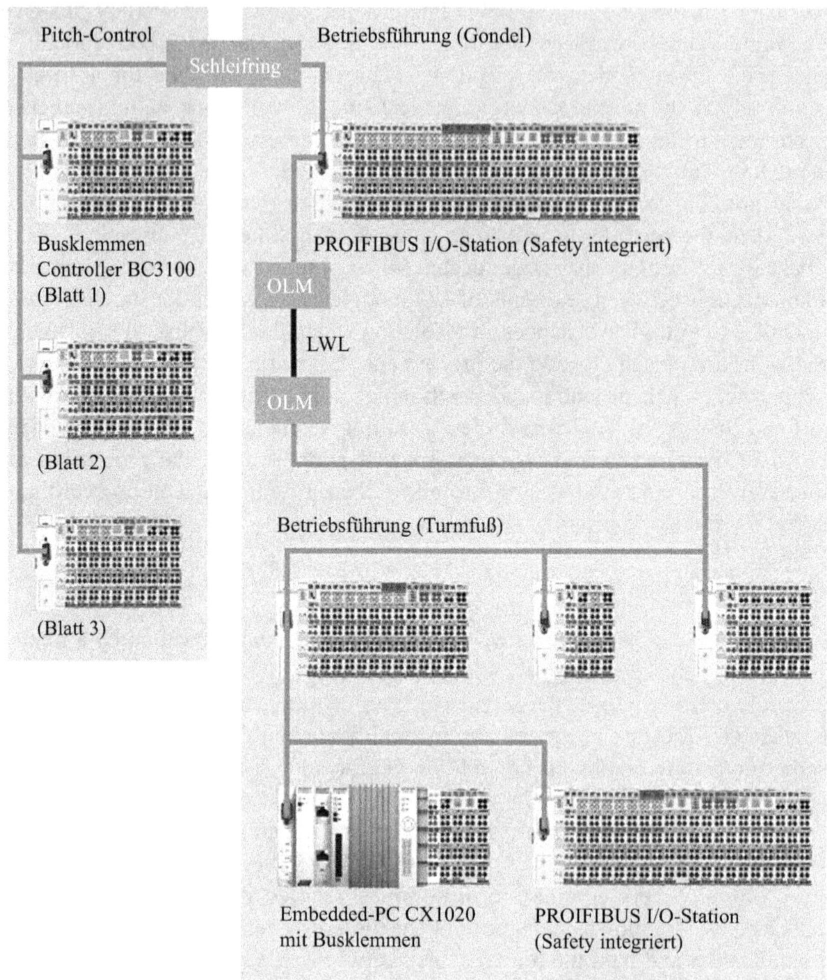


*Figure 7.4 A control system topology of Vensys Energy AG from the year 2010 based on PROFIBUS with fiber-optical converters (OLM) [9]. Copyright © Beckhoff Automation GmbH & Co. KG.*

consume data. As PROFINET utilizes a network based on switches with management functions it allows combining PROFINET devices with common internet-protocol (IP) devices in a single network, but that also requires specific real-time capable switches. Careful planning of the addressing and design of communication channels are required for more complex topologies to ensure the reliability and real-time capability of the network [11].

The protocol "Ethernet for Control Automation Technology" or EtherCAT is maintained by the "EtherCAT Technology Group" (ETG) [12] and standardized in

IEC 61158. EtherCAT only uses the Physical and Data Link Layer from the Open Systems Interconnection Model (OSI model) of Ethernet and implements the Application Layer. That prevents the mixture of EtherCAT with others IP-based devices in a single network, However, it allows very lean and performant real-time communication without additional switches or hubs necessary. Gateways allow to integrate compatible devices over IP-based communication but also other fieldbus protocols like CANopen, PROFIBUS or PROFINET and allow communication above the master in the EtherCAT network. As EtherCAT is based on a master/slave architecture, there is only a single master and a nearly unlimited number of slaves in a single network, but multiple networks can be combined by so-called bridges. Some unique features as "processing on the fly" in the EtherCAT slaves or the full-duplex utilization by "summation-frame communication" in the EtherCAT master achieve high performance and rapid data exchange with low delays. Due to the "Distributed Clocks" in the EtherCAT slaves, signals can be sampled simultaneously as well as time-controlled outputs can be switched at the same time and several networks can be synchronized with each other. The software-based abstraction of any physical topology of the network into a ring bus topology is implemented by a software-side termination in the EtherCAT slave and thus allows any physical topology as well as cable redundancy and hot plug of stations [13].

In Figure 7.5 such an EtherCAT-based topology for a wind turbine control system is shown. The tower and nacelle cabinet are pointed out, as well as a pitch control system, yaw control system, converter system and the wind farm connection. The central control approach is implemented by an embedded PC in the tower cabinet, with additional I/O terminals and some fieldbus interfaces as well as the EtherCAT interface to a converter. By an integrated fiber-optical connection the EtherCAT network is extended into the nacelle, where further I/O modules are installed in the control cabinet. Additionally further fieldbus interfaces, as well as condition monitoring terminals and yellow terminals to implement the safety inputs, outputs and overspeed detection. A further optical feedthrough indicates the connection of EtherCAT via the slip ring into the hub and to the pitch control system. The pitch control is implemented by an independent embedded PC with EtherCAT Terminals for the I/O as well as a blade monitoring module with a potential wiring of strain gauge sensors. In the nacelle and hub, there are additional EtherCAT Box modules that allow to be installed outside of the cabinet, directly mountable on aggregates such as the gearbox or inside the rotor blades. The connection to the superior wind farm network is as well an EtherCAT network to allow real-time capable control of all turbines in the wind farm by means of a wind farm controller.

Ongoing is the development in the Institute of Electrical and Electronics Engineers (IEEE) to extend Ethernet on Layer 2 by Time Sensitive Networking (TSN). It shall allow time synchronization of all devices in the network, as well as deterministic behavior based on Time Aware Shaper (TAS) with hard timing of defined streams and queues in specific Ethernet switches. Efforts are already being made to use established but originally non-deterministic protocols as OPC UA over TSN for real-time communication [14]. Also, the already real-time capable and Ethernet-based fieldbus protocol PROFINET integrates TSN as an option in

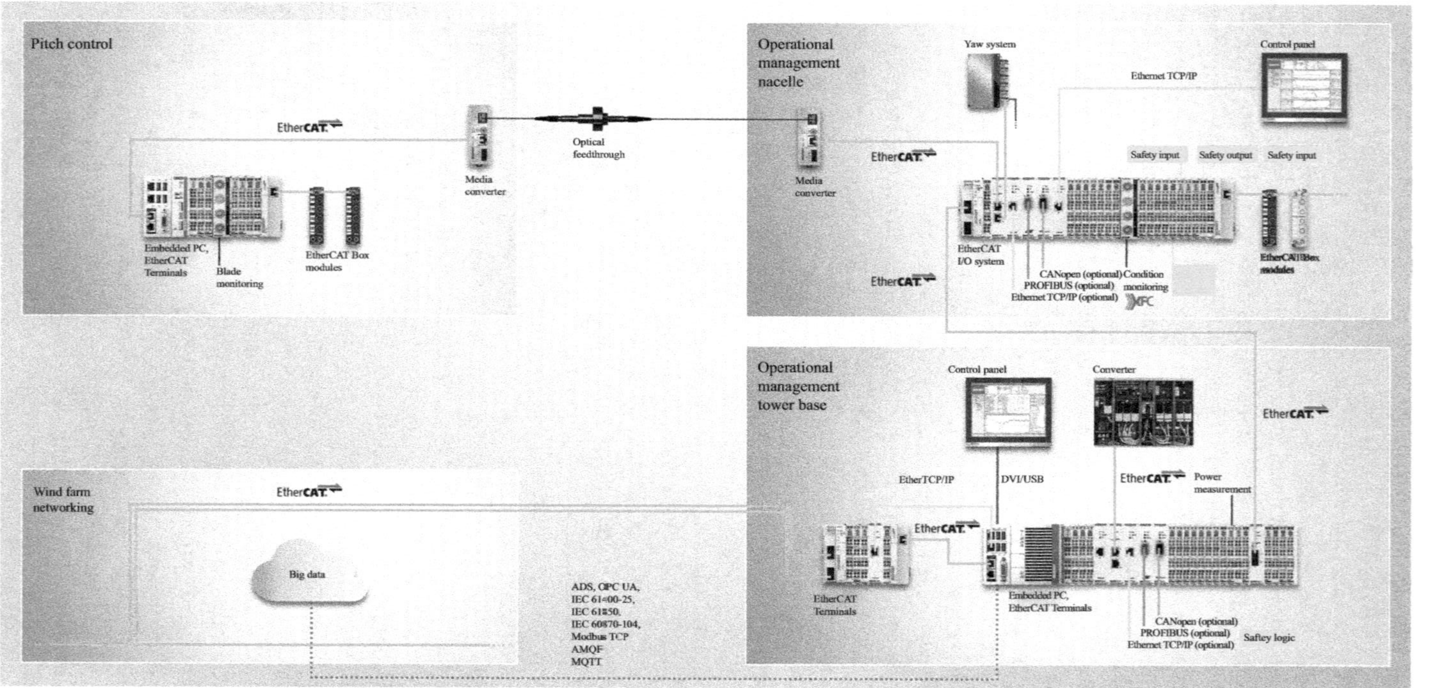


*Figure 7.5* *The control system topology based on EtherCAT with a central embedded PC in the tower cabinet, an integrated fiber-optical connection to the nacelle cabinet, and the integrated yellow terminals for the safety system. Copyright © Beckhoff Automation GmbH & Co. KG.*

version 2.4 of the specification [15]. The future will show whether this standard will prevail and whether further protocols will be based on it.

## 7.4 Safety system

The safety system of a machine typically implements measures to mitigate any accidents or injuries to persons in the workplace. For wind turbines the safety system also must protect the workers, as well as the machine from any dangers itself. But because wind turbines are often located in public areas, they additionally must protect the environment. With its size and construction some parts of the turbine could aviate in a wide range around the site and harm people or damage properties there [16].

The design of such a safety system includes risk assessments, evaluations of precautions and the development of measures to avert the risk. Even though many potential risks in the turbine can be eliminated by simple physical, mechanical, or structural measures, such as installing a cage around moving parts or simply disconnecting the power supply, many essential measures while operating the turbine are usually implemented by an electrical safety system in relation with the control system. For reasons of simplicity, safety, and security, including cyber-security, the measures should be implemented pragmatically, as close to the process as possible and intrinsically safe if possible. A physical safety component such as a pressure relief valve cannot be manipulated by cyber-attacks, and it is very unlikely that such a simple mechanical component will fail twice at the same time [17]. The measures beyond this, which are implemented typically electrically, should be independent as possible of the control system software to ensure a safe condition even without it or otherwise not allow to operate the system at all.

Since the safety of the turbine is essential, also in public interest, there are also precise standards and regulations on how the entire development should be implemented. The requirements are defined in the GL Guideline and in the IEC 61400, while both defining the ISO 13849 as the standard for the safety functions inside a wind turbine. The ISO 13849-1 defines different safety functionalities in so-called "Performance Level" (PL) from PLa to PLe. Apart from that, there are the EN 50308, ISO 13850, ISO 14121, IEC 61508 as well as the IEC 61439 for definitions regarding safety in machines and plants. For example, the EN 60204-1 describes safety and functional requirements for shutting down a machine in three categories. Category 0 requires an immediate stop by interrupting the power supply to the actuators. For category 1, it is sufficient to stop the machine, bring it to a safe state and then disconnect the power supply. While category 2 only requires a safe state, not to switch off the actuators, but to prevent an unintentional restart of the machine [18].

Apart from the technical requirements and implementation discussed here, the safety consideration also includes guidelines and training for the service and maintenance workers as well as engineers who are on site and how to behave. Such guidelines are about the behavior while being on site, but also how to use the turbine in a convenient way to allow maintenance and repair. A simple and self-evident rule could be to stop the turbine before climbing up the tower to the nacelle.

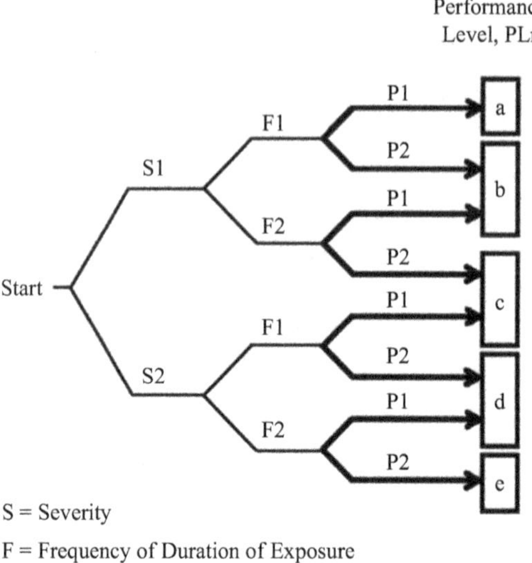


*Figure 7.6 The ISO 13849-1 defines the required Performance Level (PLr) based on the severity of injury (S), the potential frequency or exposure to happen (F), and possibility of avoiding any harm (P)*

But also beyond, locking the turbine by a service switch or like prevent a restart as well as any remote interaction to the control systems is reasonable.

In the risk assessment, possibilities and probabilities for causes are then identified, evaluated, and classified. The classification results in the requirements of implementation according to a performance level as PLa to PLe from ISO 13849 or the safety integrity level from SIL1 to SIL3 from IEC 61508. Figure 7.6 illustrates the decision-making process as to which performance level is necessary for an application. It determines the severity of injury, the frequency with which any harm might occur and the possibility of avoiding it. This level then describes whether redundancy must be built up or whether diverse evaluations must be used for the implementation. This leads to the fact that essential sensors are then present multiple time and work with different measuring methods, whose are based on either direct measurement or indirect methods such as software-based sensors. The cabling of these sensors, up to the evaluation as well as reacting aggregates or components must also be redundant and take the best possible separate routes through the turbine.

The most important property for the safety of wind turbines is the rotor speed, the most important inputs are the emergency stop buttons, and because a multi-megawatt wind turbines can only stop the rotor by adjusting the angle of attack on the rotor blades, the pitch system is the most important actuator [19]. Therefore, for example, the rotational speed is always measured on the rotor and on the generator. By using inductive sensors on the rotor and an incremental encoder on the generator, diversity is warranted. The cable from the sensors and the encoder then uses different paths through the system and the control cabinet to the evaluation unit. The evaluation unit must then compute the two signals, check their plausibility, and compare them with each other. It should not only trigger when one of the two speed signals is too high, but also when the values are different, or one value cannot be evaluated. When the unit has been triggered, the "emergency feather command" (EFC) should then also be sent to the pitch system via the most redundant possible paths to trigger an emergency stop, moving of the rotor blade autonomously out of the wind, into the feather position.

This would then be one scenario of the safety system to be considered and implemented. The usual additional triggers are overcurrent or overpower, as well as a shock caused by high vibration at the top of the tower, excessive twisting of the cables at the cross-over from tower to the nacelle and the emergency switches at different positions in the turbine. Since in the first wind turbines all these sensors were connected in series, just like all actuators where then connected to this series, the safety system was or still is referred to as a safety chain. If one element of the chain opens, the whole chain is open and therefore the safety chain is said to be opened or closed. This was implemented in practice often with a digital high signal of for example 24 V and a series of relays through which two wires were routed in parallel. If one relay in that chain opened, the 24 V signal drops. Other relays of actuators then controlled by this signal would also open when the high signal drops and thus trigger the emergency stop.

However, one difficulty in practice was connecting the 24 V signal over long distances, especially between the tower base and the nacelle, without a voltage drop due to the resistance of the long wires. The diagnosis of the interconnected relays was also often difficult, when one switch was triggered, it was not easy to find the cause of triggering the emergency stop. Therefore, in modern turbines the safety system is generally integrated into the control system, implemented using safe I/O modules, as well as a safe protocol for communication of the signals via the fieldbus and a safe evaluation unit.

The modern fieldbus protocols offer an integrated feature to allow transmission of such safe signals, as PROFISsafe for PROFINET [20] as well as FailSafe over EtherCAT (FSoE) [21]. This is implemented via an additional area in the protocol, also known as the black channel, whose data is also validated using checksums and other mechanisms. The advantage is that any inputs and outputs can be installed easily and decentral, these are evaluated redundantly and intrinsically safe according to safety standards, usually extended by a monitoring of the signals during operation. Since these protocols are routed and distributed (but not influenced) through the control system, the processing unit can still call up all information about the safety system and knows all the states of sensors, actuators as well as the connections between them always. This allows full diagnosis and easy troubleshooting.

When communicating with a modern fieldbus via the slip ring, safe communication can be integrated directly into the hub. This avoids additional phases on the slip ring to be able to exchange the redundant digital safety signals in both directions. Even if the pitch system is integrated directly into the main control system or integrated via a modern fieldbus, the safe signals can be communicated and exchanged right into the logic of the pitch system. No additional safe switches and feedback signals are required in the hub.

## 7.5 Control cabinets

The control cabinets for wind turbines are the housing and substructure for the electrical and electronic components of the control system. As the most components are not protected against dust or water and approved for an ingress protection (IP) class of IP20 after the standard IEC 60529, they cannot be installed directly in the nacelle or tower without additional protection. In addition, several components are only allowed to be operated in an environment with a temperature of for example 0 °C–55 °C and would need cooling or heating as well as protection against condensation, which can result from humidity at temperature fluctuations.

While there are other parts as sensors and actuators as well as especially designed I/O modules with a protection class of IP65 or IP67, thus suitable for the environmental conditions within the turbine and for example a temperature range from −25 °C up to +60 °C are designed to be installed in the turbine without additional enclosure. Nevertheless, these components require a power supply as well as communication, must be connected to the other parts of the control system and therefore are wired to the control cabinets. Thus, the control cabinet forms a central component where the cables are brought together and from where the distribution of the power supply takes place.

Depending on the design and philosophy of the turbine only one central cabinet in the nacelle or multiple distributed cabinets in the tower, nacelle and hub are installed. Common is one tower, one nacelle and one hub cabinet, while two or more cabinets in the nacelle as well as a single cabinet in the hub per blade respectively pitch system is not unusual. In designs where all major components including the transformer are in the nacelle, a control cabinet in the tower may not be necessary. However, while components such as the switchgear are often installed in the tower base or foundation, it is common to install one control cabinet there, which contains the hence easily accessible processing unit. Two such control cabinets are shown in Figure 7.7. The external cables are fed in and connected from below, while the internal and external signals are distributed upwards, routed via protective devices, and supplied with power. The power distribution, components for the Ethernet network, the industrial PC, the I/O modules, multiple contactors, and breakers are shown as well.

In the case of separated and distributed control cabinets, these can have separate supplies, but always require communication for the control system. In modern systems, the use of additional wires for the transmission of signals between the control cabinets has become uncommon, because the communication of all signals including the safety systems can be integrated and exchanged by a modern fieldbus communication. Communication connections between the control cabinet in the tower and in the nacelle are nevertheless necessary and in today's turbines need to cover a distance of over 120 m or more. However, the usual fieldbus protocols and their transmission mostly allow for copper cables with a maximum length of 100 m and therefore fiber-optical connections are common.

Fiber-optical connections are furthermore insensitive to electromagnetic influences. This electromagnetic compatibility (EMC) needs to be considered while planning cables and their routing through the turbine. Not only the cables that

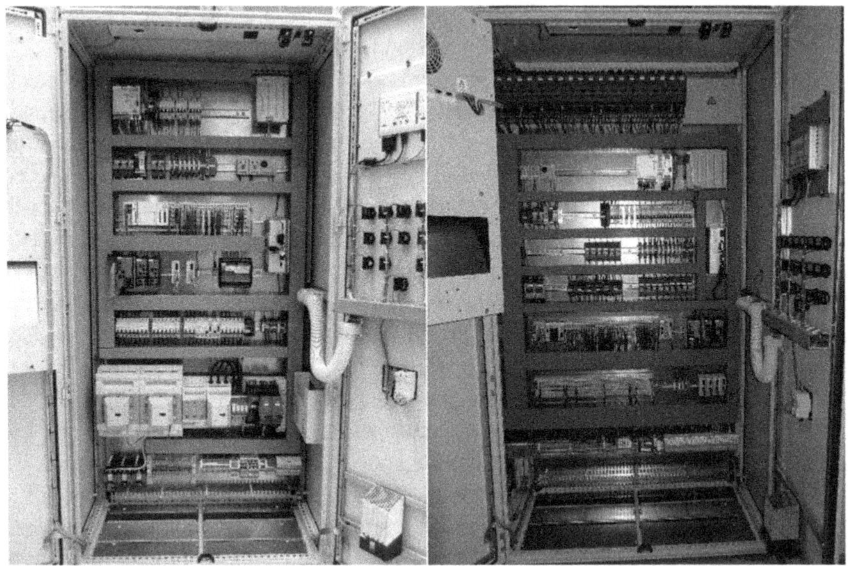

*Figure 7.7 Control cabinets for the tower base (left) and nacelle (right) of a 5 MW offshore wind turbine show the components for the automation, including power supply, protective devices, and air conditioning. Copyright © Beckhoff Automation GmbH & Co. KG.*

transmit several megawatts of power, but already connections to aggregates like fans or pumps can lead to transient currents and thus electromagnetic interference on other cables, especially when switching on. Therefore, shielding and the correct handling of it by applying it in the control cabinet, for sensor and communication cables is essential in wind turbines.

Since electrical bonding within the overall wind turbine is not always possible or well implemented, this could also lead to difficulties in the control cabinet. Sensors and signals that are based on a voltage and its adjustment can lead to incorrect results or malfunctions due to the changed potential. Also, the potential differences can lead to an unwanted current flow through the cable or its shielding and must therefore be considered in the design as well as while manufacturing and commissioning of the machine.

Although the wind turbines are constantly getting larger and the dimensions of the nacelles have meanwhile reached the size of multi-family houses, there is not always much space left for the control cabinets next to all the large components installed. This is why a compact design of the control cabinets as well as the components inside them is also important. It can also lead to the components being divided among several control cabinets to be able to accommodate them in the construction rooms at all. The cable lengths already mentioned, but here to sensors and actuators, can again become significant with these dimensions and thus it would be reasonable or necessary to divide and distribute the control cabinets or use separate modules with higher protection class.

After selecting the necessary components such as power supplies, circuit breakers, relays, and terminal blocks, the housing and enclosure of the control cabinet will be chosen depending on the available space in the turbine. The components need to be compatible with the electrical and control systems and fit in case of size and power ratings. But also, the supply power for all components inside the cabinet, as well as the aggregates connected outside of it, such as the pumps, heaters, and fans must be considered. A separate interruption of the supply by the safety systems as well as protection against overcurrent and other effects by means of fuses is necessary. The operation of large aggregates with high power consumption can often not be done directly from the I/O modules and thus require contactors, which then are utilized to switch on and off the large currents. The single line diagram (SLD) in Figure 7.8 shows such large actuators and their circuit breakers. Among them are two cooling systems, each with two heaters, one with two pumps and the other with two fans.

As with electric potential differences, other external and environmental influences can also lead to issues, malfunctions or damaging of components. Therefore, further protective measures and protection devices like lightning protection must be implemented especially for sensors outside the nacelle, as well as components in the hub. Examples are the meteorological sensors which are usually mounted on top of the nacelle as well as the components like blade bending sensors with connection into the rotor blades thus in a lightning protection zone. The standard IEC 61400-24 applies to wind turbines and defines the environment, different levels as well as risks and measures for lightning protection [22].

After all the considerations about the external conditions for the control cabinets, the mounting plate, and the placement of the components as well as the internal cabling can be planned inside the control cabinets. Here, not only the influence of cables and components on each other must be considered again, but also the behavior towards temperatures inside the cabinet. For this the functional enclosure layout in areas like high power, low power, main power as well as actuators and sensors as susceptible devices are defined. The use of so-called bus bars for power circuits is common as well as for applying the shielding of cables before the connection of sensor or communication cables to the I/O terminals. For earthing and grounding are bus bars, as well as earthing straps used for potential equalization by large-area conductive mounting with low inductive inference. The power circuits should be separated, if possible, with a separation of high and low-voltage lines, while practices as right-angled cable crossovers are used to lower electromagnetic interferences.

Finally, the control cabinet must comply by all these measures with the current standards as the IEC 50204 and IEC 61439 up to a marking of "conformité européenne" (CE) for commercial products sold in the European Economic Area (EEA). These requirements and tests are intended to affirm conformity of electronic equipment with guidelines for health, safety, and environmental protection.

Closing the note on solutions offering a complete control system without a cabinet. Depending on the level of integration, those might include all major

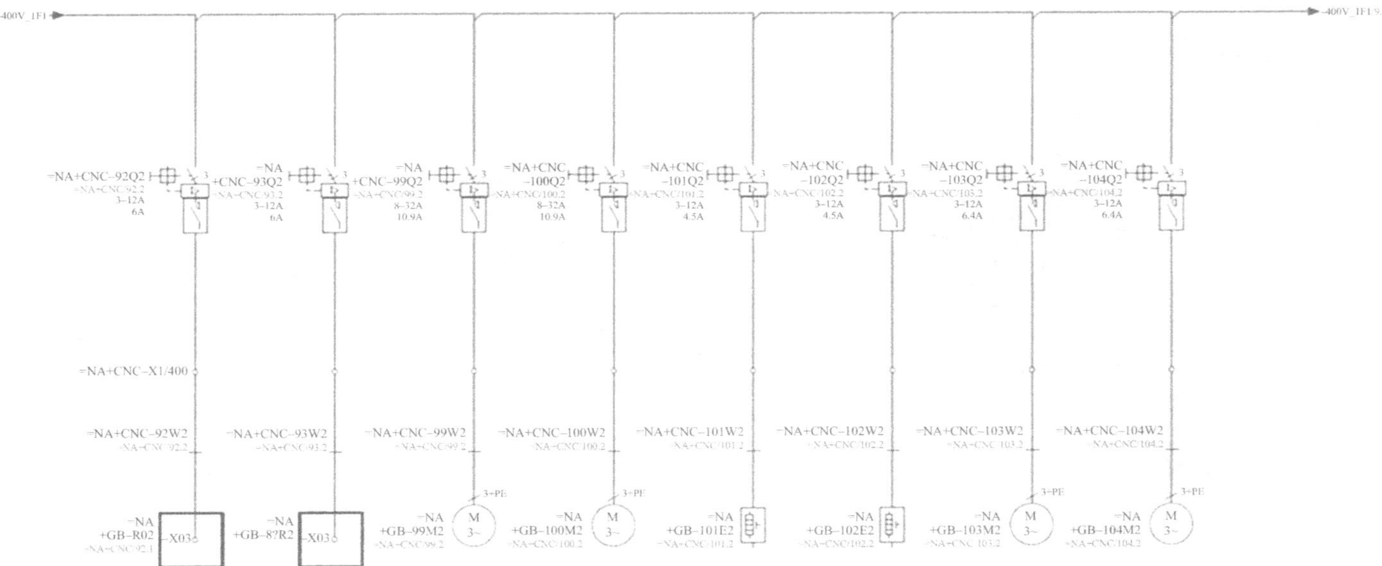


*Figure 7.8* *A single line diagram that shows the heaters, pumps, and fans of a cooling system. Copyright © Beckhoff Automation GmbH & Co. KG.*

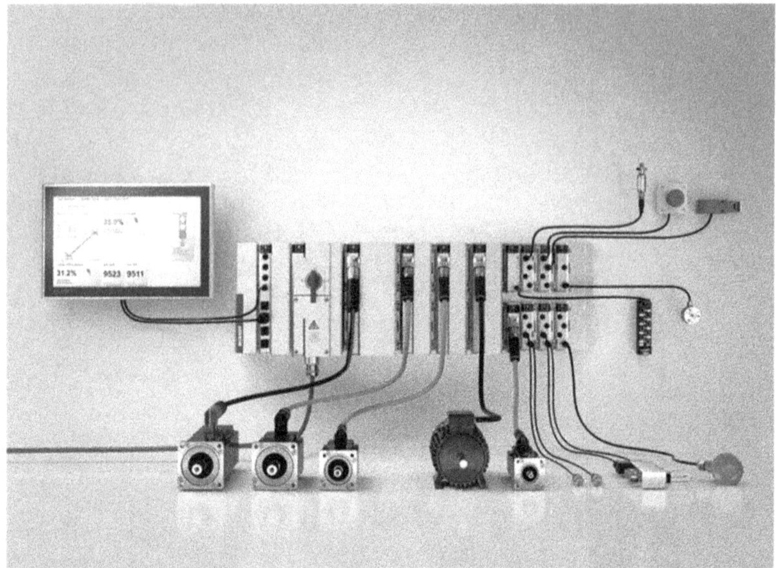

*Figure 7.9 The MX-System allows the installation of control systems without a cabinet, by integrating supply, protection, distribution, and communication with Industrial-PC, I/O modules, and drives. Copyright © Beckhoff Automation GmbH & Co. KG.*

components of the control systems such as the processing unit, I/O modules, drives as well as the power supply, communication, relays, and protection of electronics by fuses or others with a high ingress protection level. An example of such a system is shown in Figure 7.9, where an industrial PC, frequency converters for drives, direct switched motors, as well as I/O modules with different sensors, encoders, and safety signals are shown as protected modules in a metal housing with integrated power supply and protection mounted on a base plate. In this way, it can be made much easier to implement solutions for critical problems outside the control cabinet and close to a subsystem. But also, the need for large and complex cabling inside the plant can be eliminated by separating small islands of integrated I/O modules, for example directly mounted at a gearbox, generator, or mainframe.

## 7.6 Software application

The software application describes the software for a specific wind turbine type whose is usually implemented by the OEM based on the solution of an industrial control system supplier. The individual properties, functionalities, and singularities or individual wind turbines will be implemented into this software application and executed in the real-time capable runtime environment on the central processing unit (or distributed to the individual units). Thus, the software application also includes the specific know-how about the turbine and the intellectual property of the manufacturer, while the operating system and the runtime environment are usually supplied by automation and control system suppliers.

The timing requirements for this real-time runtime environment result from the dynamics of the turbine which shall be controlled. So, the usual time constant of the systems and subsystems, as well as structural oscillations and their frequencies as continuous signals must be considered. By the Nyquist-Shannon sampling theorem, the time discrete sample rate of the control algorithms must be at least twice as fast as the continuous signals. However, it is common practice to use a time step that is more than 10 times faster than the maximum transients in the system [23], also due to the performant systems available today. The Campbell diagram of the IWT-7.5 [24] shows for example oscillations up to 5 Hz and with cycle time of ten times the signal to allow accurate reaction times results in a sampling rate faster than 50 Hz or the commonly used cycle time of 10–20 milliseconds for the main control system.

Due to the autonomous, unobserved, and permanent 24/7 operation of the systems, a particularly resilient implementation of the application software is required. This should be considered in the conception and implementation with appropriate design principles. Any behavior of the systems must be acquired and monitored via the sensors, the signals must be checked for plausibility, and external influences must be eliminated. Problems such as faulty communication sensors or actuators, as well as in the communication or in the cabling should be reliably detected. If a malfunction or unexpected event occurs, the system must continue to operate or stop safely. If possible, the turbine must be able to resolve the situation and restart independently. Such strategies are also known as self-healing, allowing the application to recover from a malfunction or simply retry an operation. In computer science, this is also referred to as robustness and defensive programming to achieve a fault-tolerant system. But be aware that redundancy is not necessarily always the only or best solution after considering the error probabilities due to the increased number of components.

### *7.6.1 Basic functions*

To implement the overall software application, basic functions are usually standardized, at least within an organization. This is good practice for all major software developments, to define an abstraction of complex but reusable functions and to unify them. This not only increases the quality of the application software, because all functions and their improvements over time happen in one central place instead of being distributed and possibly not even unified and thus implemented differently, but also a uniform way of using and understanding is used across engineers and optionally across multiple projects. Such standardizations may also include directives for programming, so-called coding guidelines, as well as uniform naming guidelines like the definitions in the IEC 61400-25 for wind energy systems.

For this purpose, the usual methods of software abstraction are used, the elements are encapsulated and stored in a separate area often called as software library. This process often happens during the development and enhancements of the software project. It is recognized that elements are used more frequently or are very similar and these are then standardized, optimized if necessary and stored in the library. The library can then be easily accessed by the various other areas of the software and this function can be used. Should optimization, extensions be carried out or malfunctions be corrected over time, all users of the function automatically benefit from this change.

This leads to one or many software libraries and software functions. Suppliers of industrial automation technology often offer libraries, tools, and project templates to simplify the use of their technologies. On the one hand, these can be basic functions for the use of the operating system, the runtime environment or extending functions and services. In many cases such libraries are also offered especially for the application of a wind turbine. Depending on the scope and functionality of this prepared functionality it is then referred to as a software library of a software framework. That could include not only services that simplify implementing the application, but also interfaces to other software tools or services as well as elements for the visualization in a user-interfaces, up to completely prepared applications.

Developing the whole software application from scratch does not only include implementing the functionalities, but also structuring them in an understandable and clearly arranged layout with hierarchical relationships between the components of the software. The extension and maintenance of the software over decades while operating the turbine will otherwise suffer from complexity and make it difficult for established but also new developers to continue working on the software. Therefore, the software frameworks of the control system suppliers often include prepared applications as template, which includes the entire wind turbine application, the operation modes, stop modes, but also control and interfaces of subsystems. Every turbine design and integrated components are unique in its combination, nevertheless, the essential elements are often the same, prepared in such an application template and only must be adapted to the specific system, not developed from scratch anymore. The architecture of such a software framework is illustrated in Figure 7.10. Generic functions are provided by the framework in services and specific functions are offered in a template. All essential components of the application software are already implemented in prepared modules and the subsystems are also prepared as individual exchangeable modules with the possibility of simulation. The connections to I/O modules via the fieldbus, as well as data exchange with external systems and storage via a database, are modularly integrated.

Back to the basic software functions, these are often named differently and implemented differently in detail. However, some core properties that can be found similarly over the different implementations have prevailed. Due to the similar application and requirements, there are not a plurality of implementations as straight forward solutions possible. As well as through the engineers who implemented the solutions in the beginning of the industry, passed their knowledge and standardization to subsequent engineers, perhaps even passed it to other companies so that the solutions are similar or even the same, to the point of software parts that have existed in companies since the 1990s.

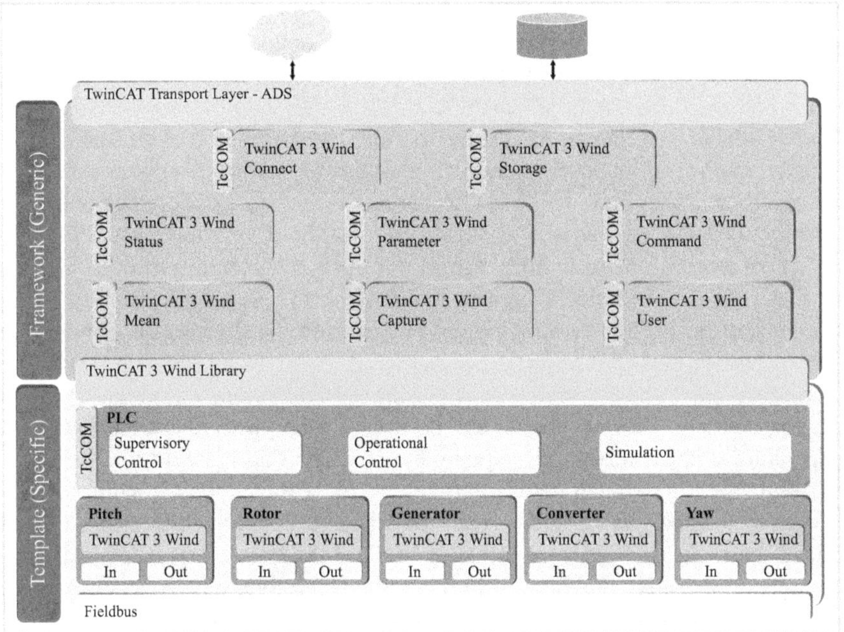


*Figure 7.10* *Overview of a software framework for wind turbine automation, including generic services as well as specific implementations for the turbine and its subsystems. Copyright © Beckhoff Automation GmbH & Co. KG.*

#### 7.6.1.1 Status codes

Such functionality is the monitoring of the overall turbine by many individual status signals, also called status codes or status objects. Such a status represents a specific situation in the system, where a monitored signal has exceeded a threshold, or its value is outside of a specific range. Thereby it is for example possible to observe an input signal from a sensor, whether it is plausible, in a physically possible range or an issue in the acquirement and processing of that signal occurred. But also, physically valid values might trigger a status signal because they are outside of an allowed operational range of that system. By triggering a status signal several reactions can follow and the trigger is handled as a kind of event. One essential reaction could be to initiate a stop of the turbine. Detecting a malfunction or impermissible condition could lead to damage and requires stopping power production. This event will then be notified as information, warning or even error that must be reported to the maintenance staff.

This functionality is often used hundreds or thousands of times in one application. Nearly every connected sensor with all its input values must be validated and approved, resulting in multiple such status signals to detect a broken sensor, broken wire, the value is invalid by any other circumstance, whether the values are too low or too high or have changed to fast are only some possible opportunities for evaluation. So one sensor for the temperature of the ambient air could lead to many types of status signals: The temperature is too low to operate, the temperature is too high to operate, the temperature could lead to potential icing and if the temperature is not available it might be necessary to stop also because it is not possible to validate that the turbine is in sufficient conditions to operate. The value of a pressure sensor could be high, low but also the rate of change could indicate an issue with pipes or other parts as well as slowly and steadily decreasing pressure without any consumer could indicate a leakage of oil. Such evaluation must be considered in the overall system, since switching on/off the heating/cooling naturally affects the temperature, just as activating pumps or opening valves affects the pressure and so the monitoring of those must be considered. This is also used to monitor such components since the system reacts to changes in an actuator through an output and, for example, the function of the heating can be checked by the increasing temperature or the function of the pump by the increasing pressure.

While there is so much status information and messages about conditions in the turbine, it is important to avoid reporting massive consequential errors or subsequent messages. It is very irritating for the user if a message is displayed which does not correspond to the root cause. If possible, therefore only the status message of the cause should be displayed, while subsequent or lower priority messages should be hidden or avoided. For example, if a cable break leads to a wrong signal of a sensor, the message should be called "cable break" and not "temperature too high." Also, not both messages should be shown, but just the original one indicating the real cause for a stop of the turbine and whose might need to be fixed by the maintenance personnel.

The term "status codes" has spread through the original use of numbers or alphanumeric codes which represent the status signals and was used as an indication of the operational status or condition of a wind turbine. By this nomenclature and grouping of such codes, the service and maintenance personnel could easily draw conclusions about the condition, its probable cause and allowed sufficient troubleshooting [25]. The numbers were then memorized and led to familiar sayings among employees such as "code 42 again" instead of "overtemperature again."

#### 7.6.1.2 Parameter

As every turbine has fixed conditions defines by its design, as rated power, cut-in and cut-out wind speed, but also allowed operational ranges, several conditions need to be adjustable while running the turbine without the need of a complete software update. Therefore, the use of parameter values is a common implementation to allow a convenient change of settings in the field. Those values are not fixed and constant within the software application but prepared to be changed externally without the need to change the software application itself to change a simple limit or behavior. Very common use cases are set values for control algorithms as the pressure a control logic for a pump must reach, the temperature for a heating and/or cooling system, the position or speed that a motor should drive. Usually, the wind turbines are all built-in series in the same way, but due to small changes, production deviations or different environmental conditions, adjustments for set and limit values become necessary.

To ensure safe operation despite the possibility of configuration changes by these variable parameters, the defined values must be checked especially. For example, there may be limits like a minimum and maximum values within a user is allowed to change the parameters or preconditions which are required to allow a change at all. In addition, authorizations are also assigned to the parameters, which users or group of users are allowed to change the parameter or whether they need to be on side to change them. The parameters are often also divided into categories, for example depending on the phase of the lifetime of the turbine in whose it is intended to change the parameter. Some Parameters arise specifically from the design of the turbine and can therefore no longer be changed after installation, other parameters arise during commissioning such as the offset of the nacelle position for aligning the nacelle to the north direction and other parameters such as fan speeds or temperature ranges that can be controlled are subject to changes due to aging and can therefore still be changed even after several years of operation.

#### 7.6.1.3 Commands

During the commissioning but also afterwards, during the operation there are many kinds of operations an engineer would like to perform but which are not necessarily intended for the automatic operation of the system. For example, an aggregate shall be switched on for testing purposes or switched off to reduce noise while staying in the nacelle. It must also be possible to stop the turbine manually at any time, to reset it in the event of an error and to manually request a start of the turbine. Manual processes may also be necessary during service and maintenance or while operation from remote.

But like the parameters, those manual processes must be started with care to continue safe operation. Therefore, the preconditions to allow a manual operation, the execution of those operations but also the permissions to run them must be carefully considered. For example, certain aggregates may only be manually controlled if the demanding person is on site or if the turbine has been stopped and is in standstill. Such processes can be executed locally by pressing a hardware button at the cabinet door or through a human-machine interface (HMI) whose is used locally or remotely, where both ways trigger a command in the control system.

Figure 7.11 shows common buttons, switches, and lights in the front door of the control cabinet. The important emergency stop button to open the safety chain and stop the turbine in case of a critical situation is apparent. But also, several buttons to reset, start the operation as well as bypassing of safety functionalities are visible. Such buttons are connected directly to I/O modules, but nevertheless, their commands require a prior check of authorizations before they are to be executed by the control system as well as later traceability.

#### 7.6.1.4 Signal processing

For many signals pre-processing is necessary. That could include the calculation of the rotational speed from a proximity sensor: The proximity sensor will have a high signal in case a bolt head or tooth passes by these high signals will be counted and

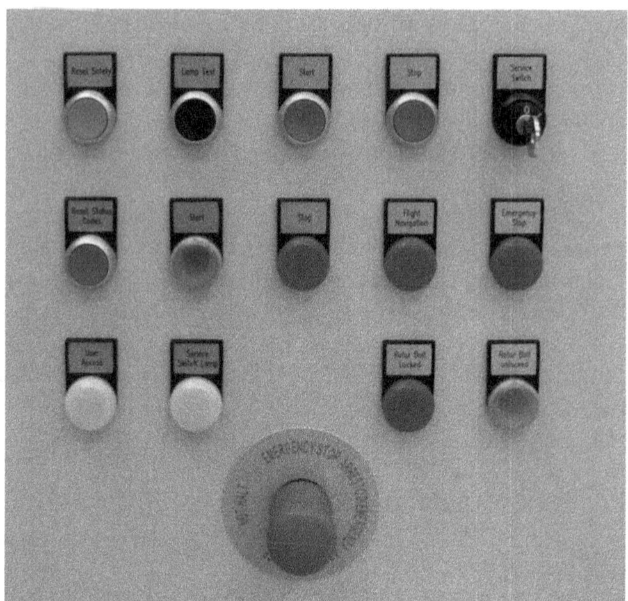

*Figure 7.11 The basic user interface of the automation system are buttons and lights on the control cabinet, which are used to display states and trigger commands. Copyright © Beckhoff Automation GmbH & Co. KG.*

timely evaluated to calculate based on the total amount of teeth per rotation a rotational speed in rounds per minute. One supposedly completely different signal is determined from another signal, often referred to as a virtual sensor. However, a value is often calculated from several other signals, such as the tip speed ratio calculated from the rotor speed and wind speed. In addition, virtual sensors or virtual encoders are used as software-based algorithms to estimate or predict a signal by the relationships of completely different signals or by using a model of a component to determine a signal. That could be a strain-gauge sensor that is used to determine movements, bending, loads and forces acting on a shaft, a carrier, or a rotor blade. Due to external influences or inaccuracies of sensors and their processing in the I/O modules, the signals may be noisy or subject to interference. Such signals must be filtered before they are used for further evaluations to reduce or eliminate interference. Also, mechanical disturbances or oscillations are often removed from signals before they are used for control purposes.

It is also common to apply statistical means to the signals, both for preprocessing but also for post-processing. This is how averages of signals are formed to eliminate higher frequency influences or to interpret or compare the behavior over a longer period. But statistical values are also used for the evaluation of the plant and its performance. The way of evaluation is part of standards such as the IEC61400 and prescribes the calculation to ensure comparability among turbines and manufacturers. The most popular example will probably be the power curve which is normally shown in every data sheet of a wind turbine and determined from a 10-minute arithmetic average of wind speed and the power fed into the grid. To eliminate environmental influences and local conditions, the wind speed is often corrected with the air density. To remove unusual operational states of the turbine from the statistics of the power curve, interaction by conditions is then added, such as that only after 10 min of a stable operation the values are taken into account for the statistics so that not during start-up and ramp-up a reduced power value affects the curve.

#### 7.6.1.5 Signal recording

Due to the remote location and non-permanent observer, but also for long-term traceability and statics, many signals need to be permanently recorded, stored, and kept available. This means that even if only 50 major 10-minute average values are retained over the lifetime of 20 years will need at least 200 MByte for a single turbine (24 hours * 6 values/hour * 50 values * 365 days/year * 20 years * 4 byte/value). In the past files on the hard disk of the processing unit were often used for this purpose and the data were stored in a text format for easy readability for users. With today's modern turbines hundreds to thousands of values are generated and recorded every second and under special circumstances even faster for specific time ranges in case of an event. For this purpose, simple text files are no longer productive and more compact, more precise, and more structured storage formats are used as, for example, databases are directly integrated or connected by the control system.

Beyond that are all the events as status signals must be recorded, the latest parameters and their changes, executed commands and their interactions, as well a special high-resolution signal recording in case of a malfunction or stop of the turbine. This high-resolution recoding is often called a trace log and is especially convenient for troubleshooting: In case the turbine has detected a malfunction that results in a stop of the turbine, the trace shall allow a much later examination of the circumstances that lead to the malfunction, but also how the turbine responded to that und how the stop operation has been proceeded. For special circumstances, the service and maintenance personal are no longer able to know the details of all the components and turbines, so an extended post-event evaluation of such malfunctions is performed by special personnel or even the development department itself.

### *7.6.2 Supervisory control*

The overall behavior of the turbine is usually defined in operational states, as well as conditions for the transitions between these states. This functionality is concluded as supervisory control and contains the overall state machine of the wind turbine application. Overall targets and logic to control and operate the turbine are defined by the different states within this state machine, while these targets could be the rotating speed of the rotor respectively generator, the minimum allowed pitch angle, or the power to be drawn from the generator by the converter. But also, higher-level changes in the states of subsystems are controlled as well. This is used for example to stop the turbine or to implement the initialization and startup of the turbine from a kind of standby state.

From the standards as the IEC 61400-1 [3] there are also basic definitions of operational states in the design load cases (DLC) for example, “power production” in the DLC 1.1 to 1.6, “start-up” in DLC 3.1 to 3.3, “normal stop” in 4.1 and 4.2 as well as “emergency stop” in DLC 5.1 or “parked” in standing still or idling with allowance to slowly spin the rotor in DLC 6.1 to 6.4. These states differentiate, for example, the speed of the pitch system moving back the rotor blades to feather position for a stop and then evaluating the resulting structural loads. Nevertheless, as stated in this section there might be specializations necessary to differentiate the behavior for other standards like the machinery directive or requirements from subsystems.

#### 7.6.2.1 Stop modes

The conditions to stop the turbine are usually any kind of malfunction that is detected. Depending on the origin and importance of this malfunction, different kinds of stop modes are common. There could be a “normal stop,” a “fast stop” as well as “emergency stop,” but also special stop modes like a “grid stop” are used. In the following list are suggestions of such modes and their names:

- Rotor Parking: Positioning and holding the rotor in a certain direction, in case of a helicopter approaching an offshore wind turbine.
- Normal Stop: A slow shutdown of the turbine with avoidance of increased loads due to non-critical malfunctions.
- Fast Stop: With marginal loads a faster stop of the turbine, in case of malfunctions that can lead to damage in a timely manner.
- Grid Stop: A stop of the plant in case of failure in the supply network, which requires the use of energy storage devices such as batteries.
- Pitch Stop: If a fault occurs related to the pitch system, an immediate stop with the help of increased torque on the generator may be necessary.
- Safety Stop: The safety chain triggers an emergency drive caused by overspeed of high vibration, but only for the safety of the machine.
- Emergency Stop: If the safety chain performs an emergency drive due to an emergency stop button with potential risks for persons.

In case of a common issue, where a safe and stable operation is not possible, but no potential damaging of any component within the near future is foreseeable, an often called “normal stop” is sufficient. The turbine will slowly reduce rotational speed and produce power without increased structural loads to bring the turbine to a standstill or idling operational state. A further cause, which initiates a stop mode but should be rather unusual, arises from the opening of the safety chain. The safety chain might already independently from the control system trigger an emergency feather command of the pitch system and if necessary, also a load throw-off on the converter. This very fast attempt to stop the turbine to a standstill resulting in the corresponding high extreme loads and shall therefore be very uncommon. In the main control system this stop might be defined as “safety stop” or “emergency stop,” depending on the cause. In case the safety chain was opened by a device like the overspeed-detection or shock-detection a “safety stop” will be initiated, stops the turbine and results in an idling of the turbine, but it might be still allowed to spin the rotor very slowly. On the other hand, the safety chain might be opened by a human pushing one of the emergency stop buttons whose could initiate the "emergency stop." Because this case does not originate from a device to protect the machine, but from a person on site so that there is a potential human safety risk to that person, it may require a different reaction. Unlike a "safety stop," an "emergency stop" would therefore require switching off actuators immediately, except the pitch system, but potentially including the power supply to aggregates such as motors, fans, and pumps that are rotating or consist of moving parts and could therefore cause harm to persons. This stop operation is still initiated by the safety chain with emergency feather and load throw-off, but then also applies the brake depending on the rotor speed. So finally, the applied brake shall hold the rotor and prevent any spinning so that a moving part could harm any person on site.

Early software applications used enumerations or integers to define the different states by a number. This number was shown in the software and user interface then also as a number and not converted into a text representation. Therefore, many control and service engineers in the wind industry also know the numbers of these stop modes as "normal stop" with 50, the "fast stop" modes starting with 100 and then "emergency stop" will start with 200. Variants or specializations of the stop modes were then sorted into this numbering according to weighting it. This also resulted in normal stop modes with the number 51 or fast stop modes with numbers like 150 or 160. Several stop modes are also differentiated by their origin which led to or triggered this mode. If a stop whose was initiated by an opening of the safety chain the "safety stop" with number 200 was used and if the stop was initiated by an emergency button, which was pressed by a human to stop the turbine is named as "emergency stop" with number 210. In today's applications, however, depending on the OEM the name of the state is more commonly used again and displayed as text in the user interfaces, while internally in the software the enumeration with numbers is still used.

This numbering can also be understood as a priority. A stop mode with lower priority is represented by a lower number and can be replaced or overwritten by a stop mode with higher priority. For instance, if a normal stop mode with the number 50 would be applied and might take some minutes to bring the turbine to standstill but in the meantime, another effect causes a safety stop with the number 200. That higher prior safety stop will replace the original normal stop mode and initiate a faster operation to stop the turbine and demand other targets.

Specific requirements whose require the invention of further stop modes might arise also by a subsystem. For example, from a stop which is initiated by a grid loss: The overall energy grid has been down over a longer period, so that a low-voltage ride-through has already been exceeded but the voltage has not recovered. Then the pitch systems had no power supply and might have applied an emergency stop by using its storage. Nevertheless, whether the storage is in a hydraulic pitch with pressure accumulator or in an electrical pitch with electrical capacitor, this storage must recover, be refilled, or charged. Such conditions must be met either during the execution of the stop mode and during the shutdown or of the turbine, because on a grid fault also other aggregates do not work and need to be switched off, or after the end of the stop mode and during the attempt to restart the turbine. In these cases, it might be reasonable to introduce such a stop mode "grid stop" and allow to implement the requirements in the software application in a simpler and more targeted manner.

#### 7.6.2.2 Operation mode

Besides the stop mode to interrupt the production and break the turbine, the operation modes are used to start, ramp up and operate the turbine. Should any malfunction occur or be initiated by the safety chain, a stop mode can be triggered at any time to interrupt during operation any of the following operation modes. This should not be the normal case and the usual operation modes therefore should outweigh the stop modes in terms of the total time of execution. Some examples for possible operation modes are:

- Initialization: First state after power on the control system to boot and initialize the system and all subsystems.
- Standby: Ready to operate but waiting for example environmental conditions to start the turbine into a productive state.
- Service Mode: Maintenance personnel are present and servicing the turbine.
- Self-test: The turbine carries out tests to verify its own safe operability and the functionality of important subsystems.
- Start Up: The start of the turbine to reach a minimum rotational speed and to switch on all operational processes and subsystems.
- Grid Synchronization: The rotational speed of the drivetrain is sufficient to start but the synchronization to the energy grid is outstanding.
- Reduced Production: Due to specific conditions or external requirements, the turbine is not operated at full power.
- Production: The turbine is fully productional and feeds electrical energy into the grid.
- Fault Ride Through: Due to a fault in the electrical grid, special functions are used to keep the turbine in operation.
- Pitch Lubrication Move: If further pitch system movements are required to lubricate the pitch bearings. There are also comparable modes to lubricate the yaw system.

The turbine will be initialized after power up and with the booting of the systems, an initialization of all components, the establishment of communication and an initial exchange of information between the devices. In this initialization mode also a configuration and parametrization of all systems might occur, based on data and information form a persistent storage as a database or file on the hard disk of the control system.

Only when all major systems are up and running the turbine will enter standby mode. Here the control system is then ready to receive manual commands and to accept local or remote interactions. Depending on the design of the system, the rotor is allowed to spin and idle or must be braked and in a standstill. Many of the aggregates might already be in operation here: Cooling and heating circuits are already used to bring or keep all aggregates at operating temperature, oil and lubrication systems can be perform greasing but also the yaw system can proceed with extended limits to align the rotor into the wind direction.

While all sensors are already being evaluated, an average value of the measured wind speed is used to determine whether it is appropriate to start the turbine If the wind is too weak or strong, the turbine remains in standby due to the environmental conditions. Corresponding limits are used for this purpose, which result from the performance curve and are stored as parameters in the control system. Only when all internal and environmental conditions are suitable, the start of the turbine is aimed. Otherwise, the turbine must wait for the wind to rise.

Before the turbine can be started to rotate the rotor with higher speeds, important systems necessary for safe operation and stopping may have to be checked. For this purpose, the so-called self-tests are required at regular intervals or after special events like an emergency stop. In this self-test mode, an emergency drive of the pitch system may be required and carried out with the goal of checking the accumulators of the pitch system. For electrical systems that might be batteries or ultracapacitors and for hydraulic systems these might be pressure accumulators, as described in Chapter 3 of Volume 1. The rotor blades are then moved by the pitch system simultaneously or successively to a position such as 60° and then an emergency feather drive is executed. The procedure to move the blades until the pitch drives come to a standstill in the feather position at 90° will be observed and evaluated by the control system, as it is operated independently by the pitch system. The pitch motors must drive the rotor blade out of the wind with a required speed, reach the end position within a certain time, and during this time the currents and voltages of the batteries, for example, are also monitored and evaluated. In doing so, you could even test the rotor brake at the same time, because at the targeted 60° and at corresponding wind speed, the rotor should already start to rotate. But if the brake is applied during the emergency drive test and there is no movement of the rotor, the rotor hold will probably have worked.

To start the wind turbine and bring it to productive operation, a start mode is operated. Since most systems cannot feed directly into the grid from zero speed, a rotational speed above the minimum speed is aimed to switch on and synchronize systems such as generator and converter. The pitch system and the pitch control are used to approach this speed and keep it stable. Because the power coefficient $C_p$ of the rotor at the common optimal pitch position $\beta$ of 0° becomes negative with a very low tip speed ratio $\lambda$, it is sufficient to implement a limitation of the pitch position to the optimal angle in relation to the actual tip speed ratio. That will always guarantee to operate at maximum power coefficient while starting up and not to choke the rotor while trying to speed it up. Then the converter can be started to magnetize the generator if necessary and synchronize it with the current phase position of the mains voltage. As soon as stable operation and power production have been achieved, the rotational speed can slowly be ramped up together with the torque applied to the generator until the nominal set-points are reached and the system switches over to the power production state. The visualization for a wind turbine is shown in Figure 7.12, where such a start and stop of the turbine is

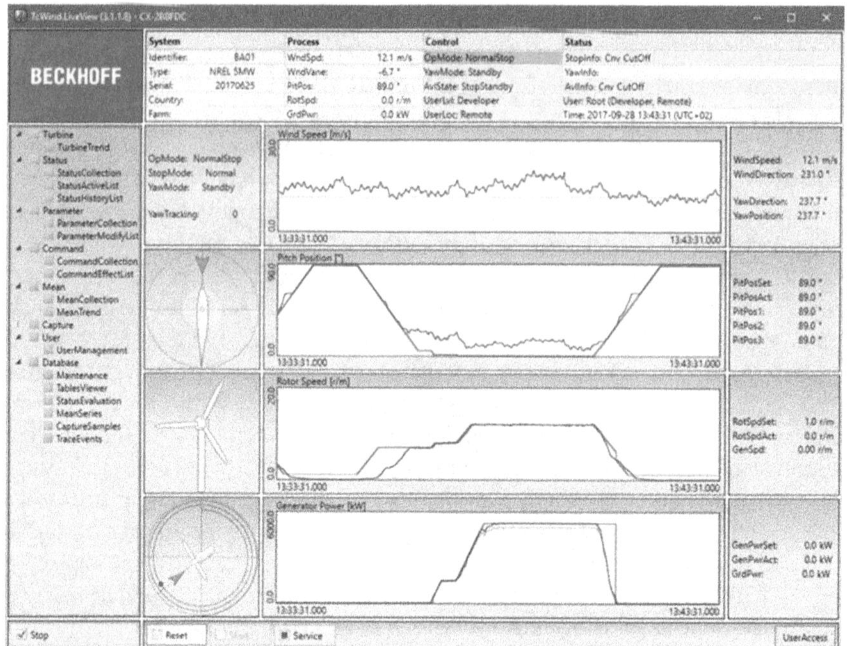


*Figure 7.12 The visualization shows the start as well as a stop procedure of a wind turbine, with the trends of wind speed, pitch angle, rotor speed, and produced power. Copyright © Beckhoff Automation GmbH & Co. KG.*

displayed. It visualizes how the pitch angle is decreasing and the rotor speed is slowly increased until the converter is switched on and power is produced.

Since the classification of these operating modes is not exclusively used for functionality, but later also for monitoring and statistical evaluation, further subdivisions of the operating modes are made. In a later calculation of the technical and operational availability of the turbine as defined in the IEC 61400-26, these specific operation modes are then considered differently in the statistics. For example, when there is any request to reduce the power, another mode is used for reduced production. Such requirements can arise operationally when the temperatures of the bearing or gearbox become too high and therefore the speed should be reduced. But there are also increasing external conditions, whether to reduce noise emissions or shading during specified periods throughout the day or to protect animals such as birds or bats from the effects of the turbine. The power production might also be reduced because there is simply no corresponding demand for this energy, the electrical grid cannot transport the quantities of energy, production is not attractive due to electricity price trading, or a reserve of power will be kept available for short-term demands.

Further operation modes arise also from technical requirements. In the event of a fault in the electrical grid such as a voltage drop, the turbine is required to support the grid, keep operative and not switch off for specified time despite the fault.

This is referred to as low voltage ride through (LVRT) as well as high voltage ride through (HVRT) and summarized as fault ride through (FRT). In the case of the operation of the turbine over a long period but without any major movement of the rotor blades the lubrication can become unfavorable. The lubricant would always be applied to the same spot, could not spread, and might leak. Therefore, it becomes reasonable to briefly pause the full load and move the pitch system. While the rotor blades move to a higher position of for example up to 40°, the turbine can stay operative in partial load and apply lubrication of the bearing while moving it. This enables the lubricant to be optimally distributed over the bearing.

The service mode is also specific as other settings may apply here. If the maintenance personnel are on site, climb up the tower and enter the nacelle, the rotor must standstill and the yaw is not allowed to move as well. That might require applying the brakes as well as switching off other subsystems or actuators. However, in many cases it can be necessary to enable the manual operation of aggregates for testing purposes or the commissioning. Through to the need to operate the system during initial tests and approve the system according to manual specifications of target values. Reducing the noise exposure and risk due to moving parts of the staff when they are present in the nacelle by switching off fans and pumps is also obvious.

#### 7.6.2.3 Yaw modes

There is also a state machine involved to control the yaw system. Its states are loosely coupled to the stop and operation modes, because even if the turbine is stopped and there are no errors in the yaw systems, it is still used to align the nacelle. This is because the oblique flow can lead to high loads, even without the rotation of the rotor and therefore any misalignment always be avoided. Possible modes are, for example:

- Automatic: The automatic tracking of the nacelle to the wind direction.
- Standby: In standby of the turbine a reduced and slowed down tracking.
- Unwinding or untwisting: Dissolve the twisting of the cables in the loop.
- Suspended: The tracking cannot take place due to a malfunction.
- Manual: A tracking based on manual demand by a person.
- Avoid: Move the rotor out of the wind by 90° or 180° to avoid any loads in special wind conditions for instance.

Because it is intended to keep the yaw misalignment low, so that the difference between the wind and the nacelle direction is always minimal, an automatic tracking of the nacelle is mostly aspired. While in the automatic mode it is intended to keep the misalignment below 4° to 8° to prevent any loss in the power performance and keep the loads withing the envelope, it is allowed to keep a higher misalignment in the standby mode. Since still in today's most wind turbines the movement from the yaw is performed against an applied brake, every movement becomes subject to loads, but especially to wear of brakes, gears, and motors. Therefore, it must always be weighed up whether a movement is appropriate. The limits used for the closed-loop control and to start the tracking are defined differently in the automatic and standby modes. While the turbine is not in operation, for example, stopped or in standby, much longer periods with higher misalignments are allowed to reduce the number of movements. On the other hand, during the operation of the turbine, tracking is started in automatic mode at smaller intervals and minor misalignments for maximum efficiency.

The cables that are routed between the tower and the moving nacelle are hung in a loop. This cable loop allows the nacelle to move against the tower while the cables are twisting. But this twisting has a limit and so the cables can usually be twisted up to two full turns in each direction. That corresponds to a total permissible movement and an end position of the yaw system at e.g., 720° in one or the other direction. Since excessive twisting of the cables can cause massive damage, this twisting is also monitored by a position sensor and limit switches from the safety system. Recently, safety encoders with safe communication and determination of the position have been used to implement limit switches by end positions in software. In the event of excessive twisting, the safety chain is thus opened and manual disengagement by a maintenance employee is necessary. To avoid this circumstance, the control system must intervene before strong twisting occurs and carries out untwisting. However, since this untwisting or unwinding requires the nacelle to rotate completely around its own, rotation during operation is not possible and would lead to high loads. Due to the relatively slow process of the yaw at a few degrees per second, a full rotation can already take half an hour or longer. It is therefore important to consider when the unwinding is to be carried out and whether the possibly productive turbine shall be stopped for this purpose. For this reason, unwinding is often carried out while the turbine has already been stopped anyway. It might be ideal to unwind at low wind speeds, also because no large loads are to be expected then due to oblique flow. Since the turbine is waiting for the wind anyway, an unwinding can already be carried out prematurely, even if the cables are only twisted one turn.

In the event of an emergency stop, the yaw system must of course also be shut down and switched off. In the event of many other faults and stops of the turbine, the yaw system can still be operated. But during stop and start-up, while the strong acceleration or deceleration of the rotor takes place, a pausing of the yawing can be useful to reduce the gyroscopic force. Furthermore, even if the end position may have been reached, it must be possible to allow manual movement by employees to resolve the situation. Since this takes place when the employees are present and visually monitored by them, manual tracking can still be allowed even in the event of faults in the positioning system.

How the yaw system is designed is described in Chapter 4 of Volume 1, as well as the closed-loop control and the subsystem from the control system point of view in Sections 7.6.3 and 7.7.7.

### *7.6.3 Operational control*

The dynamic control of the wind turbine during production can be divided into three parts: yaw control, torque control and pitch control. The yaw control aligns the nacelle into the wind direction so that the rotor is optimally aligned to the incoming wind flow. The energy taken from the wind flow and converted by the rotor blades into a rotation of the rotor and drivetrain is controlled by the pitch control. The angle of attack of the rotor blades is adjusted mainly based on the rotational speed of the drivetrain. The energy which is then taken from the drivetrain via the generator and converted from rotation into electrical energy is determined by torque control. In this way, the energy balance in the drivetrain is controlled as a function of the speed by applying the appropriate torque.

How the blade angle of the rotor is then adjusted by the pitch system or how the torque is applied and converted into electrical current from the converter is the responsibility of the subsystems and not intended here. Subsystems may have further subordinate closed-loop control algorithms as, for example, the speed control of pitch motors or the current control in the converter. However, the operational control here concentrates exclusively on the higher-level control of the drivetrain by specifying the pitch angle and generator torque based on the rotational speed.

During the design of the dynamic control and the corresponding design load calculation, the following partially competing objectives must be weighed against each other [23]. While it is intended to produce the highest possible output of electrical energy, the structural loads must be minimized to ensure the required lifetime of 20 years or more.

- Economic objectives
  - maximization of power production
  - improvement of power quality
  - maximization of profit
- Protective objectives
  - reduction of loads
  - reduction of vibrations
  - guarantee of stability

For the design of the controls, the time constants of the controlled systems are relevant. These result from the inertias of the process and delay times of an actuator but also dead times in communication, why also the cycle time of the controller, the fieldbus and its jitter are relevant. Typical times of the subsystems are [26]:

- Yaw control with 1 to 4 °/s
- Pitch control with 4 to 8 °/s
- Converter control less than 1 ms (2 to 8 kHz)

#### 7.6.3.1 Operation points

Completely autonomous and mostly independent is the yaw control. Even when the turbine is not in production, the nacelle is tracked into the wind direction to keep the optimum angle of attack on the rotor as well as low loads on the structures due to oblique inclined wind flows.

The pitch and torque control, on the other hand, interacts especially during the transition from partial load to full load. While in the partial load only the torque control tries to keep the optimum torque for the extraction of the energy and to maintain the energy balance in the drivetrain, the pitch angle is mostly kept in the optimum for a maximal energy extraction. Starting with the transition to full load and when the actual rotational speed approaches the nominal speed the pitch control does take effect. The energy drawn by the rotor is limited and the angle of attack on the rotor blades will be reduced by an increasing pitch angle. In this way, the optimum of the energy balance is left and an optimum of the drivetrain at nominal rotational speed is aimed, where the maximum energy yield of the turbine shall be produced as steadily as possible. This correlation in relation to the wind speed is shown in Figure 7.13: In Region 1 as partial load the rotor speed and power are slightly increasing, while the pitch angle is steady and Region 2 as full load where the rotor speed and power is steady while the pitch angle is decreasing.

Due to the continuously changing wind, in speed, direction and turbulence, it is not possible to stay permanently at an optimum operation point. But the dynamic pitch and torque control must continuously operate to keep the turbine as near as possible to the optimum. The transition from one operation point such as partial load, to another operation point at full load can happen fluently and continuously change back and forth. The tuning of pitch and torque control, especially the interaction at

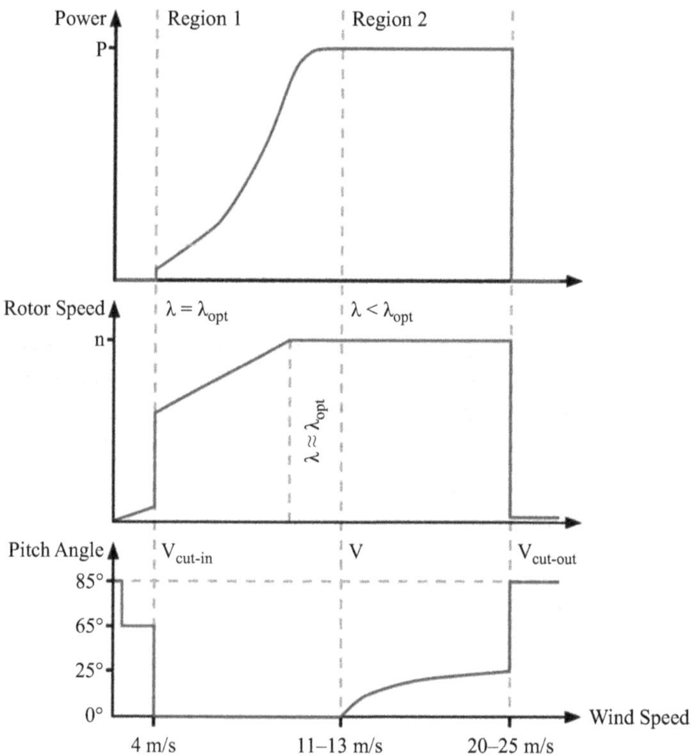


*Figure 7.13 Regions 1 and 2 show exemplary the different operating points based on the power curve, as well as the relations of speed, torque, and pitch angle to produced power. Copyright © Beckhoff Automation GmbH & Co. KG.*

this transition is relevant to ensure the stability of the closed-loop control, but it is also often a critical point where optimal power curve is not maintained.

Other special features are the operation points in modes in which no optimum and neither nominal speed is aimed. While starting the turbine often a constant rotor speed is approached to switch on the generator/converter and synchronize it with the energy grid frequency. In doing so, a rotational speed above the minimum switch-on speed of generator/converter and optimally also away from any resonance frequency as determined by the Campbell diagram is chosen. But also reduced operation points, for example by reducing the power production of a wind farm, are required, where the wind speed is above nominal, but the desired power production is below nominal. These are conditions outside the optimum operating points as usually aimed at but where a balance of energy in the drivetrain must nevertheless be maintained.

In the following, a classical approach to the closed-loop control of a wind turbine will be discussed. The approach is intended to explain the principles of a common collective pitch control and optimal mode gain torque control. While in today's turbines the usage of individual pitch control as well as model-based controllers is becoming common. However, there is a lot of research on even more specialized and extended control algorithms, as well as fuzzy based control or the use of neural networks, which will not be mentioned here, and more specialized literature should be consulted instead [23].

#### 7.6.3.2 Torque control

In partial load the generator torque $M$ is regulated to balance the aerodynamic torque, keep the optimum tip-speed-ratio $\lambda^*$ and obtain the maximum power coefficient $C_P{}^*$. This basic approach is usually referred to as optimal torque control (OTC) [23] and is implemented by the following computations. The tip-speed-ratio $\lambda$ is defined as following and based on the rotor radius $R$ depending on the rotational speed $\omega$ in relation to the wind speed $v$:

$$\lambda = R\frac{\omega}{v}$$

As already described in the first chapter of the first volume of this book, the power $P_{\text{wind}}$ extracted by the rotor area $A = \pi R^2$ from the wind with wind speed $v$ behaves as:

$$P_{\text{wind}} = \frac{1}{2}\rho v^3 A = \frac{1}{2}\rho v^3 \pi R^2$$

With the definition of the power coefficient $C_P = P_{\text{ex}}/P_{\text{wind}}$ for a ratio of the extracted power $P_{\text{ex}}$ to the available power $P_{\text{wind}}$ the power extraction is defined as:

$$P_{\text{ex}} = \frac{1}{2}C_P \rho v^3 \pi R^2$$

By means that the torque $M$ relates to the power and rotational speed as $M = P/\omega$ and using the optimal tip-speed-ratio $\lambda^*$ to replace the wind speed, the power equation can be arranged to the optimal torque $M^*$ as:

$$M^* = \frac{C_P{}^* \rho \pi R^5}{2\lambda^{*3}} \omega^2$$

After extracting the constant values, an optimum torque $M$ can be calculated for any rotational speed $\omega$ using the optimal mode gain $K_{\text{opt}}$:

$$M^* = K_{\text{opt}} \, \omega_g{}^2$$

where $\omega_g$ is the actual rotating speed of the generator, $\rho$ is the air density, $R$ is the rotor radius and $n$ the gearbox ratio, the $K_{\text{opt}}$ as optimal gain of the generator torque to speed relation can be calculated as [26]:

$$K_{\text{opt}} = \frac{\pi \rho R^5 C_P{}^*}{2(n\lambda^*)^3}$$

By considering the losses and the efficiency $\eta$ of the drivetrain, the optimal mode gain will result from [23]:

$$K_{\text{opt}} = \frac{\pi \eta \rho R^5 C_P{}^*}{2(n\lambda^*)^3}$$

The optimal gain mode is used to keep the optimal operating point with maximum power coefficient in partial load. In full load conditions a constant nominal torque, a nominal power or a combination of torque and power limitation is applied but can also be partially controlled to reduce any harmonics and other oscillations. As the drivetrain of a wind turbine includes multiple coupled stages of inertia, the twisting of shafts and interaction of gears in a gearbox result in a variety of torsional modes. Coupled frequencies of two components often lead to resonances as well, indicated by crossing lines in the Campbell diagram. Thus, torsional vibrations can occur during operation of the turbine, which are intended to be reduced by means of drivetrain damping via torque control. Using the generator as a damping component, by applying a torque oscillation, based on a band-pass filtered generator speed and is 90° phase shifted to frequencies smaller than the resonance frequency is applied to achieve sufficient damping [23]. This approach benefits from the fast torque control in modern converters. While the pitch system is highly dependent on the inertia of the blades, the generator by means of the converter can ensure fast and effective reaction on oscillations in the drivetrain [27].

#### 7.6.3.3 Pitch control

The closed-loop pitch control in pitch-regulated variable-speed wind turbines is used to keep the rotational speed of the rotor and generator coupled optionally by a gearbox in the optimal range for maximum energy production. If the wind speed is high enough and even above nominal to reach and keep the nominal rotational speed the optimal pitch position must be left to prevent the rotational speed from increasing. As the wind speed increases, the rotor blades are turned further out of the wind to reduce the angle of attack to keep the rotational speed as stable as possible at nominal. Due to the unsteady wind, this adjustment must be made continuously, and the angle of attack must always be adapted according to the current situation.

The conventional closed-loop controllers are designed by classical control theory and use proportional-integral-derivative (PID) based control algorithms. An error will be defined based on the difference of the set value, the nominal speed, and the actual value, the current speed. The proportional part $K_p$ of the controller attempts to correct the error by applying a gain factor to it and to calculate the controller output. To adjust a permanent deviation, the integral part with the gain $K_i$ is used, and a value accumulated over time is added to the controller output. A very rapid change shall be compensated by the derivative part with the gain $K_d$ and a differential value of the error over the time $T_d$. The transfer function of such a PID controller can be defined as [23]:

$$G_{(s)} = K_p + \frac{K_i}{s} + \frac{K_d\, s}{1 + T_d\, s}$$

The controller output is then used to define the pitch position or pitch angle in a range from 0° which is usually the optimal pitch position and 90° which is the feather position and thus completely out of the wind. Various limiters in the controller ensure compliance with limit values and load envelops, such as a limitation of the maximum pitch speed. An anti-windup of an integral part of the controller is required to limit the demand of the pitch angle as well as react quickly during a sudden change of the operation point in partial load or while the pitch has moved to its end positions. The setpoint is transferred to the pitch system and then approached via internal mechanisms for controlling the drives as described in Chapter 3 of Volume 1. Any delays while applying the setpoint, as well as dead times during transmission by means of a field bus, must be considered during the design of the parameters for the controller.

As the wind speed increases, so does the sensitivity of the rotor against a change in the pitch angle [28]. Therefore, a single set of parameters for the configuration of the PID controller is usually not sufficient. Specific settings and in particular gain factors are determined over the entire operating range of the plant. This mechanism of adapted control parameters is called gain scheduling and calculated based on the aerodynamic gain of the rotor.

As the wind turbines are getting continuously larger, the components must become more compact to reduce the total weight relative to the size. This results in softer, less stiff structures. Band-stop filters such as notch filters are used in the signal processing of the speed measurement to eliminate the natural frequencies that could be excited by the pitch system. Especially the towers, which are getting higher and higher, are stimulated to oscillate by the pitch angle adjustments. For this reason, tower vibration damping is often integrated into the pitch control. The acceleration of the tower head is measured, filtered, and incorporated into the control as an additional offset. In this way for instance a positive acceleration causes the rotor blades to be taken slightly out of the wind, thus counteracting the thrust force and tower movement [18].

The extension of an individual pitch control (IPC) creates an additional zero-mean demanded pitch angle for each blade to be superimposed to the collectively demanded pitch angle. A further improvement of the individual pitch control can be achieved from measured blade root out-of-plane bending moments to reduce asymmetrical load and imbalances on the rotor, produced by wind turbulence, wind shear, yaw misalignment and tower shadow.

#### 7.6.3.4 Yaw control

The classical but still used approach to control the alignment of the nacelle direction is implemented by means of the nacelle anemometer. The anemometer signals are affected by the turbulence of the wind itself, as well as because of the positioning downstream of the rotor and thus influenced by rotor-induced turbulences. Also, because the very unsteady wind field is distributed over the large surface of the rotor and the anemometer measures only a very small part of this field, large deviations must be expected. However, since the wind direction usually only changes significantly in longer periods, an attempt is made to compensate for these circumstances by a strong filtering of these signals over seconds and minutes.

During the tracking, the yaw system drives against an applied brake, on the one hand, to prevent the load, impacts, and shear forces on the gear teeth, but on the other hand, it generates high wear and loads on brakes, motors, and their gearboxes. Therefore, every tracking must be weighed up against the expenditures caused by these wears. In combination with the long-term filtering of the measured wind direction, a tracking procedure is usually started only every few minutes. The nacelle alignment is weighed up and in case of significant differences to the wind direction, the alignment is started immediately, while in the case of small differences, longer periods are waited.

This is often implemented by different thresholds and levels. For example, the deviation of wind and nacelle direction averaged over a period of 10 min leads to an alignment when this value reaches the threshold of 2°. In addition, an average over a period of 30 s leads to an alignment when the values reach 4°. The last and immediate alignment is started when the average value exceeds a threshold of 8°. The IEC 61400-1 defines in Section 6.3 the considerations of wind conditions with oblique inflow of up to 8°. This results in the usual operating range of $\pm$ 8° for the yaw system [3].

The direction of the tracking is often described from a bird's eye view of the turbine as clockwise (CW) and counterclockwise (CCW). How the different signals and values are used for positions and directions of the yaw and nacelle depend on the purpose. The anemometer indicates, for example, a wind direction as an analog or digital signal, which is measured in $-180°$ to $+180°$ relative to the orientation of the anemometer and thus usually also relative to the nacelle. So, this measured wind direction can be defined as relative wind direction to the nacelle position. But the nacelle position has no superior reference after the installation of the turbine. Since the nacelle position is often determined with an encoder, the position the encoder had during installation also determines the first position of the nacelle. A reference must be defined during the commissioning, whereby the twisting of the cables in the passage between the tower and the nacelle is often used to determine a zero position. Then the nacelle position is relative to and indicates the cable twist.

Since this position depends on the installation of the turbine and may vary from one to another turbine, the nacelle position relative to the cardinal direction north is often selected as additional value. This direction relative to the north allows comparing the nacelle position of several turbines, for example on a site in the farm. Based on the nacelle position relative to the north, a direction relative to the north can now also be determined from the wind direction and used for comparisons.

An interesting aspect of the filtering and the calculation of mean values based on the wind direction is the overflow of the values. So, the value for the direction of wind direction or relative nacelle position can be defined from $-180°$ to $+180°$ or relative to north from $0°$ as north, over $180°$ as south to $360°$. If an arithmetic average value is calculated by means of the sum of the values divided by the number of values and the values move around the overflow, this leads to erroneous results. Because with a cardinal definition from $0°$ to $360°$ and a value of $45°$ for northeast and a value of $315°$ for northwest. This results in an average value of $180°$, which corresponds to the south, while the correct direction would be $0°$ and north. To remedy this situation, modulations by means of the sine and cosine of the direction can be used. In the following equations, a direction $\gamma$ will be separated into $\gamma_{\sin}$ and $\gamma_{\cos}$ including the conversion from degree into radians:

$$\gamma_{\sin} = \sin(\gamma * 2\pi/360)$$

$$\gamma_{\cos} = \cos(\gamma * 2\pi/360)$$

After calculating the average values of $\gamma_{\sin}$ and $\gamma_{\cos}$ over a period of, e.g., 30 s into $\overline{\gamma_{\sin}}$ and $\overline{\gamma_{\cos}}$ the following equation calculates the average of the direction $\overline{\gamma}$:

$$\overline{\gamma} = \tan^{-1}(\overline{\gamma_{\sin}}/\overline{\gamma_{\cos}}) * 360/2\pi$$

Due to the circumstances of the nacelle anemometer, turbines are often operated with a permanent deviation of several degrees to the wind direction. As the power consumption of the rotor is reduced with the cosine square of the wind direction deviation, this leads to losses of 2% and more [29]. For compensation, the deviations of the anemometer to a reference measurement from a wind measuring mast are often determined to make a correction by factor and offset to the values.

In contrast to the approach of an open-loop control by thresholds, intelligent controls are increasingly being used. In modern systems, the rotor itself is used as a better means of measuring wind speed as well as direction. By adjusting the yaw system only a few degrees, it can be observed whether the power extraction increases or decreases as a result. Thus, the regular method for wind tracking can be used to determine a correction factor and offset to the wind direction determination. In this way, to determine a correction, explicit tracking procedures, but also the regular tracking procedure for yaw alignment can be utilized. This approach is referred to as hill climb searching (HCS) or hill climb control (HCC) using an adaptive gradient algorithm [30].

#### 7.6.3.5 Predictive control

As the wind speed is very important to the pitch and torque control of the wind turbine, the wind speed measurement is often very inaccurate, placed behind the rotor and influenced by its induced turbulence. Therefore, an effective wind speed measurement or estimation in advance to the turbine would allow improved control algorithms. A variety of special measuring sensors, e.g., for mounting on the front of the spinner, attempts are made to replace the nacelle anemometer with a better measurement. From the nowadays frequently used rotor blade sensors, which use the strains, vibrations, and loads of the rotor blades for damage detection and other purposes, oblique inflow can be determined as imbalances. The blade root moments indicated by such sensors can also be used to draw conclusions about loads or imbalances as well as the wind speed and turbulence. Another approach to optimize the yaw alignment would be the application of a modern drive system, which allows to evaluate the torques of the motors as loads and thus determines an oblique inflow on the rotor.

A Light Detection and Ranging (LIDAR) system can be used to measure the wind speed in front of the rotor and predict the wind speed on the rotor. Unfortunately, the assumption that the same wind speed measured at a distance in front of the wind turbine also reaches the rotor with the same conditions is not valid under all circumstances. Nevertheless, using state-space models as state observers based on Kalman filters is a common approach to optimize the estimation, but the usage of neural-networks and machine learning algorithms also finds application not only for the prediction of the wind speed, but also the reaction of the rotor and turbine [31].

Thereupon, so-called feedforward controllers are used to optimize the control by means of the prediction, but also model-predictive controllers (MPC) are increasingly used. Especially the research around nonlinear model predictive control (NMPC) in combination with LIDAR and artificial neural networks (ANN) [32] for the modeling and prediction of the wind turbine behavior shows promising results [33]. While some are still working on the implementation and challenged due to the required computing performance of the industrial PCs or the lack of efficiency in the execution of the NMPC and ANN [34], others already show successful applications in the field [35]. Individual optimization criteria can be applied to compute an optimal control trajectory for the corresponding prediction. It would become possible to dynamically optimize to maximum power, minimum loads, or other criteria on the fly, depending on the strategy for the operation and actual conditions of the turbine. But apart from the promising research, the price of LIDAR systems is still too high to be used as standard equipment on series turbines.

## 7.7 Subsystem control

The following section will introduce the common subsystems with typical sensors, actuators, and key objectives from the viewpoint of the control system. The major subsystems can be listed according to the energy flow in the turbine, such as rotor, drivetrain, generator, converter, transformer and grid. In addition, there are also plenty of components not included in the primary energy flow but perform auxiliary tasks and nevertheless provide an important support. Such components are the control cabinets, cooling systems or hydraulic systems as well as measurement systems like meteorological and vibration measurements.

The use of modular software structures and the orientation on real objects in the architecture of the software is state of the art. Thus, the subsystems should also be implemented as objects or modules in the software. This has many advantages, such as the independent development and testing of the modules as well as their reusability.

### *7.7.1 Control cabinets*

The control cabinets are part of the control system in which the electronics are installed, and all sensors and actuators are brought together by means of wiring. This includes not only the electronics like the industrial PC and I/O modules but also the power supply of aggregates, own climatic air-conditioning system and electronic protection devices. Some common components are as follows:

- 400 V/230 V AC power supply
- 24 V DC power supply
- temperature sensors
- air heater and cooler
- supply protection
- motor protection
- lightning protection
- circuit breakers and fuses
- uninterruptable power supply (UPS)
- Ethernet infrastructure as switches
- manual operation controls

The power supply to and in the control cabinet itself must be protected by circuit breakers and fuses. The electrical power for larger units such as pumps, motors and fans are taken from the 400 V power supply and operated by contactors. These units must be protected against overcurrent. The 400 V power supply is then used to generate the 24 V supply for smaller units as sensors and the control system components like the I/O modules. The 24 V supply of all those units, collected into groups, must be protected from overcurrent as well. States of all those contactors as well as the protection devices must be reported back into the control system via feedback signals to allow the control system to diagnose and report any malfunction. In the event of a power supply failure, an uninterruptible power supply (UPS) is used to ensure a further supply of usually a few minutes to bring the systems into a safe state before they are switched off. Which in turn results in further control and feedback signals from the UPS into the control system.

Cables of sensors inserted from the outside must be connected and special signals must also be protected, for example, by means of overvoltage protection, lightning protection, surge, and burst protection. The supply to the sensors themselves can be checked whether current is applied, thus whether cables are connected or faulty. Information about the states of connected components is often included in the diagnostics of the I/O modules and needs to be evaluated as well. But also, internal components and those necessary for the control cabinet must be controlled as well as evaluated and diagnosed. This includes heating and cooling for the air conditioning of the control cabinet. For offshore installations, the air conditioning might include filtration that avoids ingress of salty and humid air.

A complete diagnosis of the control itself should of course also be implemented. Starting with the industrial PC, a diagnosis of the CPU and RAM utilization as well as the remaining free space on the hard disk should be evaluated to prevent a too high or complete utilization of such resources. The real-time capability must be monitored and kept, like the cycle time in the execution of the software but also the response times in communication, for example over fieldbus interfaces.

To enable communication and diagnostics via the widespread Ethernet interfaces for the increasingly intelligent systems, additional IT infrastructure is provided in the control cabinets. These are usually industrial Ethernet switches, but wireless communication might be used as well as routers, firewalls, or other gateways for communication to higher-level networks, including the internet.

### *7.7.2 Hydraulic and cooling systems*

To be able to dissipate the heat of many components, several cooling systems are used in a wind turbine. These are usually cooling circuits of oil as for a gearbox or of water as for the converter, where not necessarily pure water but a coolant is used. These cooling systems often consist of the following components:

- pumps and valves
- heaters and coolers
- temperature sensors
- level sensors
- pressure sensors
- pressure accumulators
- over pressure relief valves

The temperature of the component to be cooled must be measured, as well as the temperature of the coolant. So-called PT100 sensors are commonly used as robust resistance thermometers. Pumps are used to move the coolant in the circuit to absorb the waste heat from the component and can often be operated at different speeds to influence the liquid flow rate. The resulting pressure in the pipes is monitored by pressure sensors and used to control the pumps to maintain a specific pressure, but also to check the function of the pumps, the leak tightness of the pipes, and finally to prevent overpressure. A change in pressure or temperature over time can also be used to conclude on whether there is a blockage in case the pressure rises too slowly, just as the difference between the pressure upstream and downstream of a pump or the temperature upstream and downstream of a heater can be used to check their functionality and efficiency. A sufficient quantity of coolant must also be checked via level sensors. Again, the coolant must then be cooled to dissipate the heat it has absorbed and heated up.

Equally important are the heaters to warm the coolant. In the case of water cooling, but especially with oil, a minimum temperature must be maintained before the coolant can be used and fed through the pipes by means of the pumps. This ensures the viscosity of the fluid, as oil can become very sticky at low temperatures, which can cause damage to pumps and pipes. If the turbine is installed in cold climate regions or has been switched off for several days in winter times, such low temperature may well occur. Heating up large quantities of oil, sometimes hundreds of liters, can take many hours and must be carried out urgently before the system is operated.

Hydraulic systems are used to generate and transmit high power by means of pressurized liquids. These usually supplies to other system, such as:

- rotor brakes
- rotor bolts
- yaw brakes
- pitch systems

The pressure is generated by pumps, corresponding the liquid pressure and level are measured, the temperature is measured as well as adjusted by means of heating and cooling. Find more details in Chapter 7 of Volume 1.

To enable the different speeds of pumps and fans, but also to prevent the inrush current by switching them on, the connection of the supply voltage is changed during operation. They are started in star circuit or Y connection for a low speed with low current consumption and then switched forth and back to a triangle circuit or delta connection. This concept is referred to as the start-delta or wye-delta transformation and used to avoid the tripping of overcurrent protection devices due to the otherwise high starting current in a delta connection. Other concepts are the Dahlander pole changing motor that allows to utilize two speed motors or the use of much more flexible and conserving frequency converters.

Such load peaks of inrush currents might result in high costs for the auxiliary power supply and can be avoided by peak load management with the use of peak shaving. This implements a delayed and coordinated switching of the large actuators to separate the inrush currents from each other and thus avoid high current peaks. To avoid a high switching frequency of digitally controlled aggregates hysteresis controls are used. For the hysteresis, two levels are defined: One for switching on, and one for switching off. By the difference of those two levels a corresponding interval between switching on and off is caused. But also, time delays are common for the throttling of the switching frequency of the aggregates. In addition, the use of frequency converters for continuous control of any speed of the motor is possible, but due to the higher costs not so common. This continuous control of the motor, however, has the advantage of not leading to high wear of pumps and fans as well as the contactors due to a high switching frequency.

### *7.7.3 Tower, nacelle, and hub*

Tower, nacelle, and hub are merely passive parts of the turbine. Nevertheless, they determine the ambient conditions for other systems installed in them. It is therefore important to control and necessary to influence these conditions. Thus, the ambient air is primarily monitored and controlled because the systems installed in the system mostly emit their waste heat into this ambient air, which in turn should be dissipated to the outside. It is mainly the task of air conditioning, its temperature, and offshore also its humidity. Used for this purpose are for example following devices:

- temperature sensors
- cooling and ventilation systems
- dehumidifier

Since the outside air is used for such heating, ventilation, air-conditioning, and cooling (HVAC) systems, it is advisable to include the outside conditions in the control logics. In this way, for example, cooler outside air can be used efficiently to cool and ventilate the interior. However, if the outside temperature exceeds the inside temperature, the installed cooling systems must be used. In the case of offshore installations, the outside air cannot be used directly due to the moisture and salt content and must be purified beforehand using treatment systems like dehumidifiers.

### *7.7.4 Pitch and rotor*

The rotor absorbs the energy of the wind and converts it aerodynamically into mechanical rotational energy. The rotor blades are used for this purpose and their angle of attack is adjusted by the pitch system. Since the pitch system is either a hydraulic or an electrical drive system, the components used must be monitored and controlled. Even at a low speed of the rotor, the energy contained in it is so high that only changing the rotor blade position by the pitch system is still able to decelerate the rotation. This makes the pitch system the primary and most important system in the turbine to stop it. It needs to ensure a safe stop in case of an emergency and must therefore be implemented intrinsically safe and equipped with redundancies. However, since at least two, usually, three rotor blades are used and even a single rotor blade in feather position has a sufficient braking effect on the rotor to decelerate it, redundancy is already provided by the individual axis. In addition, important sensors such as the positions of the rotor blades are usually installed redundantly via encoders. The following sensors and actuators are usually installed in the rotor:

- encoders for redundant positions of each axis
- drives, gearboxes, and motors (electrical pitch system)
- valves and cylinders (hydraulic pitch system)
- ambient, motor and cabinet temperatures
- bearing lubrication with pump and level sensor
- slip ring with power and communication transmission from the nacelle.

Since each rotor blade allows movement relative to the hub via its bearing, the lubrication must be controlled, and its condition monitored. A detailed overview of the pitch system can be found in Chapter 3 of Volume 1.

The power supply for all the systems in the hub is managed from the nacelle and transmitted to the hub via the slip ring. About the same slip ring, the communication to the pitch system and other systems, as well as the emergency stop signals for the safety system in both directions must be transmitted. The lubrication and hub ventilation system might be supplied and controlled by the main control system or by the pitch system. In case a safety communication of the primary fieldbus and main control system is used, the emergency stop signals as well as feedback are not needed to be physically transmitted via separate lines on the slip ring but could easily be communicated together with the fieldbus over its lines, as already stated in Section 7.3.

As the rotor is mounted to the shaft and held there by the main bearing, several loads across the rotor react to the main bearing. Thus, these bearings face special conditions in comparison to usual bearings, as the aerodynamically induced loads from the rotor and the loads induced by the rotating components further down the drivetrain. Because it is typically not possible to replace the bearing rollers or raceways on site, any damage will result in a replacement that requires the complete removal of the main bearing. Therefore, this critical component is often subject to vibration monitoring, especially on direct-drive systems. Using individual vibration sensors, where two sensors are mounted offset by 90° to each other, damage in the bearing, as well as any imbalances and other loads, can be detected and monitored.

Another aggregate in relation to the rotor, but usually installed in the nacelle, is the rotor bolt. This bolt is used to fix the rotor and ensure a safe working inside the hub and blades for the service and maintenance staff. The bolt can be operated mostly hydraulically, either manually or sometimes automatically, using the following components:

- main, lock and unlock valves
- bolt positioning sensors

Most important is the feedback signal whether the bolt has been unlocked or not arranged correctly. This signal is often also included in the safety system to prevent any start of the turbine while the rotor is locked by the bolt. Otherwise, the high load of the rotor presses on the bolt and could lead to damage.

### *7.7.5 Gearbox, generator, and shafts*

The rotation and torque of the rotor is then passed to the low-speed shaft or also called the main shaft. Depending on the type of turbine, a gearbox and high-speed shaft are followed, as well as the generator. This mechanically interconnected series of rotating systems is referred to as the drivetrain. The drivetrain units are supported by multiple bearings, each of which requires a lubrication system.

The gearbox in a wind turbine usually has multiple stages and requires many sensors and actuators in total for the operation. An oil circuit is required to lubricate the gearbox, but also to cool it and to dissipate the waste heat via the oil. This oil circuit is then in turn supported by a water circuit as a further cooling system of the oil. This results in the use of the following units:

- gear stage temperature sensors
- shaft bearing temperature sensors
- oil circuit with pumps, heaters, and coolers
- oil temperature, pressure, and level sensors, with limit switches
- oil valves, filters, and pressure relief valves

- water cooling circuit with pumps, heaters, and coolers
- water temperature, pressure, and level sensors

Both the oil circuit and the water circuit can be implemented like conventional cooling systems described in Section 7.7.2. Due to the huge size of a gearbox for modern multi-megawatt plants, as well as the hundreds to thousands of liters of oil, very high time constants must be provided for the cooling circuits. It can take hours to days to heat the oil to liquefy it and pump it. but even cooling it down to a few degrees can take several hours. This is often done in a permanent attempt to keep the oil at a good operating temperature. whether it is to continue pumping and heating the oil when the system is switched off or to continue cooling the heated gearbox after operation. By operating the turbine slowly while the oil is sufficiently liquid and with a limited torque, the oil can be heated up as well or respectively cooled down after a high load. If the oil and gearbox temperature increase steadily beyond certain limits during operation, this leads to a necessarily reduction of the power output of the turbine. For instance, the load demanded via the gearbox must be reduced, the power extracted in the generator and thus the torque reduced.

Depending on the generator, whether slow running, directly driven or fast running and driven by the rotor via a gearbox, either air cooling is sufficient or water cooling is necessary. It follows that another cooling system must be implemented in the control system. Depending on the structure of the generator, the following additional sensors and actuators might be used:

- rotational speed sensor or encoder
- stator temperatures
- bearing temperatures
- bearing lubrications with pump and level sensor
- heater, cooling fans, and air temperatures
- water cooling with pumps, fans, and temperatures
- earth brush limit sensor

Additional to the rotor speed measurement, an encoder or similar kind of speed sensor is used at the generator. While an absolute encoder with position measurement is often used for the rotor, an incremental encoder with speed measurement is used for the generator. This also means that the speed measurements, which are very relevant for the safety system, are directly diversified. It is important to compare and validate the speed of the generator with the speed of the rotor. This allows the plausibility of the speed signal to be evaluated and any difficulties in the drivetrain to be identified.

The brake disc can usually be found on the high-speed shaft as well. The brake is actuated via the hydraulic system, as described in Chapter 7 of Volume 1. Since this is a holding or parking brake, the brake may only be applied at low rotational speeds, for which a locking or control in the safety chain is often implemented. This leads to the common usage of following parts:

- hydraulic supply unit with pumps, heaters, coolers
- hydraulic pressure, level, and temperature sensors
- valves to apply brake at different pressure levels

If the brake shall be applied, the thresholds for the speed are observed. But if no deceleration is detected to avoid the risk of overheating, in case of any doubt the brake might be released again.

### *7.7.6 Converter, transformer, and switchgear*

After the generation of electrical power in the generator, this is conducted by the converter and fed into the grid. Depending on the generator, the current conduction must be set, and a torque determined from the dynamic control be applied and demanded from the generator. In addition, external demands or limitations from a superior control system and the grid utility in order to control the active and reactive power of the whole wind farm must be taken into account. Further details are described in Chapter 4 of this volume. The control system mainly interfaces the converter system by means of a fieldbus interface. The converter is also supported by water cooling, which is operated by the control system as well. That leads to an integration of sensors and actuators as follows:

- communication interface
- safety system contacts
- cooling system with pumps, heater, coolers, etc.
- water inlet and outlet temperatures

The real-time capable communication between the converter and the control system is nowadays often already implemented by Ethernet-based fieldbuses such as PROFINET or EtherCAT. The control systems demand different states in addition to the applied torque from the converter system, to start it up, charge the internal DC bus or let it synchronize with generator and grid. The feedback of the converter systems determines the actual status and internal signals of voltages, currents, as well as temperatures but also errors and warnings. While the turbine is being started and has already reached a minimum rotational speed, the converter is initiated to switch on, synchronize to the voltages and then feed current into the grid.

Depending on the design of the converter and the use of a circuit breaker, switching it off and on again must be considered. Because such a circuit breaker only has a limited number of switching cycles and should not be used several times a day. The connection to the mains is often only opened in an emergency or service case and otherwise the converter tried to keep in operation on the mains side.

The case of an emergency is interfaced usually by two redundant 24 V signals from the safety system to the converter systems as well as by two feedback signals from the converter into the safety system to indicate a safe operability of the converter. Even if safety communication via a modern fieldbus is possible to save such additional hardware signals and components, this is still not yet common to implement PROFIsafe or Safety over EtherCAT into the converter control system.

If the turbine is directly connected to the electrical grid, but also if connected to the grid of a wind farm, the voltage of the grid is usually higher than the voltage generated by the converter. Therefore, the voltage is increased via a transformer and the currents are reduced accordingly to feed into the grid. Most transformers are monitored by the temperature and conclude a cooling system, what results in following sensors and actuators to interface:

- winding temperature sensors
- air cooling fans or water-cooling system
- ground fault sensor
- arc detection sensor
- Buchholz relays

To detect any current flows to earth, a ground fault detection sensor, as well as an earth switch could be integrated in the switchgear to protect the electrical equipment. Additionally, arc detection sensors might be used to detect an intensive light which can result from an electric arc flash.

The three phases of voltages and currents are evaluated via power measurements to determine following signals:

- voltages, currents, frequency, power
- detection of faults or asymmetries
- detection of overpower
- auxiliary measurements

As many detections regarding the electrical grid are already integrated into the converter, the power measurements implement a reference measurement as well as optionally additional measurements on different levels of the turbine. So might be an additional power measurement of the generator and before the converter used to analyze the electrical behavior of the generator from which conclusions can also be drawn about its mechanical conditions. Measurement of auxiliaries can be used to determine their consumption as well as potential malfunctions. While the measurement to the grid then records the generated energy, as well as the conditions of the grid, faults of high or low voltage, rapid changes of frequencies, or the case of too high power or current induced by the turbine.

### 7.7.7 Yaw system

To move the nacelle and align it with the direction of the incoming wind, the yaw system is used. For this purpose, different types of drive systems are implemented in the various turbine designs, but mostly a hydraulic brake system. This results usually in the following components of a yaw system:

- drives, motors, and gearboxes
- soft-starter, frequency converter or servo drives
- motor protection devices
- electrical brakes on the motors
- hydraulic brakes on the gear ring
- hydraulic supply unit with pumps, heaters, coolers, ...
- position sensors with limit switches
- bearing lubrication system

Apart from the operational control discussed earlier in Section 7.6.3, the control system is concerned with implementing the timing between opening the brake, starting the drives, partially closing the brake until the end of the movement, and then completely breaking the azimuth. Because in many turbines, the drives are still moved against an active brake to tension the drives against the gear rim to prevent any movement caused by inclined wind or oblique gusts. To start the movement, the breakaway torque by the inertia must be overcome and to not induce an additional load, the brakes are completely released here.

If the motors are connected directly to the three-phase supply, they already generate a very high starting current themselves additionally to the current used to overcome the breakaway. Such high inrush currents should be avoided or reduced, to prevent any overcurrent protection to trip. Soft starters or motor starters are electrical devices used to protect the motor against such an overcurrent and reduce the starting current. More flexible control of the motors is possible via frequency converters, which allow the control of the speed of the motor and thus completely freely adjustable a motor can be started slowly, and the speed can be increased smoothly. With such dynamically controllable drive systems, intelligent couplings of the motors can be used to protect the gear rim. Mechanical stresses such as backlash or impacts on the teeth can thus be largely avoided.

There are also concepts to reduce or eliminate the use of the brake by dynamically controlling the motors before, during, and after the movement of the azimuth. For this purpose, some motors create a counter-torque to tension the gear rim and prevent any unintended movement while backlash or other loads on the gears are reduced. The overcoming of the initial inertia of the yaw during start-stop can be optimized by dynamic drive control, as well as the resulting gyroscopic forces can be reduced. An even more dynamic control but also overload capability is offered by servo drives. This also allows to use several small motors rather than a few large motors. With today's heavy turbines, it is still possible to exchange the smaller motors and gears on site, while for the large drives the use of a crane in the nacelle makes such work massively more difficult.

A position sensor is used to determine an absolute position when the yaw is at standstill, but also to be able to record the speed from the azimuth while moving the yaw. An absolute position with multiple turns evaluated in the encoder allows the position and alignment of the system to be determined again even after a power failure. The additional safe evaluation of limits and end positions is necessary to protect the cable passage from tower to nacelle and excessive twisting as well as tearing. This is realized either by mechanical end positions as digital switches, but modern encoders also allow to communicate a safe position value via secure fieldbus protocols to the safety system.

### *7.7.8 Meteorology*

The meteorological measurement is usually located on top of the nacelle. Even if not included in the safety chain, the current standards require redundant wind speed and direction measurement. Furthermore, it must be protected against the accumulation of ice, which is why a heater is usually integrated into these sensors. Due to the placement on the nacelle, the signals must be protected by lightning protection before they can be connected to the I/O modules.

During the operation of the turbine, these wind measurements on the nacelle are rather estimates due to the aerodynamic turbulence caused by the rotor and are therefore evaluated by filters in the software. They are used, among other things, for wind tracking or to shut down the turbine when wind speeds are too high, as well as in the standstill to evaluate the minimum wind speed for starting the turbine. Common components outside on the nacelle are:

- wind speed and direction sensors
- outdoor temperature sensors
- barometric pressure sensors
- global positioning system
- obstruction and flight navigation lights.

Barometric pressure is useful to estimate the density of the air, what has a high impact on the power contained in the inflowing wind. Therefore, the barometric pressure is allowed to correct the statistically calculated power curve, to obtain a uniform statistical curve, mostly independent of location and environmental conditions. In addition, there are systems available that determine the weather, detect rain, snow, or ice to take precautions if necessary.

Sometimes are navigation, gyrospopic or compass systems used to get the orientation of the nacelle to the sky direction, and global positioning system (GPS) to get a very accurate time source for the comparability of the data recorded later and their statistical evaluations. Depending on the local regulations, an aviation lighting system is installed on the turbines, to prevent any collision with approaching airplanes or helicopters. Therefore, they are also known as flight navigation lights, anti-collision lights, beacon lights or strobe lights. For offshore installations, there is a similar concept for ships. Here, the visibility must also be determined, which may be reduced by fog and air humidity, to send light signals accordingly.

### *7.7.9 Monitoring and protection systems*

There are several monitoring systems for monitoring and optimizing turbine operation. A list of potential monitoring systems as follows:

- tower vibration sensor and shock detection
- rotor blade bending measurement
- structural health monitoring
- gearbox condition monitoring
- bearing condition monitoring
- generator power monitoring
- fire protection system

Some of them are safety-critical and must therefore be integrated into the safety system, like the shock detection sensor. By vibration or acceleration sensors, it is mandatory to detect excessive loads caused by a jerk on the tower head, defined as shock. Bearings, gears, or structures are also monitored based on vibration/acceleration sensors to implement condition monitoring or structural health monitoring. Such monitoring systems are usually for lifetime monitoring and early detection or prediction of any damage. In the case of rotating systems, the vibrations are then evaluated by means of an order analysis in combination with a frequency analysis for instance. In the order analysis, the vibrations recorded at variable speeds of the systems are normalized to a uniform speed to be able to use a better evaluation of the frequencies occurring. Based on a comparison of the analyzed frequencies with the frequencies generated by the mechanics, as in the case of tooth meshing frequencies, deviations and sidebands can thus be identified as potential damage.

The monitoring of the rotor blades by means of stains or vibrations in the blade root can also be used to detect potential damage. But also, to determine ice buildup, as well as oblique wind inflow, rotor unbalance, and other asymmetric loads. These evaluations are then also useful to improve the individual pitch control. Such structural analyses are often carried out by determining natural frequencies. The natural frequencies under different operating conditions allow conclusions to be drawn as to whether, for example, the mass or strength of the structure has changed, and thus whether any damage might have occurred.

Fire alarm systems serve to prevent worse damage by applying an extinguishing system. Since the nacelle at height cannot be extinguished, but only let burned down in a controlled manner, an extinguishing system is often the only way to contain a fire. The installation of cameras in the turbine to visually inspect and record any activities becomes increasingly common. The focus is not only to detect faults or fires, but also on preventing theft and damage to property.

## 7.8 Monitoring and maintenance

The automation and control systems of wind turbines are designed to operate autonomously as possible and without human intervention. As turbines are frequently in volatile locations they rarely should be visited for service. Each service intervention requires a stop of the turbine with corresponding losses, as well as the costs for travel and personnel. Instead, it should be possible to access as much information as possible and influence the system from remote.

Intelligent computer systems can be diagnosed remotely via an Ethernet interface over a network. These devices often can be configured or reset remotely via such an interface, records of important signals and events retrieved from which errors can be traced. Modern control systems offer extensive facilities for displaying signals, states, and conditions, as well as recordings, retrieving and displaying data over long-term.

Nevertheless, there are always tasks that cannot be automated and require service personnel on site. Refilling of lubricant and renewal and replacement of oil are just some of these regular tasks. Finding and fixing faults, so-called troubleshooting is not always possible from remote. In addition, there are fault conditions that cannot be reset remotely. Somebody must be on site and carry out a visual inspection in advance. For example, the safety system must be reset and closed manually again locally after a violent shock occurs.

### 7.8.1 Remote access

If the service personnel access the control system even remotely or locally, it must be clearly defined which authorizations and permissions this access currently has. Such permissions are often defined according to groups, such as operators, maintenance, commissioner, or developers. For example, a developer is allowed to change important parameters from the design of the system, while the operator only has rights to monitor, reset, start, and stop the turbine. But also, the location of access is important in the case of wind turbines. In this way, one person on site can stop the turbine and then no one else from remote is allowed to restart it. This is primarily required for safety reasons, because if a person is on site and climbing in the turbine, maybe on top of the nacelle or inside the hub, the turbine must not be operated, and remote access could lead to serious consequences. But also, several manual interactions, for example the manual operation of actuators, are often only allowed locally, because it will ensure that the person on site is inspecting this operation.

To assign permissions and localities to accesses, it must be possible to identify them. A variety of user management tools can be used to identify the user to the system with a username and password, or via a unique identification tag on a small electronic device with a transponder. To control local access, there are often service switches on the control cabinets. These switches can sometimes be protected by means of a key or a lock, so that no other person can open them again while the originator is still working within the turbine. But even here there are electronic locking systems, up to digitalized locks in the main door that regulate access to the turbine.

### 7.8.2 Data recording

If a malfunction in the turbine occurs, it must still be possible to figure out the causes afterward. Not only the misbehavior, but also operating states and selected signals must be permanently recorded, as e.g., the 10 minute average values of some characteristic signals like wind speed and rotor speed must be stored over the total lifetime. In many cases, the monitoring of the plants takes place live via recorded information sent to the control rooms. Nevertheless, the information is recorded in parallel to serve for later evaluations and statistics.

The most common example for a statistical evaluation and comparison of wind turbines is the power curve, which can be found in almost every data sheet for a turbine type. The 10-min arithmetic average of the power a turbine produced is plotted against the wind speed. Figure 7.14 shows such a power curve as defined in the IEC 61400-12: a standard especially written for power performance measurements of wind turbines [36].

To evaluate such statistics, the 10-min average values of the relevant signals are recorded permanently. In addition, relevant signals are also recorded more quickly, so there are some values such as wind speed, wind direction, rotor speed, produced power, which are recorded every second. This requires exact time synchronization within the control system for comparability, acquiring the measured values simultaneously and using uniform time periods for post-processing as averaging or statistics. The recorded signals are used to evaluate the general

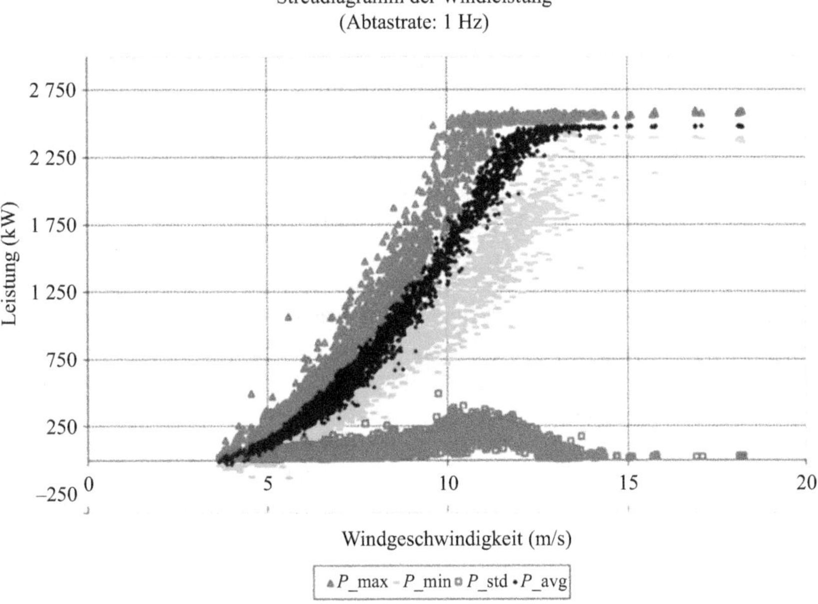


*Figure 7.14 Scatter diagram of the produced power over the wind speed as defined in the IEC 61400-12*

behavior of the turbine, as well as the most important conditions. After 20 years of operation to estimate for example, the loads absorbed due to the wind can be estimated from such data and thus enable a possible life-time extension to operate the turbine for further years.

Rather slowly changing signals such as temperatures are not exclusively stored cyclically. Instead, the values are regularly monitored for changes and only then recorded if, for example, the outdoor temperature has changed by 1° C, whose corresponds to the accuracy of common PT100 temperature sensors. Thus, a value is no longer generated every second but only every few minutes or hours to save considerable storage space of data that should be kept on the local hard disk of the controller for as long as possible. All events that occur, the status messages, as well as relevant changes to parameters or the execution of commands, as well as other actions by a user who logs in, logs out are also recorded. On the other hand, high-resolution recordings are often stored for a short period around an event. Such trace logs or event logs could store the most important 100 values in the cycle time of the controller, several minutes before an event, the pre-trig, and several minutes after an event, the post-trig. An event could then be a relevant error of the turbine which leads to an emergency stop. The trace log then enables a very precise and targeted analysis of the cause of such behavior by means of the pre-trig, but also an evaluation of the reaction of the turbine to the error and whether it was then stopped correctly by means of the post-trig.

All this information is then usually first saved to the local hard disk of the industrial PC. Traditionally, text files are written in the format of comma-separated values (CSV). This CSV format is not uniformly standardized and can therefore be implemented in different ways. Furthermore, the conversion of the digital values into a text is complex and lossy. Thus, floating point values, which usually use 4 bytes and are accurate to the sixth decimal place, are written into the file using a text of eight characters with only three decimal places. So, in the worst case, twice the amount of memory for half the resolution. However, the CSV format is still very common because it is easy to read for humans and tools, it is supported almost everywhere. To limit the size of the files, these files are often created daily, so that a new file is created for each day.

However, such text files are unsuitable for continuous management of data over thousands of turbines and decades. This leads sometimes to the use of proprietary binary files or other binary formats of countless variations. In the end it is an attempt to build a database with such individual files, where the files are mostly converted and inserted into the enterprise environment anyway. Therefore, it is a good idea to write the data from the control software directly onto a database. There are the usual Structured Query Language (SQL) databases or newer non-relational, often called NoSQL databases. Due to the mostly time-occurring information consisting of timestamp, meta-data and the values, the use of time-series databases is practical. These time-series databases are specially developed for such time-related series of values and offer high performance and compression of such data.

Since the storage space on the local systems is often limited, just to save costs for the hard disks, the local data is kept in a time-limited frame. Older data is then deleted to make space for new data. This way, the local data is kept for a limited time in an often-called ring buffer or circular storage. To nevertheless be able to store the data over the entire lifetime of the turbine, the data is transferred to superordinate servers and collected there. Servers installed within the wind farm, in the company of the manufacturer, operator, or even in server farms or the cloud can then be used. For example, at Siemens Gamesa the "turbines use over 300 sensors which transmit over 200 gigabytes of data per day to state-of-the-art remote diagnostic centers" via cloud services [37]. In these servers, large and redundant hard disks are then often installed, which can be used and maintained centrally more easily and efficiently. There are technologies such as data warehouse carried out to utilize intensive analyses of large amounts of data.

### *7.8.3 Farm communication*

Within a wind farm consisting of several turbines, an Ethernet-based network is installed as well. Due to the high distances between the turbine and the limitation of Ethernet over copper lines to a maximum of 100 meters, fiber-optical lines are used to establish Ethernet connections over several kilometers. At the nodes and interconnection of the network, industrial and managed switches are used. Within the turbine, Ethernet switches are also used, often installed in tower and nacelle to connect all computer systems. These switches also allow connecting the cables to rings of networks to increase availability and implement redundancy, what is not defined by the Ethernet standard. The Ethernet standard provides a spanning tree protocol (STP) whose does not allow fast convergence times and redundant topologies, why are proprietary solutions used to implement more resilient Ethernet networks, while some Ethernet based fieldbuses implement similar mechanisms for redundancy purposes. This way, if one line is interrupted, the connection can still be maintained via another line.

To separate the networks that are used by different users for different purposes, the configuration of virtual local area networks (VLAN) is common. In this way, a certain level of data and system security can be established, while the hardware only needs to be installed once. If special use cases require a separate connection and exclusive access, physically separated networks will also be used. Thus, for the control in the wind farm, which may require high bandwidth and real-time capable communication, a separation is made via the hardware. But there are also technologies available to prioritize real-time communication and then allow the other data exchange to take place in a subordinate manner. Independent of established real-time protocols such as EtherCAT or PROFINET, such possibilities are also developed in the context of the time sensitive networking (TSN) in the IEEE 802.1 working group.

The SCADA system then uses this communication to retrieve or receive information about the turbines and provide it to the operators. For this purpose, a web interface is often provided by the SCADA system to serve as a human-machine-interface (HMI), enable observation and interaction with the turbines. However, the SCADA system usually only allows manual interactions but is not intended to implement automated control of the turbines.

The wind farm controller instead serves to control and coordinate the energy production of all turbines so that an optimal feed-in takes place at the point of common coupling (PCC). Many of the grid connection rules are implemented in this way, which includes the control of active power, reactive power, and frequency stabilizing measures, but also the reaction to grid faults as voltage dips [18]. The wind farm controller is then a separate control system that executes an application and dynamic control in real time, which in turn communicates with the turbines in real time and thus implements the closed-loop control. Depending on the network and communication or whether a fieldbus is used, reaction times can be as short as just a few milliseconds. This allows the reaction to a fault in the grid to be communicated very quickly to the turbines, thus enabling a collective reaction of the wind farm to support the grid. Any dead times, jitter and other influences of poorly executed communication can have a particularly strong effect on such a distributed control loop of many plants and the long lines for transmitting the energy [38].

### *7.8.4 Remote communication*

Communication to external systems is usually implemented by Ethernet as well. Apart from the IEC 60870-5-101, where serial communication is still commonly used by grid utilities. But the IEC 60870-5-104 is already based on IP and TCP as common Ethernet protocols. As a successor to this standard for communication with grid utilities, the IEC 61850 has been published 2003 and is just starting to find a wider use twenty years later. Especially with the extension IEC 61400-25 to the IEC 61850, a description of the data model of a wind turbine, a common standard for wind turbines is provided.

In the meantime, the OPC UA standard IEC 62541 is becoming more widespread in the machinery industry and finding further application areas due to the extension to the server/client protocols by publisher/subscriber architectures. In addition, there is a lot of work on so-called companion specification, as for instance a working ground on porting the data models from IEC 61400-25 to OPC UA [39].

To enable communication between the wind farms and the operators mostly located in service and support centers, but also accessed from their notebooks anywhere mobile, the internet was already used at an early stage of the industry. Web interfaces for the service staff were implemented to enable the essential operations of the turbine. For the protection of the system from unauthorized access, see Section 7.8.6, the connection over the internet was in the meantime established usually via virtual private networks (VPN). A VPN allows an encrypted and authenticated connection between the network of the user and the network of the wind farm and turbines. In this way, the service staff can access the systems over the internet with all the applications that might not be secured by themselves but preferably without the possibility of use or insight by third parties.

With the invention of cloud-based services and the increasing number of things that communicate with them, new technologies have become established. Where there is talk of the Internet of Things (IoT), the systems no longer communicate directly with each other but via brokers. The message broker is a central intermediary at which the telegrams between the things are communicated. The advantage is the reduced effort required to configure the systems, because they no longer need to know each other. Only the configuration with the address of the broker is necessary to share information with all other connected services. Protocols used for this purpose, such as the Message Queue Telemetry Transport (MQTT), offer encryption and signing of the communication, services to ensure the quality and transmission of telegrams, as well as a management of the information whose are published to so-called topics.

Such message brokers can be placed on the internet and accessible to all via a public address, as well as used in local networks. If in a wind farm or in the network of an operator with hundreds of turbines all addresses and communication channels have usually to be set individually, the effort is immense. A local and central message broker can avoid this and even improve security. This is because it is no longer necessary to open ports for incoming connections in the firewall of each system and open ports for communication. The systems only establish connections outbound to the message broker. Thus, only the firewall of the broker must open a single port, and only this must be protected by measures. The system for the broker can be placed in a demilitarized zone (DMZ) and thus completely separated from the network, the software process of the message broker can also be protected in the operating system by means of containers.

All further services that work with the information and data of the systems must no longer poll and query these directly from all the individual systems. They can simply connect to the message broker, register there, and subscribe to topics for the receipt of new data. This allows to easily establish services for further data-processing, store it and save it in databases or use it for visualization for operators. However, potential connection issues and temporary disconnects must always be taken into account so that this is not accompanied by any loss of data. A local storage should be integrated as fallback option, which stores the data locally in the event of a disconnect and then sends the data after the reconnect.

### *7.8.5 Predictive maintenance*

To reduce the occurrence of maintenance and to schedule service not at fixed time intervals, but according to the needs of the turbine and components, predictive maintenance is used. There are different concepts and approaches to use additional or existing measurement technology to evaluate the information about the turbine and to make a prediction. Among other things, the vibrations of bearings or gears are evaluated to detect emerging damage at an early stage before it leads to damage that would result in the shutdown of the turbine and the replacement of the component. With delivery times of large components of several months to years, early detection is necessary to be able to procure and replace the component, with the organizational effort, procurement and planning to optimally avoid or at least minimize the downtime of the plant.

A more recent approach is the use of machine learning (ML), a discipline from artificial intelligence (AI) and the use of artificial neural networks (ANN). ML uses algorithms that are not individually programmed, but learning algorithms that are taught by the data and signals recorded or simulated by a turbine. These algorithms then model the turbine or components, which are used to calculate predictions or detect anomalies. During anomaly detection, the results of the models are compared with the real measured values and deviations are detected. Such deviations are then referred to as anomalies and used to detect errors, faults, and specific events or to implement a health monitoring of the components. With transfer learning it is also possible to transfer the model from one system to another [40]. This way a model can be pre-trained very fast with the help of a simulation, because virtually a lot of training data can be generated in a short time. The result is a generic model which physically represents the system. With the real data measured on the system the model can be further trained to perform an adaptation to the real system and to transfer the model specifically for this system. In this way, continuous improvement of the models can also be carried out or adaptations to changes.

Due to the ever-increasing fleet of turbines that need to be monitored and maintained, there is no longer sufficient manpower available to handle the tasks in view of the impending shortage of skilled workers. Therefore, it is a goal to automate the evaluation of the data and to reduce the service operations, or to optimize and perform them more efficiently. ML approaches help to evaluate the vast amount of data from the fleet [37]. However, anomaly detection can only recognize and point out deviations, but not determine their causes. This would again require trained and experienced personnel. To generate meaningful information for the operators from the anomaly, the anomalies are classified. Further learning algorithms are used, which determine the possible causes via probabilities. If an anomaly is detected and reported, a cause is entered as feedback into the system by the service personnel after troubleshooting and intervention. The classification of that feedback can then be used to determine the most frequent causes for the types of anomalies that have already been learned. In this way, the next time the anomaly occurs, suggestions for causes are already made and, finally, perhaps even a very specific source is given with instructions on how to rectify it.

### *7.8.6 Computer security*

The term computer security or also named cyber security, aims to protect and secure computers and networks from unauthorized access, tampering or other attacks. Since the control systems in the wind turbines today largely correspond to the common personal computer systems for business and personal use, the practices for implementing security are also the same. The objectives are often defined as confidentiality, integrity, and availability (CIA). In the widespread computer systems and networks where users communicate and exchange personal data over the internet, the protection of information, confidentiality is often a top priority. On the other hand, for machines and especially for critical infrastructure with energy production as wind turbines, the availability of systems is the top priority. It is more important to keep the turbine in operation and to supply people with energy than protecting the operating data of the turbines from being viewed by third parties. For this reason, operators are also afraid of software updates, which are supposed to fix issues and vulnerabilities, but can lead to unforeseen side effects and instabilities.

Especially in the first decades of wind energy, when the share of turbines in the mix of energy production plants still played a subordinate role, the security of the systems was not yet so much in the focus. But today with the large share of wind energy in the total energy supply, as well as after recent incidents of attacks on power plants or the communication interruption of hundreds of turbines due to a disrupted satellite communication, the topic is coming under strong attention. The requirements of the grid utilities, operators of wind farms but also public facilities regarding the security of the systems in a wind turbine are increasing.

Common requirements are derived from standards as the ISO/IEC 17799 and ISO/IEC 27001, which describe the cyber security management for information technology systems. But the new standard IEC 62443 is especially for the security of industrial automation and control systems (IACS), often also called the operational technology (OT) in distinction to the information technology (IT) as used in business and enterprise. For the specific area of IACS the IEC 62443 describes not only specific requirements for the components, but also for processes like the development and maintenance at the company as well as their design and integration into the machine or plant. The whole product development lifecycle shall be audited and improved continuously. Like the requirements for functional safety, risk analyses are carried out to determine a required level of maturity and security. These include not only the consideration of intentional violation, but also the risk from unintentional or accidental misuse. While, on the other hand, protection from intelligent agencies with specific knowledge, high motivation, and extensive resources is not completely possible [41].

In IT the most communication, especially via the internet, is now secured by means of transport layer security (TLS). TLS is a cryptographic protocol to implement confidentiality by encryption, but also integrity by authentication of the communication between computer applications by means of certificates. These digital certificates are used to identify the devices or applications and offer a public/private key pair together with the signature of a superior certificate authority. The key pair is used in such a way that the public key is distributed and used to encrypt data sent to the application, while the private key remains a secret but is the only one that allows to decrypt of this data. The certificate authority (CA) is used as a common basis of trust between the otherwise unknown applications by signing the certificates. This way, mutual certificates can be used to implement complete encryption and security of communication. However, a distinction must also be made between transport encryption and end-to-end encryption. Because with some protocols, such as MQTT, the transport from client to the broker is encrypted, but the data is then processed unencrypted on the broker and will only be re-encrypted for the transport when communicating with another client. Therefore, if an intermediary unit such as a message broker or proxy server is used within the communication, it must be evaluated whether it is trustworthy or whether end-to-end encryption is preferred.

In the networks of the critical infrastructure, a complete encryption of the communication is often even an impediment. Unlike in IT, in OT the availability is the highest priority and how to detect any unusual communication in a network in case all communication is encrypted. Most information as machine data must be protected from manipulation rather than from inspection. But the source of the information becomes especially important in case of control signals, settings and parameters are changed, or new set values are demanded, then authentication and authorization is required. Meanwhile, so-called intrusion detection systems (IDS) are used to observe communication within a network, as also recommended in the IEC 62351. Therefore, in local networks the operators of OT are moving back to unencrypted but only signed communication. That does not implement confidentiality, but integrity and identifiability of the systems. The usual protocols and ways of communication are taught to the IDS and in case of anomalies an alarm will be raised. Thus, the IDS and its functionality is largely dependent on unencrypted communication to be able to evaluate its internals. Nevertheless, sensitive information that needs to be protected or communication outside the local and private network can then continue to be encrypted [42].

Apart from the necessary and relevant protection of computers against attacks, the physical systems, be they mechanical, hydraulic, or pneumatic, should also be implemented as intrinsically safe as possible. If a system is self-protecting and fail-safe by design, utilizing physical means to prevent any damage or destruction, should be used to implement a tamper-proof and secure prevention. The probability of a malfunction in mechanical components like valves is very low and the attacks via cyber space can be excluded [17].

## 7.9 Conclusions and outlook

Nowadays, what is referred to as digitalization, Industry 4.0, and the IoT already has the automation of wind turbines in its genes. This is because the distributed turbines in the remotest corners of the world made communication via the internet for remote operation and maintenance inevitable. It was only called SCADA until now and implemented with proprietary means. Thus, the new technologies from IT, be it cloud services, IoT protocols, or edge devices are finding rapid acceptance and implementation in the industry [43]. However, this is often based on proprietary rather than standardized protocols or open interfaces. In addition, concerns, and requirements regarding the cyber security of the wind turbines, farms, and remote operating stations (ROS) are increasing.

Unfortunately, the initial uses of latest technologies such as PC-based control systems, modular I/O systems and Ethernet-based fieldbuses have not been continued by further innovations for control systems in the wind industry. Under the cost pressure, the partly weak local markets in Europe, and the following market shakeout in the 2010s, the manufacturers set their focus on continual larger turbines [44]. This further development happened primarily through a form of upscaling, the components simply became larger and larger. Solving the resulting difficulties did not lead to any fundamental changes, as well as a lack of quality and reliability. Components specially designed for the turbines were not produced in large series, but rather in manufactories, and thus no continuous optimization of the systems or reduction of production costs through scaling was realized [45]. The control systems in some new designs of turbines today mostly look the same as they did 20 years ago. Maybe new components were added left and right, then new tasks were gladly implemented by closed and proprietary systems, but no comprehensive revision of the concept was carried out. This results in several difficulties, obscure black box systems, various special interfaces between systems, each system with its own hardware and infrastructure components, lack of time synchronization, different formats for data and protocols with the loss of information due to data concentration and lack of communication. This also leads to a significant cyber security risk: Each subsystem must be maintained individually, vulnerabilities due to diversity and combination of subsystems may remain undetected, every component requires its specific software, tools, updates, as well as individual expertise and trained personnel. Finally, this leads to high complexity, difficulties in integration, lifetime maintenance and troubleshooting, and thus high costs both at installation and spread over the lifetime. Standardizing protocols and interfaces, as well as unifying systems and integrating the various tasks into a centralized system, would circumvent such problems and provide new opportunities for optimization. This would also enable utilization of modern sensors and measurements for modal analysis and holistic control algorithms, as well as modern learning algorithms and predictive control for the comprehensive wind turbine system.

An exemplary project would be a 5 MW offshore wind turbine designed in 2014 with approximately 1500 input signals and 500 output signals, as well as a further 10 fieldbus interfaces. That project utilized a major task for the execution of control logics and algorithms in a 10 millisecond cycle time as well as a minor task for the execution of concurrent jobs, such as asynchronous communication, the interface for storing in a database and so on with a 1-ms cycle time. On the used processing unit Intel® Atom™ E3827 with 1.6 GHz on two cores, the two tasks could be executed like threads separately on the cores. In the application, based on a software framework, a total of about 2000 status objects, 1000 parameter objects, 250 command objects, about 50 mean objects and 200 recorded signals were used for implementation. Due to the modern platform, the efficient runtime and the high-performance framework, the CPU was only utilized by 15% for the execution of the application. This means each cycle takes 1.5 ms to execute and is then executed again after a further 8.5 ms due to the 10 ms cycle time of the task. At the same time, the EtherCAT fieldbus with 100 MBit/s bandwidth was only utilized to 8% and otherwise unused.

There are still plenty of resources available to integrate further functions and features into the application. The CPU used is in the medium to lower price range and is also available as a variant with 1.91 GHz and four cores as Intel Atom® E3845. With such processors or even from higher performance classes, advanced tasks such as the evaluation of vibrations or rotor blade bending on a third core, as well as the execution of models or machine learning algorithms on the fourth core, can thus be implemented almost eligibly. However, such applications require a rethinking and paradigm shift by the manufacturers, away from thinking in separate hardware implementations to separate software components that are integrated on a common hardware technology.

In addition, many manufacturers are reluctant to provide data from the turbines by offering open interfaces. Usually, only a few 10-minute averages are made available via the SCADA system and by proprietary protocols. The data of all the sensors is available in the control system and could be offered, the owners and operators should have the right to use this information. In doing so, the high-resolution data would allow better root cause analysis (RCA), failure mode and effects analysis (FMEA) [46], as well as promote immense progress in data-driven developments [47], if it were possible for owners, operators, maintenance companies as well as data scientists and research to make use of the data. In addition, the interconnection and communication of systems in energy production becomes necessary for the implementation of a smart grid in the all-electric future [48]. It would be possible to provide the data every second by a standardized interface and a data model such as defined in the IEC 61400-25. In many other industries, data models are now established for this purpose via OPC UA [49]. The current developments to port the data model from IEC 61400-25 to OPC UA would be suitable to become such a standard for the wind industry [39].

The need for improved dynamic control algorithms, enhanced monitoring and thus necessary computing performance will continue to increase, as required by floating wind turbines. Enabled by the continuous advances in computing and communication technology, there is an opportunity for further improvements in the control systems. The integration of all information on a central platform would then allow the use of higher-level algorithms and data analytics. Advanced control strategies, such as model predictive control, adaptive control, and advanced pitch control algorithms, up to swarm intelligence can be used to optimize power harvesting and reduce stress on critical components. Machine learning techniques are being further developed to augment digital twins with fault and anomaly detection algorithms, leading to predictive maintenance strategies and other data-driven solutions to improve turbine performance and reliability. These advances will further drive the growth and efficiency of wind energy as a clean and sustainable energy source. In addition, greater integration with the energy grids, through the combination with batteries, hydrogen production and sector coupling requires an open connectivity and performant data exchange by standardized communication protocols. Due to the continuously decreasing rotational inertia in the energy grid, the provision of virtual inertia and ever shorter reaction times on events by the wind farm are required for a comprehensive all-electric energy society, as described in Chapter 10.

## References

[1] International Electrotechnical Commission. *Programmable controllers – Part 3: Programming languages*. 3rd ed. No. IEC 61131-3 in International Standard. Geneva, Switzerland: International Electrotechnical Commission; 2013.

[2] Sayed K., Abo-Khalil A., and Eltamaly A. ′Wind Power Plants Control Systems Based on SCADA System′ in *Control and Operation of Grid-Connected Wind Energy Systems*. Berlin Heidelberg: Springer; 2021. Available from https://www.researchgate.net/publication/349807006 [Accessed May 2023]

[3] International Electrotechnical Commission. *Wind Energy Generation Systems – Part 1: Design Requirements*. 4th ed. No. IEC 61400-1 in International Standard. Geneva, Switzerland: International Electrotechnical Commission; 2019.

[4] MathWorks. *Industrial Automation and Machinery*. Available from https://mathworks.com/solutions/industrial-automation-machinery.html [Accessed May 2023]

[5] The Modelica Association, *Modelica*. Available from https://modelica.org/ [Accessed May 2023]

[6] Moore G., *Moore's Law*, 1975. Available from https://en.wikipedia.org/wiki/Moore%27s_law [Accessed May 2023]

[7] Wörn H., and Brinkschulte U. *Echtzeitsysteme*. Berlin: Springer; 2005

[8] German Institute for Standardization. *DIN 44300*, 1988

[9] *High-Quality, Proven Control Components for Vensys Wind Turbines*. PC-Control 03/2010. Available from https://www.pc-control.net/pdf/032010/solutions/pcc_0310_goldwind_e.pdf

[10] PROFIBUS and PROFINET International (PI). Available from https://www.profibus.com/ [Accessed May 2023]

[11] PROFIBUS and PROFINET International (PI), *ProfiSAFE*. Available from https://us.profinet.com/profisafe-profinet-profibus/ [Accessed May 2023]
[12] EtherCAT Technology Group (ETG). Available from https://www.ethercat.org/ [Accessed May 2023]
[13] EtherCAT Technology Group (ETG). *Safety over EtherCAT (FSoE)*. Available from https://www.ethercat.org/en/safety.html [Accessed May 2023]
[14] Goller V. Industrie 4.0 – Überblick zu Echtzeit-Ethernet. 2019. Available from https://www.elektroniknet.de/automation/industrie-40-iot/ueberblick-zu-echtzeit-ethernet.162383.html [Accessed May 2023]
[15] PROFIBUS and PROFINET International (PI), *Update of PROFINET Specification V2.4*. Available from https://www.profibus.com/newsroom/press-news/first-maintenance-update-of-profinet-specification-v24-is-complete/ [Accessed May 2023]
[16] Hau E. *Windkraftanlagen*. 4th ed. Berlin: Springer; 2008. Ch. 10
[17] Atug M. Interview mit Manuel Atug, AG Kritis. 2022 Available from https://www.lanline.de/it-security/grundschutz-statt-cyberwar-gerede.254110.html [Accessed May 2023]
[18] Kursiek A. *Windenergieanlagen*. München: Carl Hanser; 2022. Ch. 30–41
[19] Burton T., Jenkins N., Sharpe D., and Bossanyi E. *Wind Energy Handbook*. 2nd ed. Chichester: John Wiley; 2011. Ch. 8
[20] PROFIBUS and PROFINET International (PI), *PROFINET*. Available from https://de.profibus.com/index.php?eID=dumpFile&t=f&f=82431&token=056ec23a1f6860dc926064f796d49f75c7d72a76 [Accessed May 2023]
[21] EtherCAT Technology Group (ETG), *EtherCAT – The Ethernet Fieldbus*. Available from https://www.ethercat.org/download/documents/ETG_Brochure_EN.pdf [Accessed May 2023]
[22] International Electrotechnical Commission. Wind turbines – Part 24: Lightning protection. No. IEC 61400-24 in International Standard. Geneva, Switzerland: International Electrotechnical Commission; 2010
[23] Gambier A. *Control of Large Wind Energy Systems*. Cham: Springer; 2022
[24] Popko W. and Thomas P. *IWES Wind Turbine IWT-7.5-164*. Rev 4. Bremerhaven: Faunhofer IWES; 2018. Available from https://publica.fraunhofer.de/handle/publica/299323 [Accessed May 2023]
[25] Kusiak A. and Li W., The Prediction and Diagnosis of Wind Turbine Faults in Renewable Energy, vol. 36, issue 1, 16–23, 2011. Available from https://www.sciencedirect.com/science/article/abs/pii/S0960148110002338 [Accessed May 2023]
[26] Gasch R. and Twele J. *Windkraftanlagen*. 5th ed. Berlin: Springer; 2007. Ch. 12
[27] Wang G., Yin S., and Karimi H.R. Health monitoring of wind turbine: data-based approaches. In Karimi H.R. (ed.), *Structural Control and Fault Detection of Wind Turbine Systems* (pp. 169–191). Stevenage: The Institution of Engineering and Technology; 2018
[28] Meng F., Wenske J., Neshati M., and Bartschat A. Advanced control of wind turbine system. In Karimi H.R. (ed.), *Structural Control and Fault Detection*

*of Wind Turbine Systems* (pp. 113–148). Stevenage: The Institution of Engineering and Technology; 2018

[29] Tegtmeier M., Effects of Yaw Misalignment on the Performance of a Wind Turbine. Available from https://www.turbit.com/post/effects-of-yaw-mis-alignment-on-the-performance-of-a-wind-turbine [Accessed May 2023]

[30] Xin W., Yanping L., and Wei T., *Modified Hill Climbing Method for Active Yaw Control in Wind Turbine*. Available from https://ieeexplore.ieee.org/abstract/document/6391113 [Accessed June 2023]

[31] Snleckus D., *Seal of Approval for World's First Laser-Aided Turbine Control*. Available from https://www.rechargenews.com/wind/seal-of-approval-for-worlds-first-laser-aided-turbine-control/2-1-701472 [Accessed June 2023]

[32] Schild A., Ritter B., and Gros S., *High Performance Nonlinear Model Predictive Control for Wind Turbines*, Wind Europe Summit 2016, Hamburg. Available from https://windeurope.org/summit2016/conference/allfiles/93_WindEurope
2016presentation.pdf [Accessed June 2023]

[33] Luna J., Falkenberg O., Gros S., and Schild A., Wind turbine Fatigue Reduction Based on Economic-Tracking NMPC with Direct ANN Fatigue Estimation, *Renewable Energy*, vol. 147, issue P1, 1632–1641, 2020. Available from https://www.sciencedirect.com/science/article/abs/pii/S0960148119314235 [Accessed June 2023]

[34] Wintermeyer-Kallen T., Basler M., Konrad T., Zierath J., and Abel D., *Challenges of Applying Model-Based Predictive Wind Turbine Control in the Field*, CWD DSEC 2023, Aachen. Available from https://link.springer.com/article/10.1007/s10010-023-00634-1 [Accessed June 2023]

[35] eco4wind – Echtzeitbetriebsführung für moderne Windenergieanlagen: Abschlussbericht. Available from https://www.tib.eu/de/suchen/id/TIBKAT:1733103201/eco4wind-Echtzeitbetriebsf%C3%BChrung-f%C3%BCr-moderne-Windenergieanlagen [Accessed June 2023]

[36] International Electrotechnical Commission. Wind turbines – Part 12: *Power performance of electricity-productin wind turbines*. No. IEC 61400-12 in International Standard. Geneva, Switzerland: International Electrotechnical Commission; 2014

[37] Gutierrez J., *Wind is Going Digital: New Technologies are Driving Increases in Operational Efficiency and Excellence*. Available from https://www.linkedin.com/pulse/wind-going-digital-new-technologies-driving-increases-juan-gutierrez [Accessed June 2023]

[38] Hau M. and Shan M., Windparkregelung zur Netzintegration. *Symposium Energy System Technology*. Kassel 2011.

[39] USE61400-25, AWEA 2019. Available from https://use61400-25.com/wp-content/uploads/2019/05/AWEA2019_Flyer.pdf [Accessed June 2023]

[40] Tegtmeier M., *Transfer Learning: A Guide to Utilizing Pre-Trained Models for Wind Turbine Data*. Available from https://www.turbit.com/post/

transfer-learning-a-guide-to-utilizing-pre-trained-models-for-wind-turbine-datatransfer-learning [Accessed June 2023]

[41] International Electrotechnical Commission. *Security for Industrial Automation and Control Systems*. 1st ed. No. IEC 62443-3 in International Standard. Geneva, Switzerland: International Electrotechnical Commission; 2013.

[42] Ritter H. and Klien A., *Infrastruktur sicher schützen, aber wie?*, ETZ 12/2022. P. 51–43

[43] Reed J., Nordex on Their Push Towards Real World IoT at Scale. Available from https://diginomica.com/conxion-2020-nordex-their-push-towards-real-world-iot-scale-software-ag-use-case [Accessed June 2023]

[44] Henseter F. and Dehoorne D., It's Time to Slow Down on Turbines if We Really Want to Scale Up the Wind Industry, RECHARGE. Available from https://www.rechargenews.com/wind/its-time-to-slow-down-on-turbines-if-we-really-want-to-scale-up-the-wind-industry/2-1-1439317 [Accessed June 2023]

[45] Durakovic A., Ever-Growing Offshore Wind Turbines Bring Unsustainable Market Risks. Available from https://www.offshorewind.biz/2023/05/04/ever-growing-offshore-wind-turbines-bring-unsustainable-market-risks-gcube/ [Accessed June 2023]

[46] Zickert B., 10 Years of Alpha Ventus with PC-Based Wind Turbine Control, PC-Control 03/2020. Available from https://www.pc-control.net/pdf/032020/interview/pcc_0320_alpha-ventus_e.pdf [Accessed June 2023]

[47] Tegtmeier M., *Advantages of High-Resolution Wind Turbine Data with 60-Second Data*. Available from https://www.turbit.com/post/unleash-the-power-of-high-resolution-wind-park-ai-monitoring-with-turbit-s-60-second-data [Accessed June 2023]

[48] Bent R., *Schlüssel zur Energiewende*. Available from https://www.computer-automation.de/feldebene/vernetzung/schluessel-zur-energiewende.204526.html [Accessed June 2023]

[49] Krischke I., *Ein Pfad durch den Zoo der Standards*. Available from https://www.computer-automation.de/feldebene/vernetzung/ein-pfad-durch-den-zoo-der-standards.200356.html [Accessed June 2023]

# Chapter 8
# Structural health monitoring

*Carles Colomer Segura*[1], *Carsten Ebert*[1] *and Peter Kraemer*[2]

Continuous structural health monitoring (SHM) of structures – rotor blades, towers and foundations of wind turbine generators (WT) – is still a relatively young discipline in engineering. For several decades, scientists have been investigating possible methods and the directly associated outcomes as well as achievable sensitivities. After the condition monitoring of drivetrains has become the standard in wind energy since the beginning of the 2000s, the monitoring of structural components has increasingly become the state of the art – especially for offshore but also for onshore wind turbines.

The authors have been successfully involved in the research and further development of SHM solutions for supporting structures for many years and have extensive experience in deploying these solutions in commercial environments. They see SHM as a data-driven process capable of providing reliable diagnoses and assessments of the structural performance with the main goal of ultimately adding value to the operators of wind turbine assets.

## 8.1 Motivation

### *8.1.1 Goals of structural health monitoring*

Wind turbines are designed with a strict eye on the costs of construction and operation, CAPEX and OPEX, respectively. Accordingly, only systems that are indispensable acc. to the authorities for safe operation, such as need-based night marking or rotor blade ice detection, or have a commercial benefit for the turbine manufacturer or operator are integrated into the WT.

In this sense, SHM systems can offer a wide range of benefits for turbine operation. Some of them are listed below exemplarily:

- Fulfillment of monitoring requirements by authorities.
- Reduction of maintenance work and costs.

[1]Wölfel Wind Systems GmbH + Co. KG, Germany
[2]Department of Mechanical Engineering, University of Siegen, Germany

- Optimization of the wind park operation.
- Improving asset management.
- Enabling lifetime extension.
- Continuous validation against design assumptions.

It is not possible to present the commercial benefits of SHM systems on a general basis. A case-by-case evaluation is always required, as the performance of the systems available on the market, the interests of the involved stakeholders and the calculation assumptions for CAPEX and OPEX costs are too different between projects. The rated power of the wind turbines in which the systems are integrated, site-specific boundary conditions (such as icing periods), the accessibility of the turbines for inspection and maintenance and many other boundary conditions have an influence on the commercial viability of an SHM solution.

In addition to the investment costs for the SHM systems, the maintenance and repair costs – which are also necessary for these systems – must not be neglected, as these are usually decisive for economic efficiency over the planned service life of at least 20 years (today usually 25 years).

However, one thing is important to remark: systems that offer no economic advantage for turbine manufacturers or turbine operators will not be integrated into the WT.

### *8.1.2 What is the difference between condition monitoring and structural health monitoring?*

In the literature and in regulations, different definitions are given for "Condition Monitoring" (CM) (see [1] for example) and "Structural Health Monitoring" (SHM). The terms are not uniformly defined and overlap each other in some situations. In the wind energy sector, the term CM or CMS (condition monitoring system) is often understood as a generic term for the continuous monitoring of mechanical components as well as structural elements. In this view, systems and methods for monitoring machinery parts (especially the drivetrain) as well as systems and methods for monitoring structures (rotor blades, towers and foundations) are included under the term CM.

One possible reason for this confusion is probably that both CM and SHM are usually based on continuous measurements, data evaluation and system assessment. Both disciplines, CM and SHM, share the same objective, but refer to different monitoring objects.

**Condition monitoring systems (CMS)** are systems for the continuous monitoring of rotating machinery components, especially drivetrains.

Main and rotor bearings, gearboxes and generator bearings are regularly monitored. The objects to be monitored are precisely manufactured and standardized machine parts with tolerances in the range of micrometers or less. In the case of gearboxes, for example, the distribution of masses and stiffnesses is very well known. This means that natural frequencies, rollover frequencies, and/or other modal features of such machines are well known. These objects are associated with geometric and kinematic conditions that are very closely linked to certain damage features. For example, in classical CM, the damage features are often associated with "rollover frequencies" (e.g., rolling bearing damage) or "gear meshing frequencies" (gear damage). This enables data-driven localization and identification of the type of damage. Thus, for example, the occurrence of a ball pass frequency at the outer raceway can be associated with outer ring damage of a roller bearing via the kinematics, geometry and number of rolling elements.

In the context of machine learning, damage detection in CMS can often be associated with model-based supervised learning, since the frequencies that occur can be directly assigned to one or another type of damage.

Most CM applications for damage detection today are based on the interpretation of parameters such as temperature, oil quality, torsional load (e.g., in the case of shafts), and in particular on the evaluation of kinematic-related frequencies.

**Structural Health Monitoring (SHM)** is the continuous monitoring of support structures, especially rotor blades, towers, and foundations, but also hubs and main frames of the nacelle. One of the main objectives of SHM is to assess the load-bearing capacity, the stability, and fatigue loads of these elements.

However, usually, no information about any damaged state is directly available, so SHM systems are often designed to detect structural changes in relation to the original design or the initial state after commissioning.

The production of the components with a load-bearing function that are to be monitored is partly manual (in wind turbines this applies to the rotor blades, but in the civil engineering it also applies to bridges, towers, etc.), which results in a comparatively low production accuracy in the range of millimeters to centimeters. Accordingly, the masses and stiffnesses of the structures also vary in the single-digit percentage range.

As a rule, the measured structural responses to ambient and to operational excitations are used for monitoring. These can often only be described stochastically, but not deterministically, which makes the identification and evaluation of eigenfrequencies, damping ratios and vibration patterns difficult. Furthermore, in most cases, the damage-related changes cannot be predicted or localized based on expected specific feature changes. This is different from CM where, for example, kinematic-related rollover frequencies can be determined analytically before the installation of monitoring systems.

It follows from the above that every load-bearing structure monitored with SHM has a certain individuality. In order to be able to make statements about the condition of such an "individual," the behavior of the object must first be learned. This means that reference values must first be determined. In this context, damage detection in SHM is often combined with algorithms for unsupervised learning and change detection (novelty detection) during the monitoring process.

Why is this distinction between CM and SHM so important from the authors' point of view? The most significant reason is the substantially different methods and algorithms that are applied and used for both areas! Furthermore, the measurement hardware used also differs (measurement and frequency ranges of sensors, sampling frequencies of data acquisition, continuous data acquisition vs. cyclic data acquisition with snapshots). As a result, one usually finds companies in the market that specialize in either CM for drivetrain monitoring or SHM for monitoring support structures.

In this chapter, the structural health monitoring of wind turbines is dealt with specifically. For a description of methods, algorithms, and application examples from the field of drivetrain monitoring, we reference to the extensively available technical literature (see [2,3] for example).

### *8.1.3 Relevant standards and guidelines*

There are many standards and regulations that deal with the topics of vibration measurements, condition monitoring and structural health monitoring. Only the basic ISO guidelines ISO 2041 [4], ISO 13372 [5], and ISO 17359 [1] are listed here as examples.

For the design of onshore and offshore wind turbines, the international standards IEC 61400-1 (see [6]) and IEC 61400-3 (see [7,8]) initially apply, but they do not contain any explicit requirements regarding structural monitoring. However, these standards are essential for understanding the functioning and design of wind turbines and thus also for assessing monitoring results. Other thematically related regulations from certification bodies (e.g., DNV guidelines) are also available.

For some time now, regulations have been available that describe how vibration measurements should be carried out on wind turbines and how the measured mechanical vibrations can be assessed (VDI 3834, part 1 and 2, see [9,23]). These guidelines focus on the assessment of the drivetrain, however also contain additional information on the vibration assessment of the supporting structure. It should be taken into account that the zone boundary values recommended within these guidelines are not directly suitable for use as limit values in monitoring systems. The guidelines recommend defining turbine-specific limit values on the basis of reference measurements and taking into consideration any operating parameters that influence vibration. Very similar rules and assessments are described within the corresponding international standard ISO 10816-21 (see [10]).

In Germany, a comprehensive guideline has been drawn up which documents and describes in textbook fashion the current state of knowledge for the monitoring of onshore and offshore wind turbines as well as for offshore stations (VDI 4551, see [11]). The guideline provides information on the objectives and tasks of structural health monitoring, on inspection and service life forecasts, on model preparation, on periodic inspections, on measurement concepts and other aspects.

The Norwegian Geotechnical Institute (NGI) has prepared Guidelines for the structural health monitoring for offshore wind turbine towers and foundations on behalf of BSEE Environmental Enforcement Division (see [12]). This report describes objectives and solutions for industrial application in a very specific and practice-relevant way.

Furthermore, there are superordinate regulations that describe how monitoring solutions should be organized. The DNV service specification DNV-SE-0439 (see [25]) is a representative example. This service specification documents basic requirements that must be fulfilled if monitoring systems are to be certified. For the suppliers of monitoring systems, it is often important to have their systems certified by independent certification bodies, as the approval authorities require the use of certified systems for numerous applications (e.g., for ice detection on rotor blades).

In the recent past, the importance of asset management for wind farm projects has been increasing. In this context, a central component is lifetime management and the associated lifetime extensions. The basic principles and procedures for this are listed in IEC TS 61400-28 Ed. 1 (see [26]), among others.

## 8.2 Methods and monitoring principles

In general, SHM methods and algorithms can be divided into different classes according to several criteria.

A first classification can be made according to the physical measuring principle. A distinction is made between vibration-based methods (low-frequency range), acoustic methods (medium frequency range), and ultrasonic methods, such as acoustic emissions (high-frequency range).

The excitation also plays a role: if the excitation is known, we speak of methods with known excitation, if the excitation is unknown, we speak of output-only approaches.

The procedures can be further divided into global and local procedures. Global methods are usually used to monitor the entire structure, require only a few sensors that do not necessarily have to be placed close to the damage and do not require a priori knowledge about so-called hotspots. Since global methods are based on the observation of relatively low frequencies, they are also not very sensitive to "small" damages. In contrast, local methods require a dense sensor network, the sensors must be placed close to the damage and knowledge about hotspots is often necessary. As local methods are based on the observation of high frequencies, these methods are highly sensitive to small structural damages.

Another classification is based on the inclusion of numerical models in damage detection. In this case, the procedures are called model-based. Otherwise, if only measured data are used for evaluation, the procedures are called data-driven.

A very well-known classification in the literature refers to different levels of damage identification. According to [30], several levels of damage identification can be defined: damage detection, damage localization, determination of the extent of damage, prediction of remaining service life.

Since only very limited design and model information is usually available when implementing SHM systems, and for cost reasons, mainly vibration-based, output-only, data-driven, global methods have prevailed in industrial implementation in the wind energy sector to date. Thesc are more suitable for damage detection or early detection of changes in the load-bearing capacity of structural components.

The vibration-based methods are based on the structural dynamics of a non-linear, time-variant excited system [20]. The stochastic vibration excitation can often not be measured directly (forces from the wind field on the rotor blades for example), so that output-only methods must be used for system identification. The description is possible with the help of the following equations of motion and damage evolution (see [15–17]):

$$\boldsymbol{M}(\boldsymbol{\theta}_d,\boldsymbol{\theta}_e,\boldsymbol{x},t)\ddot{\boldsymbol{x}}+\boldsymbol{g}(\boldsymbol{\theta}_d,\boldsymbol{\theta}_e,\dot{\boldsymbol{x}},t)=\boldsymbol{f}(\boldsymbol{\theta}_d,\boldsymbol{\theta}_e,t) \tag{8.1}$$

$$\dot{\boldsymbol{\theta}}_d=\boldsymbol{\Gamma}(\boldsymbol{\theta}_d,\boldsymbol{\theta}_e,\boldsymbol{x},\dot{\boldsymbol{x}},t), \tag{8.2}$$

$$\boldsymbol{y}(t)=\boldsymbol{h}(\boldsymbol{\theta}_d,\boldsymbol{\theta}_e,\boldsymbol{x},\dot{\boldsymbol{x}},t). \tag{8.3}$$

The matrix $\boldsymbol{M}$ describes the mass, the matrix $\boldsymbol{g}$ contains the elastic forces, the damping forces and so on and the vector $\boldsymbol{f}$ describes the excitation forces; $\ddot{\boldsymbol{x}},\dot{\boldsymbol{x}},\boldsymbol{x}$ describes the acceleration-, velocity, and displacement-vectors. A damage parameter $\boldsymbol{\theta}_d$ is depending on the time $t$; the time-depending environmental influences are denoted with $\boldsymbol{\theta}_e$ (e.g., changes of temperatures, boundary conditions). The generally non-linear function $\boldsymbol{\Gamma}$ describes the evolution of the damage parameter $\boldsymbol{\theta}_d$ – a loss of stiffness, a change in mass or a crack propagation for example. A temporary increase in mass, e.g., due to ice accretion, is formally assigned in this context to $\boldsymbol{\theta}_d$, as it is not part of an expected (normal) environmental or boundary condition change. The measurement equation (8.3) represents the measured structural response $\boldsymbol{y}$ related to the non-linear and time-depending vector $\boldsymbol{h}$.

According to [17], a feature vector $\boldsymbol{f}_y$ can be determined by means of a feature extraction procedure (FE):

$$\boldsymbol{f}_y(\boldsymbol{\theta}_d,\boldsymbol{\theta}_e)=FE(\boldsymbol{y}(\boldsymbol{\theta}_d,\boldsymbol{\theta}_e,t)) \tag{8.4}$$

The sensitivity $s$ of the feature $j$ to a structural change (type, place, and extent of change are included) indicates that the dynamics of the structure and thus the damage-sensitive features are simultaneously dependent on damage and on the environmental and operational conditions (EOC) of the structure:

$$s_j=\frac{\partial \boldsymbol{f}_y(\boldsymbol{\theta}_d,\boldsymbol{\theta}_e)}{\partial \boldsymbol{\theta}_{dj}} \tag{8.5}$$

Ideally, under laboratory conditions, the EOC do not change. However, in real applications, this is the case because of changes in soil properties, temperatures, humidity, mass moments of inertia, clamping stiffness, etc.

**Compensation of environmental and operating condition influences in practice:** The aforementioned task of EOC compensation (in the context of damage diagnosis) has increasingly come into focus in recent years, so that several methods have been developed for this purpose. According to [18], these approaches can be divided into two categories:

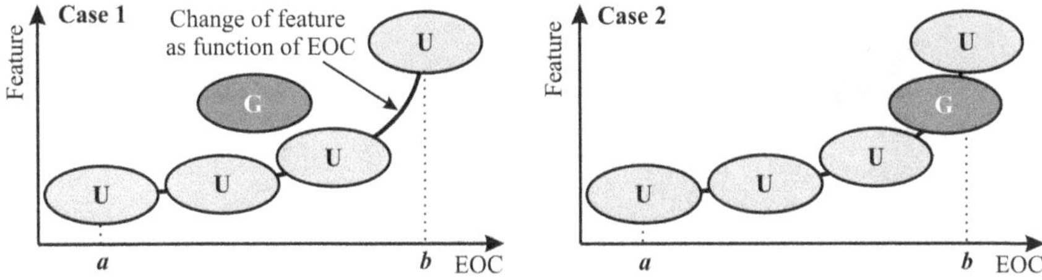


*Figure 8.1 Two cases for consideration of changing EOC according to [15,18]*

- Case I: Methods that do not require measurements of EOC variables,
- Case II: Methods that use direct measurements of the EOC variables.

Both mentioned classes of methods for EOC compensation assume that the characteristics of the undamaged state can be described as linear or non-linear functions of the EOCs.

**Case I**: If the features of a damaged state (G) differ from the function of the undamaged state (U), see Figure 8.1 (left), the EOCs are not needed. In this case, e.g., statistical dependencies between the features of the undamaged state are used to build a reference function. A damage indicator is built using the significance of the distance of the features to the reference function. If the predefined distances between the current features and the reference state exceed an established limit from a statistical point of view, one talks about the detection of a "new," "unexpected" state of the system. Different methods of pattern recognition and data mining can be used for the construction of reference functions, see [19,27] for example.

**Case II**: If the features of a damaged state coincide with the reference function, see Figure 8.1 (right), a separation between the states is no longer possible without knowledge of the EOC variables. In this case, only the observation of the EOCs shows that the same environmental and operating states of a system cause different features. Suitable methods for constructing references based on a feature-EOC space can be found in [15,18,27].

The two classes of methods are applicable for functions of the undamaged state in a considered EOC interval, see the interval [$a$, $b$] in Figure 8.1. Outside this interval (where the reference function is no longer valid), it is not possible to predict how the features evolve/behave depending on the EOCs. Exceptions to this are methods based on algorithms that are able to adaptively learn new states outside the interval and expand the reference (see [15]).

Practical experience confirms the above theoretical assumptions. Monitoring of hundreds of wind turbines in-situ [21] as well as large-scale tests in the laboratory [28] have shown that compensation of the EOC on the damage features (automatically extracted structural natural frequencies, modal damping, etc.) is absolutely necessary, as these change more with EOC than as a result of damage. A simple example of how the dynamics of a rotor blade change with temperature is shown in Figure 8.2.

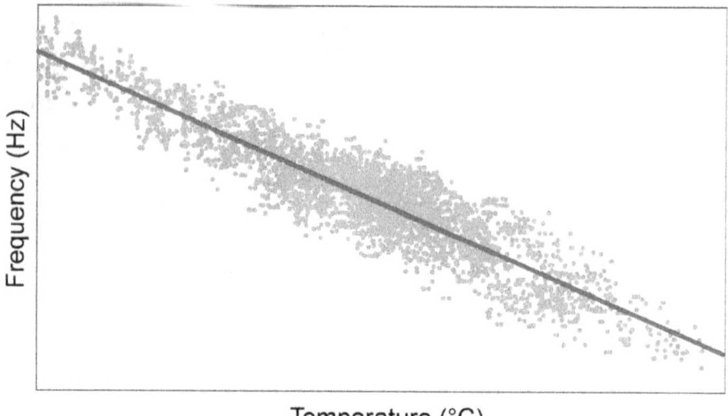


*Figure 8.2 Temperature dependence of a natural frequency of a rotor blade; similar behavior (although not always linear) exists between the blade dynamics and pitch angle as well as rotor speed. Courtesy © Prof. Dr. Peter Kraemer.*

## 8.3 Types and goals of monitoring

In this section, the authors provide a general overview of systems and their related monitoring goals that are commonly considered in modern WT. An actual implementation of these systems is mostly related to site-specific and commercial considerations.

### *8.3.1 Rotor blades and pitch bearings*

Rotor blades are major and expensive components of a wind turbine. They have reached considerable sizes, are exposed to extremely complex load situations and at the same time are very individual in terms of both the material and the manufacturing process as well as the associated tolerances.

Nevertheless, for a long time rotor blades have been the only essential components of a wind turbine that have not been monitored by sensors. Even today, the number of sensors within the rotor blades is quite small compared to other components of a WT. However, different monitoring techniques and goals have been established in recent years and are described below.

#### 8.3.1.1 Blade load monitoring

While designing new rotor blades, it is common to equip individual rotor blades on the test bench and on turbine prototypes with a comprehensive set of strain and acceleration sensors. The measured loads are then compared with design loads and used for model adjustment and turbine certification. Such extensive sensor equipment is not feasible in mass-produced wind turbines for cost reasons.

The current state of the art is to measure the bending moments in edgewise and flapwise direction at the rotor blade root on all modern wind turbines. The blade root bending moments measured in real time are used to control the turbine and to limit the loads. Often, an individual pitch control system is used, which prescribes the pitch angle for each individual rotor blade depending on the position and measured load. Other use cases – such as monitoring the clearance between the blade tip and the tower or detecting excessive torsional vibrations – can also be realized in real time. In some cases, the recorded bending moment time series can also be evaluated with regard to fatigue loads – for example by means of rainflow classification.

Various solutions with different physical operating principles are available on the market for measuring the blade root bending moments.

The most common approach consists of instrumenting the root area of the rotor blade with strain gauges at four measuring positions around the circumference. This can be done either with resistance-based electrical strain sensors or fiber-optic strain sensors (FBG sensors).

The main advantage of optical strain sensors is their high fatigue strength. On the other hand, the sensor connection cables in optical systems are comparatively sensitive and thus less robust.

In the case of electrical strain gauges, one of the main challenges is to prove fatigue resistance for the whole lifetime of the wind turbine, in which the sensors on the blade shell will undergo high strain rates and a high number of cycles.

It should also be noted that both strain sensors types must be bonded to the rotor blade shell. Here, the temperature and humidity conditions applicable to the adhesive must be observed. Moreover, in most cases, special equipment and trained personnel is required, which complicates the application in blade production and limits the applicability during maintenance in the field.

In addition to strain sensors, other measuring principles are available for recording the blade loads:

- An alternative to strain gauges consists of using inductive displacement sensors in conjunction with cantilever beams, which can be mounted in the blade and used to measure strains and bending moments.
- Occasionally, cameras have also been used in conjunction with optical targets. Image recognition software then detects the movement of the targets relative to the mounting position of the camera. It is advantageous that several targets/measuring positions can be set up. On the other hand, contamination must be taken into account, which can limit durability.
- The rotor blade loads are transmitted to the pitch bearing via the rotor blade bolts, so that in principle sensors on the blade bolts can also detect the rotor blade loads. In the course of data preparation and analysis, it should be noted that the stress behavior is non-linear due to the pitch-bearing properties and the static pretension of the blade bolts.

#### 8.3.1.2 Structural integrity of blades, blade icing, and blade imbalances

A number of vibration-based systems for monitoring rotor blades with regard to rotor blade damage, rotor blade icing and aerodynamic imbalances are established on the market. These systems usually have a bi- or triaxial acceleration sensor inside each of the rotor blades – usually at a distance of about 1/3 of the rotor blade length from the blade root for lightning protection reasons. The aim of such global monitoring approach is to detect changes in the structural dynamics in order to be able to recognize serious blade damages at early stages. In most cases, the rotor blade's natural frequencies are monitored for changes. It is particularly important to simultaneously detect and compensate for environment conditions (e.g., blade temperature) and operating conditions of the wind turbine (e.g., pitch angle), as these have a relevant influence on the dynamics of the blades and thus on the sensitivity of the monitoring systems.

A safety-critical task is the detection of rotor blade icing at locations with roads, trails or industrial areas in the vicinity of wind turbines. Systems certified by independent certification bodies are required by approval authorities in many cases. In the course of certification, it is tested whether the systems are capable of reliably detecting critical icing. Critical icing is defined as the hazard potential when ice is thrown off a wind turbine in production. Since icing is concentrated at the leading edge of the blade, the thickness of the ice is usually the underlying criterion. The hazard potential varies depending on the rotor speed, tower height and rotor blade length. As a rule, however, ice thicknesses of approx. 15 mm shall be detected reliably.

The ice detection systems require a communication interface to the turbine controller to control the turbine accordingly. Vibration-based systems with accelerometers inside the rotor blades have the advantage of not only being able to stop the wind turbine. Since the sensors can also detect rotor blade vibrations at very low wind speeds, such systems also allow the detection of icing when the wind turbine is at a (ice-induced) standstill and thus the automatic restart without on-site inspections by service personnel. This is a decisive criterion for the economic operation.

Moreover, some systems are able to detect aerodynamic induced blade imbalances additionally. Such imbalances occur if one or two blades have a different pitch angle in opposite to the remaining blades. Imbalances can have far-reaching consequences, as the fatigue load and reduction in component lifetime is proportional to the fifth power of the increase in vibration level. Therefore, even small pitch angle misalignments with deviations smaller than 0.3° should be detected and corrected.

#### 8.3.1.3 Pretension of bolts

Rotor blades are mounted to the hub structure by high-strength friction grip bolts. Accordingly, the bolts have to withstand very high static forces and additionally the superimposed dynamic force components. The pre-tensioning force is essential to achieve an adequate service life. Depending on the turbine manufacturer and turbine type, different assembly methods (torque method; hydraulic pre-tensioning) can be used. In addition, the cross-sections of wind turbines are very large, and therefore, characterized by imperfections, which in turn result in complex load transfer and relaxation behavior.

As a rule, these connections are therefore subject to regular inspection and maintenance work. So far, there are only a few solutions in the field that systematically monitor bolt forces. However, there is an increasing demand for such solutions, which are currently not being used more frequently for cost reasons.

In the course of the ever-increasing blade lengths, two-piece blades are becoming necessary more often. Monitoring of the pre-tensioning forces is also desirable for the bolted connections of the two blade parts. In addition to the market requirement to offer cost-effective systems, the long-term stability and the accessibility for installation and maintenance of the systems are particularly challenging from a technical point of view.

#### 8.3.1.4 Lightning

Rotor blades and wind turbines are often exposed to lightning strikes. Besides the direct impact of lightning strikes on blade structure and blade surface, also the lightning protection system and all of the electronic components within are turbine can be affected. In the past, there were only limited solutions available to determine lightning strikes. In addition to manual inspection of the rotor blades and the lightning protection system by service staff, flash cards were used on some turbines to collect information about the maximum seen currents. With this approach, the service staff has to enter the turbine hub with special equipment. This means that monitoring is only possible at cyclical intervals in the course of periodic inspections.

In the recent past, systems have been developed that allow permanent and continuous monitoring in combination with remote access and alarm. In addition to recording the maximum lightning current intensity, all other representative lightning current parameters can also be recorded and stored. This enables condition-based maintenance and inspection, which increases reliability and economic efficiency.

Similar to systems for strain measurement, electrical and fiber-optic measuring principles are available for systems for lightning current measurement. Optical systems have the advantage of galvanically decoupling the sensors and the data acquisition unit, which is particularly important for this application, in order to avoid damage due to overvoltage.

#### 8.3.1.5 Pitch bearing

Pitch bearings are mechanical engineering components. Accordingly, their monitoring is more in the area of condition monitoring than structural health monitoring. Nevertheless, special monitoring techniques are required because the bearings – completely atypical for other bearings in the drivetrain – do not rotate around the complete axis. In most operating conditions, they have only very small and alternating angles of rotation – outside of service and commissioning conditions, the angles of rotation are limited to approx. $-5^\circ \ldots 95^\circ$.

These are reasons why the monitoring of rotor blade bearings has been neglected so far. Due to the increasing length of blades, the increasing loads, the increasing service life expectations and the extremely high follow-up costs of undetected damage, the monitoring of blade bearings is also becoming more important.

Most systems for monitoring uses accelerations sensors for vibration monitoring or uses distance sensors to measure and observe the gap between inner and outer ring.

### *8.3.2 Drivetrain and nacelle*

So far, no monitoring systems are established to monitor the support structure of the main frame of the nacelle. There is currently little need for this, so that the results of research and development work in this area are only published sporadically.

In contrast, the monitoring of the drivetrain including the associated bearings of the main shaft, the gearbox and the generator has been state of the art for a long time. As already mentioned at the beginning, own measurement concepts, special analysis functions, and very extensive empirical experience in the industry are available for this. Since this area is assigned to condition monitoring and is not dealt with further here, we refer here to the extensive specialist literature on the subject (see for example [1,3,5,10]).

### *8.3.3 Tower and foundation*

The tower and foundation of a WT are often considered just the supporting elements of the machine and, therefore, have not received much special focus from a monitoring perspective on the onshore WT market (unlike for offshore applications as shown in Sections 8.3.4 and 8.5). Nevertheless, with the increasing height and power yield of modern WTs, the tower and foundation have become deeply interconnected to the machinery components of the turbine from a global dynamic perspective. Therefore, monitoring the status of the supporting elements has become more relevant to validate general dynamic assumptions under which modern WTs are designed.

#### 8.3.3.1 Vibration monitoring

Monitoring the vibrations at the tower top offers a simple but effective approach to validating design assumptions regarding the global dynamic behavior of the WT. Additionally, if the vibration levels among WTs within the same fleet or turbine model can be compared, it allows for a simple way to pinpoint suspicious turbines, which deserve further detailed analyses.

A single accelerometer at the tower top is rarely sufficient to identify any damages occurring at a flanged connection or in the foundation. But in combination with a finite element model of the structure, it is often used as a base input to virtualize sensors at other locations of the structure.

For these reasons, monitoring vibrations at the tower top is a versatile, robust, and economic way to gain knowledge about the structure.

#### 8.3.3.2 Verticality monitoring

The large concentration of masses and loading at the top of a WT sets high requirements of the underlying support structure with regard to its verticality.

An efficient way to monitor the verticality of a WT consists of placing an inclinometer at the foundation level of the tower.

This type of monitoring is standard in an offshore environment, where the foundation elements and the soil structure interaction are of big interest.

In onshore applications, this type of monitoring is rarely seen, unless prior foundation damages have raised concerns and explicitly requested monitoring of the verticality of the WT.

#### 8.3.3.3 Load cycles and lifetime monitoring

The continuous monitoring of internal loads (mostly bending moments) occurring in the structural components of the foundation system is a necessary input for their health assessment during extreme events (ULS) and for the proper evaluation of cumulative fatigue damage throughout their operation lifetime (FLS). Given the large number of structural components and fatigue details present in a foundation system (especially in an offshore scenario), it is common practice to focus on the monitoring of a single cross-section (typically at the transition between tower and foundation), as this "reference" cross-section is well characterized in the design documents with regards to ultimate and fatigue loads and allows for a quantitative comparison between design and measurements. While the assessment in the ULS is based on the direct comparison between measured and allowable maximum bending moments at this reference cross-section. The assessment of the FLS requires a derivative magnitude, the so-called Damage Equivalent Loads (DEL), which accounts for the loading cycles and fatigue accumulation, by including rainflow counting and Palmgren-Miner's accumulation in its formulation. The biggest advantage of using DEL both during design and monitoring is that it provides a normalized fatigue damage, easy to compare quantitatively, which can be referred to any fatigue detail a posteriori given a specific Wöhler curve.

There are two common approaches for the load determination at relevant cross-sections:

- *Direct measurement with strain gauges:* It is the most accurate approach for the measurement of bending moments at the reference cross-section. It consists of the direct measurement of longitudinal strains along its circumference. This methodology requires the calibration of the strain gauge offsets by means of a nacelle rotation in non-operational, low-wind conditions. Despite being highly accurate, this measuring methodology is strongly limited in its practical application due to the relatively high installation cost of the gauges and measuring equipment. Additionally, this method does not provide any information about the loads at non-instrumented locations.
- *Indirect measurement with virtual sensors:* The determination of bending moments is possible based on virtual sensors derived from accelerometer measurements. To do that, a finite element (FE) model of the system must be used, which provides knowledge about the mechanical coupling between the different locations of the structure. Due to the introduction of an engineering

model in this concept, the results are less accurate than the ones from a direct measurement as described above. On the other hand, the use of a model introduces an additional flexibility, allowing to virtualize an arbitrary number of measuring locations.

In practice, both measurement approaches are combined to offer the most cost-effective solution. Due to their easy installation and low cost, accelerometers can be found in almost all turbines, providing an excellent base for park-wide load monitoring. To reduce implementation costs, it is common to use a common FE model for all turbines. This approach yields an optimal ratio between implementation and operational hardware costs and achieved load monitoring coverage. Nevertheless, the exclusive use of virtual load monitoring techniques relies too heavily on the model assumptions of the FE model, leading to a relatively large uncertainty of the results.

Therefore, wherever possible an SHM application should rely on direct strain gauge-based measurements to establish a reference benchmark for the calibration of the virtual loading sensors and their FE model. During the operational life of the strain gauges, their results are used by the SHM application both for monitoring purposes and for training the virtual sensors. The results of this training are regularly transmitted to all locations, i.e., all turbines of a park profit from the knowledge of few turbines with "expensive" strain-gauge instrumentation.

#### 8.3.3.4 Other monitoring goals

**Relative displacement between concrete foundation and tower:** All onshore wind turbines are founded on concrete foundations. A steel built-in part is integrated into the foundations to transfer the tower loads into the foundation. At the upper end of the built-in part is a steel flange onto which the first tower segment is bolted. Permanent and reliable anchoring in concrete is challenging due to the large dynamic loads. In the past, construction faults have become known on some foundation types, and in some cases, faults also occur in the field when pouring the concrete. In these cases, the foundation insert loosens during operation, so that larger relative displacements between the insert as well as the tower base flange and the concrete foundation are observed. As a rule, however, the turbines should not be dismantled but should continue to be operated. First of all, the foundation must be strengthened by concrete refurbishment. However, displacement sensors are often installed to permanently record the relative displacement between the tower flange and the concrete foundation. With three sensors offset at 120° to each other, the vertical movements can be reliably determined in both spatial directions of interest. If required, horizontal relative displacements and strains can also be recorded.

For the evaluation and analysis of the measurement data, the environmental and operating conditions of the turbine must be taken into account in any case, as the occurring displacements are dependent on the acting loads. Permissible limit values are to be determined in consultation with the foundation designers. In [13] a traffic light based alarm concept is listed. Up to 0.5 mm, an intact connection is assumed (green), between 0.5 and 1.5 mm, regular monitoring and an assessment by experts is recommended, from 1.5 mm, short-term immediate measures should be implemented and a refurbishment concept planned.

**Measurement of bolt pretension:** Bolts are one of the most stressed parts of a wind turbine. For a safe and reliable operation of the wind turbine, every single bolt must be intact. Of particular interest are the bolts of the tower flange bolting. These connections are highly complex. Imperfections due to manufacturing tolerances occur as well as insufficient pretensioning forces. The state-of-the-art is to check the bolt connections regularly (manually) in the course of the periodic inspection. Particularly in the case of offshore wind turbines with limited accessibility, however, interest in structural monitoring is increasing, as damage to the bolt connections can result in high follow-up costs.

The central challenge in monitoring the bolt connections is the amount of bolts within a wind turbine. For cost reasons, it is not possible to permanently monitor all bolts of the tower flanges. It is therefore only possible to monitor individual bolts of a flange, which, however, cannot generate a reliable statement for the entire connection. Various systems are available on the market for permanent monitoring. Forces can be measured with load cells (a type of washer), alternatively, bolt stress can be measured with strain sensors or, for some time now, bolt lengths can be measured with ultrasound-based length measurements.

As an alternative to the permanent monitoring of individual bolts, cyclical testing has become established in the course of periodic testing at offshore wind turbines. Mobile ultrasonic systems are used here with which the service personnel can record and document the length of all bolts, and thus also the pretensioning force, in a short time by placing an ultrasonic sensor on the bolt head. However, bolts specially prepared for this technology must be used (flat surface on the bolt head, exact documentation of the bolt length in the untensioned state for each individual bolt), so that this inspection method must already be provided for and taken into account when planning the project.

### *8.3.4 Additional considerations for offshore foundations*

In an offshore environment, the foundations of the wind turbines are complex systems, whose design, construction and maintenance have a decisive impact on the operation of the wind farm. In this section, the authors focus on some key aspects relevant to the monitoring of fixed-bottom monopiles and jackets, as both foundation types dominate the offshore market to date. Future or special technologies such as floating foundations, mono bucket foundations and gravity-based foundations are not explicitly considered below. For this, please refer to further technical literature (e.g., [12]).

#### 8.3.4.1 Soil structure interaction

Knowledge and realistic modeling of soil-structure interaction is essential in the design of wind turbines. The stiffness of the soil has a significant influence on the first natural frequencies, which in turn is an important design criterion to avoid or minimize operation in resonance (1P/3P excitation). Due to the complex foundation situations of offshore wind turbines, many projects monitor after commissioning at least at the beginning of the operating phase the "as-built" soil structure interaction. Depending on the deviation between design and "as-built" natural frequencies as well as the distance of the natural frequencies to wave excitation spectra a relevant impact to tower and foundation vibration amplitudes may arise. Increases in vibration stresses in turn lead to a disproportionate reduction in service life, in accordance with the S–N curves for steel.

Currently, two monitoring methods have established themselves on the market to assess the soil structure interaction:

**Monitoring of the natural frequencies** for the foundation/tower structure with acceleration or inclination sensors: This method is very easy to implement on any wind turbine within the offshore project, as the natural frequencies can be determined with only one acceleration or inclination sensor. The sensors are usually installed in the area of the transition piece or the tower base. This makes them easily accessible for installation and maintenance work. The natural frequencies can then be determined from the measured acceleration or inclination time series using various methods. Normally, methods in the frequency domain (e.g., Fast Fourier Transformation) or in the time domain (e.g., output-only modal analysis by stochastic subspace identification) are used.

**Monitoring of bending moments on monopiles through strain measurements**: In the case of monopile foundations – which currently dominate the offshore market – the distributed horizontal stiffness of separate soil layers are essential for global foundation stiffness. The local horizontal stiffness is dependent on the soil parameters and thus height-dependent. Various design approaches are available. One common method linearizes the non-linear stiffness behavior by the so-called p-y-curves. This allows the modeling of soil as linear elastic springs. Because of non-linearities, the design process is interactive to find the correct linearization point. If we look at the global load-bearing behavior of the wind turbine tower and foundation in a very simplified way, we have a cantilever beam clamped in the soil, which is stressed by horizontal wind and wave loads. The maximum bending moment thus results at the seabed (in reality somewhat below). Due to the horizontal support of the soil, the bending moment is reduced with increasing depth until it is completely reduced to zero. The stiffer the subsoil, the faster (at shallower depths) the moment is reduced: The softer the subsoil, the greater the depth required.

The knowledge of this load-bearing behavior can be used for structural monitoring. If strain sensors are installed at different height levels of the monopile, bending moments at the measurement levels can be determined. Subsequently, it is possible to determine the bending moment distribution over the height within post-processing by means of quite complex analyses and thus inversely adapt the ground model with the layered foundation soil to reality.

In principle, the strain measurements can be recorded with electrical foil strain gauges in conjunction with a Wheatstone bridge. A well-established, reliable and robust method under "normal" environmental conditions. However, since the measuring positions are below the salt water level, the sensors are extremely difficult to protect against water ingress and are still affected by frequent failures that cannot be repaired.

For this reason, fiber optic strain sensors (so-called fiber Bragg gratings, FBG) have recently been used for such applications. Compared to electrical strain sensors, this is a fairly new and complex technology. The advantage is the much higher resistance to water penetration (this does not interfere with the measurement technology, only adhesives should be resistant) and the ability to integrate several strain and temperature sensors in one fiber line. The disadvantage is the high sensitivity to fiber rupture (it is glass fibers that are usually used) and the extensive special know-how required, which is only available from very few companies in the market.

In general, it should be noted that the costs for the installation of strain sensors on monopiles and the associated data evaluation are comparatively high, so only individual projects decide to provide such extensive monitoring technology. In addition, for cost reasons, only (significantly) less than 10% of all sites are equipped.

#### 8.3.4.2 Scour monitoring

In the past, some projects have experienced considerable scour depths. As a result, it is state of the art that foundations are protected against scouring. Independently of this, three technologies have become established that can be used to monitor the actual depth of the grout:

**Sonar 3D-scanning of the seabed by vessel:** At least for projects in the German exclusive economic zone with regulations from Bundesamt für Seeschifffahrt und Hydrographie (the responsible German authority) it is required to inspect the seabed surface during periodic inspections. Within BSH-Standard Design (see [22]) a scour inspection is required during the first two years of operation annually, then depending on the conditions, recommended every 4 years. Usually, this requirement can be most easily and cost-effectively realized by vessel-based sonar scanning, where the vessel sails successively through the entire wind farm and records the seabed geometry.

**Single- and multipoint echo sounder mounted on subsea foundation structure:** Different solutions are available in the market to have individual information about scour origin on single foundations. Single-point echo sounder are able to measure the (vertical) distance between the sensor position and seabed at one single path. To have more information available multipoint systems are available which provide distance information on different paths – e.g., four different paths with different angles to the vertical. Newer solutions allow also 2D and 3D scanning modes to get information that is more detailed. The measuring arrangement and result images are shown as examples in [12]. All these solutions require multiple sensor heads if the entire perimeter of a foundation is to be monitored. The advantage of these solutions compared to sonar scanning by vessel is that continuous, uninterrupted monitoring can take place. However, for cost reasons, this is only possible at individual locations and not throughout the entire park. Vessel-based solutions have clear advantages here.

**Identification of natural frequencies by acceleration/inclination sensors:** An easy to realize scour detection can be implemented via the observation of natural frequencies – for monopile as well as for jacket foundations. Analogous to the procedure for monitoring the soil structure interaction (see the previous section) the sensors can be mounted in the easily accessible transition piece/tower base area and the natural frequencies extracted from the measurement data. Among other things, it could be shown that the sensitivity is sufficient to detect critical scour depths (see [29]) and that the method is well suited to realize permanent monitoring at all foundation locations. Especially in combination with vessel-based sonar methods from the inspections of the first two years of operation, this provides a very reliable and cost-effective monitoring option.

#### 8.3.4.3 Further relevant aspects in offshore environment

*Corrosion monitoring*

Steel structures of offshore wind turbines are subject to a high risk of corrosion. Accordingly, extensive measures are taken to protect the structures against corrosion and to inspect them regularly as part of the periodic inspections.

Monitoring systems that directly record the corrosion protection or the corrosion that has occurred in the foundation are currently not available on the market. An exception to this is foundations whose corrosion protection is realized by Impressed Current Cathodic Protection (ICCP) systems. With these systems, it is possible to permanently measure and document the impressed current and the corrosion protection potential. This also proves that the structure to be protected has sufficient corrosion protection.

*Grouted connections*

Grouted connections are well established in the oil and gas sector and are also widely used in offshore wind energy projects. In some of the first large offshore projects with monopile foundations, damage occurred during the operating phase, partly because the dynamic load components are much greater in the foundation of wind turbines. In the meantime, the design regulations have been adapted and improved. In addition, it is now possible to install monopiles so precisely vertically that grout connections are now rarely used for monopile foundations.

Occasionally there was a request to monitor the grout connection between the monopile and the transition piece. Due to the very complex assembly on-site offshore, this is a great challenge. From an engineering assessment point of view, it is best to know the concrete and steel strains inside the grout connection. Therefore, in the early phase of the offshore industry, strain sensors were partly installed in the grout joint area. However, practice shows that the survival rate of the sensors is very low. Alternatively, the relative vertical displacement between the monopile and transition piece can be measured. This works reliably. However, the costs are very high due to the necessary offshore installation. The corrosive environment must also be considered, so suitable sensors (e.g., eddy current displacement sensors) are selected and installed.

## 8.4 Requirements to common monitoring sensors

An overview of general available measurement principles and sensor as well as data acquisition requirements is provided within the relevant set of standards (see [9,11,23], for example). In the following, we focus on most often used acceleration, inclination and strain sensors.

### *8.4.1 Acceleration sensors*

Acceleration sensors are used for very different use cases for the monitoring of wind turbines. There are multiple examples of this type of sensor integrated into the safety chain in the nacelle of every wind turbine to detect overloads, installed in the foundation and tower area to assess service life, at the drivetrain to detect damages and in the rotor blades to detect ice.

Due to the low natural frequencies of towers and rotor blades (in all cases significantly lower than 1 Hz), acceleration sensors capable of measuring down into the quasi-static range must be used for monitoring applications. Capacitive MEMS sensors, able to measure down to 0 Hz, have become established on the market.

It is important to know that the acceleration amplitudes decrease significantly with decreasing frequency and are usually very small. For structural monitoring tasks, it is therefore essential to use extremely low-noise MEMS sensors. Simple low-cost sensors are not suitable for this purpose.

Table 8.1 lists some selected technical characteristics of accelerometers that should be evaluated and considered for structural monitoring tasks at wind turbines.

*Table 8.1 Technical requirements for acceleration sensors*

| **Measurement axis** | **Three-axis recommended; at least two-axis** |
|---|---|
| Measurement range | $\pm 2$ g sufficient for most applications |
| Frequency range | 0 Hz to $\geq$25 Hz |
| Noise density | $\leq$50 $\mu g/\sqrt{Hz}$ at least;<br>$\leq$25 $\mu g/\sqrt{Hz}$ recommended |
| Protection grade | $\geq$IP 65 recommended |
| Operation temperature | −40 °C–60 °C |
| Sensor output | 4–20 mA or digital bus recommended for electromagnetic immunity |
| Overvoltage protection | Following IEC 61400-24 requirements |

*Table 8.2 Technical requirements to inclination sensors*

| Measurement axis | 2-axis |
|---|---|
| Measurement range | ±3° sufficient for most applications |
| Resolution | ≤0.001° recommended |
| Frequency range | 0 Hz to ≥10 Hz |
| Temperature behavior | Drift of zero point, drift of sensitivity as low as possible; temperature compensated sensors strongly recommended |
| Aging behavior | In most cases, despite the great importance, no information is available from the sensor manufacturers |
| Protection grade | ≥IP 65 recommended |
| Operation temperature | −40 °C–60 °C |
| Sensor output | 4–20 mA or digital bus recommended for electromagnetic immunity |
| Overvoltage protection | Following IEC 61400-24 requirements |

### *8.4.2 Inclination sensors*

Inclinometers are used to monitor the inclination of the tower and foundation. The main task of the sensors is therefore to reliably detect the (quasi-) static inclination over the entire operating period of usually 25 years with an accuracy of less than 0.1°.

There are no suitable sensors available on the market to meet these extreme requirements of durability. As a rule, cost effective inclination sensors are used, which cannot fulfil these requirements, as temperature and the aging behavior of the sensor often cause sensor drifts surpassing the range of 0.1°.

In order to reliably monitor the verticality misalignment, the initial misalignment must be recorded and documented using alternative reference methods (e.g., by geodetic measurements) when the monitoring system is commissioned. Periodic repeated reference measurements can then be traced back to the initial misalignment. These are necessary when sensors are recalibrated or when defective sensors are replaced. Sensor calibration at least every three years is recommended despite the costs involved.

Table 8.2 lists some selected technical characteristics of inclinometers that should be evaluated and considered for structural monitoring tasks at wind turbines.

### *8.4.3 Strain sensors*

Strain sensors are usually used in structural monitoring to calculate internal forces at interfaces (rotor blade root; tower base) and compare them with design values.

#### 8.4.3.1 Electrical strain gauges

Electrical strain sensors have been established in experimental stress analysis for decades and are also frequently used on wind turbines. With regard to the selection of sensors, the electrical wiring and the application, reference is made to the extensively available technical literature. It is important to mention that for the application of foil strain gauges in the sensor application area, the corrosion protection must be removed locally. However, this is not a disadvantage as the sensors are subsequently covered in several stages using proven processes.

Especially when only a few strain sensors are to be applied (<10 pcs.), strain measurements with electrical sensors can be implemented more cost-effectively and robustly than fiber-optic sensors. For this reason, these are used as standard to monitor the tower base bending moments. In general, the high number of load cycles in wind turbines must be taken into account. Special fatigue-resistant foil strain gauges are available. Of course, the fatigue strengths depend on the amplitudes of the load cycles. With fiber composites at the rotor blade root, the strains are much greater than with steel. Therefore, directly applied foil strain gages in the rotor blade can only withstand the stresses for a limited time.

#### 8.4.3.2 Fiber Bragg grating

In the recent past, fiber-optic strain sensors – so-called fiber Bragg grating (FBG) – have established themselves in the market. Just like electrical strain sensors, these are to be connected to the structure at the measurement positions. As a rule, this is done by bonding, in the case of steel structures partly by spot welding. With fiber-optic strain sensors, several sensors can be integrated into one optical fiber so that sensor chains can be installed. Fiber optic sensors have particular advantages when measurements have to be taken in environments with electromagnetic fields (e.g., generators), under water (offshore foundations), or with a very large number of sensors (cost advantages over electrical sensors). It should be noted that the principle of measurement technology makes it sensitive to dirt and mechanical stress (e.g., bending of the optical fibers) and thus less robust in some cases.

## 8.5 Practical considerations for implementing SHM – example for monitoring of offshore foundations

While there is no general recipe for a successful implementation of SHM, the authors can offer some insight into practical aspects to consider when designing and implementing an SHM solution for a customer. Despite the large variety of monitoring needs and implementation scenarios, the authors focus here on the practical analysis of an SHM solution for an offshore foundation, where they have gathered extensive practical experience over the course of the last decade.

For German offshore projects, the responsible licensing authority (Bundesamt für Seeschifffahrt und Hydrographie, BSH) has required structural monitoring of 10% of all foundations since 2007 (see [24]). This legal requirement has contributed to the establishment of foundation SHM as a standard component in all German offshore wind projects. Similar requirements have found their way in an international context, so that foundation monitoring is now established as a state-of-the-art component in the offshore wind industry, required by wind farm project planners, operators, and designers in most international offshore wind projects.

Despite the broad presence of SHM in the offshore foundation market, the definition of the monitoring scope is still handled on a project-specific basis, ideally in cooperation between geotechnical experts, the foundation designer, the SHM provider, and the project owner. The goal is to develop a monitoring concept to support the operator during the operating phase of the wind park. The focus is normally set in monitoring the overall behavior of the foundation elements. Parameters, such as shifting, deformation, component stress, and frequencies, shall be measured and recorded (see [22]). After several years of experience providing commercial SHM solutions, the authors have identified four monitoring goals, which yield the most value-to-investment ratio in an SHM system for offshore wind turbine foundations and towers:

| Monitoring of... | Helps to detect... |
|---|---|
| loads and lifetime consumption | • problems in operation<br>• deviations from load assumptions in design<br>• anomalies in the fleet<br>• critical operation states |
| dynamic state | • anomalies in the global behavior<br>• scour |
| vibrations at the tower top | • problems in operation<br>• critical operational states |
| verticality | • ground settlements<br>• soil degradation |

Most projects require the monitoring of 10% of all offshore foundations. A "standard" monitoring sensor setup, as shown in Figure 8.3 (left), has become a common solution. Acceleration sensors in the tower top and in the transition piece/tower bottom record the movements and vibrations from which the dynamic properties are determined. For larger wind turbines, an additional acceleration sensor in a middle tower section can be recommended in order to reliably detect higher natural frequencies. At the transition piece/tower bottom area, inclination sensors monitor the verticality and strain sensors enable the monitoring of extreme loads and cycle counts, which are required for fatigue assessments.

In some situations, the operator requests a full monitoring coverage of the whole fleet. In these cases, it is common practice to equip all other foundations in the wind farm with a basic sensor setup consisting of a single accelerometer and inclination sensor, which uses advanced algorithmic methods, often known as load virtualization techniques, to reconstruct loading states despite the missing instrumentation. These concepts rely on the transfer of knowledge from the standard equipped turbines to the basic equipped turbines. This approach provides additional value for the owner and operator of the wind farm – especially for optimization of

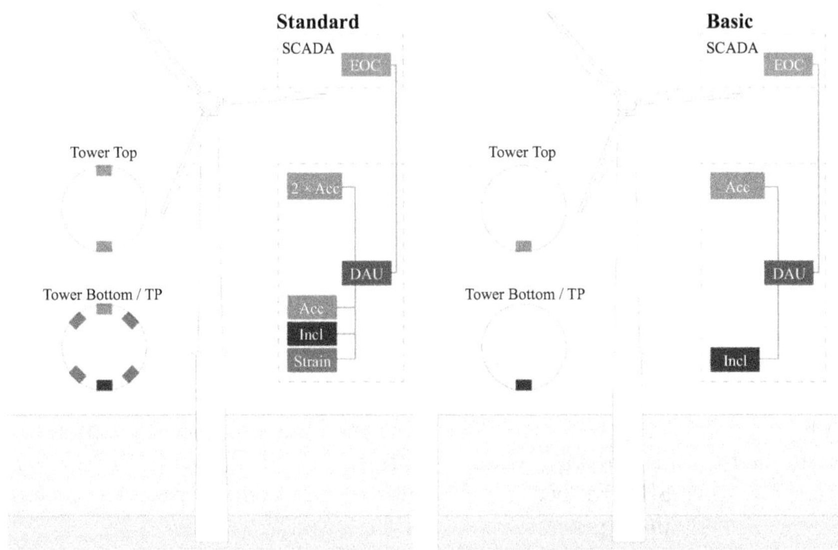


*Figure 8.3 Typical sensor configurations for the monitoring of offshore WT. Courtesy of © Wölfel Wind Systems GmbH & Co. KG.*

operation and for a reliable assessment of lifetime extension in all turbines of the wind park.

## *8.5.1 Definition of valuable KPIs*

The success of a monitoring solution is determined by its ability to provide valuable and understandable insights to the operator of the structure. These types of results are often referred to as Key Performance Indicators (KPI).

KPIs are the interface for information exchange between the SHM provider and the operator. A good SHM solution shall provide a comprehensive amount of KPIs to the operator. Here the principle of quality before quantity applies strictly.

Any SHM application generates a large number of features, such as eigenfrequencies, inclinations, loads, etc. These features are constantly evolving over time, mostly due to their dependence on the operating and environmental conditions (EOC). It is the task of the SHM application, to implement reliable algorithms or models to compensate the effect of the EOC on the available features to extract the valuable information hidden within the data. Answering the question whether a variation in the monitored features is related to a change of the EOC or due to a change in the structure, is the key to a successful KPI definition.

Thus, the calculation of KPIs is often linked to solving an anomaly detection problem. At this stage, machine learning methods are frequently used to train models to establish a baseline of expected behavior and detect anomalies during operation.

Following this approach, KPIs are easy to understand and can be linked to thresholds. Only after crossing these thresholds, further investigative actions are triggered, offering a starting point for the definition of condition-based monitoring approaches by the operator of the structure.

### *8.5.2 Comparing monitoring results to design*

The interpretation and evaluation of monitoring results, even after the successful definition of KPIs, remains a challenging issue for any SHM application.

While it is generally accepted that the absence of relevant changes in the monitored KPIs, can be interpreted as a good sign, such a simplistic approach fails to actually generate significant value for the operator of the wind park. Furthermore, in the case of some KPI becoming noticeable, big questions arise as to how to interpret this and what actions or effects shall be derived from it. It is for these reasons that SHM has been often perceived as a system only able to "generate problems," rather than "generate value" for the operator.

In the authors view, many SHM systems' lack of added value originates in the fact that SHM is often completely decoupled from the design of the structure. Considering that the real behavior of any structure always deviates from its initial design, SHM systems identify these differences and inevitably pose challenging questions to the operator.

For these reasons, allowing the designer of the structure to become a stakeholder in the evaluation of SHM results has proven very meaningful to deliver valuable and actionable insights to the operator of the structure.

In this context, the authors in collaboration with leading experts in the design of offshore foundations, have introduced the concept of *agile design*.

Agile design is an evolution of the current design process of structures, which is focused on the continuous delivery of value throughout the lifetime of the structure. While traditional design processes focus on the one-time comprehensive planning and specification before the construction of the structure, agile design builds on top of this process to enable its update in regular iterations, based on the inputs from a structural health monitoring system.

One of the most important iterations in agile design occurs soon after the commissioning of the SHM system. After few months of operation, SHM is able to deliver an accurate image of the individual as-built situation of every WT. By providing the right information to the designer (dynamic properties, loading behavior at given operating states), they can update the original design, often uncovering lifetime reserves, but also identifying new relevant KPIs to be tracked, which would have otherwise remained undiscovered.

Once *agile design* is setup, it can be updated (or iterated) on a regular basis or on a condition base (e.g., after KPI threshold exceedances), thus providing the operator with short feedback loops and actionable engineering assessments ratified by the designer and the SHM expert.

## 8.6 The lifecycle of an SHM solution

The success of an SHM system relies on the optimal planning and execution of a wide spectrum of tasks from choosing the right hardware and sensors, designing software, implementing algorithms, and ultimately providing engineering support to take the right O&M decisions. You will need a dedicated team of experts in hardware design, software development, data scientists and structural engineers to support all stages of the SHM process from beginning to end. In addition, comprehensive cooperation across all phases of a project is required between the various project stakeholders. These are for example monitoring solution providers, project developers, operators, designers, EPCI contractors and their subcontractors, geotechnical experts, certification bodies, and many others.

Since successful implementation of SHM systems requires interactions between stakeholders at all stages of the project, these stages are listed in the following:

**Phase 1 – Planning of SHM solution**: The challenge of providing valuable structural insights begins with the optimal choice of sensors, their meaningful positioning, and a reliable data acquisition. At the same time, budgeting and non-technical constraints are important to find the best SHM solution.

In addition to the design of the measurement technology, it is also important to define interfaces to other IT systems and SCADA systems, to plan the installation of the components with all stakeholders involved in the project and to prepare method statements and risk assessments. In this project-phase, it should also be determined how the monitoring data will be analyzed and who will be responsible for the analysis. As a rule, the monitoring concepts drawn up must be submitted to the authorities for approval or at least agreed with the wind farm operator.

**Phase 2 – Delivery, installation, and commissioning**: First of all, a distinction can be made as to whether the monitoring solution is installed as a series system by OEMs directly in production with their own production personnel or whether it is a retrofit solution or a third-party system. Often, monitoring solutions are retrofitted during turbine construction or even later after turbine construction and commissioning by companies specializing in structural monitoring. In the case of monitoring offshore support structures, this is standard, as the foundations – and thus also the associated monitoring solutions – are required and purchased by the operator and not by the WT manufacturer.

In all cases, it is essential to have extensively trained and experienced engineers and technicians for installation and commissioning. In addition to knowledge about Health Safety Environment (HSE) it is crucial to know the – in most cases very specific – technology for sensor installation, sensor testing and measurement system commissioning. If the systems are to be installed all over the world, which is common for offshore applications at least, early planning and allocation of the required well-trained engineers and technicians is important for the success of the project. Measurement technology that is easily adaptable to all kind of common electrical and network interfaces makes integration much simpler. If network remote access is available during system commissioning, technicians on site can also be supported by special trained engineers via remote control.

**Phase 3 – Deployment of the SHM software and baseline definition**: After installation and commissioning of the SHM hardware (sensors, servers, IT connectivity), the collection of raw data begins. As a rule, the results and key performance indicators identified from the raw data after commissioning cannot be used directly for monitoring, as the properties of the structure must be referenced first. Results such as vibration intensities, inclinations, natural frequencies, and other quantities can be determined immediately after commissioning. These initial results are usually used as a reference and then continuously monitored for structural changes in relation to this initial reference. Depending on the monitoring system this type of referencing can be automatic (e.g., by machine learning algorithms) or individual by monitoring experts. During the commissioning and referencing period, data consistency and signal quality should be taken into account. Sensor offset corrections as well as other required calibrations are carried out in this phase.

The results of these initial evaluations provide a baseline for the future operation and are generally used as basis for the definition of meaningful alarm criteria. This period needs typically some days to a couple of weeks for well established procedures and methods, but can take up to 1 year for pure machine learning methods without any pre-information.

**Phase 4 – Live monitoring**: After the deployment and configuration of the SHM software the operational phase of the SHM begins. From this moment on the SHM will be providing regular updates of all defined key performance indicators (KPI) and checking against defined thresholds. The responsibilities for data storage, analysis, visualization of results, and reporting should be clearly defined between the operator and the monitoring system provider in order to be able to produce reliable data and results at any time.

System maintenance and the development of a concept for regular sensor recalibration is important in order to generate reliable and usable measurement data. Often, sensor re-calibration traceable to national standards is not given sufficient attention or importance, which can be disadvantageous or critical for future applications – especially for the evaluation of fatigue loads and remaining lifetime.

**Phase 5 – Periodic reporting:** The reporting requirements of SHM systems also depend on the expectations of the end user or client and the monitoring solution. For some applications, it is sufficient if the monitoring system transmits the monitoring results to the turbine control or a SCADA system in real time (e.g., systems for the detection of rotor blade icing). In other applications (e.g., foundation monitoring of offshore wind turbines), operators and sometimes licensing as well as certification authorities expect detailed engineering reports – analyzing the structural behavior and assessing it against design assumptions – on a regular monthly or annual basis.

**Phase 6 – Engineering support:** Sometimes things do not go as planned and SHM systems identify assets that require further attention. In these situations, experienced engineers shall stand side-by-side with the operations team of the wind park – and sometimes the certification authorities – to locate of the problem, recommend measures and adapt the SHM system to new requirements.

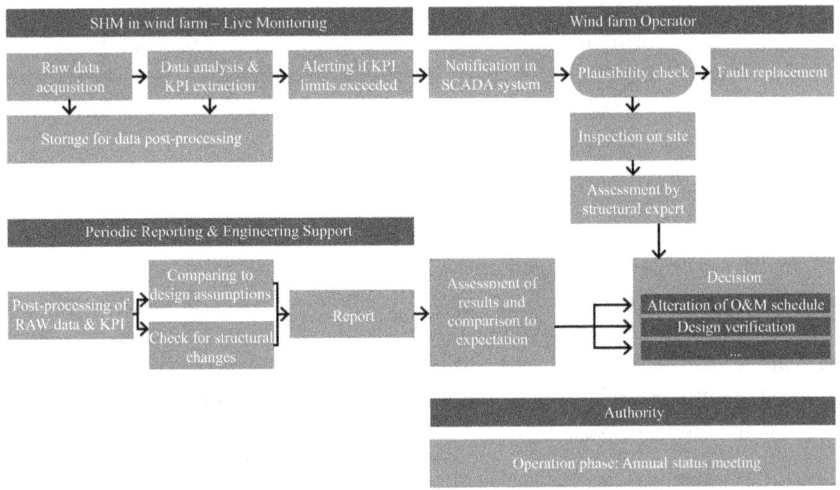


*Figure 8.4 Flow chart of SHM process from data acquisition to approving authority on example for an offshore foundation monitoring. Courtesy of © Wölfel Wind Systems GmbH & Co. KG.*

In Figure 8.4, a typical flow chart for an individual monitoring project with scheduled interaction between the operator, monitoring body, and authorities is shown. Of course, such flow charts can vary between different projects but the reader can see the major responsibilities and interfaces.

### *8.6.1 Formal requirements for monitoring solution providers*

Due to the importance of what is to be achieved with the monitoring results – often it is about the fulfillment of conditions from the building permit or commercially extremely important decisions such as the lifetime extension of wind farms – formal requirements should also be placed on the solution providers, which must be demonstrated in addition to technical competence.

The following formal aspects should be addressed.

**Quality management:** The solution provider for SHM should have introduced a quality management system in order to be able to ensure and further develop the delivered quality at all times. The proof of a successful implementation in the company can best be provided by a certification according to ISO 9001.

**Environmental management:** The installation, operation and maintenance of monitoring systems also consume resources that should be conserved with regard to the environment. The authors also recommend considering this aspect when selecting service providers and checking their environmental management system. The easiest way to illustrate this is with an ISO 14001 certification.

**Management of IT security:** In the course of advancing digitalization, the number of attacks on companies' information and operation technology (IT and OT) is constantly increasing. A large number of companies have already been affected in the past. Since the reliable operation of wind turbines is usually essential, they generally belong to the critical infrastructure, which on the one hand is particularly worthy of protection, but on the other hand, is also a particularly frequent target of attackers. It is therefore advisable for providers of monitoring solutions to have their IT security management certified according to ISO 27001.

**Health and safety management:** The solution provider for SHM should have introduced a health and safety management system in order to ensure that all processes and methods are reviewed and optimized for occupational health and safety. The main objective must always be to minimize risks to employees, prevent accidents and maximize health protection. The proof of a successful implementation in the company can best be provided by a certification according to ISO 45001.

## References


[1] ISO 17359:2018: Condition Monitoring and Diagnostics of Machines – General Guidelines

[2] Kolerus, Jh., and Becker, E., Condition Monitoring und Instandhaltungsmanagement, *Expert*, 2022

[3] Randal, R. B., Vibration-based Condition Monitoring: Industrial, Aerospace and Automotive Applications, Willey, 2010

[4] ISO 2041:2018-10 Mechanical Vibration, Shock and Condition Monitoring – Vocabulary

[5] ISO 13372:2012-09 Condition Monitoring and Diagnostics of Machines

[6] IEC 61400-1:2019, Wind Energy Generation Systems – Part 1: Design Requirements

[7] IEC 61400-3-1:2019, Wind Energy Generation Systems – Part 3-1: Design Requirements for Fixed Offshore Wind Turbines

[8] IEC 61400-3-2:2019, Wind Energy Generation Systems – Part 3-2: Design Requirements for Floating Offshore Wind Turbines

[9] VDI 3834, Part 1: Measurement and Evaluation of the Mechanical Vibration of Wind Turbines and Their Components, Wind turbines with Gearbox; Guideline of Verein Deutscher Ingenieure; August 2015

[10] ISO 10816-21:2015 Mechanical Vibration – Evaluation of Machine Vibration by Measurements on Non-Rotating Parts – Part 21: Horizontal Axis Wind Turbines with Gearbox

[11] VDI 4551: Structure Monitoring and Assessment of Wind Turbines and Offshore Stations; Guideline of Verein Deutscher Ingenieure; January 2020

[12] BSEE Offshore Wind Recommendations: Guidelines for Structural Health Monitoring for Offshore Wind Turbine Towers & Foundations; NGI Report; Doc. No. 20160190 16-1036; Rev. No. 3; 2017-06-15

[13] BWE-Ratgeber: Umgang mit Schäden an Fundamenten von Windenergieanlagen – Onshore, Inspektion – Bewertung – Sanierung; Bundesverband Windenergie, Deutschland, www.wind-energie.de

[14] IEC 61400-13:2015+AMD1:2021: Wind Turbines – Part 13: Measurement of Mechanical Loads

[15] Kraemer, P., Schadensdiagnoseverfahren für die Zustandsüberwachung von Offshore-Windenergieanlagen, Dissertation, Universität Siegen, 2011.

[16] Balageas, D., Fritzen, C.-P., and Güemes, A., *Structural Health Monitoring*, Hermes Science Publishing, 2006, ISBN 9781905209019.

[17] Fritzen, C.-P., Klinkov, M., and Kraemer, P., Vibration-based damage diagnosis and monitoring of external load, in *New Trends in SHM*, Springer, 2013.

[18] Sohn, H., Effects of environmental and operational variability on structural health monitoring, *Phil. Trans. of the Royal Society (Band 365)*, S., page 539–560, 2007.

[19] Sohn, H., Worden, K., and Farrar, C. R., Statistical damage classification under changing environmental and operational conditions, *Journal of IMSS*, Vol. 13, page 561–574, 2002

[20] Kraemer, P., Friedmann, H., and Ebert, C., Vibration-based ice detection of rotor blades in wind turbines—The industrial realization of an SHM-system, *10th International Workshop on Structural Health Monitoring*, Stanford University, USA, page 2841–2848, 2015.

[21] Kraemer, P., and Friedmann, H., Vibration-based structural health monitoring for offshore wind turbines structures – Experimental validation of stochastic subspace algorithms, *Wind and Structures, An International Journal*, Vol. 21(6), page 693–707, 2015.

[22] BSH Standard Design – Minimum Requirements Concerning the Constructive Design of Offshore Structures Within the Exclusive Economic Zone (EEZ); Bundesamt für Seeschifffahrt und Hydrographie; 1. Update 28 July 2015, corrected as of 1 December 2015, upgrade from 1 June 2021; BSH-No. 7005

[23] VDI 3834, Part 2: Measurement and Evaluation of the Mechanical Vibration of Wind Turbines and Their Components, Wind turbines Without Gearbox; Guideline of Verein Deutscher Ingenieure; June 2022

[24] BSH: Konstruktive Ausführung von Offshore-Windenergieanlagen; Standard des Bundesamt für Seeschifffahrt und Hydrographie; 12 June 2007; BSH-No. 7005

[25] DNV-SE-0439: Certification of Condition Monitoring; Service Specification of DNV; June 2016

[26] IEC TS 61400-28, Edition 1: Wind Energy Generation Systems – Part 28: Through Life Management and Life Extension of Wind Power Assets

[27] Kraemer, P., Büthe, I., and Fritzen, C.-P., Damage Detection under Variable Environmental Conditions Using Self Organizing Maps, *Proc. of 11th IMEKO TC, Kraków, Poland*, 2010

[28] Kraemer, P., and Ebert, C., *Schwingungsbasierte Erkennung von Strukturveränderungen in der Theorie und Praxis*, VDI-Berichte 2227, VDI-Fachtagung "Schwingungsüberwachung," Leonberg, 2014

[29] Lendve, S.B., Enss, G.C., Tsiapoki, S., Ebert, C., and Asmussen, J., Probabilistischer Ansatz zur Detektion von Strukturveränderungen an Monopile-Gründungsstrukturen mit Messdaten aus einem Structural Health Monitoring System. Stahlbau, Vol. 89, pages 542–550, 2020.

[30] Rytter, A., *Vibration Based Inspection of Civil Engineering Structures*, dissertation, Aalborg, 1993

# *Chapter 9*

# Advanced concepts for control of wind turbine and wind farm systems

*Tobias Meyer*[1] *and Niklas Requate*[1]

Although wind turbine technology is comparatively young, with large-scale series production starting in the 1980s, it has reached a high degree of maturity. The early years of wind turbine development saw many iterative improvements, with some proving themselves worthy of further usage; others are now regarded as design experiments. The basic concept has remained mostly unchanged since the mid-1990s: a three-bladed upwind rotor and a variable-speed generator. Turbine operation is controlled to keep the rotor speed at its aerodynamically optimal setpoint below rated power and to limit the power by controlling the pitch angle of the blades to feather. This concept has reached a high degree of perfection. The turbines work autonomously, produce power with low maintenance, and support grid operation.

The above-mentioned concept is driven by wind physics and the limits of the selected generator size. Thus, there is a very small margin for further improvements of control regarding the primary objective: power generation. With the mature wind turbine controller technology as a basis, advanced control concepts can put a focus on secondary objectives and wider system scope. With such a wider scope, a single turbine is not an independent operating system serving the primary objective, but it acts as part of a larger system: a full wind farm that is controlled as a whole within the surrounding ecosystem composed of the environment, the market, and the grid. Within this context, the possibilities for advanced control concepts become vast. We give an overview on these possibilities focusing on improving the ecologic impact of wind energy from load-related usage of materials.

We start by taking a brief look into the ecological impact of wind energy in Section 9.1, from which we derive potentials for economic, ecologic, and operational improvements by structuring control concepts for these purposes in Section 9.2. With these concepts, we present the architecture for a system-wide control of an entire wind farm in Section 9.3. Within this architecture, operational planning over the entire lifetime is a key concept that combines turbine-level control with global lifetime objectives. Details of an optimal planning approach are presented in Section 9.4. We conclude by summarizing the challenges and required methods for evaluating advanced control concepts in Section 9.5.

[1]Fraunhofer IWES, Fraunhofer Institute for Wind Energy Systems, Germany

## 9.1 Ecological impact of wind energy

Renewable energy sources, one of which is wind energy, have a very low ecological impact. According to [1], the production of 1 kWh electricity from wind energy emits approximately 8 g $CO_2$eq. Comparing this with natural gas (374 g $CO_2$eq./kWh), coal (815 g $CO_2$eq./kWh), or lignite (1142 g $CO_2$eq./kWh) [2], the importance of wind energy in a future climate-neutral energy system is clear.

Traditional large fossil-fuel power plants have low up-front costs, but high operating costs, where the marginal cost of an additional kWh is driven by the cost of fuel itself. This is contradictory to wind turbine operation, where the fuel (wind) is free and the marginal cost is zero. The ratio of capital expenses to operational expenses is highly skewed towards capital expenses. This also holds true for the $CO_2$eq. emissions over the entire lifecycle of a wind turbine. In [2], the emissions are broken down into individual components and life cycle phases. While the exact values differ greatly depending on wind conditions and site, all cases have in common that most emissions are caused by the production of the turbine itself. Grid connection, logistics, installation, and operations and maintenance have a minor contribution. Dismantling of the turbine feeds some of the material back into the recycling processes, which effectively yields a negative $CO_2$eq.

We can conclude that, unlike conventional plants which burn fossil fuel at each conversion process, wind turbines convert the effort that was spent during production into electric energy. Luckily, they do so with high efficiency. Vestas calculated that its V150-4.2MW low wind power plant has a breakeven time of 7.6 months [3]. The calculation was done using the net-energy approach, which relates the life cycle energy requirement of the wind plant to electrical energy output. This assumes electricity as primary energy, but in view of the entire European energy system with its currently still large share of fossil-fuel-based electricity generation, the primary energy that the wind turbine replaces is even greater. Computing the payback with regard to primary energy, it yields a return of only 2 months. While these low values are only valid for this specific turbine model and the site assumptions that were used, values for other turbines and other manufacturers do not differ greatly.

While wind energy is already a source of low-emissions electricity and approximately 8 g $CO_2$eq./kWh is a great value, wind energy is not devoid of emissions. To reduce these emissions, the main leverage we have is to convert what's being built in the field into more electric energy. This means that turbines must be operated more efficiently.

The efficiency of a wind turbine is commonly defined as aerodynamic efficiency, and the term power coefficient, denoted as $C_p$, is used. Betz's law states that the upper limit to the power coefficient is 59.3%, with commercially available turbines reaching a maximum power coefficient of 45%–50%. However, the maximum power that is produced is still limited by the wind itself, over which we have no control. To get more electric power, a second turbine could be erected, and the aerodynamic efficiency would not be impacted negatively. However, this neglects the fact that a second turbine would have to be built and that material effort would have to be spent. So instead of relating efficiency to theoretically available wind power, we need to address material efficiency. This is also directly related to $CO_2$eq. emissions. It can be improved by extracting more energy out of a given wind turbine structure.

The lifetime of a wind turbine is, among other factors, limited by fatigue loads imposed on load-bearing components. These cyclic loads over time reduce the load-bearing capacity of the structure until a critical point is reached, at which the turbine has to be dismantled to mitigate catastrophic failure. Turbines are designed to last for 20–25 years; with the possibility of lifetime extension, the entire lifetime can be well above 30 years. The process of lifetime extension is characterized by re-evaluating site conditions and relating them to design conditions, the so-called analytical approach, but also by a detailed inspection on site [4]. Lifetime extension directly reduces the ecological impact by increasing the total energy production of a turbine without additional material effort.

During the lifetime of a turbine that is erected today, the energy system will undergo drastic changes. According to most studies [5], wind energy will play a major role by contributing well over 50% of the power demand by 2050. The transformation from a traditional grid formed by large generators of conventional power plants to a system of inverter-based renewable energy sources is a continuous process, but the solutions to stabilize a grid with a high percentage of inverter-based generation are not fully certain yet, e.g., in Germany, the discussion of future grid services just started [6]. Therefore, wind farms built now must be ready for a future energy system, where they must respond to new requirements and demands that are not fully known yet.

In the past, wind energy was highly subsidized with fixed feed-in tariffs that were well above the actual cost of producing electricity. Once this was changed to bidding schemes for new wind farm projects, zero cent-bids became common, indicating that wind farms can be operated competitively in the current energy system. However, this makes the operators of wind farms highly dependent on market prices. The ongoing deregulation of the energy market requires a response to low or even negative prices. This is a big challenge because weather conditions are similar over large areas. Here, wind power generation among wind farms is strongly correlated and electricity prices are lowest when generation conditions are most favorable. To get an economic benefit over competitors within the same region, wind farm operators must adapt their turbine operation such that the turbines can feed in more power at times of unfavorable wind conditions, but high price, and save turbine lifetime at other times.

These changes will happen within the expected lifetime of wind farm projects being developed right now. Those will be in operation for at least two decades with options for further extension, easily reaching well into the 2050s. With a single wind farm consisting of more than 1 GW of nominal electric power, the shift from a simple collection of multiple turbines towards a large power plant acting as a system of interacting systems has accelerated in recent years and remains an ongoing process.

In effect, we need to expect that it will be necessary to augment turbines with additional control features, but at the same time, we must monitor their lifetime and operate them optimally within their load envelope. Ideally, turbines produced today are reconfigurable to account for future requirements, changing operating conditions, and variable prices.

Wind farm control combines multiple turbines into one system and aims at controlling them together such that overarching objectives are achieved to a higher degree. It is gaining traction for increased power across the entire wind farm. However, future wind farm control needs to create the best long-term value for a wind farm. It will integrate value metrics with information about the state of the wind farm, its environment, the market, and the energy system into an optimal long-term operating strategy and short-term reaction to new information. It will build on reconfigurable turbine controllers and wind farm control to distribute loads among turbines and turbine components, in turn, synchronizing and extending the lifetime.

## 9.2 Potential for reduced $CO_2$eq. emissions from wind farms

As outlined in the introduction, the emissions are driven by production, whereas energy is generated over the entire operating lifetime. This yields two ways to reduce the emissions: Reduce material effort in production or increase energy production in operation. These can be achieved in multiple ways, as depicted in Figure 9.1.

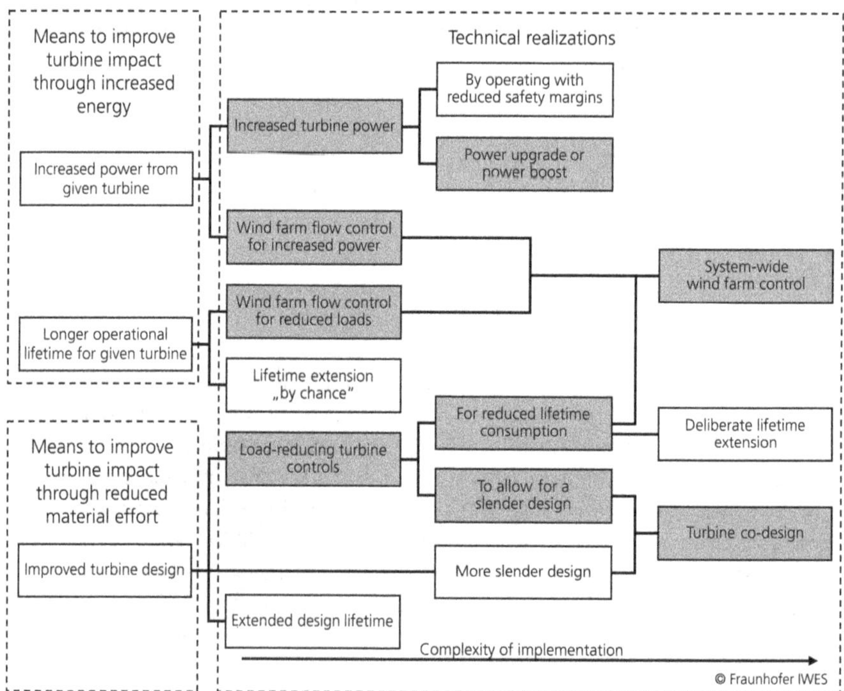


*Figure 9.1 Means to improve turbine impact and associated technical realizations. Gray boxes indicate means that are made possible through advanced control techniques.*

### *9.2.1 Reducing $CO_2$eq.: decrease the material effort of a turbine*

#### 9.2.1.1 Improved turbine design

The basic design process of a wind turbine relies on load simulations for specific load cases. This builds on a model of the entire wind turbine with all component masses. An introduction to design load cases is given in Chapter 1 of Volume 1. One goal during design is to reduce the material effort by building turbines as light as possible. A major contribution to different weights is the drivetrain topology, several examples of turbines of different sizes are given in [7]. A reduced tower top mass allows for a design with lower mass of tower, substructure, and many other components as well, thus in total reducing the material effort greatly.

These are currently mainly economic considerations by OEMs. Recently, new metrics beyond levelized cost of energy (LCOE) have been discussed. With these also comes the suggestion to design turbines to different objectives such as ecological objectives [8].

A slender, light design also comes with reduced stiffness, which can not only lead to visible deformation but also to increased vibrations and increased fatigue loads. To alleviate these problems, contemporary turbines are equipped with load-reducing controller features. These aim at introducing artificial damping into the structure to reduce vibrations.

#### 9.2.1.2 Load-reducing turbine control

The wind turbine controller aims at extracting power from the wind as efficiently as possible. To do so, it controls the torque of the generator to keep the turbine at its aerodynamically optimal operating point, thus keeping the power coefficient as high as possible. At rated power, it reduces the efficiency of power conversion by pitching the blades. This basic principle is augmented with additional requirements. In Chapter 7, the current industry standard for closed-loop wind turbine control and automation is described. The operational controller, cf. Section 7.6, is part of the overall automation system and the capability to reduce loads is one of its most important objectives. More details on the theory and design of wind turbine controllers can also be found in [9]. Load-reducing features for the classic pitch and torque controllers include active tower damping, individual pitch control, setpoint smoothing, or peak shaving.

The literature provides many turbine control techniques [10,11], each aiming for a specific goal with the main topics of power maximization and load reduction. Examples of these are model predictive control (MPC) with LiDAR measurement for inflowing wind. MPC combines multiple control objectives into an optimal control scheme and has been used effectively for load reduction [12] or improving operation in derating situations [13]. LiDAR uses laser measurements of wind speed for predicting incoming wind speed, which can be incorporated in a feed-forward control to reduce component loads [14]. While this has been researched extensively, this technology has reached maturity to a point where it is employed in industrial applications [15].

The combination of load-reducing controllers and a highly load-dependent wind turbine design allows for combined design, which is referred to as control co-design. A general introduction can be found in [16]. The principle is based on combining the sequential process of structural design and control design into a common design process, where load-reducing features of a controller are designed in conjunction with the load-bearing strength of the turbine components. According to [17], this has a great potential for reduced cost of energy. One of the greatest challenges for the application of control co-design is the computation cost of this multidisciplinary problem which usually combines approaches from reliability-based design optimization (RBDO) with control optimization [18–20]. There is a high chance, that this challenge can be solved in the near future.

While all these additional works on turbine control can reduce loads and allow for more material-efficient and less conservative designs, they are still limited to greedy control on the turbine level. Each wind turbine is controlled as a singular system, without cooperation with surrounding turbines, at the cost of less-than-ideal operation of an entire wind farm. Wind farm control aims at solving this problem, but even for a single turbine, there are more means to increase the lifetime energy production.

## *9.2.2 Improving kWh: increase the lifetime-energy production*

There are two ways to increase the energy yield of a turbine, as is already apparent in its unit: either increase the power (kW) or increase the operating time (h). Both are equally important but require entirely different methods and most often even contradict one another.

### 9.2.2.1 Improving kW: increased turbine power

The nominal or rated power of a turbine is limited by the maximum loads of the entire chain of load-bearing turbine components, by the maximum power converted in the generator, and in the power electronics. All these components have their individual maximum power rating, where the lowest power rating limits the total turbine power. However, not all components experience the same maximum loads in the same environmental conditions. For example, mechanical components are highly influenced by turbulence, whereas electrical components are more limited by heat dissipation due to inefficiencies.

Under specific conditions, these individual power ratings can be increased during the lifetime of the turbine. A permanent increase is sometimes possible, e.g., through a reduction of safety margins, through a retrofit of higher-rated components, through reduced losses, or even by exchanging some components with newly developed aftermarket solutions. Component exchanges are offered by OEMs themselves, e.g., by Siemens Gamesa for their legacy turbines [21], but also by aftermarket companies such as by Deif for Vestas V80 and Senvion MM82/92 turbines [22]. A temporary power increase, which is triggered if environmental conditions allow so, is being offered as a regular product by several OEMs, e.g., PowerBoost by Siemens Gamesa [23]. This is purely a software feature that monitors critical measurements such as ambient temperature, main component temperatures, wind speed and turbulence, or grid voltage. Based on these measurements and pre-determined safe envelopes, power can be increased. Siemens Gamesa claims a temporary increase up to 5%.

#### 9.2.2.2 Improving h: longer lifetime

The selection of 20 years as the design lifetime according to IEC61400-1 seems, from today's perspective, to be an arbitrarily selected value. In the German bight, site concessions for offshore wind farms are granted for an operating time of 25 years, which was in stark contrast with the design lifetime of the first offshore turbines. Only in 2017, the design lifetime of then-erected turbines was increased to 25 years. This was shortly after the first offshore wind farms were erected [24]. Thus, to make the best use of the site concessions, extending the lifetime beyond the design lifetime is almost mandatory for these early offshore wind farms. But also for newer wind farms, German legislation was recently changed to allow for an extension of the site concession for up to ten years, allowing for a total operating time of 35 years.

However, lifetime extension is only possible if the load-bearing capacity of the structure is not used up at the end of the designed-for operating period. This can be due to differences between site conditions and design conditions, e.g., a turbine designed for wind class 1 being erected at an onshore site with lower mean wind speed. It can also be due to excess strength, where the turbine is designed with overly high safety coefficients. In this view, having the possibility of lifetime extension can also be regarded as excessive material use in the first place.

While lifetime extension is generally regarded as economically beneficial, the required excess material increases the initial investment. Viewing this as a decisive trade-off gives rise to the concept of optimal lifetime, where additional investments for greater strength and resulting increased lifetime yield the best economic benefit. This was conducted in [25], and it was found that the optimal lifetime can deviate, depending on assumptions, quite greatly from the 20-year default value.

Aside from the technical requirements, mainly the need for a sufficiently high remaining load-bearing capability, legal processes need to be conducted which are highly country-specific [26]. They have in common that the design lifetime is compared to a site-specific individual lifetime. These two values are computed with the so-called analytical approach according to DNV-0262 [4]. This is basically a two-lane computation of the load-bearing capacity: In one lane, design conditions are assumed; in the other lane, site conditions are used. Additionally, site conditions can be augmented with operational data about actual turbine usage or operation, e.g., to take extended times of idling or curtailment into account. Both lanes result in a specific value of the loads. One lane yields the loads that individual components can experience; the other one yields the loads which they have experienced. These two are then used to compute the remaining load-bearing capacity. If no changes in operation occur, this result can be directly used as an extended lifetime. The one component that has experienced the highest ratio of site loads compared to design loads is the one limiting the lifetime. Additional works might be required, e.g., practical evaluations on site.

A major drawback to the currently used process for lifetime extension is the one-time evaluation in a late stage of the wind turbine lifecycle. The turbines are operated at their intended nominal operating mode, usually in greedy single-turbine power maximization control. As a result, the lifetime is extended for turbine operation as-is. This means viewing the operation regime as a given input and the time-to-failure as a result. Instead, switching to viewing the existing load-bearing capacity as a given input and then optimizing operation to make the best use of this would be beneficial.

#### 9.2.2.3 Wind farm flow control for increased power

All prior-discussed approaches to increase power production rely on extended or higher-rated operation of a single turbine. Wind farm control aims at operating multiple turbines, usually one wind farm, together such that the total wind farm power is increased. This is only possible with coordinated action across all turbines.

In a wind farm, individual turbines extract power from the wind at the cost of reduced wind speed and increased turbulence behind the rotor. This is called the wake. The wake extends many rotor diameters downwind; far longer than the turbines are usually spaced apart. With surrounding wind mixing into the wake, it gradually decreases. This means that if a turbine is standing behind an operating turbine, it experiences the wake and needs to cope with lower wind speed and increased turbulence. In effect, it will produce lower power.

Wind farm power can be increased by reducing the wake effect, such that the impact on downwind turbines is reduced. The basic principle is depicted in Figure 9.2. There are several ways to do so: increased wake mixing, wake steering, or wake reduction (also called low induction turbine control). The high level of attention paid to this topic is also reflected in various overview papers published in recent years [27–30]. All these references give an insight into the use of control mechanisms on turbine level for farm-level optimization. On farm control level, setpoints to the individual turbine controllers are defined.

Wake mixing focuses on an individual turbine and aims at increasing the stream of undisturbed air into the disturbed airstream behind the turbine. Examples are, among others, the Helix strategy or pulsing the turbine thrust [31]. Wake steering and wake reduction aim at using existing inputs into a turbine and reducing the wake effect on a downstream turbine directly. Wake steering uses a deliberate yaw misalignment, which deflects the wake slightly sideways and can direct it away from a downstream turbine. Wake reduction is based on reducing the power of a turbine, usually by pitching, in turn reducing the thrust and the resulting wake. All these methods improve a turbine's wake for downstream turbines but at the cost of reduced power of the upstream turbine. The aim of wind farm control is then to offset individual power losses with global power gains across all turbines.

A good overview of the status on wind farm flow control was recently published [32]. Here, the technology readiness levels are clearly defined for wind farm control and examples for implementations of yaw-based wake steering are given. According to it, wind farm flow control has already reached TRL 6 or even 9 in individual projects. Clearly, wind farm flow control is a mature technology, on

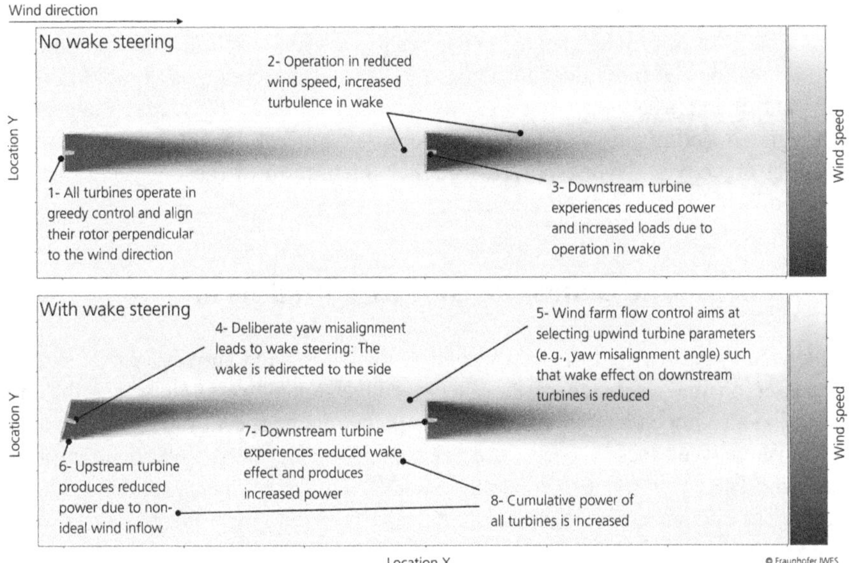


*Figure 9.2 Wake effect and basic working principle of wind farm flow control with wake steering (flow visualization taken from foxes yawed_wake example, https://github.com/FraunhoferIWES/foxes/tree/main/examples/foxes/yawed_wake)*

which higher-level controls can build upon. Concepts that have reached a high TRL are currently mainly focusing on power maximization, as this was the main objective for wind farm flow control in the past. In some cases, load reduction was considered as a secondary objective. An influential dissertation [33], that was further developed into the first commercial solution [34] also focused on power maximization or load mitigation but did not plan for the optimal long-term trade-off between both.

Despite the high TRL in some projects, the research and development of wind farm control with respect to power maximization is far from completed. There is still a large uncertainty with respect to the overall gain in annual energy production (AEP) because the benefit is highly dependent on the uncertain inflow of the wind and its characteristics. In addition, [32] also name prospects and challenges for the further development of wind farm control beyond power maximization. This includes the influence of the above-mentioned control mechanisms on loads directly on the upstream turbine which changes its setpoint, and indirectly on the downstream turbine by influencing the wake.

Therefore, an integrated perspective for a system-wide control of entire wind farms where load management is a primary objective for the overall optimization of the wind farm operation brings a high potential and can be augmented with some existing concepts.

Just like on the turbine level, a control-co design for wind farms provides another path for improvements [32]. For design, the main optimization parameters are the locations of individual turbines, which in effect creates a combined wind farm control and layout optimization problem. In [35], the AEP is maximized with an integrated optimization of layout and control. For a fully integrated approach including objectives beyond power, a higher integration of these aspects into the control of wind farms is required first.

## 9.3 System-wide control for an entire wind farm

While wind farm flow control already controls the entire wind farm as a single system, it focusses on the wind-side of the energy conversion process. However, the grid connection also imposes requirements on the power plant aspect of the wind farm. To distinguish between these aspects, a recent paper [36] suggested a clarification of the term Wind Farm Control, as it is commonly used in different meanings depending on context. The authors suggest distinguishing between Wind Power Plant Control, Wind Farm Control, and Wind Farm Flow Control. They define Wind Farm Control as an Umbrella Term that unifies all aspects and interfaces with surrounding influences, e.g., changes in the energy price. It builds on Wind Farm Flow Control, which aims at controlling the flow of air inside a wind farm by changing the operation of individual turbines, and on Wind Power Plant Control, which interfaces the entire wind farm with the power grid and controls it according to grid requirements. While both individual aspects are not new, the combination is becoming more important due to high interdependencies. This yields several challenges, among them *Balancing control objectives* and *Access to wind turbine control by wind farm control* [37].

In this context, it becomes of great importance to have clearly defined system boundaries in order to clearly identify and define the inputs and (required) outputs. It is important to know not only wind conditions at a specific site but also how they are influenced by turbines or wind parks in the vicinity, possibly even how they will be impacted in the future by planned or expected projects. While wind farm flow control is a comparatively slow process (compared to turbine dynamics), it relies on individual turbine controllers to be reconfigurable according to current requirements. Additionally, wind farm flow control, as it is currently employed, is open-loop control, e.g. [38]. True closed-loop control must evaluate the effect that control actions, e.g., changed setpoints of an individual turbine, have on the controlled system, i.e., the wind flow within a wind farm. While this is being researched, e.g., [39], it is much more involved and not yet used in applications.

Common to all wind farm flow control mechanisms is that knowledge of the wind conditions is crucial. A single turbine usually relies on a cup anemometer for wind speed measurement and a wind vane for directional measurement, both of which are not sufficiently precise for wind farm flow control, where the measurement error of these devices is in the same order as the small yaw misalignment angles required to deliberately steer the wake. Some turbines are now being equipped with nacelle-mounted or hub-mounted Lidar that measures the inflow wind conditions, but this is far from standard equipment and not universally usable for wind farm flow control. Instead, additional data postprocessing or measurement equipment is required. The most accessible means to improve wind measurements is to unite all turbine measurements. This was done in, e.g., [38]. More sophisticated methods include observers for wind turbine inflow [40]. Also, Lidar on some key turbines has been used to measure undisturbed reference inflow conditions [41].

## *9.3.1 Structuring the multiple levels of control*

Currently, the layers of control present in a wind farm are clearly distinguished and are developed separately. Wind turbine control and grid-side control are ubiquitous on modern turbines, and wind farm control for increased power production is gaining traction. However, combining these is expected to yield great potential beyond what is currently possible.

A first step towards organizing the interaction and collaboration across all three control stages has been published in [36], which defines wind power plant control to be the intermediate layer between wind farm flow control and wind turbine control. It is accompanied by a reference implementation in MATLAB® Simulink® [42]: This does not include the controllers per se, but instead organizes the layers into separate blocks and thus facilitates further developments.

For autonomously adapting systems, so-called self-optimizing systems, a generalized control and communication architecture was suggested as operator-controller module [43,44], which builds on concepts for modeling cognitive processes [45]. It orders the information processing into three levels: controller, reflective operator, and cognitive operator, as depicted in Figure 9.3. On the lowest level, a real-time capable adaptable wind turbine controller stabilizes the system in the motor loop. Adaptation of the controller is achieved by switching between different controller configurations, which might be composed of individual setpoint values, trajectories, or entirely different control laws (depicted as individual controllers A, B, C) in the reflective loop. The intermediate level, the reflective operator, includes a transition between hard real time and soft real time for higher-level adaptations. Thus, a typical wind farm controller is situated on this level. Information about the situation is provided by additional wind and state estimators, and emergency routines are implemented. The highest level provides optimal configurations to use in specific situations through the cognitive loop. These decisions are made within the cognitive operator using model-based self-optimization. In this optimization process, variables are adapted, and optimal system parameters are determined. Therefore, a global maximization of wind farm value can be performed on this level.

We now describe requirements, recent developments, and prospects on each of these levels, to allow for such an advanced system-wide wind farm control.

## *9.3.2 Turbine control as actuator for higher-level wind farm system control*

All techniques for system-wide wind farm control rely on the turbine reconfiguring itself to specific setpoints that are transmitted from a higher-level supervisory

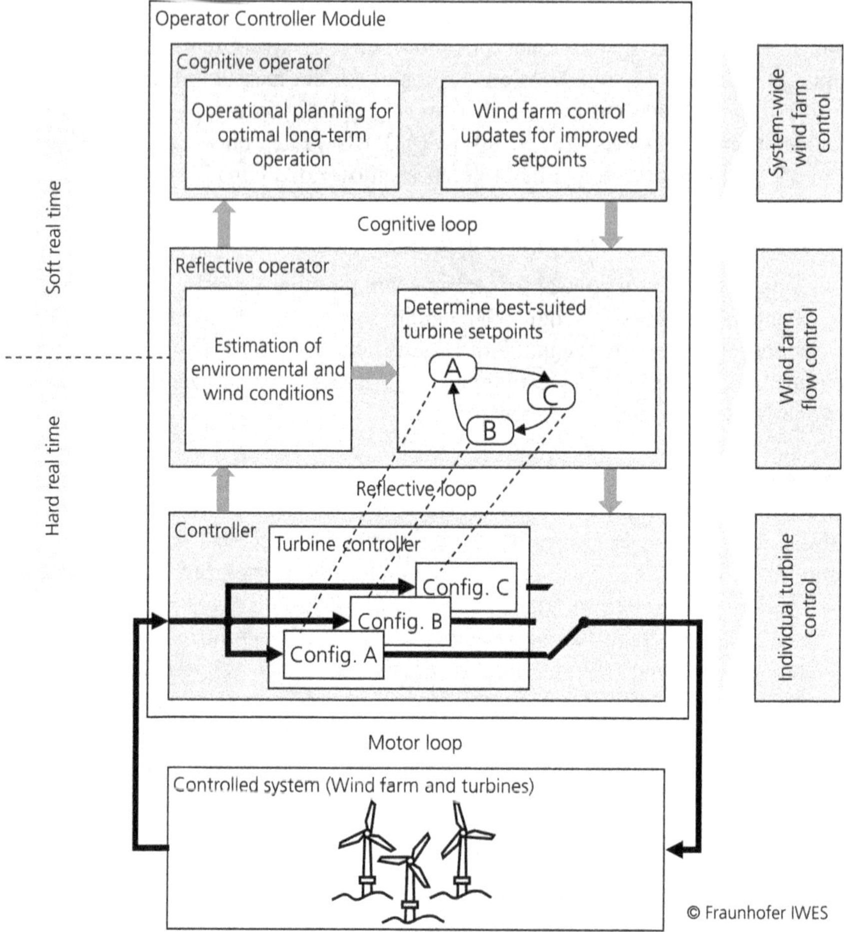


*Figure 9.3 Control levels for wind turbines and wind farms structured according to operator controller module (original OCM graphic in [43])*

instance. The principal setup of modern wind turbine controllers provides such a setup because it reacts to different operational states, e.g., required from the grid operator or simply to the cut-in and cut-out wind speed. For the basic setup of wind turbine control, we refer to Chapter 7. In Section 9.2.1.2, we already gave an introduction to load-reducing turbine control. Thus, we take the wind turbine controller as a required and existing prerequisite for advanced system-wide control which principally can be used as an actuator for higher level control. In the simplest form, this builds on already-existing inputs, e.g., curtailment setpoints, or it adds a new control input by altering sensory measurements. This has successfully been done for wake redirection through wind vane measurement value overrides [38].

While these basic turbine controllers allow for some degree of wind farm flow control, they don't offer up all possible means for the full interaction of turbines inside a wind farm. This is only possible with additional controller features that are specifically implemented to allow for more sophisticated control [31]. A major benefit in the long term can be achieved if the trade-off between load reduction for a specific failure mode and energy production is actively balanced across various environmental conditions [46].

Such turbine controllers designed for wind farm control must be reconfigurable controllers that are capable of continuous trade-offs between secondary objectives, e.g., specific failure modes, and energy production. Especially, the controller features must be reconfigurable not only to a specific site but also during operation. Thereby, an optimal performance can be achieved over the lifetime of a wind turbine within an environment with changing constraints. To support this, a fast communication system such as those introduced in Chapter 7 is required.

### 9.3.3 Value optimization

Currently, wind farm operation is focused on maximizing short-term monetary gains, where a maximum gain is defined as selling the greatest amount of energy at the highest possible price at that moment. Contrasting with this is the idea of continuously adjusting the operation to obtain a higher long-term gain. The main driver and most-researched aspect is operation control depending on the electricity price [47,48]. However, this does not consider that operation also comes at the cost of turbine degradation, which highly depends on the environmental situation.

The value itself is usually limited to monetary value summarized as levelized cost of energy (LCoE). This is limited though; more complex metrics that take the changing price of electricity into account are currently discussed [49]. Governments have also started using non-price criteria in their wind auctions. The Dutch lead the way with two successful auctions for the Hollandse Kust West offshore wind farms, one using system integration and the other biodiversity criteria [50], but other countries are following.

Generally, for an operator, the decision if a unit of electricity is produced depends solely on the remuneration. In contrast, the margin of profit attainable at a particular moment should consider the actual load-bearing capacity of turbines in the wind farm. The profit is attenuated by the cost of the negative impact of production on the lifecycle value of the wind farm. This consideration becomes possible by integrating surrogate models for fatigue damage into an optimization for creating a planned lifetime strategy for optimal material usage [51].

Thus, a business-culture transition from instantly gratifying short-term transactions to long-term position profitability is needed. Each unit of electricity to be produced carries a cost consisting of its direct production cost and of the indirectly incurred future cost of maintenance. The planning needs to build on detailed modeling of the long-term economic development of a wind farm. In addition, the lifetimes of various turbines and components need to be aligned with each other and with the economic objectives.

## 9.4 Optimal operational planning as input for wind farm flow control and turbine control

With the aim of using control as an instrument for optimal usage of the invested materials, pre-planning of operation over the entire lifetime becomes important. For planning to be successful, the operational plan must be implemented into an operational strategy that is adapted at regular intervals. This strategy then forms the basis of a continuous adaptation of the turbine controllers over the entire lifecycle. The various operating conditions, which occur during this lifetime, differ greatly in the amount of fatigue lifetime that is consumed and the energy output that is obtained. Under some conditions, e.g., at high turbulence but low wind speeds, the loads of a turbine are high, but energy production is low. Therefore, it is desirable to reduce loads during this situation, and in turn to be rewarded with an extended lifetime and more low-load, high-yield situations later during the extended lifetime of the turbine. This becomes ecologically beneficial if the saved additional lifetime allows for production of more energy over the entire lifetime than would have been possible otherwise. While optimally balancing lifetime consumption and energy yield will in principle lead to the lowest possible value for invested material per unit of energy ($CO_2$eq./kWh as introduced in Section 9.1), i.e., ecologic cost, the economic cost per unit of energy can only be balanced when considering the variable market and including economic assessments into the planning. An illustrative example of this conflict is negative electricity prices, which usually occur in very favorable wind conditions when many turbines operate at high power levels, and energy supply exceeds energy demand. During these wind conditions, loads on the turbine are comparatively low, energy yield is high, and the resulting ecologic cost is low. However, economic cost is extremely high with operators even having to pay for the operation of their turbines!

Overall, the relationship between lifetime consumption, energy production, and economic value is highly nonlinear. To make use of this for an optimized operational planning, all influencing factors need to be considered:

- *Operation-influencing conditions:* Under this term, we summarize all conditions that the wind turbine or wind farm controller can use or needs to react. They include:
  - (Natural) environmental conditions (e.g., wind speed, wind direction, turbulence)
  - Events or faults, e.g., from the grid
  - Required reactions to certain situations, e.g., to the grid
  - Market prices

  Under each of these conditions, variable setpoints of the turbine controller can be used. The combined influence of operating condition and control setpoints on energy production and lifetime consumption of the wind turbine components or failure modes needs to be known.
- The frequency of occurrences for each of these conditions is highly relevant for the distribution of induced damage over lifetime.

- The financing concept combines energy yield with a model or forecast for market development. This finally determines the economic costs.

Therefore, each of these aspects needs to be modeled to determine optimal planning. The planning procedure will thus determine an operational strategy which is the input for the subsequent operation of the wind energy system. Figure 9.4 shows how the planning interacts with the operating stage. In the context of the multi-level control architecture with an operator controller module (cf. Figure 9.4), those two stages represent a more specific implementation of the first two levels, i.e., the cognitive operator (slow adaptation using methods as in the planning stage) and the reflective operator (fast re-adaptation during the operating stage based on operational strategy as main planning result). The operational strategy is used as an input to the operating stage of the wind energy system. Within the operating stage on this level, we refer to a time scope between minutes and hours, i.e., on the scope of wind farm flow control. Therefore, a supervisory controller or operational management is implemented on this level. It provides a setpoint for the configuration of the real-time controller for each wind turbine or potential other controllers of components for a hybrid system depending on actual inputs on this time scope. In both stages, feedback of the system's performance is required, to close the loop between planned operation and desired behavior. The direct readaptation will be implemented within the supervisory control loop on the minutes scope. The readjustment of the planning can integrate major changes in the forecast or the desired objectives on a very large time annual time scope.

The optimized operational planning forms the basis for the operation with these control loops at different stages. The process for this was introduced in [51]. We explain the basic concept for this process in the following. For full details, please refer back to the paper.

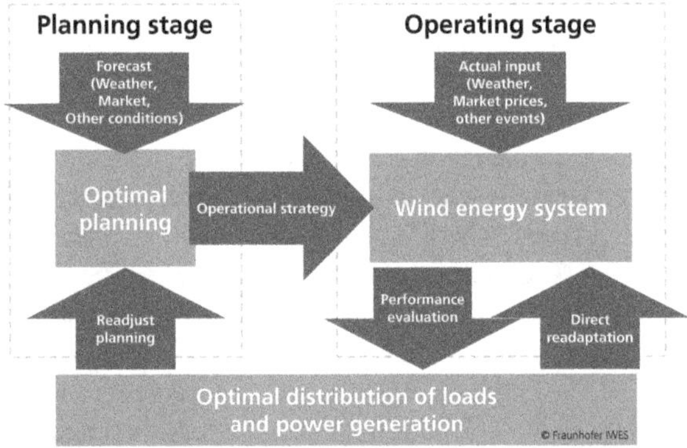


*Figure 9.4 Overview of the separation of planning and operating stage of wind farm planning and control (adapted from [51])*

### 9.4.1 Optimization

The key part for the optimized operational planning from [51] is setting up a mathematical optimization problem, with which system operation can be adapted through operation-parameterizing optimization variables such that lifetime objectives are met while influencing factors are being considered.

To enable the mathematical optimization, it is first necessary to define the problem unambiguously and to describe it via mathematical relationships. This is achieved by defining suitable objective functions that depend on certain adjustable-value parameters, which are used as optimization variables. For a problem with different conflicting objectives, in general, any number of objectives that depend on the same optimization variables and parameters can be considered. In this context, the objective functions often cannot be formulated as a closed mathematical equation, but each objective function evaluation may involve performing different simulations and evaluating them. Thus, setting up the optimization problem for a complex engineering problem is a major part of the process for automated optimization and is at least partially detached from the targeted solution method. However, the solution method is crucial under the specific boundaries and limitations of a specific problem and setup.

For operational optimization, the system model must cover the entire value chain from changed operational parameters to long-term gains. This implicitly requires the use of a model for the behavior of the system itself, and for long-term behavior such as the aging of components or monetary value considering the cost of capital, credit repayments, and O&M activities. The objective function values are derived using predefined metrics that aggregate one or several simulation results into a single value like lifetime energy production or lifetime damage. Simulation results can consist of time series or contain statistical distributions for the relationship between in- and outputs.

Figure 9.5 shows the principal setup for the system model that is used in the optimal operational planning process. The operational strategy is determined as a set of setpoints for the real-time controller of the considered wind energy system. These setpoints of the operational strategy define the optimization variables of the optimization problem. Thus, they are input to the annual scope calculation which contains a system model and needs a forecast of environmental conditions as an input. This yields the cumulative annual values, which in turn can be accumulated into total lifetime values. While total energy and total damage define technical objectives that don't need any additional input, the additional economic or other influences are critical for the value calculation. The lifetime of the system is determined by the accumulated damage value and varies depending on the operational strategy. This relationship is crucial and allows to make use of saved damage later during an extended system lifetime.

From the setup of the optimization process, requirements for the system model and the forecast can be derived. The annual scope calculation needs to be computed as part of the objective function and thus many times within an optimization loop. Thus, it needs to be a fast-running model that can still cover the influence of

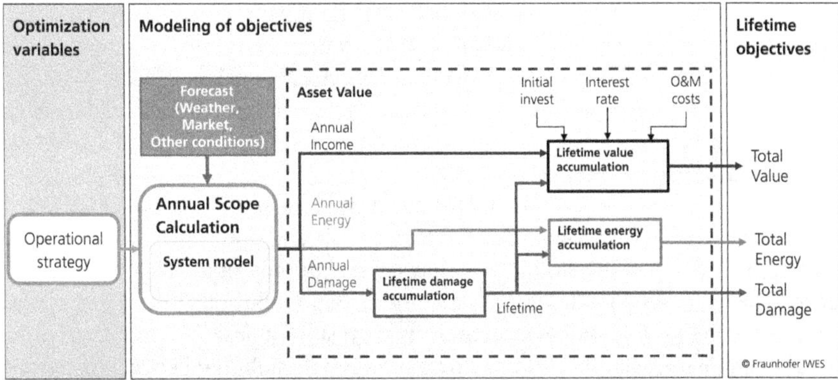


*Figure 9.5 Process for computing the lifetime objectives dependent on operational strategy. The setpoints of the operational strategy define the optimization variables (adopted from [51]).*

control setpoints under each of the operating conditions. Lifetime damage accumulation is driven by fatigue loads. These can only be computed from the results of an aeroelastic simulation of a wind turbine, which requires small time steps and high computational effort. For this purpose, surrogate models, which are usually derived from more complex simulations, are a suitable choice that we take as a required and existing prerequisite for a four-step process for operational planning (see Figure 9.6).

## *9.4.2 Surrogate model setup*

In general, the direct use of an aeroelastic simulation model is not possible within an optimization loop of the size and complexity required for operational optimization. In addition to the high computational effort comes the uncertainty of simulation results. This wind-energy-specific problem comes from the use of stochastic wind fields as simulation input. Slight changes in wind turbine operation mean that the blades sweep over a different section of the wind field, thus experiencing different input values. This leads to great differences in the simulation result for wind fields with the same nominal conditions. A long simulation duration or a large number of fixed-length simulations with random seeds for random wind

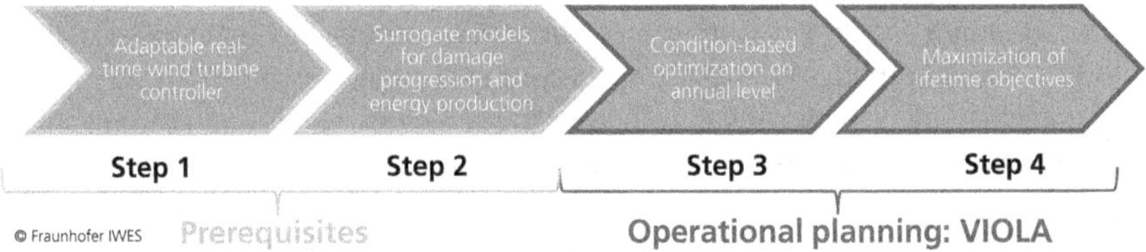


*Figure 9.6 Four-step process for the optimization of operational planning*

generation reduce this problem but cannot mitigate it entirely. Mozafari *et al.* [52] have researched this problem in detail and found that the minimum recommendation for the number of seeds in IEC-standard, six, is only sufficient for low-priority bins, while more seeds are required in bins where a large amount of damage is induced. However, this makes the use of an aeroelastic simulation model within an optimization loop even more challenging!

Instead of running more simulations and running them during optimization, surrogate models can be used. This exploits the fact that during an optimization, we're not interested in the simulation time series, but only in the final value of our metric. The basic idea is then to map the influencing input conditions and control setpoints to energy and damage increments as result values. The most accessible surrogate model is a simple lookup table, possibly interpolating between individual data points. More sophisticated surrogate models combine the values of neighboring data points by approximating all data points with a multidimensional function, e.g., using polynomial regression. In [53], different machine-learning models were trained and evaluated. Within the context of operational planning, a surrogate model is merely a means to achieve fast-running optimization models. As such, it should be as simple as possible, but cover all relevant effects and be suitable for the utilized optimization method, i.e., smooth enough or even differentiable.

The surrogates are created by defining in- and outputs for models on different time scopes (white rectangles). The white arrows describe a transition from input to output with a corresponding model. The rectangle in the center contains the prerequisites for computing energy and damage on the annual scope, an thus for the operational planning. The creation process of surrogates is depicted by the bright blue arrows, starting from the pool of input samples and with the creation of surrogate models which can be evaluated on the minutes scope.

Within the annual scope, the surrogate models are used to compute the annual value with the frequency distribution as an additional input. The whole process needs to be well defined for the optimization of operational planning. To optimize the operational planning on a technical level, the natural environmental conditions build the crucial part of the influencing operating conditions. They determine power production and loads during normal power production operation. Thus, those are required inputs for the surrogate models. Their specific choice and the utilized control setpoints are derived from the definition of system boundaries. For each combination, a simulation of the aeroelastic turbine model is run for a duration of multiple minutes. Afterwards, loads and power are evaluated and damage or energy increments for operation with the respective inputs are obtained. Many simulations are run to create a data set from which the surrogate model is created. Later on, only the surrogate model is evaluated, no further simulations are run.

To evaluate the benefit of a specific operational strategy, the site-specific annual wind conditions and the operational strategy form a set of conditions on the annual scope, i.e., with a time step of 1 year. Those conditions are mapped to environmental input conditions and corresponding control setpoints where the surrogate models are evaluated, and energy and damage increments are accumulated with their corresponding frequency distribution to form the annual energy and damage. Since the surrogate model is usually stateless, i.e., each result depends only on inputs on the minute scope, but not on other time steps, all time steps can be evaluated in any order or even parallel, thus allowing for efficient and fast simulations.

### *9.4.3 Classification of operation-influencing conditions*

Until now, we assumed that environmental conditions were known. Normally, they are derived from measurement data and then postprocessed to form, e.g., a Weibull distribution of wind speeds. These parameters are then used for the load calculation process of an individual turbine to assert that the site-specific loads are lower than the design loads. This process can directly be used for operational planning.

At first, specific environmental conditions are identified and measured. The main parameter is wind speed, but secondary or derived parameters, e.g., turbulence intensity, must be considered as well. For state-based optimization, individual bins are required. Within operational planning, the operation is adapted for each combination of specific bin value for wind speed, turbulence intensity, wind direction, price of electricity, and all other environmental conditions. This requires a careful selection of the bin width, as with increasing dimensionality and small bins, the resulting number of states and thus the number of optimization variables increases greatly. However, a too-small number of states would neglect important aspects. A vivid example is the wind direction: Ideally, different wake overlap situations can be differentiated, i.e., there are multiple different direction bins for no wake hitting a downstream rotor, a wake hitting a downstream rotor partially and the downstream turbine standing fully in the wake. This usually requires a bin width of around 1°.

To cover the whole range of operation-influencing factors, man-made conditions need to be considered, e.g., the price of energy. The price will have no influence on the damage increments and is thus not an input to the surrogate models, but it directly influences the relationship between cost and utilized material. In addition, it also depends on the global wind conditions due to the availability of electric power in the grid [47].

Overall, the classification process results in a combined probability or frequency distribution of all operation-influencing conditions which is valid for a specific time horizon.

### *9.4.4 Scheduling for long time horizon*

With the surrogate models and combined probability or frequency distribution of all the environmental conditions, the optimization problem can be formulated. Ideally, a robust optimization approach that directly maximizes the value as the main objective under all potential cost factors together with their probabilistic uncertainty would be performed. Principally, it would be possible to setup such a problem if there were no limitations on computational power and time. Such an approach would also require a high modeling effort because all partial influences would be influencing the results in a single objective function evaluation. Therefore, interpretability would be difficult. To overcome these issues, intermediate steps that yield several operational strategies pursuing partial objectives can be made.

In [51], only the technical aspects are covered in the first step. The energy production is maximized under a given target fatigue budget $D^{target}$:

$$\max_{\overline{u}} \sum_j P(x_j, u_j) h_{\Delta\tau}(x_j) \tag{9.1}$$

$$\text{subset to} \sum_j d_{fm}(x_j, u_j) h_{\Delta\tau}(x_j) \leq D^{target}, \forall fm,$$

where $\Delta\tau$ is usually selected to be one year, as is the commonly used design basis in wind turbine development. The environmental input conditions $x_j$ are assumed to be known, while the control setpoints $u_j$ are used as optimization variables $\overline{u}$. For each individual combination $j$ of input conditions $x_j$ and control setpoints $u_j$, all surrogate models are evaluated at once. Then they are combined to compute total energy yield from individual power values $P$ and the frequency distribution $h_{\Delta\tau}$ over time span $\Delta\tau$. The target fatigue budget is computed similarly from individual damage increments $d_{fm}$, which is done for each failure mode $fm$. This process is also depicted within the annual scope of Figure 9.7.

This results in a specific value for each control parameter that should be set under a specific combination of environmental conditions, to obtain maximum power, but restrict damage progression such that the target fatigue budget is met. However, the target fatigue budget would, at this point, be an arbitrary decision without being backed by sound decision-making. Instead, we suggest running multiple optimizations with different target fatigue budgets, thus essentially using the $\epsilon$-constraint method for multiobjective optimization. This results in multiple different solutions for different target fatigue budgets, but also energy yield. Some examples of optimized operating strategies are shown in Figure 9.8. For each optimization scenario, percentage power setpoints are optimized for two different sets of environmental conditions. The left side shows solutions for wind direction and wind speed as environmental condition input, whereas the right side shows solutions for wind speed and turbulence intensity. In all cases (a) to (c), the optimization is performed considering a different failure mode for both environmental setups. The operational strategy reacts to the high load situations of the selected failure modes and reduces the damage by finding the optimum global setpoints. The three failure modes, which are selected here, represent fatigue damage failures at two major components: the blades and the tower. The blades are represented by two separate failure modes: bending moments in flapwise and edgewise direction at the blade root. For the load calculation procedure and their representative loads refer to [54]. Typically, the tradeoff between energy and fatigue loads results in several unique points on a Pareto front for each of these failure modes or as a Pareto front with many objectives (one per failure mode plus energy production).

A similar approach can be pursued when market prices are included as another variable environmental condition. Then, the annual income can be maximized while adhering to a predefined target fatigue budget. This way, a reaction to low-price and high-damage situations becomes possible, in turn increasing annual income at possibly even reduced annual energy yield.

After this first phase, multiple possible operating strategies exist, but the decision which one to use is still open. To find the truly optimal operating strategy,

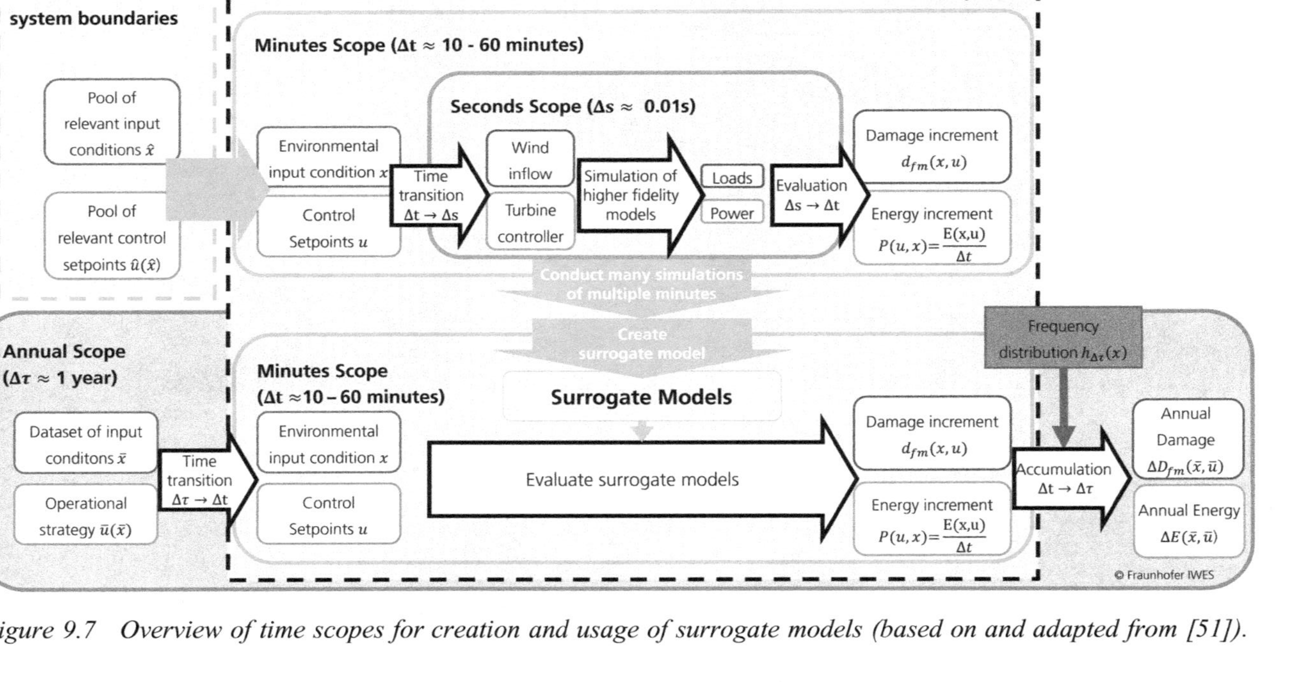


*Figure 9.7 Overview of time scopes for creation and usage of surrogate models (based on and adapted from [51]).*

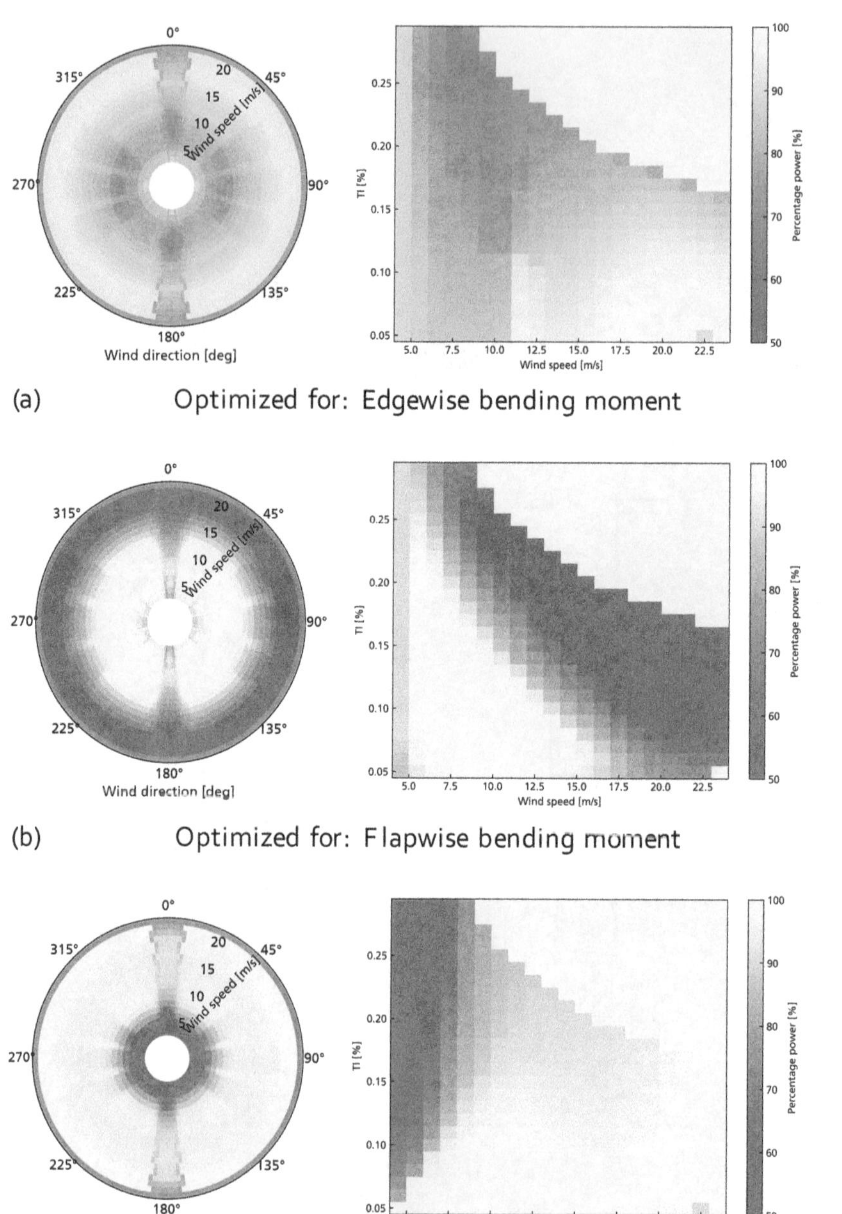


*Figure 9.8 Example operation strategies optimized for specific failure modes, as indicated in the subtitles. Strategies are depending on environmental conditions: wind speed and wind direction (left figures from [51]), right figures from [55]).*

economic long-term aspects must be brought into the decision-making. To do so, we perform a second optimization process. Now it is crucial to keep in mind that a changed target fatigue budget also results in a changed time to failure – or, vice versa, a potential for lifetime extension. We assume that this extended lifetime is utilized and transformed into more operating hours, which in turn also allows for more beneficial low-load and high-yield operating hours. With a very generalized value function $V\left(\tau^{\text{life}}(\overline{u}), E\left(\overline{u}, \tau^{\text{life}}(\overline{u})\right), q_V\right)$ depending on the total lifetime of the wind energy system $\tau^{life}(\overline{u})$, the energy production within that time $E\left(\overline{u}, \tau^{\text{life}}(\overline{u})\right)$ and the input parameters which are needed for the cost model $q_V$, the best operational strategy $\overline{u}$ can be found by maximizing the objective $V$. The optimization process can range from a simple selection process from several pre-computed strategies over a second mathematical optimization based on a mathematical description of the Pareto-front of annual energy and annual damage up to including a probabilistic reliability model. A first step towards this was published in [55]. In all cases, the relationship between annual damage progression and total lifetime needs to be known. In the simplest case, a deterministic linear relationship can be assumed. This yields that the lifetime of each failure mode $\tau_{fm}^{life}$ is reciprocal to the cumulative annual damage, i.e., $\tau_{fm}^{\text{life}} = \frac{1}{\Delta D_{fm}}$. The total lifetime can be assumed as a minimum value from all failure modes, i.e., $\min\left\{\tau_{fm}^{\text{life}}\right\}\forall fm$. Independent of the approach for computing the total lifetime, both the total lifetime as well as the total energy production depends on the selected operational strategy and can by augmented with a cost model.

A potential metric for the economic value is the net present value (NPV). By assuming a fixed, credit-financed CAPEX, time-variant cost for O&M, and fixed interest rate, the net present value at the end of the lifetime can be computed and the economic optimum can be selected.

To reach this economic optimum in actual operation, the optimal operating strategy is used to provide control parameters that depend on environmental conditions. The process is then to measure the environmental conditions as precisely as possible, to match them with those assumed during the optimization of the operating strategy, and to find the dataset that fits most closely. This dataset is composed of environmental conditions and corresponding control parameters. These matching control parameters are then set in the turbines, thus adapting their behavior.

### *9.4.5 Continuous update of operational planning*

The planning process can be conducted at any time during the operation of an asset, not just prior to or during commissioning. The only aspect that needs to be changed is that for a mid-lifetime planning, the load history must be used as a baseline that cannot be altered anymore. The available fatigue budget is reduced accordingly. The process itself is not changed, as it relies on optimization on a yearly scale, which is expanded to the entire lifetime. In the case of mid-lifetime-planning, it is expanded to the remaining lifetime only.

This makes it also possible to replan once new information about the current system state is available. Ideally, recurring updates would be used. Intervals for replanning could be fixed, e.g., yearly, monthly, or weekly, or variable, e.g., to consider major events that were not anticipated. These can include natural events such as storms, which occur so seldom that a single occurrence significantly alters the wind distribution, or system-inherent events, e.g., grid congestion or non-scheduled downtime due to failures.

Another driver for re-planning can be the monitored system state. If structural health monitoring, condition monitoring, or digital twins are used to assess the current system state in near real-time, this can be regarded as the current value, whereas the theoretical damage progression is the desired value. Deviations lead to differences between the two, that a closed-loop control can compensate for by adapting future operation.

Such a closed-loop process between monitoring the current status and replanning for future damage progression captures the essential aspects of MPC, as documented in [46]. This not only allows for long-term adaptation but also for short-term replanning and regular model updates.

Model predictive control is based on an observation of the current system state and an open-loop feedforward control, to steer it away from the current system state and back onto the desired state trajectory. Usually, this is employed for a time-varying desired trajectory, where it has the benefit of bringing the future trajectory into the feedforward control. The time horizon for feedforward control is usually longer than the control timestep. In every timestep, the feedforward action up until this time step is employed, but then a new feedforward control for the future is computed and the old one is discarded. This process is repeated continuously, thus forming closed-loop control. The main advantage is that a time-varying setpoint can be followed well, but also time-varying environmental conditions are considered during feedforward control computation. In the context of operational planning, this allows to compensate short-term increases in damage progression by reducing damage contribution before or after the critical event.

### *9.4.6 Short-term adaptation and override of planning*

Planned operation prescribes the behavior of the turbines in a wind energy asset according to the current environmental conditions. But what if these environmental conditions do not capture the complete environment? What if there are further restrictions on reconfiguration, which are not captured by the planning process? Such further restrictions could be, e.g., a minimum power to be delivered, e.g., to fulfill a power purchase agreement, or a maximum power that can be exported, e.g., due to grid restrictions and deliberate curtailment imposed by the grid operator. It must be possible to fulfill such restrictions, and thus to override the control parameters from planning.

Short-term readaptations can be implemented in two ways: either using alternative optimal operating points or switching to arbitrary control parameters, thus negating optimized operating points. It is usually desirable to select between optimal operating points on the Pareto front of compromises between damage contribution and power. This allows to adjust the tradeoff between different objectives, e.g., to restrict total power output at the cost of any objective value for damage progression and at operating at an optimal compromise, which is not the economic optimum. However, this process relies on the established chain for transferring control parameters on a slow timescale. For fast emergency reactions, this is not feasible. For fast reactions, for complex reactions that go beyond what was expressed as mathematical objective functions, or for true emergency reactions, e.g., on grid events that demand immediate changes, dedicated routines are required. Especially, emergency reactions usually rely on the reaction of each individual turbine, which overrides its own prescribed planned control parameters.

If the entire planning process is regarded and setup as a model-predictive control loop, it aims at controlling degradation according to a predetermined damage progression. The setpoint damage progression can be derived from the original planning without non-optimal reactions. Any deviations from this due to short-term readaptations act as perturbation to the closed loop controller and will be compensated for. However, such reactions violate the assumptions made during planning and thus also inhibit the validity of the operation strategy. These violations are not problematic, as they don't directly influence the integrity of the wind energy asset operation. However, the long-term effects lead to a changed time to failure or energy yield and thus directly contradict the original purpose of planning. Business strategies based on planned operations would be impacted as well. To mitigate these negative effects, short-term overrides mandate continuous replanning according to Section 9.4.5.

## 9.5 Evaluating the technical and economic benefit of advanced control concepts

Common to all advanced control concepts is that the action and the reward are further removed than in classical control. If a single turbine controller is tuned to operate at a higher power output, the effect is directly measurable. If, on the other hand, it is tuned for secondary objectives, e.g., to reduce loads, the benefit will only be observable with comparably complex measurements of component loads, possibly during rare events or over a long time period.

For advanced control concepts, action and reward can be removed temporally or spatially. In wind farm control, even for simple objectives such as power increase, the action on one turbine is rewarded on another turbine. This makes it difficult to isolate the effect of wind farm control from other effects and thus to assess the complete effect of wind farm control. A common evaluation approach to measure the benefit despite uncertain inflow conditions is to compare a wind farm with and without wind farm control to itself. This is achieved by splitting it into a controlled and a non-controlled part. For wind farms with simple boundary conditions and largely the same inflow conditions, e.g., offshore, this split can be a physical split into two groups of turbines, both of which experience the same inflow conditions. For more complex terrain, this approach is not feasible. Here, toggling between two operating modes has been employed successfully [56,57]. A common approach is to toggle after a fixed interval, leave some time to approach a stationary operating point, then to record measurement data. After this fixed interval, the operation mode is toggled again. The total test phase duration must be long enough to acquire sufficient data to average out any secondary effects. Also, toggling should be at an odd interval so as not to lead to synchronization with natural intervals, e.g., daily weather patterns.

Measuring the benefit of operational planning is more complex. The main difficulty is the desired long-term effect, which could only truly be observed over this long term, i.e., the entire lifetime. If operational planning works as desired, a control group of turbines would have accumulated less damage than a group without operational planning, in turn allowing for a longer lifetime extension. This means that the true benefit becomes apparent after approximately 25–30 years. This is far from realistic, thus faster evaluating concepts are required.

Since the planned optimal operation is a combination of individual turbine control, wind farm control, fatigue damage extrapolation, reliability analysis and ideally structural health monitoring, the respective established methods for evaluation should be employed. In addition, there are the complex influences of the future energy market, which are based on uncertain forecasts. With the planning of operation ahead of more than 25 years, it is thus required to quantify the risk of each planned strategy with respect to reliability, economic value, and environmental impact. This way, a risk-informed decision for all these aspects can be made when all influencing factors can be assessed.

Under all of these aspects, market considerations form a separate field for the further development and application of appropriate methods. From this, further influences on operational management can be derived. Reactions to varying electricity prices have already been considered in the previous chapter, but the reaction to the future energy market with further influences will be considered separately below.

In Section 9.4, the market price was already introduced as an operation-influencing condition which can be included within the operational planning process. Within the future energy system, a reaction to market prices and operator demands will be more and more required. Many wind farms installed today already generate income by directly selling electricity at the market without relying on fixed feed-in tariffs. In the future, a wind energy system will probably operate under zero-subsidy schemes. Already today, the price of electricity depends on the available power capacity of the wind. At times with high availability of wind power, electricity prices are lower, and vice versa. With a smart and planned reaction to current prices, the amount of income per incurred damage can be improved. While such planning can be beneficial, when trading electric energy at the current market conditions, the further transition of the energy system will demand further unknown requirements which need to be planned and optimized. On the one hand, the regulatory framework could be adapted to remunerate retaining power reserves for grid support, so that this also becomes attractive for wind farms. On the other hand, hydrogen production and storage of energy will play a crucial role in the future. Such storage of energy allows for a temporary uncoupling of the production from the market. In all of these cases, trading energy or even energy reserves at different markets provides possibilities of higher monetary income from the same amount of energy yield. In this case, the operational planning needs to be adapted to new requirements and objectives while still aiming for the highest conversion of material into energy and total value.

Overall, the temporal removement from technical influence on the system (i.e., load reduction in the presence) and reward (increased energy yield in the future) is exaggerated by uncertainties in economic forecasting and market behavior. The more trading options exist, the better these uncertainties can be turned into profits.

## References

[1] J. Hengstler, M. Russ, A. Stoffregen, A. Hendrich, M. Held, and A.-K. Briem, "Aktualisierung und Bewertung der Ökobilanzen von Windenergie- und Photovoltaikanlagen unter Berücksichtigung aktueller Technologie entwicklungen," May 2021. [Online]. Available: https://www.umweltbundesamt.de/publikationen/aktualisierung-bewertung-der-oekobilanzen-von

[2] P. Icha, "Entwicklung der spezifischen KohlendioxidEmissionen des deutschen Strommix in den Jahren 1990 – 2018," Apr. 2019. [Online]. Available: https://www.umweltbundesamt.de/sites/default/files/medien/1410/publikationen/2019-04-10_cc_10-2019_strommix_2019.pdf

[3] P. Razdan and G. Peter, "Life Cycle Assessment of Electricity Production from an onshore V150-4.2 MW Wind Plant," Hedeager 42, Aarhus N, 8200, Denmark, Nov. 2019. [Online]. Available: https://www.vestas.com/content/dam/vestas-com/global/en/sustainability/reports-and-ratings/lcas/LCA%20of%20Electricity%20Production%20from%20an%20onshore%20V15042MW%20Wind%20PlantFinal.pdf.coredownload.inline.pdf

[4] *Lifetime Extension of Wind Turbines*, DNVGL-ST-0262, DNV GL, 2016.

[5] I. Tsiropoulos, W. Nijs, D. Tarvydas, and P. Ruiz, *Towards Net-Zero Emissions in the EU Energy System by 2050: Insights from Scenarios in Line with the 2030 and 2050 Ambitions of the European Green Deal.* Luxembourg: Publications Office of the European Union, 2020.

[6] Bundesministerium für Wirtschaft und Klimaschutz (2023): Roadmap Systemstabilität. In preparation.

[7] J. Wenske, "Drivetrain concepts and developments," in *Wind Turbine System Design. Volume 1: Nacelles, Drivetrains and Verification*, J. Wenske (eds.). Stevenage: Institution of Engineering and Technology, 2022, pp. 207–296.

[8] H. Canet, A. Guilloré, and C. L. Bottasso, *The Eco-Conscious Wind Turbine: Bringing Societal Value to Design*, 2022.

[9] A. Gambier, *Control of Large Wind Energy Systems: Theory and Methods for the User*, 1st ed. Cham: Springer International Publishing, 2022. Accessed: Jun. 20, 2023.

[10] J. G. Njiri and D. Söffker, "State-of-the-art in wind turbine control: Trends and challenges," *Renew. Sustain. Energy Rev.*, vol. 60, pp. 377–393, 2016, doi: 10.1016/J.RSER.2016.01.110.

[11] E. J. N. Menezes, A. M. Araújo, and N. S. Bouchonneau da Silva, "A review on wind turbine control and its associated methods," *J. Clean. Prod.*, vol. 174, pp. 945–953, 2018, doi: 10.1016/j.jclepro.2017.10.297.

[12] D. Schlipf, D. J. Schlipf, and M. Kühn, "Nonlinear model predictive control of wind turbines using LIDAR," *Wind Energy*, vol. 16, no. 7, pp. 1107–1129, 2013, doi: 10.1002/we.1533.

[13] J. G. Silva, R. Ferrari, and J.-W. van Wingerden, "Convex model predictive control for down-regulation strategies in wind turbines," 2022.

[14] A. Scholbrock, P. Fleming, L. Fingersh, *et al.*, "Field testing LIDAR-based feed-forward controls on the NREL controls advanced research turbine," in *51st AIAA Aerospace Sciences Meeting including the New Horizons Forum and Aerospace Exposition*, Grapevine (Dallas/Ft. Worth Region), Texas, 2013.

[15] Sowento, *Important Milestone: 1000 Wind Turbines with Lidar-Assisted Control.* [Online]. Available: https://www.sowento.com/important-milestone-reached-for-lidar-assisted-control-of-wind-turbines-1000-wind-turbines (accessed: Jan. 6, 2023).

[16] M. Garcia-Sanz, "Control co-design: An engineering game changer," *Adv Control Appl*, vol. 1, no. 1, 2019, doi: 10.1002/adc2.18.

[17] L. Y. Pao, D. S. Zalkind, D. T. Griffith, *et al.*, "Control co-design of 13 MW downwind two-bladed rotors to achieve 25% reduction in levelized cost of wind energy," *Ann Rev Control*, vol. 51, no. 8, pp. 331–343, 2021, doi: 10.1016/j.arcontrol.2021.02.001.

[18] T. Cui, J. T. Allison, and P. Wang, "Reliability-based control co-design of horizontal axis wind turbines," *Struct Multidisc Optim*, vol. 64, no. 6, pp. 3653–3679, 2021, doi: 10.1007/s00158-021-03046-3.

[19] D. R. Herber and J. T. Allison, "Nested and simultaneous solution strategies for general combined plant and control design problems," *J Mech Des*, vol. 141, no. 1, 2019, doi: 10.1115/1.4040705.

[20] Q. Zhang, Y. Wu, L. Lu, and P. Qiao, "A single-loop framework for the reliability-based control co-design problem in the dynamic system," *Machines*, vol. 11, no. 2, p. 262, 2023, doi: 10.3390/machines11020262.

[21] Siemens Gamesa, *Energy Thrust Energy Output Upgrade.* [Online]. Available: https://www.siemensgamesa.com/en-int/products-and-services/service-wind/energy-thrust (accessed: Jun. 9, 2023).

[22] DEIF, *Wind Turbine Retrofit.* [Online]. Available: https://www.deif.com/wind-power/turbine-retrofit/ (accessed: Jun. 9, 2023).

[23] Siemens Gamesa, *Siemens Gamesa Power Boost Function: Increase Your Power Production.* [Online]. Available: https://www.siemensgamesa.com/en-int/-/media/siemensgamesa/downloads/en/products-and-services/services/asset-optimization/power-boost-function.pdf (accessed: Jun. 9, 2023).

[24] M. Dörenkämper, T. Meyer, D. Baumgärtner, *et al.*, "Weiterentwicklung der Rahmenbedingungen zur Planung von Windenergieanlagen auf See und Netzanbindungssystemen: Endbericht," Fraunhofer IWES, Jan. 2023. [Online]. Available: https://www.bsh.de/DE/THEMEN/Offshore/

Meeresfachplanung/Flaechenentwicklungsplan/_Anlagen/Downloads/FEP_2023_1/Endbericht_FEP_2023_Beratung.pdf

[25] C. Hübler, J.-H. Piel, C. Stetter, C. G. Gebhardt, M. H. Breitner, and R. Rolfes, "Influence of structural design variations on economic viability of offshore wind turbines: An interdisciplinary analysis," *Renew. Energy*, vol. 145, no. 3, pp. 1348–1360, 2020, doi: 10.1016/j.renene.2019.06.113.

[26] L. Ziegler, E. Gonzalez, T. Rubert, U. Smolka, and J. J. Melero, "Lifetime extension of onshore wind turbines: A review covering Germany, Spain, Denmark, and the UK," *Renew. Sustain. Energy Rev.*, vol. 82, pp. 1261–1271, 2018, doi: 10.1016/j.rser.2017.09.100.

[27] R. Nash, R. Nouri, and A. Vasel-Be-Hagh, "Wind turbine wake control strategies: A review and concept proposal," *Energy Convers. Manag.*, vol. 245, p. 114581, 2021, doi: 10.1016/j.enconman.2021.114581.

[28] L. E. Andersson, O. Anaya-Lara, J. O. Tande, K. O. Merz, and L. Imsland, "Wind farm control – Part I: A review on control system concepts and structures," *IET Renew. Power Gener.*, vol. 37, no. 11, p. 1703, 2021, doi: 10.1049/rpg2.12160.

[29] *Bestimmung der elektrischen Eigenschaften von Erzeugungseinheiten und -anlagen, Speicher sowie für deren Komponenten am Mittel-, Hoch-und Höchstspannungsnetz*, FGW-TR3, FGW e.V. Fördergesellschaft Windenergie und andere Dezentrale Energien.

[30] D. R. Houck, "Review of wake management techniques for wind turbines," *Wind Energy*, vol. 25, no. 2, pp. 195–220, 2022, doi: 10.1002/we.2668.

[31] J. A. Frederik, B. M. Doekemeijer, S. P. Mulders, and J.-W. Wingerden, "The helix approach: Using dynamic individual pitch control to enhance wake mixing in wind farms," *Wind Energy*, vol. 23, no. 8, pp. 1739–1751, 2020, doi: 10.1002/we.2513.

[32] J. Meyers, C. Bottasso, K. Dykes, *et al.*, "Wind farm flow control: prospects and challenges," *Wind Energ. Sci.*, vol. 7, no. 6, p. 2271–2306, 2022, doi: 10.5194/wes-7-2271-2022.

[33] P. M. O. Gebraad, "Data-driven wind plant control," *Delft University of Technology*, 2014, doi: 10.4233/uuid:5c37b2d7-c2da-4457-bff9-f6fd27fe8767.

[34] P. M. O. Gebraad, "Controlling Wind Turbines in Presence of Wake Interactions." Applied for by Siemens Gamesa Renewable Energy as [DK] on 6/8/2018. App. no. EP20180176703 20180608. Patent no. EP3578808 (A1). F03D7/04;F03D7/02. Priority no. EP20180176703 20180608, 2018.

[35] M. M. Pedersen and G. C. Larsen, "Integrated wind farm layout and control optimization," *Wind Energy Sci.*, vol. 5, no. 4, pp. 1551–1566, 2020, doi: 10.5194/wes-5-1551-2020.

[36] I. Eguinoa, T. Göçmen, P. B. Garcia-Rosa, *et al.*, "Wind farm flow control oriented to electricity markets and grid integration: Initial perspective analysis," *Adv. Control Appl.*, vol. 26, no. 2, p. 100381, 2021, doi: 10.1002/adc2.80.

[37] K. Kölle, T. Göçmen, P. B. Garcia-Rosa, *et al.*, "Towards integrated wind farm control: Interfacing farm flow and power plant controls," *Adv. Control Appl.*, vol. 4, no. 2, 2022, doi: 10.1002/adc2.105.

[38] J. Schreiber, D. Coimbra, and C. L. Bottasso, "Demonstration of AEP boosting by wind farm control at a commercial wind farm," 2023.

[39] B. M. Doekemeijer, D. van der Hoek, and J.-W. van Wingerden, "Closed-loop model-based wind farm control using FLORIS under time-varying inflow conditions," *Renew. Energy*, vol. 156, no. 4, pp. 719–730, 2020, doi: 10.1016/j.renene.2020.04.007.

[40] J. Schreiber, C. L. Bottasso, and M. Bertelè, "Field testing of a local wind inflow estimator and wake detector," *Wind Energy Sci.*, vol. 5, no. 3, pp. 867–884, 2020, doi: 10.5194/wes-5-867-2020.

[41] P. Fleming, J. King, K. Dykes, *et al.*, "Initial results from a field campaign of wake steering applied at a commercial wind farm – Part 1," *Wind Energy Sci.*, vol. 4, no. 2, pp. 273–285, 2019, doi: 10.5194/wes-4-273-2019.

[42] T. Hestermeyer, O. Oberschelp, and H. Giese, "Structured information processing for self-optimizing mechatronic systems," in *Proceedings of the First International Conference on Informatics in Control, Automation and Robotics*, Setúbal, Portugal, Aug. 2004–Aug. 2004, pp. 230–237.

[43] Thanasis Barlas, Qian Long, Abhinav Anand, Filippo Campagnolo, Kaushik Das, and Tuhfe Göçmen, "Open access integrated WFC platform on Simulink."

[44] J. Gausemeier, F. J. Rammig, W. Schäfer, and W. Sextro, *Dependability of Self-Optimizing Mechatronic Systems*. Berlin: Springer, 2014.

[45] G. Strube, "Modelling motivation and action control in cognitive systems," in *Mind Modeling*, U. Schmid, J. F. Krems, and F. Wysocki (eds.) Berlin: Pabst, 1998, pp. 111–130. [Online]. Available: http://cognition.iig.uni-freiburg.de/team/members/strube/ActionControl98.pdf

[46] N. Requate and T. Meyer, "Active control of the reliability of wind turbines," *IFAC-PapersOnLine*, vol. 53, no. 2, pp. 12789–12796, 2020, doi: 10.1016/j.ifacol.2020.12.1941.

[47] K. Kölle, T. Göçmen, I. Eguinoa, *et al.*, "FarmConners market showcase results: Wind farm flow control considering electricity prices," *Wind Energy Sci.*, vol. 7, no. 6, pp. 2181–2200, 2022, doi: 10.5194/wes-7-2181-2022.

[48] P. Loepelmann and B. Fischer, "Lifetime extension and opex reduction by adapting the operational strategy of wind farms," *J. Phys.: Conf. Ser.*, vol. 2257, no. 1, p. 12014, 2022, doi: 10.1088/1742-6596/2257/1/012014.

[49] E. Loth, C. Qin, J. G. Simpson, and K. Dykes, "Why we must move beyond LCOE for renewable energy design," *Adv. Appl. Energy*, vol. 8, no. 4, p. 100112, 2022, doi: 10.1016/j.adapen.2022.100112.

[50] M. James, K. Jannusch, K. Koenig, J. McGowan, J. Small, and E. Yahr III, "Using Non-Price Criteria in State Offshore Wind Solicitations to Advance Net Positive Biodiversity Goals," 2023. Available: https://www.vermontlaw.edu/sites/default/files/2023-06/iee-tnc_offshore-wind-report_20230606_1644.pdf (accessed: Dec. 18, 2023).

[51] N. Requate, T. Meyer, and R. Hofmann, "From wind conditions to operational strategy: optimal planning of wind turbine damage progression over its lifetime," 2023, doi: 10.5194/wes-8-1727-2023.

[52] S. Mozafari, K. Dykes, J. M. Rinker, and P. Veers, "Effects of finite sampling on fatigue damage estimation of wind turbine components: A statistical study," *Wind Eng.*, vol. 10, no. 4, 0309524X2311638, 2023, doi: 10.1177/0309524X231163825.

[53] N. Dimitrov, M. C. Kelly, A. Vignaroli, and J. Berg, "From wind to loads: Wind turbine site-specific load estimation with surrogate models trained on high-fidelity load databases," *Wind Energy Sci.*, vol. 3, no. 2, pp. 767–790, 2018, doi: 10.5194/wes-3-767-2018.

[54] P. Thomas, M. Leimeister, A. Wegner, and M. L. Huhn, "Load calculation and load validation," in *Wenske (Hg.) 2022 – Wind Turbine System Design*.

[55] N. Requate and T. Meyer, "Optimal lifetime operational planning of wind turbine fatigue reliability under stochastic wind uncertainty," in ESREL 2023; Accepted for Publication.

[56] E. Bossanyi, R. Ruisi, G. C. Larsen, and M. M. Pedersen, "Axial induction control design for a field test at Lillgrund wind farm," *J. Phys.: Conf. Ser.*, vol. 2265, no. 4, p. 42032, 2022, doi: 10.1088/1742-6596/2265/4/042032.

[57] E. Simley, P. Fleming, N. Girard, L. Alloin, E. Godefroy, and T. Duc, "Results from a wake-steering experiment at a commercial wind plant: Investigating the wind speed dependence of wake-steering performance," *Wind Energy Sci.*, vol. 6, no. 6, pp. 1427–1453, 2021, doi: 10.5194/wes-6-1427-2021.

# Chapter 10
# Integration of local energy systems

*Christoph Kaufmann*[1,2], *Carlos Cateriano Yáñez*[1,3], *Aline Luxa*[1,4], *Marina Nascimento Souza*[1] *and Georg Pangalos*[1]

## 10.1 Introduction

Up until two decades ago, electrical energy was mostly generated centrally in large power plants, such as nuclear-fueled or coal-fired, in the largest power networks in the world. Their power was distributed from high-voltage transmission grids down to the low-voltage distribution networks to the consumers. This predominant practice changed with the emergence of distributed energy resources (DERs) of mostly solar and wind-powered resources. DERs are often integrated at the medium-voltage or low-voltage level depending on their size. Recent developments show that the power flows locally from generation to load site, for example, from wind turbines to directly coupled consumers, and in the near future, electrolyzers [1], or photovoltaic plants connected to electric vehicle charging stations [2]. This ongoing trend of decentralization of power networks, and the energy system as a whole, led to the form of local energy systems. This chapter explains the concept of local energy systems (LESs) and addresses the following questions:

- What are future requirements for the integration of wind turbines in low-inertia networks, and how can LES help? (Section 10.2)
- What is an LES, how is it defined, and what are current examples? (Section 10.3)
- How can an LES be designed and sized? (Section 10.4)
- What are suitable control strategies for an LES? (Section 10.5)
- Which stability issues can be expected in LES, and how can they be analyzed? (Section 10.6)

Finally, this chapter is concluded with an outlook to future developments in Section 10.7

[1]Fraunhofer Institute for Wind Energy Systems IWES, Hamburg, Germany
[2]Universitat Politècnica de Catalunya, Department of Electrical Engineering, Barcelona, Spain
[3]Universitat Politècnica de València, Instituto Universitario de Automática e Informática Industrial, Valencia, Spain
[4]Leibniz University Hannover, Faculty of Electrical Engineering and Computer Science, Hannover, Germany

## 10.2 Low-inertia systems and future requirements for the integration of wind turbines

Modern power systems are undergoing a fundamental shift. The generation landscape changes from predominantly fossil-fueled synchronous generators (SGs) toward renewable energy-powered converter-interconnected generators (CIGs). This change requires new concepts in the planning and the operation of power systems [3–5].

The first wind turbines, included in the collective term CIGs, were integrated based on the assumption that the power network is behaving like a stiff voltage source. In this so-called grid-following mode, the CIGs synchronize with the grid voltage, often by means of a phase-locked loop, to inject current, thus acting as a current source. Up until now, most CIGs operate in grid-following mode [6]. However, the assumption about a stiff grid does not necessarily hold anymore depending on the instantaneous ratio between SGs and CIGs, the network topology, size, and interconnectivity [5,7]. To adjust to the behavior of a variable voltage source, additional grid-supporting functionalities provided by CIGs were developed over the years and are now partly required by grid codes, for example, reactive power provision, or fault-ride-through capabilities [8].

However, the shift toward converter-dominated networks that are low in rotational inertia leads to significant changes of the network dynamics, which were greatly studied in the past decade, e.g., [3,4,9,10]. The loss of inertia from SGs affects the frequency stability leading to larger rates of change of frequency and frequency nadirs, and generally faster frequency variations [11,12]. As a result, networks such as Great Britain, or that of Ireland and Northern Ireland, changed the required capabilities of connected units to withstand larger ROCOFs [12]. That is partly owed to the fact that CIGs were not required to contribute to these system dynamics, which, for instance, started a debate about providing virtual inertia by CIG. Thus, future requirements for integrating wind turbines are expected to increase significantly as the penetration level of CIGs rises and the system becomes low in rotational inertia [7]. Before outlining some future requirements, the differences between SGs and CIGs are highlighted.

The inherent dynamics of SGs and converters are fundamentally different. The impact on the power system dynamics by the loss of SGs are summarized as [5]:

- the loss of the rotational kinetic energy of SGs where the inertia acts against disturbances;
- loss of stable and robust nonlinear synchronization mechanism that inherently results from the rotational generation;
- the loss of robust voltage and frequency control of SGs including the provision of stabilizing ancillary services.

As opposed to SG, converters have almost no energy storage, do not inherently self-synchronize, and have a fragile voltage and frequency control. Combined with weather-dependent renewable energy sources, CIGs are variable and distributed in their generation in contrast to dispatchable SGs that are often centrally located. However, the control of CIG is very fast and flexible to achieve the desired dynamic behavior. However, owing to the use of power electronics, CIG typically has a significantly lower current overload capacity than SGs. Besides the converter, the dynamics of the CIG are determined by the power source itself [5].

We would like to make the reader aware that the provided review of the arising issues in power systems related to the transition to converter-dominated power network is based on the common viewpoint from the operation of the legacy power systems with only SGs. For example, as known from the operation of converters in microgrids [13], converters can form a stiff voltage, thus, load changes in converter-only networks may not lead to fast frequency changes if they do not have such a dynamic response programmed, e.g., a droop behavior. Therefore, the phenomena of *low-inertia* systems described here reflect the status quo of the transition period to converter-dominated networks. However, the operation of converter-based, or converter-dominated networks, opens the door to new possibilities to operate the system, where some aspects will be touched upon in Section 10.5.2.2, which might result in a more efficient way of operating the energy system than it is currently thought of.

The flexibility in designing the dynamic response of the converter and the issues that arise from the lack of SGs motivated to develop, for example, a virtual synchronous machine (VSM) control that has been applied to 23 wind turbines in Scotland in 2019. The development of the VSM control follows from the simple logic to emulate the dynamic behavior of the well-studied SGs to compensate for its phase out. Despite the flexibility of the converter response, it should be noted that the grid-forming control strategy can translate to the generator-side of the wind turbine leading to additional mechanical stresses [14]. Later in Section 10.5.2, some grid-forming control algorithms are presented.

Consequently, the flexibility of the CIG's dynamic behavior allows for a number of different response types to stabilize the power system. Therefore, grid codes are changing requiring for new services from CIG, such as, fast frequency response (FFR), or black-start capabilities. An overview of the required adjustment of the grid code depending on the system size and the amount of CIG, is given in a recent study by the International Renewable Energy Agency (IRENA) depicted in Table 10.1 [7].

For the integration of CIG at large more flexibility and the provision of services, such as, reactive power provision for voltage control, will be required to ensure a stable operation of the power system. The number of requirements for the dynamic behavior will depend on the degree of interconnection and the penetration level of CIG. For example, in smaller systems, such as island grids, the generators are likely to must withstand a wider range of voltage and frequency fluctuations than the European synchronous grid [7]. However, there is still no consensus amongst transmission system operators (TSOs) on what grid code requirements of CIG in grid-forming mode might be. This also interrelates to the lack of widespread grid-forming solutions from the manufacturers other than in small scale [7]. As a result, for example, the Irish TSO EirGrid currently limits the instantaneous penetration level of CIG to 75 [15]. EirGrid's strategy is to minimize the output of SG to keep them connected to secure the system operation [7]. In some areas of the

*Table 10.1 Suggestions by IRENA for grid code adjustment depending on the penetration level of CIG and the network size [7]*

| % VRE | Small system size | Medium system size | Large system size |
|---|---|---|---|
| High | • Storage facility integration<br>• Full frequency and voltage control capabilities<br>• Grid-forming and black-start services from storage | • Grid-forming inverters for stability issues in regions without hydropower<br>• Frequency control and active power control performance suitable for AGC integration required | • Grid-forming services and black-start functionality to be provided by new assets connected to high-voltage levels (e.g., CIG power plants or large-scale storage) |
| Medium | • LFSM-U and active power control performance suitable for AGC integration<br>• Requirements for enabling technologies (e.g., storage) | • FRT capability and active power control requirement extends to new low-voltage connections<br>• Requirements for enabling technologies (e.g., storage) | • FRT capability and active power controllability required for low-voltage connections<br>• New requirements for larger facilities<br>• Requirements for enabling technologies (e.g., storage) |
| Low | • Assets must withstand a wider frequency and voltage range<br>• Need for controllability and FRT capabilities (including small DER) | • Requirements must align with the state of the art, standards, and rules of the CIG industry<br>• Power quality, protection, suitable frequency operating ranges, and LFSM-O must apply to all newly connected CIG facilities and enabling technologies<br>• For medium-voltage connections, requirements for power remote control and FRT are needed but are not yet crucial for low-voltage connections | |

AGC: Automatic generation control; FRT: Fault ride through; LFSM-O: Limited frequency sensitive mode for overfrequency; LFSM-U: Limited frequency sensitive mode for underfrequency.

European synchronous grid, distribution system operators (DSOs) are limiting the amount of CIG due to weak grid conditions, restricting any power injection of from CIGs, which ultimately limits the increase of CIGs connected to system. Therefore, future requirements for the integration of wind turbines will require for more system services.

An aggregation of different CIGs, e.g., wind turbines co-located with battery energy storage systems, allow to provide a combined response, which has the potential to lower the requirements on a single component. Already, small-scale systems, such as microgrids, are known to be capable to support the operation of the power systems, especially for services such as black start CIGs [7,16].

Therefore, integrating CIG as part of a so-called LES, where the LES as a whole provides the system service, is a promising alternative for low-inertia systems as it may have less demanding future requirements.

The next section introduces and defines the concept of LES and how it relates to the ongoing energy transition.

## 10.3 Definition of local energy systems

Power systems historically emerged from local energy systems. Thomas Edison built the first complete DC electric power system in 1882, which consisted of a generator, cable, fuse, meter, and loads [17]. While looking for a suitable method to interconnect LESs in the 1890s, the advantage of AC systems overweight DC networks mainly because in AC networks voltage levels are easily transformable and AC generators and motors are simpler [17]. Over time, more and more LESs were built and eventually connected to the large AC networks we know today. As mentioned in the previous section, the legacy power system is dominated by large power-generating units, like conventional coal or nuclear power plants. The growing number of distributed renewable energy sources connected to the grid naturally breaks with the centralized structure. This section deals with the definition and future challenges of the integration of LES answering the following questions:

1. How is an LES defined?
2. What is the difference between an LES and a microgrid?
3. What are the potential challenges of LES integration?

Contrary to the centralized legacy power system emerging from a top-down and monopolistic approach, in a decentralized system, energy sources can be connected locally to all voltage levels [18]. Here, various loads, such as electrolyzers, batteries and heat pumps, can be used to integrate renewable energy (RE) production to different sectors (sector coupling). In this scenario, a consumer uses the energy produced by itself (a prosumer), having then a key role in the energy transition, as its investment decisions, openness to new technologies and sector coupling solutions, as well as social acceptance, will play a large role in the design and operation of an LES.

The name LES conveys through the word "local" instantly the concept of an energy system with a focal point in a geographic location. As to energy, it can manifest in various forms, such as chemical, electrical, or thermal. A system can be defined as a network of interconnected elements, ideally working together in harmony. Despite these fundamental descriptions, there can be varying interpretations regarding the precise definition of an LES within the context of an energy grid and the integration of renewable energy sources.

In the field of energy transition, there can be very different ideas about the actual definition of an LES in relation to the energetic grid and integration of RE. These go beyond the basic characterization related to geographical locality. Definitions can then include the system goal, interfaces, boundaries, dynamics, energy types, interaction rules and much more. Therefore, to create a uniform understanding of LES, in this section several descriptions and definitions are gathered from literature and interpreted. Some of those might not use the term LES itself, but still address the phenomenon.

The Agora's study [18] on European energy transition identifies ten mega-trends that will shape the energy system until 2030, including decentralization. We could assume the term "decentralized" as an analogy to "local," but is rather used as a contrary to a central and conventional overall top-down energy distribution system. According to this study, it is worth noting that bottom-up energy streams from lower to higher voltage levels are also possible. This means that energy can flow from smaller, decentralized sources, such as LES, to larger, centralized ones. Additionally, the study recognizes the existence of prosumers. Hence, we understand a decentralized energy system in the sense of an overall network structure and as a compound, which includes LES. However, it is important to note that while this study emphasizes the significance of an LES and some of its characteristics, it falls short of providing a comprehensive definition of an LES.

The trend of decentralization due to the energy transition is also pointed out in the study conducted by [19]. It suggests that the surge in decentralization can be attributed to the availability of local energy resources or the existence of local energy demands. In simple terms, an LES can be defined as the opposite of centralized production. Additionally, it also encompasses both electric and thermal forms of energy, highlighting the advantages provided by LES of optimizing energy utilization.

This description, however, is quite general. A definition of LES is given in an earlier publication, namely in the EU Directive 2009/72/EC. There it was named distributed generation (DG). The directive defines DG as generation plants that are connected to the distribution system, which comprises the high-voltage, medium-voltage, and low-voltage networks, but not the extra high and high voltage transmission system, so defining in a way the boundaries of the system. The author acknowledges the variations in the definitions of LES (or DG) but emphasizes the consensus that LES units are low-scale, connected to the distribution grid, and operated by prosumers. Additionally, it states that the asset owner is usually a relatively small actor in the electricity market. However, this definition does not include any mention of LES's control strategies concerning its overall efficiency and its contribution to the grid (regarding grid stabilization).

In [20], decentralization is also recognized as a major trend of the transition to a renewable power system. It argues that a major driving force of decentralization is the belief that only communal and small-scale prosumers can ensure sustainability. Additionally, the decentralization trend can be driven by opposition to the construction of new high-voltage lines, leading to local production and consumption of energy. The author however disagrees and also considers large-scale profit-oriented energy providers and big companies in a decentralized scenario. In addition, it is pointed out how decentralization strategies are not necessarily centered on energetic efficiency, but on cost optimization. The author does attempt to define a decentralized system in terms of energy type, control and power levels, focusing only in the geographical distribution, and so providing a definition of system boundaries that is geographically based.

The report [21] also acknowledges the decentralization trend, addressing specifically smart local energy system (SLES). However, it does not deliver a single and precise definition of SLES, but rather a framework, to help stakeholder explore how an LES delivers benefit to specific cases. It does focus on defining the meaning of "local" and "smart" in the context of an LES, as well as how to define system's boundaries. The report defines that the system should be composed of multiple energetic vectors, and should cover production, conversion, transmission, storage, distribution and consumption. It should include political, economic and social dimensions in its design, as well as consider local regulations and laws.

Regarding the "smart" label of the SLES, the report requires that the system contains communication technologies that enable data gather and usage in real time, the ability to self-regulate and to learn system dynamics (some degree of machine learning or artificial intelligence embedded in the energy system allows it to regulate itself in accordance with wider dynamics and user-set preferences). This definition gives us more precise requirements regarding control strategies. In addition, it also requires that the system is able to make smart decisions, which also falls into the category of smart modeling and control.

Regarding the "local" characteristics of an SLES, the authors address not only the geographic position, but also stakeholders, decision-making and asset ownership process from a decentralized perspective. It states that the local community and stakeholders should have an impact in the decisions on design and operation of SLES. It also states that local decision-making should improve efficiency, flexibility and sustainability and that local asset ownership increases control over its operation. Local ownership helps to reinforce the two previous points, as it enables local communities and decision-making.

The authors also give several options for defining the boundaries of an SLES, according to geographic localization (boundaries are on a map), generation resource (boundaries according to proximity to generation), network infrastructure (boundaries are defined by network segment) and social (according to the community that benefits from the system). This report is valuable, in the depth it goes to, in addressing and characterizing an LES. However, once again the literature does not provide a universal definition.

In summary, an LES is described in existing literature as a decentralized energy system that can encompass energy production, storage, transformation and transportation. These systems cater to small to medium-sized energy consumers for both domestic and commercial purposes, as well as mobility needs. Overall in the sources cited in this chapter, the terms LES and SLES are characterized, but no clear and universal definition is given. This highlights the gap in the literature for a clear definition of LES that goes beyond the basic intuitive notion of the opposite of a central power plant. A definition stating its purpose and boundaries, as well as considering voltage level and control.

As the decentralization trend might drive a reader naturally to the well-established concept of microgrids. Before introducing a definition of LES, it is important to point out its similarities and differences to microgrids. A microgrid is a local energy grid with control capability, which means it can disconnect from the traditional grid and operate autonomously. It is worth considering how it fits to and how it differs from an LES. The basic difference between the two concepts is the inclusion of multiple energy vectors in LES, which emphasizes the aspect of sector coupling and increasing energy efficiency over all forms of energy.

**Definition 1**. *A microgrid is defined as a group of distributed energy resources (DERs), including renewable energy sources (RES) and energy storage systems (ESS), plus loads that operate locally as a single controllable entity* [16].

A definition of an LES is necessary to assist policymakers, the industry, and private investors to understand how local systems bring forward the energy transition. In this context and for providing a definition, it is easier to first focus on LES consisting of electrical energy only, which is usually referred to as microgrid.

Building from the concept of microgrid:

**Definition 2**. *An LES is defined as a group of DERs, including RES and ESS, plus loads that operate locally as a single controllable entity. Additionally, the control strategy should serve efficient interaction with other LES and service to the overall grid stability. Nevertheless, an LES should be flexible to be operated as well as an island (that is, as a microgrid), in case this operational mode suits the system's requirements. An LES can exist at every power level (low, medium and high voltage). It can include chemical, electrical and thermal forms of energy.*

A few of the sources stated above address the sociological aspects of LES. Nevertheless, this chapter should not include social aspects due to the limited scope. We still want to point out the importance of the future role of communities for the energy system transition. In this section, we are seeking a more technical definition of an LES.

According to Definition 2, Figure 10.1 shows a decentralized energy system (DES), which is constructed of LES. Every LES has an individual controller, with an overlaying control layer for the optimal orchestration of all physical entities in the network. It can be seen that different energy flows can be part of a single LES, which can also leave the system boundaries. In Figure 10.1, in LES II there is heat supply of an industrial consumer to domestic consumers and also a chemical energy stream leaving the subsystem. The chemical energy can, e.g., be hydrogen produced by an electrolyzer or further synthesized hydrogen based products, like methane, ethanol or ammonia. This process is an example of sector coupling, since electric energy leaves the energy sector into other applications like transport or the chemical industry. It is not mandatory for an LES to include several energy types. The term LES should be a comprehensive category, which is also including concepts like virtual power plants (VPPs) [22].

As microgrids are a more commonly used term to describe the concept, it is used here to evidence growing examples on energy decentralization. There are

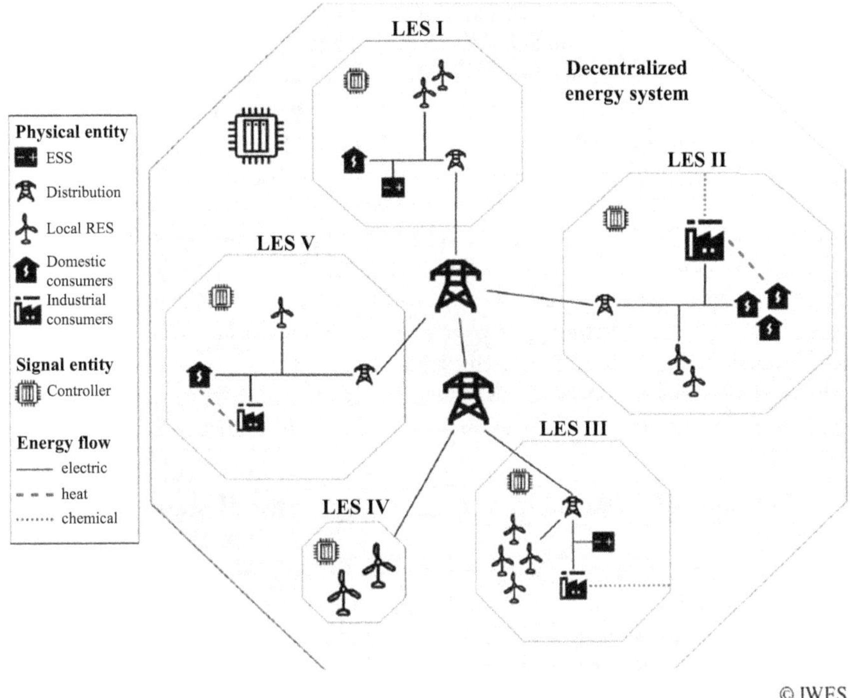


*Figure 10.1 Depiction of a decentralized energy system*

several examples of large-scale microgrid projects, such as the 1 MW microgrid in Warstein-Belecke, Germany, which is run by AEG Power Solutions [23]. As well as the energetic self-sufficient village of Feldheim in Germany [24]. Offshore wind energy production is a significant focus on reaching renewable growth targets in Europe. This is seen in the initiatives of the Danish government on investing in offshore energy islands, with the aim of using green energy to produce hydrogen [25]. In Germany, the "H2MARE" project is also exploring offshore energy and hydrogen production [1]. Offshore production is also an example of LES, according to Definition 2. As this transition occurs, it is accompanied by several technical and regulatory challenges. The paper from [26] focuses on the decentralization development of electricity production, focusing on the energetic supply of communities. This study does not however approach the technical challenges of LES integration. It shows that when communities increase their energy autonomy, they contribute less to total network costs. As the total cost of grid fees and subsidies is borne by all consumers, a significant additional cost for consumers can occur, if many communities become energy autonomous. While a higher degree of energy autonomy has clear microeconomic benefits for consumers, the net macroeconomic effect could be detrimental. Decentralized energy systems therefore pose a major challenge for energy policy and research.

The work of [27] states the potential for an increase in the efficiency of energy systems, if they get decentralized and if they include heat and biogas networks, expecting its rapid development. This shift will require substantial regulatory and policy reform, including heat network as well as appropriate incentives and a compensation model for energy services. This aligns with the outcome from [28], which concludes that challenges are not technical, but regulatory-related.

In [29], the barriers to the dissemination of decentralized renewable energy systems (DRESs) are classified in technical, economic, institutional, socio-cultural and environmental challenges. The analysis of [29] is focused on household applications and describes technical issues as poor design, lack of standardization and certification, and poor reliability. This misses the essential aspect of the control problem for power electronics on different voltage levels. Before addressing the control problem of LES, we need to size the system in question. The following sections will deal with the sizing of an LES, Section 10.4, and its control design Section 10.5.

## 10.4 Design of local energy systems by optimal sizing

Considerations regarding the components to be selected in an LES often come from the project description. A more general approach to defining the goal, e.g., the production of green hydrogen could lead to a reduced cost when an optimization over the lifetime of the LES is considered. Such an optimization can be performed using a model-based approach. Therefore, one interesting question in this context is the abstraction level of the model. When choosing a model that includes many phenomena accurately the model will be probably a nonlinear one. The resulting optimization problem will thus have little structure to exploit, and a general nonlinear solver needs to be used. When choosing a linear model some phenomena will only be approximated, but the optimization problem can be solved analytically, if the cost function is quadratic.

The question of optimal sizing is closely related to the question of optimal dispatch. Assuming the components of the LES are defined, and the sizes are to be defined. In the case of a simple LES the dispatch problem might be straight forward. As soon as more than one storage type is present (battery and hydrogen storage) or more than one consumer is to be supplied (electrolyzer and another electrical load) there arises a degree of freedom in how to operate the LES. In other words the dispatch problem needs to be optimized. And this optimization is basically part of the optimization of the component sizes. For example, in an LES with a rule-based dispatch control the dimension of the storage units might have to be bigger, since the dispatch might have a worse efficiency.

In sizing the components the approach of first defining a rule-based control for the dispatch and on this basis optimizing the component sizes is widely used. On the one hand, the approach of co-optimizing the dispatch and the sizes at the same time has the advantage of optimality in the sense of the cost function. On the other hand, when dealing with nonlinear models and non-quadratic cost functions the computational burden will be excessive.

Optimal sizing of LES with solar or wind generation and hydrogen energy storage, has been the subject of many scientific contributions, e.g. [30–32]. Here the goal is to the system's total cost by changing the sizes of the components while complying with a criterion of reliability. Six components are considered in the LES in [32], which are wind and solar power plants, lithium-ion batteries and a hydrogen storage system consisting of a hydrogen storage tank, an electrolyzer and a fuel cell. There is no grid connection but there is another electrical load. The optimization is performed by defining the sizes of the components such that the life cycle cost is minimized but ensuring that the loss of power supply (load shedding/reliability) and extra power (surplus energy/efficiency) are kept in acceptable bounds. The LES in [31] is similar but without batteries. Also, the objective is to minimize the total cost. The constraint here is the loss of power supply probability (LPSP), which is a measure on reliability. The problem is optimized for four maximal thresholds for the LPSP index, i.e., 0%, 0.3%, 1%, and 2%. Less reliability, thus bigger LPSP are found to be less expensive. This is intuitively the trade-off between system reliability and total cost.

To illustrate the optimal sizing problem an LES is introduced in the following as given in [33]. Here a system without grid connection is considered, but the concept could also be adapted to a grid-connected LES. The system under investigation is depicted in Figure 10.2 and consists of the components

- Wind turbines (WT)
- Photovoltaic arrays (PV)
- a Battery energy storage system (BESS)
- a Hydrogen energy storage system (HESS) consisting of
  - an Electrolyzer (Ely)
  - Hydrogen Storage Tanks (HST) and
  - a Fuel Cell (FC).

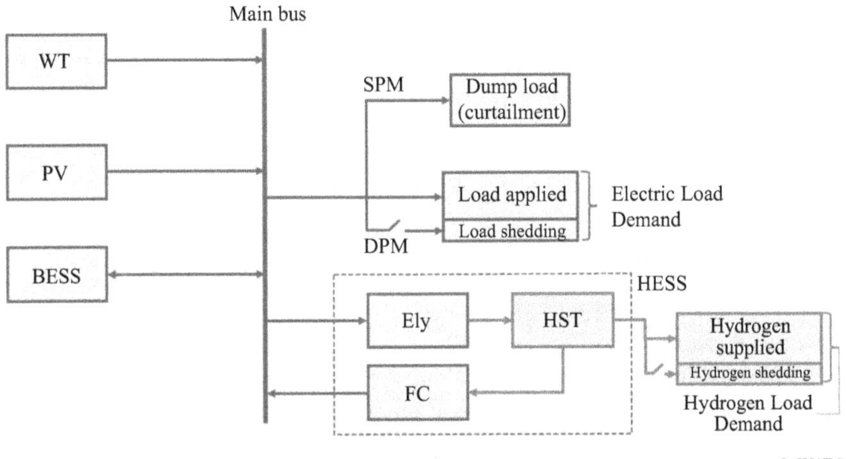


*Figure 10.2 Studied MG with six energy components [33]*

Since there is no grid connection some mechanism is needed to maintain the power balance if the components are saturated. Therefore, if too much power is produced and the storage units are at their limit, the power is curtailed by using a dump load (surplus mode (SPM)). If too little power is produced and not enough power can be drawn from the storage units, load shedding is used, i.e., the load is not supplied with power in sufficient quantity (deficient power mode (DPM)). Hydrogen shedding is used analogously when the hydrogen storage level reaches the minimum and can no longer provide the hydrogen demand. The model of the LES assumes efficiencies for its devices that are constant for all operating conditions. Also, no degradation of the devices is considered, thus, e.g., the storage capacities remain constant over the lifetime. Another assumption is that the load demand and the renewable power generation profiles are taken from one representative year. In other words, they are assumed to be the same every year. These assumptions are common in the field [30,31].

The variables of this model related to the operation are the energy levels of each energy storage component, called the state of charge for the battery energy storage system, and the level of hydrogen for the hydrogen storage tank, as well as the time series of the power to the controllable units, i.e., the energy storage system components.

The variables related to the sizing problem are the sizes of the system components given as the vector

$$\mathbf{s} = [P^{\mathrm{r}}_{\mathrm{WT}},\ P^{\mathrm{r}}_{\mathrm{PV}},\ E^{\mathrm{r}}_{\mathrm{hst}},\ P^{\mathrm{r}}_{\mathrm{FC}},\ P^{\mathrm{r}}_{\mathrm{Ely}},\ E^{\mathrm{r}}_{\mathrm{BESS}}], \tag{10.1}$$

where $P^{\mathrm{r}}_{\mathrm{WT}}$, $P^{\mathrm{r}}_{\mathrm{PV}}$, $P^{\mathrm{r}}_{\mathrm{FC}}$, $P^{\mathrm{r}}_{\mathrm{Ely}}$ are the rated power of the wind turbines, photovoltaic system, fuel cell and electrolyzer, respectively, and $E^{\mathrm{r}}_{\mathrm{HST}}$, $E^{\mathrm{r}}_{\mathrm{BESS}}$ are the rated capacity of the hydrogen storage tank and battery energy storage system, respectively.

The energy balance of the system is given as

$$P_{\mathrm{load}}(t) - P_{\mathrm{gen}}(t) = P_{\mathrm{ESS}}(t) + P_{\mathrm{LS}}(t) - P_{\mathrm{curt}}(t), \tag{10.2}$$

where $P_{\mathrm{load}}$ is the electrical demand power, $P_{\mathrm{LS}}$ is the load shedding power, $P_{\mathrm{curt}}$ is the curtailed power, $P_{\mathrm{ESS}}$ is the power from the energy storage system, and $P_{\mathrm{gen}}$ is the power from renewable generation. The two latter are given by

$$P_{\mathrm{ESS}}(t) = P^{\mathrm{dch}}_{\mathrm{BESS}}(t) - P^{\mathrm{ch}}_{\mathrm{BESS}}(t) + P_{\mathrm{FC}}(t) - P_{\mathrm{Ely}}(t), \tag{10.3}$$

$$P_{\mathrm{gen}}(t) = P^{\mathrm{r}}_{\mathrm{WT}} P^{\mathrm{ref}}_{\mathrm{WT}}(t) + P^{\mathrm{r}}_{\mathrm{PV}} P^{\mathrm{ref}}_{\mathrm{PV}}(t), \tag{10.4}$$

where $P^{\mathrm{ch}}_{\mathrm{BESS}}$ and $P^{\mathrm{dch}}_{\mathrm{BESS}}$ are the battery charging and discharging power, respectively, $P_{\mathrm{FC}}$ is the fuel cell power output, $P_{\mathrm{Ely}}$ is the electrolyzer power consumption, $P^{\mathrm{ref}}_{\mathrm{WT}}$ and $P^{\mathrm{ref}}_{\mathrm{PV}}$ are reference power profiles for 1 kW wind and photovoltaic power plants. Note that the power profiles under investigation are multiplied by the rated power, which could be in the MW or GW range. The levels of the two energy storage systems are determined by the differential equations

$$\frac{dx_1}{dt} = \frac{1}{E^{\mathrm{r}}_{\mathrm{BESS}}}\left(\eta_{\mathrm{b}} P^{\mathrm{ch}}_{\mathrm{BESS}} - \frac{1}{\eta_{\mathrm{b}}} P^{\mathrm{dch}}_{\mathrm{BESS}}\right) - \sigma_{\mathrm{b}} x_1, \tag{10.5}$$

$$\frac{dx_2}{dt} = \frac{1}{E^{\mathrm{r}}_{\mathrm{hst}}}\left(\eta_{\mathrm{Ely}}P_{\mathrm{Ely}} - \frac{1}{\eta_{\mathrm{FC}}}P_{\mathrm{FC}} - (P_{\mathrm{hl}} - P_{\mathrm{HS}})\right), \tag{10.6}$$

where $x_1$ and $x_2$ are the state of charge of the battery energy storage system and the level of hydrogen of the hydrogen storage tank. The hydrogen load power is $P_{\mathrm{hl}}$, the hydrogen shedding is $P_{\mathrm{HS}}$, the efficiencies of the battery energy storage system, the electrolyzer, and the fuel cell are $\eta_{\mathrm{b}}$, $\eta_{\mathrm{Ely}}$ and $\eta_{\mathrm{FC}}$, respectively.

The charging and discharging efficiencies of the battery energy storage system were assumed to be the same [34], and a self-discharge rate ($\sigma_{\mathrm{b}}$) was considered. The hydrogen storage tank is assumed to have an efficiency of 1.

Additionally, the operation variables are constrained by the set of operational conditions

$$\begin{aligned} x_{1,\min} \le x_1(t) &\le x_{1,\max} \\ -C_{\mathrm{r}}^{\max}E^{\mathrm{r}}_{\mathrm{BESS}} \le P_{\mathrm{BESS}}(t) &\le C_{\mathrm{r}}^{\max}E^{\mathrm{r}}_{\mathrm{BESS}} \\ x_{2,\min} \le x_2(t) &\le x_{2,\max} \\ 0 \le P_{\mathrm{Ely}}(t) &\le P^{\mathrm{r}}_{\mathrm{Ely}} \\ 0 \le P_{\mathrm{FC}}(t) &\le P^{\mathrm{r}}_{\mathrm{FC}}, \end{aligned} \tag{10.7}$$

where $P_{\mathrm{BESS}} = P^{\mathrm{ch}}_{\mathrm{BESS}} - P^{\mathrm{dch}}_{\mathrm{BESS}}$ is the battery power. The nominal C-rate is assumed to be $C_{\mathrm{r}}^{\max} = 1$. The lower limit of the electrolyzer and fuel cell power is assumed to be zero.

This defines a model of an LES, which can include a wide set of systems by setting the specific size of the power or energy to zero in the model if the component is not included in the system. In order to define the optimization problem, design constraints need to be set up. Choosing a system to be perfectly reliable will lead to huge storage units and oversized generation units. Thus, this system will be very expensive [31]. A generalization of the constraints can mostly be brought down to a concept of reliability opposing a concept of cost. Here the concepts of

- Loss of power supply probability

$$LPSP = \frac{\sum P_{\mathrm{LS}}}{\sum P_{\mathrm{load}}} \tag{10.8}$$

- Loss of hydrogen supply probability

$$LHSP = \frac{\sum P_{\mathrm{HS}}}{\sum P_{\mathrm{hl}}} \tag{10.9}$$

- Relative excess power generated

$$REPG = \frac{\sum P_{\mathrm{curt}}}{\sum P_{\mathrm{load}}} \tag{10.10}$$

are used. Using (10.8) and (10.9), the reliability index is defined as

$$RI = 1 - \frac{1}{2}(LPSP + LHSP). \tag{10.11}$$

This method was adapted from [35] when there are several load demands. Note that the relative excess power generated is growing with the size of the generation units. Therefore, restricting the excess power will also reduce the cost, since bigger plants are no longer allowed.

The goal of the optimization is to minimize the cost of the LES. Thus, a cost model is needed and will be given in the following.

The CAPEX*-OPEX† life cycle cost model given in [30,31] is used. This model includes replacement costs $RC$ over a project lifetime of 25 years. Annual OPEX is converted to total lifetime cost $C$ using the present worth annuity $W$ and an interest rate $i$ given by

$$\begin{aligned} C(\mathbf{s}) &= \sum_{j=1}^{6} \left( CAPEX_j + W \cdot OPEX_j + K_j \cdot RC_j \right) \\ &= \sum_{j=1}^{6} \left( C_j(s_j) \cdot s_j + W \cdot O_j(s_j) \cdot s_j + K_j \cdot RC_j \right), \end{aligned} \tag{10.12}$$

with

$$W = \frac{(1+i)^n - 1}{i(1+i)^n}, \tag{10.13}$$

$$K_j = \sum_{k=1}^{Y_j} \frac{1}{(1+i)^{kL_j}}, \tag{10.14}$$

$$Y_j = \left\lceil \frac{n}{L_j} \right\rceil. \tag{10.15}$$

Following an economy of scale effect, the specific costs were modeled by the logarithmic-based function

$$C_j(x_j) = \frac{C_j^{\text{ref1}}}{\left(1 + \log\left(1 + A\left(x_j - x_j^{\text{ref1}}\right)\right)\right)^y}, \tag{10.16}$$

where $y$ is a factor controlling the strength of the scaling effect, and $A$ is chosen such that

$$C_j\left(x_j^{\text{ref2}}\right) = C_j^{\text{ref2}}. \tag{10.17}$$

The specific costs of reference for each device are $C_j^{\text{ref1}}$ and $C_j^{\text{ref2}}$. For the component's lifetimes, assumptions are given in Table 10.2.

*Capital expenditure.

†Operating expense.

As mentioned in the beginning the optimal sizing problem cannot be solved without solving the operation problem. If the operation is suboptimal, the sizing problem might be suboptimal in case the operation is improved. Then the system could be oversized. Thus, both optimization problems should be solved at the same time. This is called the co-optimization approach. An example is given in [36]. However, since this method is computationally expensive, the approach of a rule-based dispatch is used in the following.

The dispatch is based on an "electricity first" rule-based strategy, which prioritizes the electrical load, as shown in Figure 10.3.

*Table 10.2 Model parameters [33]*

| Component | Parameters |
|---|---|
| WT | $L_{WT} > 25$ years, $O_{WT} = 35$ €/kW/year. |
| PV | $L_{PV} > 25$ years, $O_{PV} = 20$ €/kWp/year. |
| BESS | $x_{1,\min} = 0.2$, $x_{1,\max} = 0.9$, $\eta_b^{ch} = \eta_b^{dch} = 93\%$, $\sigma_b = 0.1\%$/day $C_r^{max} = 1\ h^{-1}$, $L_{BESS} = 15$ years, $O_{BESS} = 1\%$ of $C_{BESS}$/year. |
| HST | $x_{2,\min} = 0.05$, $x_{2,\max} = 0.99$, $\eta_{HST} = 100\%$, $L_{HST} = 15$ years, $O_{HST} = 2\%$ of $C_{HST}$/year. |
| Ely | $\eta_{Ely} = 70\%$, $P_{Ely}^{min} = 0$, $L_{Ely} = 15$ years, $O_{Ely} = 2\%$ of $C_{Ely}$/year. |
| FC | $\eta_{FC} = 50\%$, $P_{FC}^{min} = 0$, $L_{FC} = 15$ years, $O_{FC} = 2\%$ of $C_{FC}$/year. |

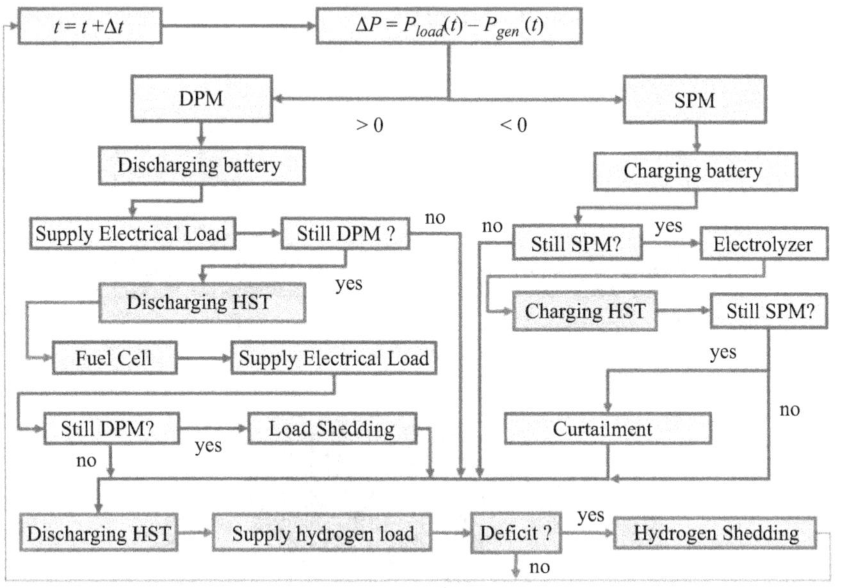




*Figure 10.3 Electricity first strategy flowchart [33]*

The minimization problem that is solving the sizing problem is given as

$$\min_{\mathbf{s}\in\mathcal{D}} (C(\mathbf{s}) + P(\mathbf{s})), \text{subject to} \begin{cases} RI_{\min} - RI \leq 0 \\ REPG - REPG_{\max} \leq 0 \\ x_1(0) - x_1(N) \leq 0 \\ x_2(0) - x_2(N) \leq 0, \end{cases} \tag{10.18}$$

where $N$ is the final time of the simulation ($N = 8760$), and $x_1(0)$ and $x_2(0)$ are the initial state of charge of the battery energy storage system and the level of hydrogen of the hydrogen storage tank, respectively. The initial values $x_1(0)$ and $x_2(0)$ were both set to $x_1(0) = x_2(0) = 0.5$. The search space $\mathcal{D}$ is defined by the upper and lower limits of the component sizes given by the application.

To enforce the design constraints, the variable penalty factor method was used

$$P = K\sum_{i=1}^{4} \max(D_i, 0), \tag{10.19}$$

where $K = 10^{12}$, and $D_i$ is the left-hand side of the four inequality constraints in (10.18). Thus, the constant $K$ is added to the objective function if a constraint is violated. The variable penalty $P$ was found to give better results than a constant penalty. This is not surprising because of the smoothness of the term opposed to a step with a constant penalty. The optimization problem was solved using a particle swarm optimization. The parameters were set as recommended in [37].

The example from [33] will illustrate the procedure. For an LES with

- an electrical load (Figure 10.4(a)) of 237 kWh/day, with a residential profile,
- a hydrogen load profile (Figure 10.4(b)) of 81.9 kWh/day,

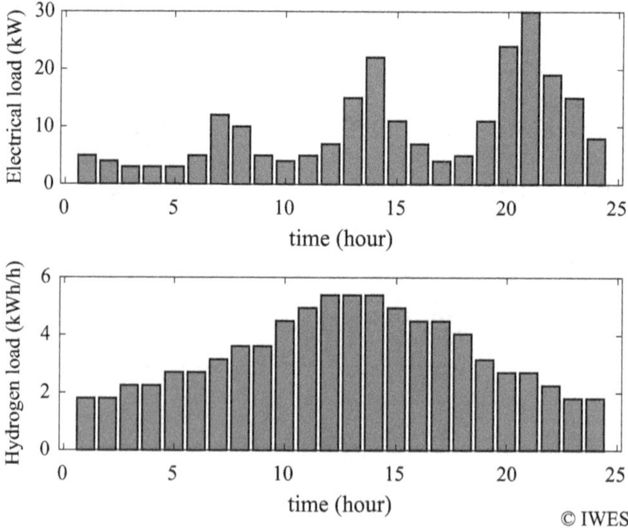


*Figure 10.4 Electrical and hydrogen load demand profiles*

- yearly renewable power generation reference profiles obtained on the platform renewable.ninjas, for the region of Hamburg

the sizes of the components are to be optimized.
The parameters of the components are given in Table 10.2.
Using the design constraints

$$RI_{\min} = 99\%,$$

$$REPG_{\max} = 60\%,$$

the minimization problem (10.18) was solved with a PSO algorithm. The algorithm was run with 80 particles and 250 iterations. The optimization history is shown in Figure 10.5. The PSO converges toward the best solution. The best cost, found at the end of the process, is a cost of $C = 682\,662$ € and called global best cost. The optimization variables (in kW or kWh) for this solution are

$$\mathbf{s}_{\text{best}} = [48.1,\ 49.3,\ 1521,\ 14.2,\ 0,\ 189.5].$$

Note that since the fuel cell is very expensive, the best solution is to set the size of the fuel cell to zero, $s_{\text{best}}(5) = 0$. Thus, the LES consists of a wind and a photovoltaic power plant, a battery and a hydrogen energy storage system. The computed reliability index for this system is $RI = 99.08\%$, and the relative excess power generated index for surplus energy is $REPG = 56.43\%$.

Replacing the rule-based dispatch with an optimized one will influence the reliability and the excess power. To state the optimal dispatch problem a state-space model formulation was used. The states of the system are energy levels of the energy storage system $x_1$ and $x_2$, and the control inputs are the powers of the controllable units

$$[\mathbf{u}_1, \ldots, \mathbf{u}_7] = [P^{\text{ch}}_{\text{BESS}}, P^{\text{dch}}_{\text{BESS}}, P_{\text{Ely}}, P_{\text{FC}}, P_{\text{LS}}, P_{\text{curt}}, P_{\text{HS}}].$$

The parameters vector of the optimal dispatch problem is given as

$$\mathbf{d} = \left[\mathbf{x}_1,\ \mathbf{x}_2,\ \mathbf{u}_1, \ldots,\ \mathbf{u}_7\right]^{\text{T}}, \tag{10.20}$$

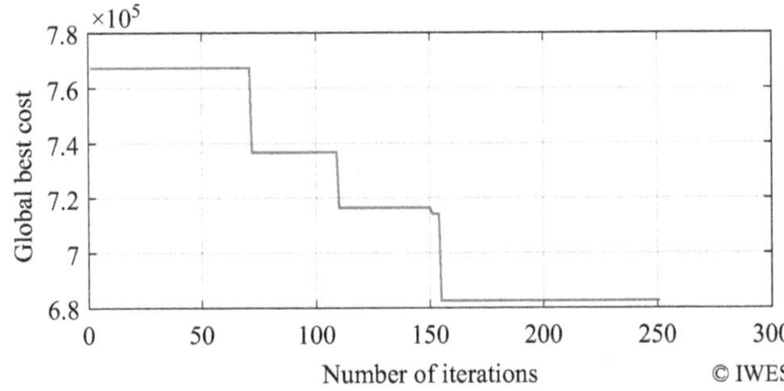


*Figure 10.5 PSO history [33]*

where

$$\mathbf{x}_i = [x_i(1), \ldots, x_i(N)], \tag{10.21}$$

$$\mathbf{u}_i = [u_i(1), \ldots, u_i(N)]. \tag{10.22}$$

The objective for optimal operation avoid load shedding and curtailment, and use as little power as possible. This can be defined as a function as

$$\mathbf{f}^{\mathrm{T}}\mathbf{d} = \sum_{t=1}^{N}(u_5(t) + u_7(t)) + \alpha\sum_{t=1}^{N} u_6(t)$$

$$+ \alpha\beta\sum_{t=1}^{N}(u_1(t) + u_2(t) + u_3(t) + u_4(t)), \tag{10.23}$$

with $\alpha$ and $\beta$ being weights to find a trade-off between the objectives.

Some constraints need to be met for the operation. The states at the end of the simulation must be greater or equal to the initial states

$$x_i(1) - x_i(N) \leq 0, \;\; i = \{1, 2\}. \tag{10.24}$$

Furthermore the states and the control inputs have lower and upper limits given as

$$x_{\mathrm{i,\,min}} \leq x_i \leq x_{\mathrm{i,\,max}}, \tag{10.25}$$

$$0 \leq u_i \leq u_{\mathrm{i,\,max}}. \tag{10.26}$$

For the sheddings and curtailment control inputs, the upper limit $u_i^{\max}$ was set to infinity. The maximal power is therefore not constrained.

Since charging and discharging of the battery are represented with two different variables, simultaneous charging and discharging is possible in the model but need to be avoided. Here this is done by tuning the weights $\alpha$ and $\beta$ in (10.23) such that overlap of battery charging and discharging occurs.

Using Euler forward discretization with a step size of 1, i.e., $\dot{x}_i = x_i(k+1) - x_i(k)$, $i = 1, 2$ the optimal dispatch problem is given as

$$\min \mathbf{f}^{\mathrm{T}}\mathbf{d}, \text{subject to } (10.2 - 10.6), (10.24), (10.25), \text{ and } (10.26). \tag{10.27}$$

Applying the optimal dispatch to the example results in the values for the indices of the constraints given in Table 10.3.

Recap that the sizing was done to have a reliability of at least 99% and a relative excess power generated of maximal 60%, one can clearly see the advantage of the optimal dispatch over the rule-based approach.

As shown, the sizing of a local energy system and the dispatch in local energy systems with multiple degrees of freedom (e.g., different storage assets) are closely connected. The sizing can only be carried out if there is an underlying dispatch strategy. The computational effort of co-optimization of both problems is quite high, such that in this approach a rule-based dispatch strategy was used. Since the

*Table 10.3 Values of the constraint indices [33]*

| | **RBS – solution** | **LP – solution** |
|---|---|---|
| LPSP (%) | 1.7 | 1.79 |
| LHSP (%) | 0.14 | 0 |
| RI (%) | 99.08 | 99.1 |
| REPG (%) | 56.43 | 53.04 |
| Computational time (s) | <0.1 | 3 |

goal is to minimize the cost of the LES a cost model was introduced. After the optimization of the component sizes, on the basis of the rule-based strategy, an optimal dispatch strategy was found. Since the difference in the rule-based strategy and the optimal dispatch solution are minor, the use of the rule-based strategy in the sizing is justified.

Obviously, load-dependent efficiencies and degradation effects could only be represented by a nonlinear model, leading to the classical trade-off between accuracy and computational effort. Here good results were found with a linear model, by averaging efficiencies and degradation effects. The high abstraction level of the model has multiple advantages. On the one hand, it is easily adaptable to different LES, on the other hand the model is linear, such that the resulting optimization problem can be solved efficiently.

## 10.5 Control of local energy systems

### *10.5.1 System level control of local energy systems*

A key aspect on the design of an LES is the selection of the right control architecture to ensure an efficient and stable operation on every timescale. The control architecture provides an overview on how different component controllers interact at different timescales and complexity levels. In the case of general microgrid literature, such structures are often referred to as microgrid supervisory/central controller or energy management structures [38]. An overview of the main LES control architectures can be found in Figure 10.6. The following subsections will cover these concepts in more detail.

#### 10.5.1.1 Centralized control architectures

Centralized architectures typically rely on a single central controller (CC) as the main unit that oversees the whole system functionality, giving the set points to the local controllers (LCs) in a master-slave setup. The main advantage of this configuration is that the CC is able to monitor, collect, and analyze the whole system data in real time, e.g., energy production/consumption, market prices, storage state of charge, weather conditions; essentially having complete system observability. This enables the CC to optimize the LES operation set points of all the LC simultaneously, considering the requirements of all its components, while keeping all the information secure on a central point [38,39].

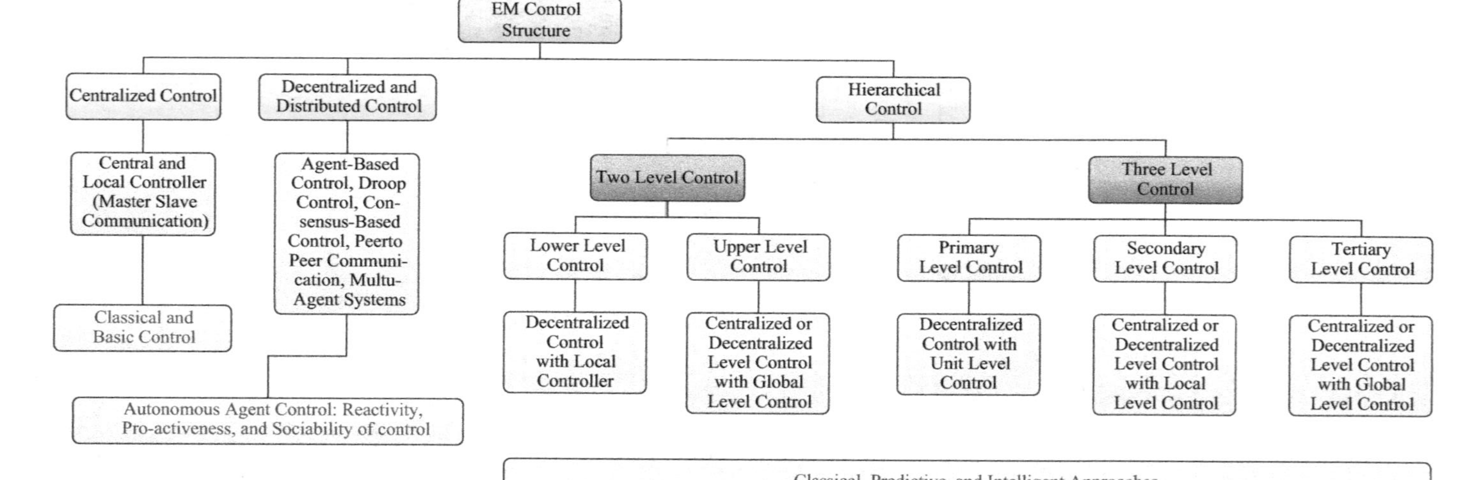


*Figure 10.6 Local energy system control architecture overview. Adapted from [39].*

However, to realize such optimization, the real-time implementation is a hard requirement that demands considerable computational resources and a fast and reliable communication network that rapidly scales with the system size, thus limiting the LES expandability. Moreover, the accumulation of all functionality onto the CC makes a clear single point of failure [38,39]. Therefore, central control architectures are only recommended in small-scale LES with a fixed structure to keep the computation and communication requirements low.

#### 10.5.1.2 Decentralized control architectures

In contrast to centralized approaches, decentralized control architectures aim at removing the CC as single point of failure by enhancing the LCs capabilities for decision-making [38]. This is possible due to recent advancements in communication technologies, e.g., Wi-Fi and Zigbee; and information exchange algorithms, e.g., Peer-To-Peer, Gossip, and Consensus [38].

It is important to note that in many fields the terms "decentralized" and "distributed" are used interchangeably. In this context, distributed approaches allow for information exchange between LCs, while centralized approaches restrict the LCs decision-making to local measurements only. This restriction considerably reduces the computation and communication requirements [39].

To ensure coordination toward a mutual goal, monitoring, processing, and data visualization are key. Thus, industry standards such as IEC-61968 and IEC-61850 are typically used as reference for LES management and communication respectively. Based on the communication setup, a distributed controller can range from fully dependent, i.e., local decisions with information share through a CC, partially independent, i.e., local decisions with only central decision information through a CC, and fully independent, i.e., direct LC communication with no CC involvement. Typical decentralized control strategies are, e.g., agent-based control, droop control, consensus-based control, multiagent systems control [39].

While decentralized architectures clearly address some major concerns with centralized approaches, e.g., single point of failure and scalability, it brings some challenges and drawbacks. By not having a central hub for decision-making, the local decisions taken by the LCs are done with incomplete information of the overall system, thus not arriving at globally optimal decisions. Moreover, it brings major implementation challenges, particularly for effective synchronization of the coordinated action of the LCs while keeping the information secure [38,39]. Naturally, these drawbacks are overweight in large LES configurations, where centralized data collection is challenging or belongs to different entities, or system reconfigurations are expected [38].

#### 10.5.1.3 Hierarchical control architectures

Hierarchical control architectures come as a compromise when the LES computation/communication complexity is too high for centralized and the LCs coupling/coordination is too critical for decentralized approaches [39]. To achieve this, hierarchical approaches split their structure into three levels, i.e., primary, secondary, and tertiary, see Figure 10.7.

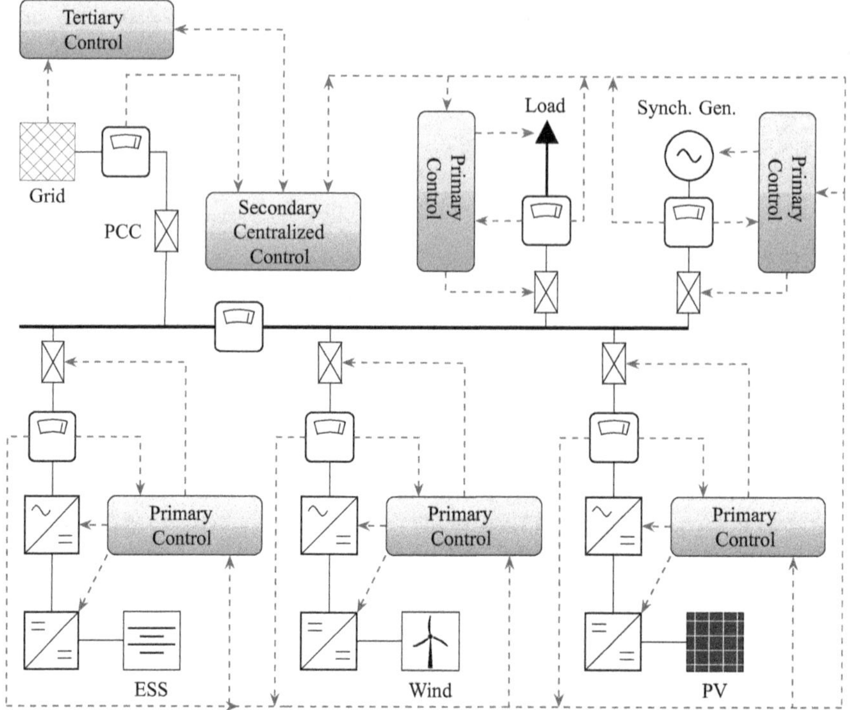


*Figure 10.7 Microgrid basic structure*

The primary level is usually tasked with the control of power converters, overseeing voltages and currents, so that the upper level set points and operating standards limits are followed. In the case of the secondary level, the main focus is to ensure the required power quality standards, which also enables the synchronization and power exchange with the main network. Typical tasks involve, e.g., voltage and frequency restoration, power imbalances, and harmonic compensation. Finally, the tertiary level is aimed at global system optimization, with multiple criteria, e.g., economic performance, technical efficiency, power quality and grid support, and social and climate impact. This multi-objective optimization requires different internal and external information sources, e.g., market pricing, weather forecast, and assets aging [38,39]. Each control layer's bandwidth is typically separated by at least an order of magnitude, ranging from milliseconds to seconds in the lower levels up to minutes and hours in the upper ones [38]. These differences have an impact on the communication infrastructure, which is normalized by the standards IEC 61850-7420 and EN13757-4 [39].

Moreover, hierarchical approaches can combine centralized and decentralized architectures at different levels in parallel. This enables them to harness each method's advantages, tailored to the particular LES setup. Therefore, hierarchical approaches can harness central optimal decisions at the higher levels, while operational constraints are ensured at lower levels in a decentralized manner [39]. Naturally, it is still exposed to risks, like the medium/long term planning dependency in higher level's forecast data, as well as the coordination with adjacent layers. While a failure from the lower levels can be isolated, a failure at the higher levels can have downstream repercussions, by limiting the information and thus the energy transfer in the lower levels [39].

### *10.5.2 Component level control of LES*

The next subsections' structure and content are mainly based on [5], which is recommended for a more exhaustive overview of the following topics.

Most of the power system generation is still done by SGs. As such, legacy component level control is geared toward SG dynamics. A standard SG model is shown in Figure 10.8. Here, the rotor with rotational inertia $M \in \mathbb{R}_{>0}$ and damping $D \in \mathbb{R}_{>0}$, is steered by a controllable torque $\tau_{\mathrm{m}} \in \mathbb{R}_{>0}$, with angle $\theta \in \mathbb{R}$ and angular velocity $\omega \in \mathbb{R}$ as states. The SG converts its mechanical energy to electrical energy through its rotating magnetic field, which interacts via the rotor and stator currents $i_{\mathrm{r}} \in \mathbb{R}$ and $\mathbf{i}_{\mathrm{s}} \in \mathbb{R}^3$, and its inductances $\mathbf{L}_\theta \in \mathbb{R}^{4\times4}$, to generate the terminal voltage $\mathbf{v}_{\mathrm{t}} \in \mathbb{R}^3$. Assuming a $dq$ frame rotating with nominal angular frequency $\omega_0$, the SG dynamics are given as

$$\frac{\mathrm{d}\theta}{\mathrm{d}t} = \omega - \omega_0, \tag{10.28}$$

$$M\frac{\mathrm{d}\omega}{\mathrm{d}t} = -D\omega + \tau_{\mathrm{m}} - L_{\mathrm{m}}i_{\mathrm{r}}\begin{bmatrix} -\sin\theta \\ \cos\theta \end{bmatrix}^{\mathrm{T}}\mathbf{i}_{\mathrm{s}}, \tag{10.29}$$

$$L_{\mathrm{s}}\frac{\mathrm{d}\mathbf{i}_{\mathrm{s}}}{\mathrm{d}t} = -(R_{\mathrm{s}}\mathbf{I} + \mathbf{J}\omega_0 L_{\mathrm{s}})\mathbf{i}_{\mathrm{s}} + L_{\mathrm{m}}i_{\mathrm{r}}\begin{bmatrix} -\sin\theta \\ \cos\theta \end{bmatrix}\omega - \mathbf{v}_{\mathrm{t}}, \tag{10.30}$$

which introduces the mutual inductance $L_{\mathrm{m}} \in \mathbb{R}_{>0}$, the stator inductance $L_{\mathrm{s}} \in \mathbb{R}_{>0}$, the stator resistance $R_{\mathrm{s}} \in \mathbb{R}_{>0}$, the identity matrix $\mathbf{I} \in \{0, 1\}^{2\times2}$, and the 90 rotation matrix $\mathbf{J} = \begin{bmatrix} 0 & -1 \\ 1 & 0 \end{bmatrix}$ [5].

Contrary to legacy systems, in modern low-inertia LES, the generation and consumption will transition toward CIGs and smart loads. Therefore, when talking about the component level control at LES, the focus is on VSC. A simple two-level VSC can be seen in Figure 10.9.

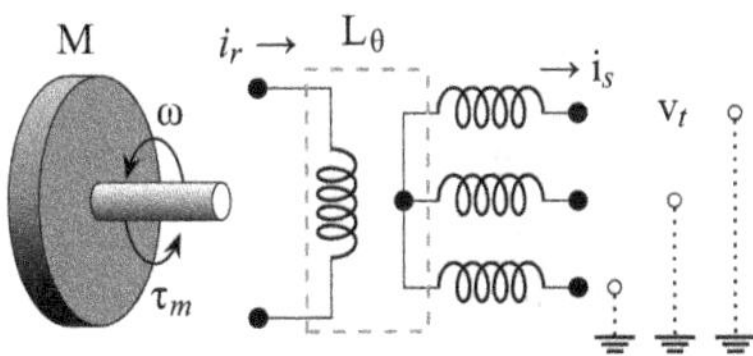


*Figure 10.8 Synchronous machine*

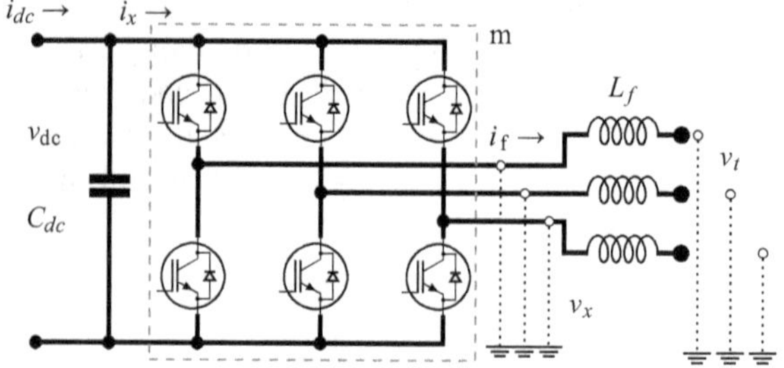


*Figure 10.9 Two-level voltage source converter (VSC)*

The main purpose of a VSC is to convert dc voltage $v_{\mathrm{dc}} \in \mathbb{R}$ and current $i_{\mathrm{x}} \in \mathbb{R}$ into the three-phase $dq$ – frame voltage $\mathbf{v}_{\mathrm{x}} \in \mathbb{R}^2$ and filter current $\mathbf{i}_{\mathrm{f}} \in \mathbb{R}^2$ via

$$i_{\mathrm{x}} = \frac{1}{2}\mathbf{m}^{\mathrm{T}}\mathbf{i}_{\mathrm{f}}, \tag{10.31}$$

$$\mathbf{v}_{\mathrm{x}} = \frac{1}{2}\mathbf{m}v_{\mathrm{dc}}, \tag{10.32}$$

with averaged duty cycle ratio $\mathbf{m} \in [-1,1] \times [-1,1]$ [40, Chapter 5]. Which leads to the averaged model

$$C_{\mathrm{dc}}\frac{\mathrm{d}v_{\mathrm{dc}}}{\mathrm{d}t} = -G_{\mathrm{dc}}v_{\mathrm{dc}} + i_{\mathrm{dc}} - \frac{1}{2}\mathbf{m}^{\mathrm{T}}\mathbf{i}_{\mathrm{f}}, \tag{10.33}$$

$$L_{\mathrm{f}}\frac{\mathrm{d}\mathbf{i}_{\mathrm{f}}}{\mathrm{d}t} = -(R_{\mathrm{f}}\mathbf{I} + \mathbf{J}\omega_0 L_f)\mathbf{i}_{\mathrm{f}} + \frac{1}{2}\mathbf{m}v_{\mathrm{dc}} - \mathbf{v}_{\mathrm{t}}, \tag{10.34}$$

with dc-link capacitance $C_{\mathrm{dc}} \in \mathbb{R}_{>0}$, filter inductance $L_{\mathrm{f}} \in \mathbb{R}_{>0}$, lumped switching, charging, and conduction losses $G_{\mathrm{dc}} \in \mathbb{R}_{>0}$ and $R_{\mathrm{f}} \in \mathbb{R}_{>0}$, and controllable dc-side current $i_{\mathrm{dc}} \in \mathbb{R}$ [5].

Despite their similar roles, SGs are more resilient and can store more energy than VSC, while the later have a much faster response that is fully controllable. Thus, even though their control objectives might be similar, their design should be suited to their inherent characteristics [5].

Regarding the VSCs' control objectives, the generally agreed framework relies on a nominal steady state in the context of a primary-level decentralized LC using only local measurements [5], given by

- *synchronous frequency* [41]: balanced three-phase periodic ac signals with nominal frequency $\omega_0$;
- *power injections* [13]: active and reactive power injection set points $(p_{\mathrm{g}}, q_{\mathrm{g}}) = \left(p_{\mathrm{g}}^{\star}, q_{\mathrm{g}}^{\star}\right)$;
- *ac voltage magnitude* [41]: nominal ac voltage magnitude set point $*\mathbf{v}_{\mathrm{t}} = V_{\mathrm{t}}^{\star}$; and
- *dc voltage* [42]: nominal dc voltage set point $v_{\mathrm{dc}} = v_{\mathrm{dc}}^{\star}$.

The nominal operating point $\left(\omega_0, v_{\mathrm{dc}}^{\star}, p_{\mathrm{g}}^{\star}, q_{\mathrm{g}}^{\star}, V_{\mathrm{t}}^{\star}\right)$ needs to be consistent with the ac power flow equations [43], either by a system operator optimal power flow solution [44] or by local objectives, e.g., MPPT [45,46]. However, due to disturbances, the nominal operating point might not correspond to an equilibrium point of the power system [5]. In this context, VSCs provide a grid-supporting steady-state disturbance response that mimics the steady-state droop response of classical SGs [13],

$$\omega - \omega_0 = m_p\left(p_{\mathrm{g}}^{\star} - p_{\mathrm{g}}\right), \tag{10.35}$$

$$\|\mathbf{v}_{\mathrm{t}}\| - V_{\mathrm{t}}^{\star} = m_q\left(q_{\mathrm{g}}^{\star} - q_{\mathrm{g}}\right), \tag{10.36}$$

with droop coefficients $m_p$ and $m_q$; while the dc voltage $v_{\mathrm{dc}}$ is assumed to be stabilized by controlling the dc power source $p_{\mathrm{dc}}$.

There are two predominant control approaches for grid-supporting VSCs, i.e., grid-following (GFL) and grid-forming (GFM) [5]. While there is no consensus on the exact definitions of these strategies, early approaches defined GFL VSCs acting as current or power sources, i.e., controllable power injection, while GFL VSCs act as ac voltage sources, i.e., ac voltage with nominal constant amplitude and frequency [13]. Currently, GFL refers to VSCs that rely on a phase-locked loop (PLL) for synchronization and current control irrespective of grid-support; whereas GFM are VSCs that impose ac voltage with adjusted frequency for synchronization and grid-support [47]. Depending on the assumptions on a VSC's ac or dc terminal, the control definitions can be terminal specific, i.e., for GFL ac-GFL or dc-GFL, depending if the approach relies on a well-defined guaranteed ac or dc voltage respectively. Similarly, ac-GFM or dc-GFM can be defined for GFM, if the controller imposes a well-defined ac or dc voltage respectively [5].

#### 10.5.2.1 Grid-following control

Currently, most converter-interfaced RESs and ESSs are controlled via ac-GFL. The main idea is to use a PLL to estimate the terminal voltage phase angle $\angle\mathbf{v}_{\mathrm{t}}$ [48] so that the VSC current $\mathbf{i}_{\mathrm{f}}$ can be controlled in the corresponding $dq$ – frame [13]. This is achieved by feedback linearization, i.e., $\mathbf{m} = \frac{2}{v_{\mathrm{dc}}}\mathbf{v}_{\mathrm{x}}^{\star}$, and a proportional-integral (PI) controller $G_{\mathrm{PI}}(s)$ [13], such that

$$\mathbf{v}_{\mathrm{x}}^{\star} = (R_{\mathrm{f}}\mathbf{I} + \mathbf{J}\omega_0 L_{\mathrm{f}})\mathbf{i}_{\mathrm{f}} + \mathbf{v}_{\mathrm{t}} + G_{\mathrm{PI}}(s)\left(\mathbf{i}_{\mathrm{f}}^{\star} - \mathbf{i}_{\mathrm{f}}\right), \tag{10.37}$$

when introduced in (10.34) leads to

$$L_{\mathrm{f}}\frac{\mathrm{d}\mathbf{i}_{\mathrm{f}}}{\mathrm{d}t} = G_{\mathrm{PI}}(s)\left(\mathbf{i}_{\mathrm{f}}^{\star} - \mathbf{i}_{\mathrm{f}}\right). \tag{10.38}$$

This strategy requires a strongly coupled ac system where the grid and VSC terminal voltages align independently of the filter current, i.e., $\mathbf{v}_{\mathrm{t}} = \mathbf{v}_{\mathrm{g}}$ [49]. These assumptions fall short in practice, potentially leading to instability with, e.g., PLL-induced positive feedback; which has limited the success on finding stability certificates in realistic conditions [49]. Despite its shortcomings, this approach remains as the standard ac-GFL/dc-GFM control setup for dc voltage control in CIG and HVDC transmission, via the current $\mathbf{i}_\mathrm{f}^{d\star}$ and active power $p_\mathrm{x} \approx \mathbf{v}_\mathrm{g}\mathbf{i}_\mathrm{f}^{d\star}$ flowing out of the dc-link capacitor [45,46]. Moreover, there is a big focus in the literature on the ac-GFL/dc-GFL variant that assumes a constant dc voltage. Here, the reference current is built as $\mathbf{i}_\mathrm{f}^{d\star} = \frac{p_\mathrm{g}}{\mathbf{v}_\mathrm{g}}$ and $\mathbf{i}_\mathrm{f}^{q\star} = \frac{q_\mathrm{g}}{\mathbf{v}_\mathrm{g}}$, where $(p_\mathrm{g}, q_\mathrm{g})$ follow the droop characteristics in (10.35) and (10.36), while using the PLL for grid frequency and ac voltage magnitude estimation [13]. However, due to the aforementioned stability concerns and the rise of ac-GFM/dc-GFL, the research community is steering away from it [47].

#### 10.5.2.2 Grid-forming control

Contrary to GFL control, GFM is designed from the ground up to provide grid stabilization, particularly in low-inertia conditions, helping enable the energy transition [47]. In general, ac-GFM operates by measuring VSC ac current or power and adjusting its ac voltage to achieve the control objectives given at the beginning of Section 10.5.2. This is typically achieved by direct modulation of the voltage $\mathbf{v}_\mathrm{x}$ or providing an LCL voltage reference for a cascaded PI current and voltage controller [5], see Figure 10.10. Most of the literature focuses on the later approach, which while appealing, still presents challenges. For example, the fine-tuning of the inner loops is critical, particularly to account for grid coupling strength and time-scales separation [50]. Furthermore, overcurrent protection is typically addressed by limiting the ac current reference $\mathbf{i}_\mathrm{f}^{\star}$; however, this can lead to a loss of synchronization or synchronous instability [51], thus complicating model reduction [52]. In the following, multiple approaches for setting the ac-GFM voltage reference $\mathbf{v}_\mathrm{x}$ will be discussed.

*Droop control and virtual synchronous machines*

The most established GFM approaches in literature are *droop control* [53] and *virtual synchronous machines* [54], which have an ac-GFM/dc-GFL setup, i.e., a constant nominal dc voltage is assumed [5].

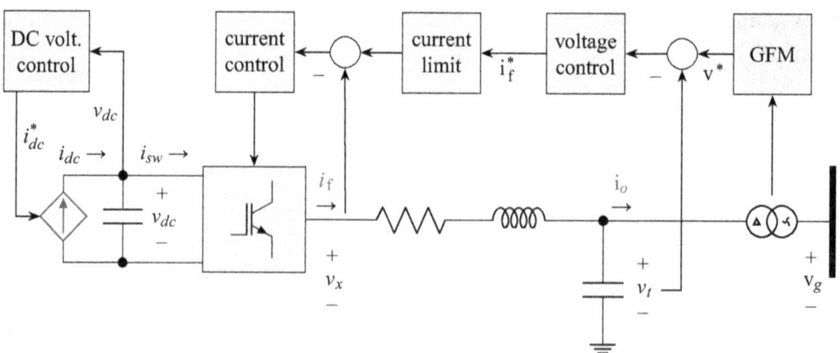


*Figure 10.10 Standard ac-GFM control architecture*

Droop control is based on the SG behavior under steady state, where frequency deviations are proportional to its active power injection and the voltage magnitude is proportional to its reactive power injection. In inductive networks, it is also observed that the active power is almost proportional to the voltage phase angle differences, while the reactive power is roughly proportional to the voltage magnitude [5]. Basically resulting in network frequency synchronization and power-sharing [55]. This is implemented by setting up feedback from the VSC active power injection $p_g$ to determine the phase angle $\theta = \angle \mathbf{v}_{gfm}$ and frequency of the ac-GFM reference voltage, and feedback from the VSC reactive power injection $p_g$ to determine the magnitude $\|\mathbf{v}_{gfm}\|$ leading to

$$\frac{d\theta}{dt} = \omega_0 + m_p\left(p^\star - \tilde{p}_g\right), \tag{10.39}$$

$$\|\mathbf{v}_{gfm}\| = V^\star + m_q\left(q^\star - \tilde{q}_g\right), \tag{10.40}$$

with droop gains $m_p$ and $m_q$, voltage magnitude setpoint $V^\star$, power setpoints $p^\star$ and $q^\star$, and low-pass filtered measured power injections $\tilde{p}_g$ and $\tilde{q}_g$ to remove switching harmonics and conflicting bandwidth interactions [5]. All the GFM approaches presented next will explicitly or implicitly exhibit a similar droop behavior [5].

Regarding VSMs, there are many approaches in literature mimicking different SG models, particularly of low order [54]. Here for reference, the focus will be on a commonly used one-axis model with exciter and swing-equation model. Assuming a null turbine time constant $T_m = 0$ and excitation winding time constant $T'_{do} = 0$, and linearizing at the nominal terminal voltage magnitude leads to

$$\frac{2H}{\omega_0}\frac{d\omega}{dt} = -D\omega + p^\star - p_g, \tag{10.41}$$

$$\|\mathbf{v}_{gfm}\| = V^\star + m_q\left(q^\star - \tilde{q}_g\right), \tag{10.42}$$

with *virtual inertia* constant $H$ and damping constant $D$ [5]. Notably, VSCs limited storage and overload capability significantly hinders the emulation of the virtual inertia constant $H$ [56]. Moreover, combined with the need of auxiliary controls for practical implementations, e.g., virtual impedance, PLL-based damping, VSMs further diverge from SGs [5].

The connection to droop control laws (10.39) and (10.40) can easily be established with a change of coordinates [57] and setting the equivalent inertia constant as $\frac{2H}{\omega_0} = \frac{\tau_{lp}}{m_p}$, with low-pass filter time constant $\tau_{lp}$ [58]. These equivalences advocate for a basic interoperability between droop VSCs, VSMs, and SGs [5]; thus, enabling the vast literature foundation of low-order SGs [59]. However, general almost or semi-global stability results are still missing [5]. Available local asymptotic stability results generally assume a lossless network and constant voltage magnitudes, which only applies to zero power flow solutions, and do not extend to dynamic network models [5].

*Virtual oscillator control*

Virtual oscillator control (VOC) is an ac-GFM/dc-GFL control approach that originated on standalone uninterruptible power supplies applications [60]. However, due to its nonlinear oscillator dynamics inherit self-synchronization, it quickly expanded on to single-phase converters network control [61] and three-phase VSC nominal operation point tracking [62]. In contrast to droop and VSM approaches, VOC offers almost global synchronization certificates for networks of VSCs [61]. Furthermore, it exhibits practical robust synchronization [61] and a clear connection to droop dynamics by averaging a cycle [63] in resistive networks [13].

Despite these promising features, out of the box, VOC is non-dispatchable, i.e., the network's operating point (power-sharing and voltage magnitudes) is not controlled, but rather depends on its parameters. This severely limits its applicability for coordinated system action and market mechanisms [5]. To address these shortcomings, the dispatchable virtual oscillator control (dVOC) was developed, which adds a synchronizing feedback and magnitude control [62]. A stationary $\alpha\beta$ coordinates formulation is given as

$$\frac{\mathrm{d}\mathbf{v}_{\mathrm{gfm}}}{\mathrm{d}t} = \omega_0 \mathbf{J}\mathbf{v}_{\mathrm{gfm}} + \eta\left[\mathbf{K}\mathbf{v}_{\mathrm{gfm}} - \mathbf{R}(\kappa)\mathbf{i}_0\right] + \eta\alpha\Phi\left(\mathbf{v}_{\mathrm{gfm}}\right)\mathbf{v}_{\mathrm{gfm}}, \tag{10.43}$$

where

$$\mathbf{K} = \frac{1}{V^{\star}V^{\star}}\mathbf{R}(\kappa)\begin{bmatrix} p_{\mathrm{g}}^{\star} & q_{\mathrm{g}}^{\star} \\ -q_{\mathrm{g}}^{\star} & p_{\mathrm{g}}^{\star} \end{bmatrix}, \tag{10.44}$$

$$\Phi\left(\mathbf{v}_{\mathrm{gfm}}\right) = \frac{V^{\star}V^{\star} - *\mathbf{v}_{\mathrm{gfm}}{}^{2}}{V^{\star}V^{\star}}, \tag{10.45}$$

with synchronization gain $\eta \in \mathbb{R}_{>0}$, magnitude control gain $\alpha \in \mathbb{R}_{>0}$, 2D rotation matrix $\mathbf{R}(\kappa)$, and $\kappa = \tan^{-}1\left(\omega_0 \frac{l}{r}\right)$, where the $\frac{l}{r}$ network ratio is assumed to be constant for transmission lines at the same voltage level [5]. Here the nominal operating point can be set through $\left(V^{\star}, p_{\mathrm{g}}^{\star}, p_{\mathrm{g}}^{\star}\right)$. Overall, dVOC can be thought of as a generalization of droop control and VOC to more general networks [5]. Connections to both approaches arise, in particular, in inductive networks with droop control [64], and in resistive networks with averaged VOC [62].

*Machine matching and dual-port grid-forming control*

In contrast to the previous ac-GFM/dc-GFL approaches, machine matching control [42] does not neglect the dc dynamics, but rather uses the dc voltage as a power imbalance indicator, akin to frequency for SGs [5]. This is achieved by using the terms

$$\frac{\mathrm{d}\theta}{\mathrm{d}t} = \omega, \tag{10.46}$$

$$\omega = k_{\omega} v_{\mathrm{dc}}, \tag{10.47}$$

$$\mathbf{m} = u_{\mathrm{mag}}\begin{bmatrix} -\sin\,\theta \\ \cos\,\theta \end{bmatrix}, \tag{10.48}$$

where $k_\omega \in \mathbb{R}$ is the angular frequency gain and $u_{\text{mag}} \in [-1, +1]$ is the controllable switching magnitude. With this formulation, the VSC dynamics in (10.33) and (10.34), closely resemble the SG dynamics in (10.28) to (10.30), by matching the inertia constant $M = \frac{C_{\text{dc}}}{k_\omega^2}$, damping constant $D = \frac{G_{\text{dc}}}{k_\omega^2}$, and torque $\tau_{\text{m}} = \frac{i_{\text{dc}}}{k_\omega}$ [42]. While structurally similar, it is important to note that the equivalent inertia and damping constants are significantly smaller [5].

The strategy of balancing power by controlling the dc voltage through the ac terminal essentially results in an ac-GFM/dc-GFM approach. More precisely, if the dc voltage is stabilized by a dc-GFM source, the ac voltage frequency is also stabilized [5]. This property is further explored in the dual-port GFM control [65] by adding a dc voltage droop term to (10.35), such that

$$\omega - \omega_0 = m_p(p^\star - p_{\text{g}}) + k_\omega(v_{\text{dc}} - v_{\text{dc}}^\star), \tag{10.49}$$

which includes typical features of ac-GFM (e.g., primary frequency control) and ac-GFL (e.g., MPPT), without the need of switching modes or using a PLL [5]. Furthermore, dual-port GFM provides an end-to-end linear stability analysis that considers ac and dc transmission, SGs and VSCs with and without controlled generation, as well as conventional and RES-generic models [65].

## 10.6 Stability of local energy systems

This section provides a brief overview of the expected characteristics of local energy systems, possible phenomena leading to instability of the system, and suitable analysis tools. A lot of these findings relate to the experiences made in microgrids that were summarized in the microgrid stability definition presented in [16] developed by the *IEEE PES Task Force on Microgrid Stability Definitions, Analysis, and Modeling*. Wherever possible, the experiences from microgrids are put into the context of the ongoing developments of LES, which strongly relate to converter-dominated systems in general. As seen in the previous Section 10.5.2, some controllers are based on the assumptions of the network that may break in the future, such as a strongly-coupled grid. Therefore, when designing an LES, these are some aspects that need to be considered.

### *10.6.1 Characteristics of LES*

Local energy systems are expected to be structurally different compared to highly interconnected bulk power networks, such as the synchronous grid of Continental Europe. The system size is comparably smaller, and the feeder lengths are significantly shorter. In addition, the voltage level is typically at medium voltage or less. As a result, the resistance-to-reactance ratio is higher, see Table 10.4 for some typical values at different voltage levels. Whether a network is inductive, inductive-resistive, capacitive, or resistive has strong implications on the relationship between active and reactive power, phase angle, and the system frequency, respectively. Therefore, the operation and control of LESs need to consider the higher $R/X$-ratio, for example, in droop-based controllers [13,66] as mentioned in Section 10.5.2.2.

*Table 10.4 Typical line parameters depending on the voltage level [13]*

| **Voltage level** | $R$ ($\Omega$ km$^{-1}$) | $X$ ($\Omega$ km$^{-1}$) | $R/X$ |
|---|---|---|---|
| Low voltage | 0.642 | 0.083 | 7.700 |
| Medium voltage | 0.161 | 0.190 | 0.850 |
| High voltage | 0.060 | 0.191 | 0.310 |

Contrary to bulk power networks, larger power fluctuations are expected. The large fluctuations result from a small number of devices and the weather dependency of the renewable source. Therefore, the production is subject to higher uncertainties. As a result, imbalances between generation and consumption can become more critical as they have a bigger impact on the system voltage and frequency. Moreover, bidirectional power flows may pose problems in control and protection. LESs are low in system inertia, where it is an open question how many units need to operate in grid-forming mode. Combined with the low short circuit capacity of the converter, large voltage and frequency variations may occur, especially in island-mode. Lastly, LES may experience unbalanced networks as it has been seen in microgrids due to unbalanced loading of the phases [16].

Thus, the different expected characteristics of LES compared to bulk power networks will lead to different phenomena seen in the network.

### *10.6.2 Stability of LES*

Stability of a power network is essentially a single problem. However, the different physical origins motivated the classification and categorization of power system stability for better analysis into first voltage, frequency, and rotor angle stability as the main categories [67]. This fundamental work was based on the legacy power system, where the network dynamics were determined by the SGs. The growth of CIGs resulted in a change of time scales of the interacting network dynamics toward the faster electromagnetic dynamics of converters compared to the slower electromechanical dynamics of SGs, as depicted in Figure 10.11. Consequently, new dynamics and interactions arise, which led to an update of the categorization by adding converter-driven stability, and resonance stability [69]. The following compares the stability definition given in [67] for power networks in general with the definition of stability of microgrids proposed in [16] to locate which is more suitable for LESs.

The stability of the LES are expected to be largely determined by the LCs, since they are acting faster than the CC to control the voltage and the frequency, balance and share power, and detect islanding [70]. When connected to the grid, i.e., the interconnected bulk power network, the grid imposes voltage and frequency. Then the stability of the LES mostly reduces to the stability of the DER, or load [16].

In the case of microgrids, [16] proposed that "a microgrid is stable if, after being subjected to a disturbance, all state variables recover to (possibly new) steady-state values which satisfy operational constraints, and without the occurrence of involuntary load shedding." Voluntary load shedding, as part of the

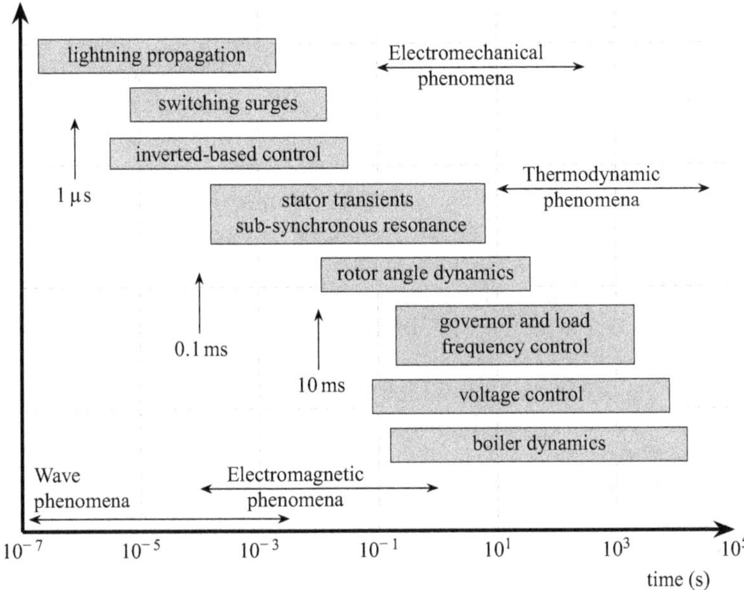


*Figure 10.11 Time scales of power system dynamics (adapted from [67])*

microgrid control, and the disconnection of elements to isolate faults, are events in which the microgrid is considered to remain stable. The proposition for microgrids assumes that the primary objective of the microgrid is to supply a few loads with power, where some of them could be critical, for example, a hospital. If after a disturbance the load is tripped, even intentionally to secure the operation of the system, the microgrid is considered to be unstable under this definition since the primary objective is not fulfilled. The proposed definition contrasts to the bulk power system, where stable operation is of utmost importance. Therefore, the intentional tripping of some elements to secure the continuous operation within the voltage and frequency constraints is acceptable.

For the general case of LES, for example, green hydrogen production sites, offshore wind power plants, the definition by [16] is not suitable because it includes an operating objective within the stability definition of a dynamical system. From our point of view, the trajectories of the system variables, i.e., mainly voltage and current, including the phase angle, and the system frequency, clearly indicate, whether a system, that prior to the disturbance operated in an equilibrium point, is able to regain an operating steady state again, i.e., an equilibrium point. Due to the physical devices including their protection, the trajectory must stay within the limits because some variables are bounded, e.g., by the required operating range given by the grid code, or the protection of the devices provided by the manufacturer. Though the system theory has a clear understanding, the complex physical nature of power systems may dilute some of these concepts for simplification. Therefore, the definition of [67] is most universal and aligned with system theory, which is: "Power system stability is the ability of an electric power system, for a given initial operating equilibrium, to regain a state of operating equilibrium after being subjected to physical disturbance, with most system variables bounded so that practically the entire system remains intact." This definition is applicable to LES.

The classification and the categorization of power system stability as in [16,67,69] follow from the physical nature of the system as mentioned before. No specific categorization of stability of LES is outlined here, as the existing ones already provide a good ground for the fault analysis by practitioners, and perhaps, even being debated due to the very complex interactions in power systems in general. Some root causes and resulting effects on the system dynamics are outlined in the following with a special focus on the interactions between CIG, indicating which aspects need to be checked when designing and operating an LES.

The disturbances in the LES can be small, for instance, operational set-point adjustments, or load changes, falling into the category of small-signal disturbances, where a linear representation of the system adequately describes the system dynamics. The disturbances are further categorized into short-term or long-term effects. For instance, heavily loaded microgrids with continuous small load changes may experience undamped oscillations in the long term. If the power-sharing between DERs is poor, undamped power oscillations could quickly exceed acceptable limits in a short time. Large disturbances relate to, e.g., the loss of generation, short circuits, or unplanned disconnection of the microgrid from the grid, switching to island-mode operation. Table 10.5 provides a summary of typical root causes and manifestations experienced in microgrids [16].

The control systems of the CIGs in Table 10.5 can experience instability due to poor tuning, or non-suitable control schemes, of one or more devices. For small

*Table 10.5 Root and manifestations of stability issues in microgrids [16]*

| **Control system stability** | | **Power supply and balance stability** | |
|---|---|---|---|
| **Electric Machine stability** | **Converter stability** | **Voltage stability** | **Frequency stability** |
| Poor controller tuning | Poor controller tuning, PLL bandwidth, PLL synchronization failure, harmonic instability | DERs power limits, inadequate reactive power supply, poor reactive power sharing, load voltage sensitivities dc-link capacitor | DERs active power limit, inadequate active power supply, poor active power sharing |
| Undamped oscillations, aperiodic voltage and/or frequency increase/ decrease | Undamped oscillations, low steady-state voltages high-frequency oscillations | Low steady-state voltages, large power swings, high dc-link voltage ripples | High rate of change of frequency, low steady-state frequency, large power, and frequency swings |

disturbances, instabilities could arise from the voltage and current control loops. For large disturbances, a special concern is the protection scheme of the CIG because the converter may trip. The requirement of a CIG to remain connected during larger disturbances, e.g., an under-frequency-ride-through, was mentioned in Section 10.2 to preserve the stable operation of the system. The interactions of the local controller result in oscillations ranging from low to high frequencies of up to several Kilohertz, which is sometimes referred to as harmonic stability [71]. The interactions of the LCs can result from the parallel connection of the CIGs in proximity. As a result, multiple resonant peaks can occur [72]. Besides the interactions due to the parallel interconnection, resonance of the LCL filter might also be triggered by the LC. Instead of adding passive filters to dampen those frequencies, using an active damping included in the LC is a typical mitigation strategy [16,72]. Another root cause in grid-following CIGs is the PLL. The PLL adds to the input admittance a negative parallel admittance deteriorating the system stability. This can, for example, affect the system voltage, which could be solved by lowering the bandwidth of the PLL which brings other drawbacks when going to low-voltage systems [16].

The category of Power Supply and Balance Stability in Table 10.5 relates to the capability to share load among the units effectively so that the power balance is maintained. Some of the effects that were mentioned before fall into this category, such as the violation of DERs active and reactive power limits, poor power sharing, or the loss of generation. The power imbalances will affect the system frequency or the system voltage.

Load and generation changes typically translate to frequency changes. Depending on the system inertia, and the participating units in the frequency control, the change in system frequency can experience large ROCOFs potentially risking the frequency stability by, e.g., causing the tripping of loads and generators. A steady-state frequency can occur if the power demand exceeds the maximum available power. Poor coordination and power-sharing can lead to sustained frequency oscillations [16]. This may also be due to higher $R/X$ ratios and the resulting strong coupling between voltage and frequency, which breaks the assumptions of some controllers that assume a decoupling of these quantities, such as droop-based control [13,66].

### *10.6.3 Stability analysis methods*

There are a variety of methods to analyze the dynamic behavior of an LES, or a power network as a whole, which depend on the kind of phenomena under investigation and the timescale of interest. Generally, these methods can be clustered into suitable tools for studying small-perturbation stability, or large-perturbation stability, where the methods for the latter are also suitable for small-signal disturbances. Prior to the perturbation, the system under study needs to be in an operating equilibrium, which is typically obtained by static power flow analysis.

The most used method for studying large perturbations are time-domain simulations based on electromagnetic transient (EMT) models or electromechanical tools known as transient stability tools [16]. Time-domain simulations based on EMT models are computationally expensive depending on the system size and the level of details of the models. Therefore, the models are simplified to speed up the simulation time. There are different approximation levels of power electronics depending on the kind of study, the relevant effects and corresponding time scales (see Figure 10.11), see [73], for example. For grid integration studies of CIGs, [74] presents phasor modeling approaches and compares these with EMT-based simulations, which can be a suitable approach if carefully considered. There are recent guidelines that recommend the kind of analysis depending on the interested phenomena as given in Table 10.6.

For grid integration studies, a technical specification such as the IEC 61400-27-1 provides generic models of wind turbines. Generally, there is a problem that the exact models of the manufacturer are intellectual properties, and thus, not openly available. One solution is that manufacturers provide their models including the controllers and their tuning as a protected component inside the software package, for example, as dynamic link libraries. However, there is a need for model validation because the provided models from the manufacturer may deviate from the actual plant behavior due to an adjustment of the controller tuning and software during different stages from pre-commissioning to the operation stage [76]. To validate the models, or to test components, sub-systems, and systems, hardware-in-the-loop (HiL) setups interfaced with real-time simulators are a suitable method [77,78] as outlined in Chapter 5 and not further explained here. However, it must be emphasized that all these methods only study one scenario at a time, requiring numerous simulation runs or tests. Contrary to time-domain simulation and testing based on HiL setups to analyze the large-perturbation stability, techniques based on Lyapunov functions provide an analytical approach to determine or estimate the region of attraction of a stable equilibrium point. The direct method by Lyapunov functions has been used in some studies, e.g., [79,80]. However, since LES are non-autonomous, time-varying systems, finding a suitable candidate for a Lyapunov function is a significant hurdle often requiring simplifying assumptions of the system [67,81], which is why it is not widely used in the industry.

To study the small-perturbation stability of an LES, the system is linearized around a stable equilibrium point. Then the linear state-space model is used to analyze the eigenvalues and participation factor. Some examples are studying the effect of the tuning of the outer power-sharing loops of the converter in critical low-frequency modes, or the converter's inner voltage and current control loop in critical high-frequency modes [16]. Depending on the system size, linear state space models can become very large, and especially for unbalanced networks, quite complex [82]. Instead of the time-domain, impedance-based methods using transfer function are suitable to study the interactions of converters at different frequencies [83], e.g., to study resonance phenomena. Impedance-based approaches have led to the concept of resonance or harmonic stability [69,71]. Small-signal perturbation analysis can access the toolbox of linear system analysis and provide useful insights into the interactions of the components of the LES. Time-domain approaches using a linear state-space representation can determine which states are involved at

*Table 10.6 Interaction studies between at least two CIGs according to Cigré WG B4-81 (adapted from [75])*

| **Type** | **Control loop interactions** | | **Interaction due to nonlinear characteristics** | | **Harmonic and resonance interaction** | |
|---|---|---|---|---|---|---|
| | **Slow control (steady state)** | **Dynamic control** | **AC fault performance** | **Transient stress and nonlinear effects** | **Sub-synchronous resonance** | **Harmonic emission and resonance** |
| Phenomena | • AC filter hunting<br>• Voltage control conflicts<br>• P/V stability | • Power oscillation<br>• Control loop interaction<br>• Sub-synchronous control interaction<br>• Voltage stability | • Commutation failure<br>• Voltage distortion<br>• Phase imbalance<br>• Fault recovery<br>• Protection performance | • Load rejection<br>• Voltage phase shift<br>• Switching and change of network<br>• Transformer saturation<br>• Isolation coordination | • Sub-synchronous torsional interaction | • Resonance<br>• Harmonics<br>• Harmonic instabilities<br>• Core saturation stability |
| Means of assessment | • Static analysis<br>• RMS time domain | • RMS time domain<br>• Small signal analysis<br>• EMT time domain | • RMS time domain<br>• EMT time domain | • EMT time domain | • EMT time domain | • Harmonic analysis<br>• Small signal analysis<br>• EMT time domain |

critical frequency modes, however, this requires a detailed knowledge of the controller and the system, which may not be available as mentioned before. Impedance-based approaches can be an alternative, as the component can be represented as a transfer function which can be obtained by an impedance scan. However, all these methods have evidently the short-coming about the unknown validity of the analysis, i.e., when is the perturbation no longer small? According to [16], the valid range of the small-signal analysis may be reduced in low-inertia systems, where fluctuations and disturbances are expected to be higher.

When studying the design and the stability of LES, there are a number of methods analyzing the transient behavior, where the ongoing discussion is about the required level of detail depending on the phenomena of interest. The methods based on the linearized models provide useful insights and are complementary to large-perturbation analysis. In the end, expertise is required to choose the right modeling approach and method, especially considering converter-based networks, such as LES.

## 10.7 Outlook

This chapter started with a discussion on the increasing requirements for the integration of wind turbines as power networks become dominated by converter-interconnected units and low in inertia. With the resulting decentralization of the power network, and the increasing combination of, e.g., wind turbines, or photovoltaic plants, and electrolyzers, local energy systems are starting to occur more often and a definition was proposed. With the coupling of different sectors, local energy systems are not limited to electric power but will interconnect different forms of energy. Therefore, it is essential to properly control the different units to achieve an efficient interaction and support the stability of the grid. Besides the control of the LES as one unit, the components of the LES need to be properly sized to achieve an efficient operation as it was shown in the example in Section 10.4. Different control architectures were discussed, starting from the central controller to the local controller. For the local controller, a review of the recent developments of grid-forming control was shown and compared with the mostly used grid-following control of CIGs. Lastly, different root causes of instabilities that could occur in LES were outlined, and suitable analysis tools were discussed.

## References

[1] German Federal Ministry of Education and Research. H2Mare: How Partners in the H2Mare Flagship Project Intend to Produce Hydrogen on the High Seas; 2023. Available from: https://www.wasserstoff-leitprojekte.de/projects/h2mare.

[2] Kumar V, Teja VR, Singh M, *et al.* PV Based Off-Grid Charging Station for Electric Vehicle. *IFAC-PapersOnLine*. 2019;52(4):276–281.

[3] F Milano, F Dörfler, G Hug, *et al.* Foundations and Challenges of Low-Inertia Systems (Invited Paper). In: *2018 Power Systems Computation Conference (PSCC)*; 2018. p. 1–25.

[4] ENTSO-E. High Penetration of Power Electronic Interfaced Power Sources and the Potential Contribution of Grid Forming Converters: Technical Report; 2020. Available from: https://eepublicdownloads.entsoe.eu/clean-documents/Publications/SOC/High_Penetration_of_Power_Electronic_Interfaced_Power_Sources_and_the_Potential_Contribution_of_Grid_Forming_Converters.pdf.

[5] Dörfler F, and Groß D. Control of Low-Inertia Power Systems. *Annual Review of Control, Robotics, and Autonomous Systems*. 2023;6(1).

[6] Kroposki B, Johnson B, Zhang Y, *et al.* Achieving a 100% Renewable Grid: Operating Electric Power Systems with Extremely High Levels of Variable Renewable Energy. *IEEE Power and Energy Magazine*. 2017; 15(2):61–73.

[7] IRENA. Grid Codes for Renewable Powered Systems. Abu Dhabi; 2022. Available from: https://www.irena.org/-/media/Files/IRENA/Agency/Publication/2022/Apr/IRENA_Grid_Codes_Renewable_Systems_2022.pdf?rev=986f108cbe5e47b98d17fca93eee6c86.

[8] Ahmed SD, Al-Ismail FSM, Shafiullah M, *et al.* Grid Integration Challenges of Wind Energy: A Review. *IEEE Access*. 2020;8:10857–10878.

[9] Collados-Rodriguez C, Cheah-Mane M, Prieto-Araujo E, *et al.* Stability Analysis of Systems with High VSC Penetration: Where Is the Limit? *IEEE Transactions on Power Delivery*. 2020;35(4):2021–2031.

[10] Markovic U, Stanojev O, Aristidou P, *et al.* Understanding Small-Signal Stability of Low-Inertia Systems. *IEEE Transactions on Power Systems*. 2021;36(5):3997–4017.

[11] Tielens P, and van Hertem D. The relevance of inertia in power systems. *Renewable and Sustainable Energy Reviews*. 2016;55:999–1009.

[12] Fang J, Li H, Tang Y, *et al.* On the Inertia of Future More-Electronics Power Systems. *IEEE Journal of Emerging and Selected Topics in Power Electronics*. 2019;7(4):2130–2146.

[13] Rocabert J, Luna A, Blaabjerg F, *et al.* Control of Power Converters in AC Microgrids. *IEEE Transactions on Power Electronics*. 2012;27(11):4734–4749.

[14] Roscoe A, Knueppel T, Da Silva R, *et al.* Response of a grid forming wind farm to system events, and the impact of external and internal damping. *IET Renewable Power Generation*. 2020;14(19):3908–3917.

[15] EirGrid Group. Electricity Grid to Run on 75% Variable Renewable Generation Following Successful Trial; 2022. Available from: https://www.eirgridgroup.com/newsroom/electricity-grid-to-run-o/.

[16] Farrokhabadi M, Canizares CA, Simpson-Porco JW, *et al.* Microgrid Stability Definitions, Analysis, and Examples. *IEEE Transactions on Power Systems*. 2020;35(1):13–29.

[17] Prabha Kundur. *Power System Stability and Control*. McGraw-Hill; 1993.

[18] Buck M, Graf A, and Graichen P. European Energy Transition 2030: The Big Picture: Ten Priorities for the next European Commission to Meet the EU's 2030 Targets and Accelerate Towards 2050: Agora Energiewende; 2019.

[19] M Altmann, A Brenninkmeijer, J -Ch Lanoix, *et al.* Decentralized Energy Systems: Directorate General for Internal Policies Policy Department A: Economic and Scientific Policy, Industry, Research and Energy; 2010.

[20] Möst D, Schreiber S, Herbst A, *et al. The Future European Energy System*. Cham: Springer International Publishing; 2021.

[21] Ford R, Maidment C, Fell M, *et al.* A Framework for Understanding and Conceptualising Smart Local Energy Systems. Publishing, UK. ISBN: 978-1-909522-57-2: EnergyREV, Strathclyde, UK. University of Strathclyde; 2019. Available from: https://www.energyrev.org.uk/media/1298/energyrev-sles-frameworkv4.pdf.

[22] International Renewable Energy Agency. Power System Flexibility for the Energy Transition, Part 1: Overview for Policy Makers. Abu Dhabi; 2018. Available from: https://www.irena.org/publications/2018/Nov/Power-system-flexibility-for-the-energy-transition.

[23] AEG Power Solutions. AEG Power Solutions Showcases a Complete Battery Energy Storage Solution in a 1MW Micro-Grid and Announces Its Partnership with Energon; 2018. Available from: https://www.aegps.com/en/company/news/detail/aeg-power-solutions-showcases-a-complete-battery-energy-storage-solution-in-a-1mw-micro-grid-and-ann/.

[24] Warneryd M, Håkansson M, and Karltorp K. Unpacking the Complexity of Community Microgrids: A Review of Institutions' Roles for Development of Microgrids. *Renewable and Sustainable Energy Reviews*. 2020;121:109690. Available from: https://www.sciencedirect.com/science/article/pii/S1364032119308950.

[25] International Energy Agency – IEA. Energy Island Project in the North Sea; 2023. Available from: https://www.iea.org/policies/11562-energy-island-project-in-the-north-sea.

[26] McKenna R. The Double-Edged Sword of Decentralized Energy Autonomy. *Energy Policy*. 2018;113:747–750.

[27] Wolfe P. The Implications of an Increasingly Decentralised Energy System. *Energy Policy*. 2008;36(12):4509–4513.

[28] Cabello GM, Navas SJ, Vázquez IM, *et al.* Renewable Medium-Small Projects in Spain: Past and Present of Microgrid Development. *Renewable and Sustainable Energy Reviews*. 2022;165:112622.

[29] Yaqoot M, Diwan P, and Kandpal TC. Review of Barriers to the Dissemination of Decentralized Renewable Energy Systems. *Renewable and Sustainable Energy Reviews*. 2016;58:477–490.

[30] Gharavi H, Ardehali MM, and Ghanbari-Tichi S. Imperial Competitive Algorithm Optimization of Fuzzy Multi-Objective Design of a Hybrid Green Power System with Considerations for Economics, Reliability, and Environmental Emissions. *Renewable Energy*. 2015;78:427–437.

[31] Maleki A, and Askarzadeh A. Artificial Bee Swarm Optimization for Optimum Sizing of a Stand-Alone PV/WT/FC Hybrid System Considering LPSP Concept. *Solar Energy*. 2014;107:227–235.

[32] Bakhtiari H, and Naghizadeh RA. Multi-Criteria Optimal Sizing of Hybrid Renewable Energy Systems Including Wind, Photovoltaic, Battery, and Hydrogen Storage with Constraint Method. *IET Renewable Power Generation*. 2018;12(8):883–892.

[33] Royer P. Optimal Sizing and Operation of a Microgrid with Hydrogen and Electricity Storage [Travail de fin d'études]. École centrale Lyon. Lyon; 2022.

[34] de Clercq S, Zwaenepoel B, and Vandevelde L. Optimal Sizing of an Industrial Microgrid Considering Socio–Organisational Aspects. *IET Generation, Transmission & Distribution*. 2018;12(14):3442–3451.

[35] Li P, Wang Z, Wang N, *et al.* Stochastic Robust Optimal Operation of Community Integrated Energy System Based on Integrated Demand Response. *International Journal of Electrical Power & Energy Systems*. 2021;128:106735.

[36] Li B, Roche R, and Miraoui A. Microgrid Sizing with Combined Evolutionary Algorithm and MILP Unit Commitment. *Applied Energy*. 2017;188:547–562.

[37] Poli R, Kennedy J, and Blackwell T. Particle Swarm Optimization. *Swarm Intelligence*. 2007;1(1):33–57.

[38] Meng L, Sanseverino ER, Luna A, *et al.* Microgrid Supervisory Controllers and Energy Management Systems: A Literature Review. *Renewable and Sustainable Energy Reviews*. 2016;60:1263–1273.

[39] Elmouatamid A, Ouladsine R, Bakhouya M, *et al.* Review of Control and Energy Management Approaches in Micro-Grid Systems. *Energies*. 2021; 14(1):168.

[40] Yazdani A, and Iravani R. *Voltage-Sourced Converters in Power Systems: Modeling, Control, and Applications/Amirnaser Yazdani, Reza Iravani*. Oxford: Wiley; 2010.

[41] Sauer PW, Pai MA, and Chow JH. *Power System Dynamics and Stability: With Synchrophasor Measurement and Power System Toolbox* 2nd ed. Hoboken: IEEE Press and Wiley; 2017.

[42] Arghir C, Jouini T, and Dörfler F. Grid-Forming Control for Power Converters Based on Matching of Synchronous Machines. *Automatica*. 2018;95:273–282. Available from: http://www.sciencedirect.com/science/article/pii/S0005109818302796.

[43] Groß D, Arghir C, and Dörfler F. On the Steady-State Behavior of a Nonlinear Power System Model. *Automatica*. 2018;90:248–254. Available from: https://www.sciencedirect.com/science/article/pii/S0005109817306441.

[44] Capitanescu F, Martinez Ramos JL, Panciatici P, *et al.* State-of-the-Art, Challenges, and Future Trends in Security Constrained Optimal Power Flow. *Electric Power Systems Research*. 2011;81(8):1731–1741. Available from: https://www.sciencedirect.com/science/article/pii/S0378779611000885.

[45] Teodorescu R, Liserre M, and Rodríguez P. Grid Converters for Photovoltaic and Wind Power Systems. Chichester, Wiley and IEEE; 2011. Available from: https://onlinelibrary.wiley.com/doi/book/10.1002/9780470667057.

[46] Blaabjerg F. *Control of Power Electronic Converters and Systems*. Amsterdam: Academic Press; 2017.

[47] Lasseter RH, Chen Z, and Pattabiraman D. Grid-Forming Inverters: A Critical Asset for the Power Grid. *IEEE Journal of Emerging and Selected Topics in Power Electronics*. 2020;8(2):925–935.

[48] Chung SK. A phase tracking system for three phase utility interface inverters. *IEEE Transactions on Power Electronics*. 2000;15(3):431–438.

[49] Dong D, Wen B, Boroyevich D, *et al.* Analysis of Phase-Locked Loop Low-Frequency Stability in Three-Phase Grid-Connected Power Converters Considering Impedance Interactions. *IEEE Transactions on Industrial Electronics*. 2015;62(1):310–321.

[50] Bala S, and Venkataramanan G. On the Choice of Voltage Regulators for Droop-controlled Voltage Source Converters in Microgrids to Ensure Stability. In: *ECCE 2010*. Piscataway, NJ: IEEE; 2010. p. 3448–3455.

[51] Paquette AD, and Divan DM. Virtual Impedance Current Limiting for Inverters in Microgrids With Synchronous Generators. *IEEE Transactions on Industry Applications*. 2015;51(2):1630–1638.

[52] Ajala O, Lu M, Johnson B, *et al.* Model Reduction for Inverters with Current Limiting and Dispatchable Virtual Oscillator Control. *IEEE Transactions on Energy Conversion*. 2022;37(4):2250–2259.

[53] Chandorkar MC, Divan DM, and Adapa R. Control of parallel connected inverters in standalone AC supply systems. *IEEE Transactions on Industry Applications*. 1993;29(1):136–143.

[54] Zhong QC, and Weiss G. Synchronverters: Inverters that Mimic Synchronous Generators. *IEEE Transactions on Industrial Electronics*. 2011;58(4):1259–1267.

[55] Dörfler F, Chertkov M, and Bullo F. Synchronization in Complex Oscillator Networks and Smart Grids. *Proceedings of the National Academy of Sciences of the United States of America*. 2013;110(6):2005–2010.

[56] Tayyebi A, Gross D, Anta A, *et al.* Frequency Stability of Synchronous Machines and Grid-Forming Power Converters. *IEEE Journal of Emerging and Selected Topics in Power Electronics*. 2020;8(2):1004–1018.

[57] Schiffer J, Goldin D, Raisch J, *et al.* Synchronization of Droop-Controlled Microgrids with Distributed Rotational and Electronic Generation. In: *52nd IEEE Conference on Decision and Control*. IEEE; 2013. p. 2334–2339.

[58] D'Arco S, and Suul JA. Equivalence of Virtual Synchronous Machines and Frequency-Droops for Converter-Based MicroGrids. *IEEE Transactions on Smart Grid*. 2014;5(1):394–395.

[59] Fouad AA, and Vittal V. The Transient Energy Function Method. *International Journal of Electrical Power & Energy Systems*. 1988;10 (4):233–246. Available from: https://www.sciencedirect.com/science/article/pii/0142061588900117.

[60] Aracil J, and Gordillo F. On the Control of Oscillations in DC–AC Converters. In: *IECON 2002*. IEEE; 2002. p. 2820–2825.

[61] Johnson BB, Dhople SV, Hamadeh AO, *et al.* Synchronization of Parallel Single-Phase Inverters with Virtual Oscillator Control. *IEEE Transactions on Power Electronics*. 2014;29(11):6124–6138.

[62] Colombino M, Groz D, Brouillon JS, *et al.* Global Phase and Magnitude Synchronization of Coupled Oscillators with Application to the Control of Grid-Forming Power Inverters. *IEEE Transactions on Automatic Control*. 2019;64(11):4496–4511.

[63] Sinha M, Dorfler F, Johnson BB, *et al.* Uncovering Droop Control Laws Embedded Within the Nonlinear Dynamics of Van der Pol Oscillators. *IEEE Transactions on Control of Network Systems*. 2017;4(2):347–358.

[64] Seo GS, Colombino M, Subotic I, *et al.* Dispatchable Virtual Oscillator Control for Decentralized Inverter-dominated Power Systems: Analysis and Experiments. In: *APEC 2019*. Piscataway, NJ: IEEE; 2019. p. 561–566.

[65] Subotic I, and Gros D. Power-Balancing Dual-Port Grid-Forming Power Converter Control for Renewable Integration and Hybrid AC/DC Power Systems. *IEEE Transactions on Control of Network Systems*. 2022;9(4):1949–1961.

[66] Li C, Chaudhary SK, Savaghebi M, *et al.* Power Flow Analysis for Low-Voltage AC and DC Microgrids Considering Droop Control and Virtual Impedance. *IEEE Transactions on Smart Grid*. 2017;8(6):2754–2764.

[67] Kundur P, Paserba J, Ajjarapu V, *et al.* Definition and Classification of Power System Stability IEEE/CIGRE Joint Task Force on Stability Terms and Definitions. *IEEE Transactions on Power Systems*. 2004;19(3):1387–1401.

[68] Hatziargyriou N, Milanovic JV, Rahmann C, *et al.* Stability Definitions and Characterization of Dynamic Behavior in Systems with High Penetration of Power Electronic Interfaced Technologies. PES-TR77; 2020. Available from: https://resourcecenter.ieee-pes.org/publications/technical-reports/PES_TP_TR77_PSDP_STABILITY_051320.html.

[69] Hatziargyriou N, Milanovic JV, Rahmann C, *et al.* Definition and Classification of Power System Stability Revisited and Extended. *IEEE Transactions on Power Systems*. 2021;36(4):3271–3281.

[70] Guerrero JM, Chandorkar M, Lee TL, *et al.* Advanced Control Architectures for Intelligent Microgrids—Part I: Decentralized and Hierarchical Control. *IEEE Transactions on Industrial Electronics*. 2013;60(4):1254–1262.

[71] Wang X, and Blaabjerg F. Harmonic Stability in Power Electronic-Based Power Systems: Concept, Modeling, and Analysis. *IEEE Transactions on Smart Grid*. 2019;10(3):2858–2870.

[72] He J, Li YW, Bosnjak D, *et al.* Investigation and Active Damping of Multiple Resonances in a Parallel-Inverter-Based Microgrid. *IEEE Transactions on Power Electronics*. 2013;28(1):234–246.

[73] de Carne G, Liserre M, Langwasser M, *et al.* Which Deepness Class is Suited for Modeling Power Electronics? A Guide for Choosing the Right Model for Grid-Integration Studies. *IEEE Industrial Electronics Magazine*. 2019;13(2):41–55.

[74] Lacerda VA, Araujo EP, Cheah-Mane M, *et al.* Phasor Modeling Approaches and Simulation Guidelines of Voltage-Source Converters in Grid-Integration Studies. *IEEE Access*. 2022;10:51826–51838.

[75] Cigré WG B4-81. Expert Group Interaction Studies and Simulation Models (EG ISSM) – Final Report; 2021. Available from: https://eepublicdownloads.entsoe.eu/clean-documents/Network%20codes%20documents/GC%20ESC/ISSM/EG_ISSM_Final_Report_211001.pdf

[76] Gomes Guerreiro GM, Martin F, Yang G, *et al.* New Pathways to Future Grid Compliance for Wind Power Plants. In: *21st Wind & Solar Integration Workshop (WIW 2022)*. Stevenage: IET; 2022. p. 115–121.

[77] Maniatopoulos M, Lagos D, Kotsampopoulos P, *et al.* Combined Control and Power Hardware in–the–Loop Simulation for Testing Smart Grid Control Algorithms. *IET Generation, Transmission & Distribution*. 2017;11(12):3009–3018.

[78] Hans F, Curioni G, Jersch T, *et al.* Towards Full Electrical Certification of Wind Turbines on Test Benches – Experiences Gained from the HiL-GridCoP Project. In: *21st Wind & Solar Integration Workshop (WIW 2022)*. IET; 2022. p. 122–129.

[79] Kabalan M, Singh P, and Niebur D. A Design and Optimization Tool for Inverter-Based Microgrids Using Large-Signal Nonlinear Analysis. *IEEE Transactions on Smart Grid*. 2019;10(4):4566–4576.

[80] Andrade F, Kampouropoulos K, Romeral L, *et al.* Study of Large-Signal Stability of an Inverter-Based Generator Using a Lyapunov Function. In: *IECON 2014 – 40th Annual Conference of the IEEE Industrial Electronics Society*. IEEE; 2014. p. 1840–1846.

[81] Anghel M, Milano F, and Papachristodoulou A. Algorithmic Construction of Lyapunov Functions for Power System Stability Analysis. *IEEE Transactions on Circuits and Systems I: Regular Papers*. 2013;60(9):2533–2546.

[82] Nasr-Azadani E, Canizares CA, Olivares DE, *et al.* Stability Analysis of Unbalanced Distribution Systems with Synchronous Machine and DFIG Based Distributed Generators. *IEEE Transactions on Smart Grid*. 2014;5(5):2326–2338.

[83] Sun J. Impedance-Based Stability Criterion for Grid-Connected Inverters. *IEEE Transactions on Power Electronics*. 2011;26(11):3075–3078.

# Index

Printed in the USA
CPSIA information can be obtained
at www.ICGtesting.com
JSHW011639210224
57814JS00003B/52